32.00

P. R. HALMOS

SELECTA
RESEARCH CONTRIBUTIONS

P. R. Halmos

P. R. HALMOS

SELECTA
RESEARCH CONTRIBUTIONS

Edited by
Donald E. Sarason
Nathaniel A. Friedman

Springer-Verlag
New York Heidelberg Berlin

P. R. Halmos
Department of Mathematics
Indiana University
Swain Hall East
Bloomington, IN 47405
U.S.A.

Editors

Donald E. Sarason
Department of Mathematics
University of California
Berkeley, CA 94720
U.S.A.

Nathaniel A. Friedman
Department of Mathematics
State University of New York
Albany, NY 12222
U.S.A.

AMS Classification (1980): 00A10

With 5 illustrations.

Library of Congress Cataloging in Publication Data
Halmos, Paul R. (Paul Richard), 1916–
    Selecta: research contributions.
    Bibliography: p.
    Includes index.
    1. Mathematics—Collected works. 2. Operator
theory—Collected works. I. Sarason, Donald.
II. Friedman, Nathaniel A. III. Title.
QA3.H245   1983   515   82-19578

9 8 7 6 5 4 3 2 1

ISBN 0-387-90755-6 Springer-Verlag New York Heidelberg Berlin
ISBN 3-540-90755-6 Springer-Verlag Berlin Heidelberg New York

TABLE OF CONTENTS

TABLE OF CONTENTS

PREFACE

This volume and its companion present a selection of the mathematical writings of P. R. Halmos. The present volume consists of research publications plus two papers which, although of a more expository nature, were deemed primarily of interest to the specialist ("Ten Problems in Hilbert Space" (1970d), and "Ten Years in Hilbert Space" (1979b)). The remaining expository and all the popular writings are in the second volume.

The papers in the present volume are arranged chronologically. As it happens, that arrangement also groups the papers according to subject matter: those published before 1950 deal with probability and measure theory, those after 1950 with operator theory. A series of papers from the mid 1950's on algebraic logic is excluded; the papers were republished by Chelsea (New York) in 1962 under the title "Algebraic Logic."

This volume contains two introductory essays, one by Nathaniel Friedman on Halmos's work in ergodic theory, one by Donald Sarason on Halmos's work in operator theory. There is an essay by Leonard Gillman on Halmos's expository and popular writings in the second volume.

The editors wish to express their thanks to the staff of Springer-Verlag. They are grateful also for the help of the following people: C. Apostol, W. B. Arveson, R. G. Douglas, C. Pearcy, S. Popa, P. Rosenthal, A. L. Shields, D. Voiculescu, S. Walsh.

Berkeley, CA

Donald E. Sarason

WORK IN OPERATOR THEORY

P. R. Halmos's first papers on Hilbert space operators appeared in 1950. Halmos was then thirty-four years old and already well known for his contributions to ergodic theory.

If Halmos was no longer a youngster in 1950, neither was the theory of Hilbert space operators. It began with the work of D. Hilbert and his school in the early part of this century. F. Riesz's fundamental contributions date from the beginnings of the subject, those of J. von Neumann and M. H. Stone from the late 1920's and early 1930's. Stone's famous book, *Linear Transformations in Hilbert Space*, was published in 1932.

In 1950, as now, the basis of the subject was the spectral theorem, and much of the work of the time can be classified as spectral theory. A few papers were appearing though, which tried to push beyond the confines of the classical spectral theory. Examples are A. Beurling's "On two problems concerning linear transformations in Hilbert space" (*Acta Math.* **81** (1949) 239–255), J. von Neumann's "Eine Spektraltheorie für allgemeine Operatoren eines unitären Raumes" (*Math. Nachr.* **4** (1951) 258–281), and M. S. Livšic's "On the reduction of a linear non-Hermitian operator to 'triangular' form" (*Dokl. Akad. Nauk SSSR* **84** (1952) 873–876). To this group belongs Halmos's "Normal dilations and extensions of operators" (1950a).

Halmos's idea in "Normal dilations and extensions" was to use normal operators to study nonnormal operators. He introduced (using slightly different terminology) the notion of a subnormal operator, that is, an operator one can obtain (to within unitary equivalence) by restricting a normal operator to an invariant subspace. He established an intrinsic criterion for subnormality, a criterion later refined by his student, J. Bram, as part of a systematic study ("Subnormal operators," *Duke Math. J.* **22** (1955) 75–94). In "Spectra and spectral manifolds" (1952a) (where the current terminology was adopted), Halmos established a spectral inclusion theorem for subnormal operators, another result that Bram refined. Interest in subnormal operators was dormant for a while after Bram's work but revived about ten years ago as people gradually became aware of the intriguing analytic problems connected with them. Recent progress has been striking, the most spectacular advance being S. W. Brown's beautiful proof that every subnormal operator has a nontrivial invariant subspace ("Some invariant subspaces for subnormal operators," *Integral Equations Operator Theory* **1** (1978) 310–333). Many of the ideas developed by Brown and others to deal with subnormal operators have proved useful in the study of more general operators.

The class of subnormal operators, while rich and fascinating, is rather special in many respects. In the hope of using normal operators to study more general operators than subnormal ones, Halmos introduced a weaker notion than extension, that of dilation. An operator B on a Hilbert space K is called a dilation of the operator A on the subspace H if the action of A coincides with that of B followed by the orthogonal projection of K onto H. Halmos proved the unexpected result that every contraction has a unitary dilation. This was the beginning of a far-reaching theory, the next and most crucial step being B. Sz.-Nagy's proof that one can produce a unitary dilation with the property that each of its positive powers is a dilation of the corresponding power of the contraction ("Sur les contractions de l'espace de Hilbert," *Acta Sci. Math.* **18** (1953) 87–92). The book of Sz.-Nagy and C. Foiaş contains a comprehensive account of the theory (*Harmonic Analysis of Operators on Hilbert Space*, North–Holland, 1970).

Historical footnote: Halmos's result on the existence of unitary dilations for contractions was anticipated, in a slightly weaker form, by G. Julia ("Les projections des systèmes orthonormaux de l'espace hilbertien et les opérateurs bournés," *C. R. Acad. Sci. Paris* **219** (1944) 8–11). The latter paper, however, seems to have been overlooked for many years.

Halmos is strongly attracted by the algebraic side of operator theory. A case in point is "Commutators of operators" (1952b) and its sequel (1954a). The starting point there was the famous theorem of A. Wintner ("The unboundedness of quantum-mechanical matrices," *Phys. Rev.* **71** (1947) 738–739) and H. Wielandt ("Über die Unbeschränktheit der Operatoren der Quantenmechanik," *Math. Ann.* **191** (1949) 21): The equality $AB - BA = 1$ is impossible for bounded Hilbert space operators A and B. If the identity cannot be a commutator, inquired Halmos, what can be? He obtained an array of tantalizing results and examples, but it was only ten years later that his basic question was completely settled by A. Brown and C. Pearcy, who showed that, in a separable infinite dimensional Hilbert space, an operator fails to be a commutator only if it is a nonzero scalar modulo the compacts ("Structure of commutators of operators," *Ann. of Math.* **82** (1965) 112–127); the converse had been noted earlier by Halmos ("A glimpse into Hilbert space," 1963b). There has been much subsequent work, especially concerning commutators in von Neumann algebras.

Shift operators and invariant subspaces form two recurrent themes in Halmos's work. I shall always have a special fondness for "Shifts on Hilbert spaces" (1961c): it is the paper Halmos gave me to read immediately after agreeing to take me on as his student. The paper continues a line of investigation initiated by Beurling in the paper already cited and advanced by P. Lax ("Translation invariant spaces," *Acta Math.* **101** (1959) 163–178) and (at about the same time as Halmos's work) by H. Helson and D. Lowdenslager ("Invariant subspaces," *Proc. Internat. Symp. Linear Spaces*, Jerusalem, 1960, Pergamon Press, 1961, pp. 251–262). Beurling's analysis revealed that the invariant subspace structure of the simple unilateral shift mirrors the factorization theory for functions in the Hardy class H^2 of the unit disk (the inner-outer factorization). Halmos's contribution was to provide a Halmosian treatment. It works for multiple shifts as well as for simple ones, and by exploiting the elementary geometry of Hilbert space and a few simple properties of unitary operators, yields the essence of Beurling's invariant subspace classification without relying on sophisticated function theory. In fact, a certain portion of the relevant function theory can be deduced in a "soft" way via the Halmos approach. The whole subject of shifts is intimately bound up with the theory of unitary dilations, as one can witness, for example, in the book of Sz.-Nagy and Foiaş.

The invariant subspace problem—whether every Hilbert space operator has a nontrivial invariant subspace—is the most basic and most challenging open question about the structure of general operators. The first progress, beyond what can be seen from the classical spectral theory, goes back to von Neumann, who in the 1930's established a positive answer for compact operators. Von Neumann did not publish his theorem, but some years later he communicated it to N. Aronszajn, who worked out a proof on his own and extended the result, first to reflexive Banach spaces, and finally in collaboration with K. T. Smith, to general Banach spaces ("Invariant subspaces of completely continuous operators," *Ann. of Math.* **60** (1954) 345–350). Smith then raised, and Halmos publicized, the invariant subspace problem for Hilbert space operators with compact squares, to which it could reasonably be hoped the Aronszajn–Smith technique would apply. The problem proved recalcitrant and was only settled in 1966 by A. Bernstein and A. Robinson, who showed that any polynomially compact Hilbert space operator has a nontrivial invariant subspace ("Solution of an invariant subspace problem of K. T. Smith and P. R. Halmos," *Pacific J. Math.* **16** (1966) 421–431). Their paper employs the viewpoint and formalism of nonstandard analysis, a circumstance which considerably raised the anxiety level among operator theorists. Halmos saved the day for operator theory by recasting the Bernstein–Robinson argument in standard terms ("Invariant subspaces of polynomially compact operators," 1966), where it could be recognized as a refinement of the Aronszajn–Smith technique. A further refinement was soon given by W. B. Arveson and J. Feldman ("A note on invariant subspaces," *Michigan Math. J.* **15** (1968) 61–64). Halmos then tried to understand what was really going on in all these arguments. The result was "Quasitriangular operators" (1968c), which singled out a class of operators suggested by one of the main ingredients in the Aronszajn–Smith proof and its variants. Halmos's idea spawned a flurry of activity which still continues, a milestone being the characterization of C. Apostol, C. Foiaş and D. Voiculescu: the operator T fails to be quasitriangular if and only if there is a complex number z such that $T - z$ is semi-Fredholm and has a negative index ("Some results on non-quasitriangular operators IV," *Rev. Roum. Math. Pures Appl.* **18** (1973) 487–514). (The result was originally conjectured by C. Pearcy and an alternative proof can be found in his joint paper with R. G. Douglas, "Invariant subspaces of non-quasitriangular operators," *Proc. Conf. Operator Theory, Lecture Notes in Math.* **345** (1973), pp. 13–57.) Thus, although it was originally hoped the general invariant subspace problem could be reduced to the case of non-quasitriangular operators, the mathematical gods were uncooperative, and just the opposite reduction was made. The notion of quasitriangularity has proved to be a most basic and fruitful one in modern operator theory.

After the birth of quasitriangular operators, V. I. Lomonosov discovered a technique for proving the existence of invariant subspaces which is essentially different from that of Aronszajn–Smith and which generally yields stronger results ("Invariant subspaces for the family of operators which commute with a completely continuous operator," *Funkcional. Anal. i Priložen.* **7** (1973) 213–214). The Lomonosov technique is remarkable in that it uses only tools available since the 1930's. It is interesting to speculate on how the development of operator theory might have been altered had Lomonosov's technique been discovered earlier.

The notion of quasitriangularity, or, more accurately, the related notion of quasidiagonality, provided a new proof of H. Weyl's theorem that every self-adjoint operator is a compact perturbation of a diagonal one ("Über beschränkte quadratische Formen deren Differenz vollstetig ist," *Rend. Circ. Mat. Palermo* **27** (1909) 373–392). The new proof can be found in the discussion of Problem 4 in "Ten problems in Hilbert space" (1970d). Problem 4 asked whether Weyl's theorem extends to normal operators. That it does was proved by I. D. Berg before "Ten problems" actually appeared in print ("An extension of the Weyl–von Neumann theorem," *Trans. Amer. Math. Soc.* **160** (1971) 365–371). Slightly later Halmos gave a more conceptual proof which reduces the normal case to the self-adjoint one ("Continuous functions of Hermitian operators," 1972a). Berg's result was an important step in a development known as BDF, the currently very active theory of extensions of C*-algebras founded by L. G. Brown, R. G. Douglas and P. A. Fillmore ("Unitary equivalence modulo the compact operators and extensions of C*-algebras," *Proc. Conf. Operator Theory, Lecture Notes in Math.* **345** (1973), pp. 58–128). Problem 8 in "Ten problems" asked whether the set of reducible operators on a Hilbert space is norm dense in the set of all operators. Voiculescu obtained a positive answer as a corollary to a generalization of the Weyl–von Neumann-Berg theorem which has had important repercussions ("A non-commutative Weyl–von Neumann theorem," *Rev. Roum. Math. Pures Appl.* **21** (1976) 97–113). An account of these interesting developments can be found in "Ten years in Hilbert space" (1979b), a sequel to "Ten problems". In fact, the two latter papers and the earlier "A glimpse into Hilbert space" (1963b) provide illuminating discussions of much of the progress that has occurred in operator theory during the past three decades.

Although I have mentioned only a fraction of Halmos's contributions, I hope that the above remarks, as incomplete as they are, indicate how some of the main developments in operator theory between 1950 and now were initiated or significantly advanced by him. His papers are not the kind that overwhelm one by their complicated proofs and startling new techniques. The depth of his papers, I believe, comes from the fundamental nature of the problems they attack and of the ideas they introduce. Halmos is continually trying to penetrate to the core, to understand the basics. He has made important contributions to operator theory through what he could not prove as well as through what he could: again and again, the problems he has raised and disseminated have focused activity in fruitful directions and resulted in unforeseen insights. His accurate overview of the subject and uncanny feeling for what the next step should be go a long way toward explaining why he is the leader of a school of operator theory. Still, to appreciate the full extent of Halmos's influence one must look beyond his publications and consider the kind of person he is. I wish to close by attempting that.

My initial encounter with Halmos came at the beginning of his first year in Ann Arbor, where I was a graduate student suffering from a classic case of post-prelim doldrums (actually, my second attack, the first having resulted in my exit from physics and entrance into mathematics). I decided to audit his course in functional analysis because I realized my understanding of the subject was weak, even though I had taken the course for credit a year earlier. A few weeks into the semester, Halmos required every student in the class—even visitors, like me—to make an appointment to meet with him in his office. It was a nontrivial commitment of his time, as there were about twenty students in all. Partly he was interested in our reactions to the course (in which he was using his version of the Moore method), but partly he just wanted to get to know us better as individuals. When, under questioning, I admitted to having solved most of the suggested problems, he expressed gratification that I was benefiting from the course. Needless to say, any student would be highly encouraged at such interest shown by a famous professor. He was lifting me from my doldrums. Some weeks later he invited me to his home for dinner. It was unheard of, to me, for a professor to seek social contact with a student.

There is no need to draw out the story. I have said enough to illustrate how Halmos, through his kindness and concern, as well as through his eventual position as my chief mathematical advisor, exerted a strong positive influence in my early development. He has helped many young mathematicians, both students and colleagues, in a similar way.

Halmos is renowned as an expositor. His writing is something he works hard at, thinks intensely about, and is fiercely proud of. (Witness "How to write mathematics," (1970a).) In his papers, he is not content merely to present proofs that are well organized and clearly expressed; he also suggests the thought processes that went into the constructions of his proofs, pointing out the pitfalls he encountered and indicating helpful analogies. His writings clearly reveal his commitment as an educator. In fact, Halmos is instinctively a teacher, a quality discernible in all of his mathematical activities, even the most casual ones.

Most of us, when we discover a new mathematical fact, however minor, are usually eager to tell someone about it, to display our cleverness. Halmos behaves differently: he will not tell you his discovery, he will ask you about it, and challenge you to find a proof. If you find a better proof than his, he will be delighted, because then you and he will have taught each other.

To me, Halmos embodies the ideal mixture of researcher and teacher. In him, each role is indistinguishable from the other. Perhaps that is the key to his remarkable influence.

Berkeley, CA
December, 1982

DONALD E. SARASON

WORK IN ERGODIC THEORY

Paul Halmos's research was primarily in ergodic theory during the period from 1939 to 1949. Halmos considered several topics such as the decomposition of a measure-preserving transformation into ergodic components, transformations with discrete spectra, roots of transformations, topologies on the class of transformations, ergodic theorems, conditions for existence of an invariant measure, and automorphisms of compact groups. Before discussing the corresponding papers, we will describe a small number of relevant results in ergodic theory obtained in the thirties. A more complete coverage is given in Halmos's 1949 review article [24].

Modern ergodic theory began early in 1931 with the astute observation of B. O. Koopman that if T is an invertible measure-preserving transformation acting in a measure space with measure m, then $Uf(x) = f(T(x))$ defines a unitary operator U acting on f in $L_2(m)$ [34]. The significance of this observation was appreciated by von Neumann and shortly thereafter he proved the first mean ergodic theorem [46]. This was soon followed by Birkhoff's proof of the first individual ergodic theorem [4]. In 1932 Khintchine extended Birkhoff's theorem to an abstract measure space [33]. The first necessary and sufficient condition for the existence of a finite invariant measure was obtained by E. Hopf [30] in 1932. In 1937 Hopf obtained the first ratio ergodic theorem [31].

There were also two significant results related to the spectrum of a transformation that were proved in 1932. The first was von Neumann's theorem that transformations having the same discrete spectrum are isomorphic [47]. The second was the theorem of Koopman and von Neumann that a transformation is weakly mixing if and only if the only proper functions are constants [35].

Von Neumann was in residence at Princeton, which soon became a center for the study of ergodic theory. Halmos arrived at Princeton in 1939 at the age of twenty-three. He stayed three years and was von Neumann's assistant for the last two years. He had just completed a postdoctoral year at the University of Illinois where he was a student of J. L. Doob. His thesis [15] was concerned with measure-preserving transformations and asymptotic independence and from there it was a short step to ergodic theory. His interest in ergodic theory was stimulated by Doob, who had also proved an ergodic theorem [11]. During Halmos's three year stay, there was a gifted group of young ergodicists at Princeton that included Warren Ambrose, Shizuo Kakutani, and Dorothy Maharam-Stone. This group would go on to develop a number of areas in ergodic theory and Halmos was to have a definite hand in this development. Moreover, due to his expository powers, he would have a definite hand in the development of generations of future ergodicists.

In Halmos's first paper in ergodic theory [16], he proved that a measure space isomorphic to the unit interval can be expressed as a direct sum of measure spaces. This result was used to give a simple proof of an extension of a theorem of von Neumann [47] on the decomposition of an arbitrary measure-preserving transformation into ergodic parts. Pertinent remarks can be found in [23], where Halmos gives a simplified proof of a decomposition theorem of Dieudonné [10]. In [2] Ambrose, Halmos, and Kakutani obtained a decomposition theorem for flows.

In [29] Halmos and von Neumann characterized measure spaces isomorphic to the unit interval with Lebesgue measure. They then proved that every ergodic measure-preserving transformation (e.m.p.t.) with discrete spectrum is isomorphic to a rotation on a compact abelian group. This result has a variety of interesting corollaries. In particular, every subgroup of the circle is the spectrum of an e.m.p.t. with discrete spectrum and every e.m.p.t. with discrete spectrum is isomorphic to its own inverse. Nine years later Anzai [3] constructed an e.m.p.t. that was not isomorphic to its own inverse.

Halmos was the first to study the root problem for measure-preserving transformations [17]. A transformation S is an nth root of T if $S^n(x) = T(x)$ for a.e. x. A square root corresponds to $n = 2$. Using results in [47], Halmos showed that if T is an e.m.p.t. with discrete spectrum on a space of finite measure, then T has a square root if and only if -1 is not a proper value of T. The same method can be used to show T has an nth root if and only if the nth roots of unity are not proper values of T. In [5] this result is used to show that, given a set P of primes, there exists a transformation T with discrete spectrum such that T has a prime pth root only for p in P.

The root problem has generated many interesting counterexamples. In particular, R. V. Chacon constructed a weak mixing transformation with no roots [7], M. A. Akcoglu and J. Baxter constructed weak mixing transformations with prime roots only for a specified set P of primes [1], D. S. Ornstein constructed a mixing transformation that commutes only with its powers and hence has no roots [43], and D. Rudolph constructed a variety of amazing examples in [45].

The transformation T constructed in [43] also has another very interesting property. Namely, the σ-algebra generated by the iterates $T^i A$, $i = 0, \pm 1, \pm 2, \ldots$, is the full σ-algebra when $0 < m(A) < 1$. Thus every non-trivial partition is a generator. This was the first example of a prime transformation.

A factor of a transformation is the restriction of the transformation to an invariant sub-σ-algebra. A transformation is prime if it has no proper factors.

It is now known that Chacon's simplest example of a weak mixing transformation that is not mixing [6] is also prime, commutes only with its factors, and is not isomorphic to its inverse [9, 12].

In [19] Halmos introduced the general problem of studying topologies on the set of invertible measure-preserving transformations G on a measure space. Since he and von Neumann had shown that the measure spaces of interest are isomorphic to the unit interval with Lebesgue measure m [29], this was the case he considered. A general conjecture is that a transformation in G can be approximated in some topology by simple transformations. A simple transformation is obtained by dividing the interval into equal subintervals and mapping a subinterval by a translation onto a subinterval. Halmos considered the dyadic intervals of rank n (n-intervals) $[(k-1)/2^n, k/2^n)$, $1 \le k \le 2^n$, of length $1/2^n$. An n-set is a union of n-intervals. An n-permutation is a transformation that permutes n-intervals by translating each n-interval onto an n-interval. A cyclic n-permutation is a permutation with one cycle of period 2^n; hence each n-interval visits each n-interval before returning to itself.

Given S in G, a subbasic dyadic neighborhood of S is a set of the form $\langle T: m(SD \, \Delta \, TD) < \varepsilon \rangle$, where D is an n-set. The weak topology is the topology generated by the corresponding dyadic neighborhoods. Halmos proved that the permutations are dense in the weak topology. In fact, he showed every dyadic neighborhood contains cyclic permutations of arbitrarily high ranks.

A transformation T is aperiodic if $T^n x \ne x$, $n \ge 1$, for a.e. x. The isomorphism class of T is the set $\langle S^{-1}TS : S \text{ in } G \rangle$. Halmos proved that in the weak topology the isomorphism class of each aperiodic transformation is dense in G. This result was used to prove the Second Category Theorem, which states that the weak mixing transformations are dense in the weak topology. Later V. A. Rohlin [44] proved the First Category Theorem which states that the mixing transformations form a set of first category in the weak topology. Together these category theorems guarantee the existence of a large supply of weak mixing transformations that are not mixing. Actually in 1941 von Neumann and Kakutani used stacking to construct an example of a transformation that was weak mixing but not mixing. This example remained unpublished until 1973 [32].

Halmos also considered the uniform metric topology on the class of transformations defined by the metric $d(T, S) = m(\langle x: T(x) \ne S(x) \rangle)$. He proved that if T is aperiodic, then for each positive integer n there exists a set E such that $T^i E$, $0 \le i < n$, are disjoint and $m(\cup_{i=0}^{n-1} T^i E) > 1 - 4/n$ [20]. Rohlin later replaced $4/n$ by $\varepsilon > 0$ [44]. This result has turned out to be very useful in a variety of applications. The result was extended to non-singular transformations in [8].

In [21] Halmos proved a ratio ergodic theorem that extended Hurewicz's ergodic theorem for transformations without invariant measure. This was the most general result at the time. Later results extended the ergodic theorem to operators.

In [18] Halmos began the study of automorphisms of compact groups. In particular, he proved that if a continuous automorphism is ergodic, then it is mixing of all orders. The ultimate extension of this result is the theorem of Miles and Thomas that an ergodic automorphism of a compact group is isomorphic to a Bernoulli shift [36, 37, 38].

In [22] Halmos extended Hopf's condition for existence of an equivalent finite invariant measure to obtain the first necessary and sufficient condition for the existence of an equivalent σ-finite invariant measure. The outstanding problem at the time was whether there exists a transformation with no equivalent σ-finite invariant measure. An example was later constructed by Ornstein [40] (a simple proof appears in [13]).

Looking back over the period that Halmos devoted to ergodic theory, one realizes that he did pioneering work in several areas that proved to be most significant in later years. In addition, one cannot over-emphasize the importance of his expository works such as his texts in measure theory [25], Hilbert space [26], and ergodic theory [27], as well as his lecture notes on entropy [28]. These notes were the first in English to present the revolutionary work of Kolmogorov and Sinai on entropy and the isomorphism problem. Many ergodicists first learned about the new ideas from these notes.

Entropy led to a result in 1969 that really accelerated the development of ergodic theory, namely, Ornstein's Isomorphism Theorem for Bernoulli shifts with the same entropy [41, 42]. Research in ergodic theory has remained extremely active up to the present, which marks the fifty-first birthday of modern ergodic theory. A recent text [48] discusses a large number of results. Also see [14, 39, 49].

Lastly, I would like to say a few words concerning my own experience as a graduate student at Brown University under the guidance of R. V. Chacon. My little bible was Halmos's ergodic theory book and I was always carrying around at least one Halmos. In fact, I have the feeling that I was raised on Halmos. No doubt this has been a common experience for many mathematicians.

Albany, NY
December 1982

NATHANIEL A. FRIEDMAN

References

[1] M. A. Akcoglu and J. Baxter, "Roots of ergodic transformations," *J. Math. Mech.* **19** (1969/70) 991–1003.

[2] W. Ambrose, P. R. Halmos, and S. Kakutani, "The decomposition of measures, II," *Duke Math. J.* **9** (1942) 43–47.

[3] H. Anzai, "An example of a measure preserving transformation which is not conjugate to its inverse," *Proc. Jap. Acad.* **27** (1951) 517–522.

[4] G. D. Birkhoff, "Proof of the ergodic theorem," *Proc. Natl. Acad. Sci. U.S.A.* **17** (1931) 656–660.

[5] J. R. Blum and N. A. Friedman, "On commuting transformations and roots," *Proc. Am. Math. Soc.* **17** (1966) 1370–1374.

[6] R. V. Chacon, "Transformations with continuous spectrum," *J. Math. Mech.* **16** (1966/67) 399–415.

[7] ——, "A geometric construction of measure preserving transformations," *Proc. Fifth Berkeley Sym. Math. Stat. Prob.*, Univ. of Cal., (1967) vol. II, part 2, 335–360.

[8] R. V. Chacon and N. A. Friedman, "Approximation and invariant measures," *Z. Wahrscheinlichkeitstheor. Verw. Geb.* **3** (1965) 286–295.

[9] A. Del Junco, M. Rahe, and L. Swanson, "Chacon's automorphism has minimal self-joinings," *J. Anal. Math.* (1980) 276–284.

[10] J. Dieudonné, "Sur le Théorème de Lebesgue–Nikodym (III)" *Ann. Univ. Grenoble* **23** (1948) 25–53.

[11] J. L. Doob, "The law of large numbers for continuous stochastic processes," *Duke Math. J.* **6** (1940) 290–306.

[12] A. Fieldsteel, "An uncountable family of prime transformations not isomorphic to their inverses," preprint (1980).

[13] N. A. Friedman, *Introduction to ergodic theory*, Van Nostrand–Reinhold, New York (1971).

[14] H. Fürstenberg, *Recurrence in ergodic theory and combinatorial number theory*, Princeton University Press, Princeton (1981).

[15] P. R. Halmos, "Invariants of certain stochastic transformations; the mathematical theory of gambling systems," *Duke Math. J.* **5** (1939) 461–478.

[16] ——, "The decomposition of measures," *Duke Math. J.* **8** (1941) 386–392.

[17] ——, "Square roots of measure preserving transformations," *Am. J. Math.* **64** (1942) 153–166.

[18] ——, "On automorphisms of compact groups," *Bull. Am. Math. Soc.* **49** (1943) 619–624.

[19] ——, "Approximation theories for measure preserving transformations," *Trans. Am. Math. Soc.* **55** (1944) 1–18.

[20] ——, "In general a measure preserving transformation is mixing," *Ann. Math.* **45** (1944) 786–792.

[21] ——, "An ergodic theorem," *Proc. Natl. Acad. Sci. U.S.A.* **32** (1946) 156–161.

[22] ——, "Invariant measures," *Ann. Math.* **48** (1947) 735–754.

[23] ——, "On a theorem of Dieudonné," *Proc. Natl. Acad. Sci. U.S.A.* **35** (1949) 38–42.

[24] ——, "Measurable transformations," *Bull. Am. Math. Soc.* **55** (1949) 1015–1034.

[25] ——, *Measure theory*, Van Nostrand, New York (1950).

[26] ——, *Introduction to Hilbert space and the theory of spectral multiplicity*, Chelsea, New York (1951).

[27] ——, "Lectures on ergodic theory," *Math. Soc. of Japan*, Tokyo (1956).

[28] ——, "Entropy in ergodic theory" (Notes) University of Chicago (1959).

[29] P. R. Halmos and J. von Neumann, "Operator methods in classical mechanics, II," *Ann. Math.* **43** (1942) 332–350.

[30] E. Hopf, "Theory of measure and invariant integrals," *Trans. Am. Math. Soc.* **34** (1932) 373–393.

[31] ——, *Ergodentheorie*, Springer-Verlag, Berlin (1937).

[32] S. Kakutani, "Examples of ergodic measure preserving transformations which are weakly mixing but not strongly mixing," *Springer Lecture Notes in Math.* **318**, 143–149, Springer-Verlag, New York (1973).

[33] A. Khintchine, "Zu Birkhoff's Lösung des Ergodenproblems," *Math. Ann.* **107** (1933) 485–488.

[34] B. O. Koopman, "Hamiltonian systems and transformations in Hilbert space," *Proc. Natl. Acad. Sci. U.S.A.* **17** (1931) 315–318.

[35] B. O. Koopman and J. von Neumann, "Dynamical systems of continuous spectra," *Proc. Natl. Acad. Sci. U.S.A.* **18** (1932) 255–263.

[36] G. Miles and R. K. Thomas, "The breakdown of automorphisms of compact topological groups," *Adv. Math. Suppl. Stud. Vol.* 2 (1978) 207–218.

[37] ——, "On the polynomial uniformity of translations of the *n*-torus," *Adv. Math. Suppl. Stud. Vol.* 2 (1978) 219–230.

[38] ——, "Generalized torus automorphisms are Bernoullian," *Adv. Math. Suppl. Stud. Vol.* 2 (1978) 231–249.

[39] D. S. Ornstein, "Ergodic theory, randomness, and dynamical systems," *Yale Math. Monographs No.* 5, Yale Univ. (1974).

[40] ——, "On invariant measures," *Bull. Am. Math. Soc.* **66** (1960) 297–300.

[41] ——, "Bernoulli shifts with the same entropy are isomorphic," *Adv. Math.* **4** (1970) 337–352.

[42] ——, "Bernoulli shifts with infinite entropy are isomorphic," *Adv. Math.* **5** (1970) 339–348.

[43] ——, "On the root problem in ergodic theory," *Proc. Sixth Berkeley Symp. Math. Stat. Prob., Vol. II*, Univ. of Cal. Press (1967) 347–356.

[44] V. A. Rohlin, "In general a measure-preserving transformation is not mixing," *Dokl. Akad. Nauk. S.S.R.* **60** (1948) 349–351.

[45] D. Rudolph, "An example of a measure preserving map with minimal self-joinings and applications," *J. Anal. Math.* **35** (1979) 97–122.

[46] J. von Neumann, "Proof of the quasi-ergodic hypothesis," *Proc. Natl. Acad. Sci. U.S.A.* **18** (1932) 70–82.

[47] ——, "Zur Operatorenmethode in der klassischen Mechanik," *Ann. Math.* **33** (1932) 587–642.

[48] P. Walters, *An Introduction to Ergodic Theory*, Springer-Verlag, New York (1981).

[49] R. J. Zimmer, "Ergodic theory, group representations, and rigidity," *Bull. Am. Math. Soc.* **6** (1982) 383–416.

BIBLIOGRAPHY OF THE PUBLICATIONS OF P. R. HALMOS†

* in front of a title denotes a book

[1938]   Note on almost-universal forms, *Bull. Am. Math. Soc.* **44** (1938) 141–144.

[1939 a]   Invariants of certain stochastic transformations; the mathematical theory of gambling systems, *Duke Math. J.* **5** (1939) 461–478.

[1939 b]   On a necessary condition for the strong law of numbers, *Ann. Math.* **40** (1939) 800–804.

[1941 a]   Statistics, set functions, and spectra, *Mat. Sb.* **9** (1941) 241–248.

[1941 b]   The decomposition of measures, *Duke Math. J.* **8** (1941) 386–392.

[1942 a]   The decomposition of measures, II, (With W. Ambrose and S. Kakutani), *Duke Math. J.* **9** (1942) 43–47.

[1942 b]   Square roots of measure preserving transformations, *Am. J. Math.* **64** (1942) 153–166.

[1942 c]   (Review) An introduction to linear transformations in Hilbert space, By F. J. Murray, *Bull. Am. Math. Soc.* **48** (1942) 204–205.

[1942 d]   On monothetic groups, (With H. Samelson), *Proc. Natl. Acad. Sci. U.S.A.* **28** (1942) 254–257.

[1942 e]   Operator methods in classical mechanics, II, (With J. von Neumann), *Ann. Math.* **43** (1942) 332–350.

[1942 f]   *Finite dimensional vector spaces*, Princeton Univ. Press, Princeton (1942).

[1943]   On automorphisms of compact groups, *Bull. Am. Math. Soc.* **49** (1943) 619–624.

[1944 a]   Approximation theories for measure preserving transformations, *Trans. Am. Math. Soc.* **55** (1944) 1–18.

[1944 b]   Random alms, *Ann. Math. Stat.* **15** (1944) 182–189.

---

†As of the publication of this volume.

[1944 c]   The foundations of probability, *Am. Math. Monthly* **51** (1944) 493–510.

[1944 d]   In general a measure preserving transformation is mixing, *Ann. Math.* **45** (1944) 786–792.

[1944 e]   Comment on the real line, *Bull. Am. Math. Soc.* **50** (1944) 877–878.

[1946 a]   The theory of unbiased estimation, *Ann. Math. Stat.* **17** (1946) 34–43.

[1946 b]   An ergodic theorem, *Proc. Natl. Acad. Sci. U.S.A.* **32** (1946) 156–161.

[1947 a]   Functions of integrable functions, *J. Indian Math. Soc.* **11** (1947) 81–84.

[1947 b]   On the set of values of a finite measure, *Bull. Am. Math. Soc.* **53** (1947) 138–141.

[1947 c]   Invariant measures, *Ann. Math.* **48** (1947) 735–754.

[1948]   The range of a vector measure, *Bull. Am. Math. Soc.* **54** (1948) 416–421.

[1949 a]   On a theorem of Dieudonné, *Proc. Natl. Acad. Sci. U.S.A.* **35** (1949) 38–42.

[1949 b]   Application of the Radon–Nikodym theorem to the theory of sufficient statistics, (With L. J. Savage) *Ann. Math. Stat.* **20** (1949) 225–241.

[1949 c]   A non-homogeneous ergodic theorem, *Trans. Am. Math. Soc.* **66** (1949) 284–288.

[1949 d]   Measurable transformations, *Bull. Am. Math. Soc.* **55** (1949) 1015–1034.

[1950 a]   Normal dilations and extensions of operators, *Summa Brasil. Math.* **2** (1950) 125–134.

[1950 b]   Commutativity and spectral properties of normal operators, *Acta Sci. Math.* (1950) 153–156.

[1950 c]   The marriage problem, (With H. E. Vaughan) *Am. J. Math.* **72** (1950) 214–215.

[1950 d]   Measure theory, *Proc. Int. Cong. Math.*, 1950, Volume II, p. 114.

[1950 e]   *Measure theory*, Van Nostrand, New York (1950).

[1951 a]   Algunos problemas actuales sobre operadores en espacios de Hilbert, *UNESCO Symp.*, Punta del Este, Montevideo (1951) 9–14.

[1951 b]   (Review) Gelöste und ungelöste mathematische Probleme aus alter und neuer Zeit, By H. Tietze, *Bull. Am. Math. Soc.* **57** (1951) 502–503.

[1951 c]   *Introduction to Hilbert space and the theory of spectral multiplicity*, Chelsea, New York (1951).

[1952 a]   Spectra and spectral manifolds, *Ann. Soc. Math. Pol.* **25** (1952) 43–49.

[1952 b]   Commutators of operators, *Am. J. Math.* **74** (1952) 237–240.

[1953 a]   Square roots of operators, (With G. Lumer and J. J. Schäffer) *Proc. Am. Math. Soc.* **4** (1953) 142–149.

[1953 b]   (Review) Introducción a los métodos de la estadística, By S. Ríos, *J. Am. Stat. Assoc.* **48** (1953) 154–155.

[1953 c]   (Review) Intégration, By N. Bourbaki, *Bull. Am. Math. Soc.* **59** (1953) 249–255.

[1953 d]   (Review) Abstract set theory, By A. A. Fraenkel, *Bull. Am. Math. Soc.* **59** (1953) 584–585.

[1953 e]   (Review) Lezioni sulla teoria moderna dell'integrazione, By M. Picone and T. Viola, *Bull. Am. Math. Soc.* **59** (1953) 94.

BIBLIOGRAPHY

[1954 a] Commutators of operators, II, *Am. J. Math.* **76** (1954) 191–198.

[1954 b] (Review) Les systèmes axiomatiques de la théorie des ensembles, By H. Wang and R. McNaughton, *Bull. Am. Math. Soc.* **60** (1954) 93–94.

[1954 c] Polyadic Boolean algebras, *Proc. Nat. Acad. Sci.* **40** (1954) 296–301.

[1954 d] (Review) Linear Analysis, By A. C. Zaanen, *Bull. Am. Math. Soc.* **60** (1954) 487–488.

[1954 e] Polyadic Boolean algebras, *Proc. Int. Cong. Math.* 1954 Volume II, pp. 402–403.

[1954 f] Square roots of operators, II, (With G. Lumer) *Proc. Am. Math. Soc.* **5** (1954) 589–595.

[1955 a] Algebraic logic, I, Monadic Boolean algebras, *Compos. Math.* **12** (1955) 217–249.

[1955 b] (Review) Mathematics and plausible reasoning, By G. Pólya, *Bull. Am. Math. Soc.* **61** (1955) 243–245.

[1955 c] (Review) Elements of algebra, By H. Levi, *Bull. Am. Math. Soc.* **61** (1955) 245–247.

[1955 d] (Review) Topological dynamics, By W. H. Gottschalk and G. A. Hedlund, *Bull. Am. Math. Soc.* **61** (1955) 584–588.

[1955 e] (Review) Theorie der linearen Operatoren im Hilbert–Raum, By N. I. Achieser and I. M. Glasmann, *Bull. Am. Math. Soc.* **61** (1955) 588–589.

[1955 f] (Review) Introducción a los métodos de la estadística (Segunda parte) By S. Ríos, *J. Am. Stat. Assoc.* **50** (1955) 1002.

[1956 a] Predicates, terms, operations, and equality in polyadic Boolean algebras, *Proc. Natl. Acad. Sci. U.S.A.* **42** (1956) 130–136.

[1956 b] (Review) Einführung in die Verbandstheorie, By H. Hermes, *Bull. Am. Math. Soc.* **62** (1956) 189–190.

[1956 c] Algebraic logic (II). Homogeneous locally finite polyadic Boolean algebras of infinite degree, *Fund. Math.* **43** (1956) 255–325.

[1956 d] The basic concepts of algebraic logic, *Am. Math. Monthly* **63** (1956) 363–387.

[1956 e] Algebraic logic, III. Predicates, terms, and operations in polyadic algebras, *Trans. Am. Math. Soc.* **83** (1956) 430–470.

[1956 f] *Lectures on ergodic theory*, Math. Soc. Japan, Tokyo (1956).

[1957 a] Algebraic logic, IV. Equality in polyadic algebras, *Trans. Am. Math. Soc.* **86** (1957) 1–27.

[1957 b] Nicolas Bourbaki, *Scientific American* **196** (1957) 88–99.

[1957 c] (Review) Logic, semantics, metamathematics, By A. Tarski, *Bull. Am. Math. Soc.* **63** (1957) 155–156.

[1958 a] Innovation in mathematics, *Scientific American* **199** (1958) 66–73.

[1958 b] Products of symmetries, (With S. Kakutani) *Bull. Am. Math. Soc.* **64** (1958) 77–78.

[1958 c] Von Neumann on measure and ergodic theory, *Bull. Am. Math. Soc.* **64** (1958) 86–94.

[1958 d]   *Finite-dimensional vector spaces, Second ed., Van Nostrand, Princeton (1958).

[1959 a]   Free monadic algebras, Proc. Am. Math. Soc. 10 (1959) 219–227.

[1959 b]   (Review) Linear operators. Part I: General theory, By N. Dunford and J. T. Schwartz, Bull. Am. Math. Soc. 65 (1959) 154–156.

[1959 c]   The representation of monadic Boolean algebras, Duke Math. J. 26 (1959) 447–454.

[1959 d]   Entropy in ergodic theory, (Mimeographed notes) University of Chicago (1959).

[1960]   *Naive set theory, Van Nostrand, Princeton (1960).

[1961 a]   Recent progress in ergodic theory, Bull. Am. Math. Soc. 67 (1961) 70–80.

[1961 b]   Injective and projective Boolean algebras, Proc. Symp. Pure Math., Lattice theory Volume II (1961) 114–122.

[1961 c]   Shifts on Hilbert spaces, J. reine angew. Math. 208 (1961) 102–112.

[1962 a]   (Review) Neurere Methoden und Ergebnisse der Ergodentheorie, By K. Jacobs, Bull. Am. Math. Soc. 68 (1962) 59–60.

[1962 b]   *Algebraic logic, Chelsea, New York (1962).

[1963 a]   What does the spectral theorem say?, Am. Math. Monthly 70 (1963) 241–247.

[1963 b]   A glimpse into Hilbert space, Lectures on Modern Mathematics, Wiley, New York, Volume I (1963) 1–22.

[1963 c]   Partial isometries, (With J. E. McLaughlin) Pacific J. Math. 13 (1963) 585–596.

[1963 d]   Algebraic properties of Toeplitz operators, (With A. Brown) J. reine angew. Math. 213 (1963) 89–102.

[1963 e]   *Lectures on Boolean algebras, Van Nostrand, Princeton (1963).

[1964 a]   Numerical ranges and normal dilations, Acta Sci. Math. 25 (1964) 1–5.

[1964 b]   On Foguel's answer to Nagy's question, Proc. Am. Math. Soc. 15 (1964) 791–793.

[1964 c]   (Review) Lectures on invariant subspaces, By H. Helson, Bull. Am. Math. Soc. 71 (1965) 490–494.

[1965 a]   Cesàro operators, (With A. Brown and A. L. Shields) Acta Sci. Math. 26 (1965) 125–137.

[1965 b]   Commutators of operators on Hilbert space, (With A. Brown and C. Pearcy) Canad. J. Math. 17 (1965) 695–708.

[1966]   Invariant subspaces of polynomially compact operators, Pacific J. Math. 16 (1966) 433–437.

[1967]   *A Hilbert space problem book, Van Nostrand, Princeton (1967).

[1968 a]   Invariant subspaces, Abstract spaces and approximation, Birkhäuser, Basel (1968) 26–30.

[1968 b]   Irreducible operators, Mich. Math. J. 15 (1968) 215–223.

[1968 c]   Quasitriangular operators, Acta Sci. Math. 29 (1968) 283–293.

[1968 d]   Mathematics as a creative art, Am. Sci. 56 (1968) 375–389.

BIBLIOGRAPHY

[1968 e] Permutations of sequences and the Schröder–Bernstein theorem, *Proc. Am. Math. Soc.* **19** (1968) 509–510.

[1969 a] Invariant subspaces 1969, *Seventh Brazil. Math. Colloq.*, Poços de Caldas (1969) 1–54.

[1969 b] Two subspaces, *Trans. Am. Math. Soc.* **144** (1969) 381–389.

[1970 a] How to write mathematics, *Enseign. Math.*, **16** (1970) 123–152.

[1970 b] Finite dimensional Hilbert spaces, *Am. Math. Monthly* **77** (1970) 457–464.

[1970 c] Powers of partial isometries, (With L. J. Wallen) *J. Math. Mech.* **19** (1970) 657–663.

[1970 d] Ten problems in Hilbert space, *Bull. Am. Math. Soc.* **76** (1970) 887–933.

[1971 a] Capacity in Banach algebras, *Indiana Univ. Math. J.* **20** (1971) 855–863.

[1971 b] Reflexive lattices of subspaces, *J. London Math. Soc.* **4** (1971) 257–263.

[1971 c] Eigenvectors and adjoints, *Linear algebra appl.* **4** (1971) 11–15.

[1972 a] Continuous functions of Hermitian operators, *Proc. Am. Math. Soc.* **31** (1972) 130–132.

[1972 b] Positive approximants of operators, *Indiana Univ. Math. J.* **21** (1972) 951–960.

[1972 c] Products of shifts, *Duke Math. J.* **39** (1972) 779–787.

[1973 a] The legend of John von Neumann, *Am. Math. Monthly* **80** (1973) 382–394.

[1973 b] Limits of shifts, *Acta Sci. Math.* **34** (1973) 131–139.

[1974 a] Spectral approximants of normal operators, *Proc. Edinburgh Math. Soc.* **19** (1974) 51–58.

[1974 b] How to talk mathematics, *Notices Am. Math. Soc.* **21** (1974) 155–158.

[1974 c] (Review) Creative teaching: heritage of R. L. Moore, By R. D. Traylor, *Hist. Math.* **1** (1974) 188–192.

[1975 a] What to publish, *Am. Math. Monthly* **82** (1975) 14–17.

[1975 b] The teaching of problem solving, *Am. Math. Monthly* **82** (1975) 466–470.

[1976 a] Products of involutions, (With W. H. Gustafson and H. Radjavi) *Linear algebra appl.* **13** (1976) 157–162.

[1976 b] American mathematics from 1940 to the day before yesterday, (With J. H. Ewing, W. H. Gustafson, S. H. Moolgavkar, W. H. Wheeler, and W. P. Ziemer) *Am. Math. Monthly* **83** (1976) 503–516.

[1976 c] Some unsolved problems of unknown depth about operators on Hilbert space, *Proc. Royal Soc. Edinburgh* **76** A (1976) 67–76.

[1977 a] Logic from A to G, *Math. Mag.* **50** (1977) 5–11.

[1977 b] Bernoulli shifts, *Am. Math. Monthly* **84** (1977) 715–716.

[1978 a] Fourier series, *Am. Math. Monthly* **85** (1978) 33–34.

[1978 b] Arithmetic progressions, (With C. Ryavec) *Am. Math. Monthly* **85** (1978) 95–96.

[1978 c] Invariant subspaces, *Am. Math. Monthly* **85** (1978) 182–183.

[1978 d] Schauder bases, *Am. Math. Monthly* **85** (1978) 256–257.

[1978 e] The Serre conjecture, (With W. H. Gustafson and J. M. Zelmanowitz) *Am. Math. Monthly* **85** (1978) 357–359.

[1978 f] *Bounded integral operators on L^2 spaces*, (With V. S. Sunder) Springer-Verlag Berlin (1978).

[1978 g] Integral operators, Hilbert space operators, Proceedings, Long Beach, California (1977) 1–15; *Lecture Notes in Math.* 693, Springer-Verlag Berlin (1978).

[1979 a] (Review) Panorama des mathématiques pures. Le choix Bourbachique. By J. Dieudonné, *Bull. Am. Math. Soc.* **1** (1979) 678–687.

[1979 b] Ten years in Hilbert space, *Integral Equations Operator Theory* **2** (1979) 529–564.

[1980 a] Limsups of Lats, *Indiana Univ. Math. J.* **29** (1980) 293–311.

[1980 b] Finite-dimensional points of continuity of Lat, (With J. B. Conway) *Linear algebra appl.* **31** (1980) 93–102.

[1980 c] The heart of mathematics, *Am. Math. Monthly* **87** (1980) 519–524.

[1981 a] (Review) The William Lowell Putnam Mathematical Competition, Problems and Solutions: 1938–1964, By A. M. Gleason, R. E. Greenwood, and L. M. Kelly, *Am. Math. Monthly* **88** (1981) 450–451.

[1981 b] Applied mathematics is bad mathematics, *Mathematics Tomorrow*, Springer-Verlag New York (1981) 9–20.

[1981 c] Does mathematics have elements? *Math. Intell.* **3** (1981) 147–153, *Bull. Austral. Math. Soc.* **25** (1982) 161–175.

[1981 d] (Review) Encyclopedic Dictionary of Mathematics, Edited by Shôkichi Iyanaga and Yukiyosi Kawada; translation reviewed by K. O. May, *Math. Intell.* **3** (1981) 138–140.

[1982 a] Think it gooder, *Math. Intell.* **4** (1982) 20–21.

[1982 b] (Review) Recurrence in Ergodic Theory and Combinatorial Number Theory, by H. Furstenberg, *Math. Intell.* **4** (1982) 52–54.

[1982 c] Quadratic Interpolation, *J. Oper. Theory* **7** (1982) 303–305.

[1982 d] Asymptotic Toeplitz Operators, *Trans. Am. Math. Soc.* **273** (1982) 621–630.

[1982 e] The thrills of abstraction, *Two-Year College Math. J.* **13** (1982) 243–251.

[1982 f] *A Hilbert space problem book*, second ed., Springer-Verlag New York (1982).

[1983] The work of F. Riesz, (to appear in *Selecta — Expository Writing*, Springer-Verlag New York (1983)).

[1984] Weakly transitive matrices (with José Barría), *Ill. J. Math.* **28** (1984).

PERMISSIONS

Springer-Verlag would like to thank the original publishers of P.R. Halmos's scientific papers for granting permissions to reprint a selection of his papers in this volume. The following credit lines were specifically requested:

[1942 e]   Reprinted from *Ann. Math.* **43**, © 1942 by Ann. of Math.
[1943]   Reprinted from *Bull. Am. Math. Soc.* **49**, © 1942 by Am. Math Soc.
[1944 a]   Reprinted from *Trans. Am. Math. Soc.* **55**, © 1944 by Am. Math. Soc.
[1944 d]   Reprinted from *Ann. Math.* **45**, © 1944 by Ann. of Math.
[1947 c]   Reprinted from *Ann. Math.* **48**, © 1947 by Ann. of Math.
[1949 b]   Reprinted from *Ann. Math. Stat.* **20**, © 1949 by the Inst. Math. Stat.
[1950 a]   Reprinted from *Sum. Bras. Math.* **2**, © 1950 by Sum. Bras. Math.
[1950 b]   Reprinted from *Act. Sc. Math.* **12**, © 1950 by the Bolyai Inst.
[1950 c]   Reprinted from *Am. J. Math.* **72**, © 1950 by Johns Hopkins Univ. Press.
[1952 b]   Reprinted from *Am. J. Math.* **74**, © 1952 by Johns Hopkins Univ. Press.
[1953 a]   Reprinted from *Proc. Am. Math. Soc.* **4**, © 1953 by Am. Math. Soc.
[1954 a]   Reprinted from *Am. J. Math.* **76**, © 1954 by Johns Hopkins Univ. Press.
[1958 b]   Reprinted from *Bull. Am. Math. Soc.* **64**, © 1958 by Am. Math. Soc.
[1961 c]   Reprinted from *J. Reine Angew. Math.* **208**, © 1961 by W. de Gruyter & Co.
[1963 c]   Reprinted from *Pac. J. Math.*, © 1963 by Pac. J. Math.
[1963 d]   Reprinted from *J. Reine Angew. Math.* **213**, © 1963 by W. de Gruyter & Co.
[1964 a]   Reprinted from *Act. Sc. Math.* **25**, © 1964 by the Bolyai Inst.
[1965 b]   Reprinted from *Act. Sc. Math.* **26**, © 1965 by the Bolyai Inst.
[1966]   Reprinted from *Pac. J. Math.*, © 1966 by Pac. J. Math.
[1968 b]   Reprinted from *Mich. Math. J.*, © 1968 by Mich. Math. J.
[1968 c]   Reprinted from *Act. Sc. Math.* **29**, © 1968 by the Bolyai Inst.
[1969 a]   Reprinted from *7th Braz. Math. Coll.*, © 1969 by the Inst. Mat. Pura e Aplicada.
[1969 b]   Reprinted from *Trans. Am. Math. Soc.* **144**, © 1969 by Am. Math. Soc.
[1970 d]   Reprinted from *Bull. Am. Math. Soc.* **76**, © 1970 by Am. Math. Soc.
[1971 a]   Reprinted from *Indiana Univ. Math. J.* **20**, © 1971 by Indiana Univ.
[1972 a]   Reprinted from *Proc. Am. Math. Soc.* **31**, © 1972 by Am. Math. Soc.
[1972 b]   Reprinted from *Indiana Univ. Math. J.* **21**, © 1972 by Indiana Univ.

Reprinted from DUKE MATHEMATICAL JOURNAL
Vol. 5, No. 2, June, 1939

INVARIANTS OF CERTAIN STOCHASTIC TRANSFORMATIONS: THE MATHEMATICAL THEORY OF GAMBLING SYSTEMS

BY PAUL R. HALMOS

Introduction. The "Regellosigkeit" principle of von Mises has been shown to correspond in the mathematical theory of probability to the fact that certain transformations of infinite dimensional Cartesian space into itself are measure preserving. It is the purpose of this paper to investigate the behavior of such transformations on more general spaces. The theorems at the basis of this work are stated in the first section and applied to obtain results concerning the existence and independence of "Kollektivs" in the second and third sections. In §§4, 5, and 6 certain invariants of the transformations considered are obtained. Previous results on these transformations are shown to be special cases of these invariance theorems.

1. **Preliminary definitions and theorems.** In this section we shall define the concepts and state the theorems which are the basis of all the work of the later sections.

DEFINITION 1. A collection $\mathfrak{F}_1$ of sets in a space Ω_1 is a *field* if $E_1 \, \epsilon \, \mathfrak{F}_1$ and $E_2 \, \epsilon \, \mathfrak{F}_1$ implies $E_1 + E_2 \, \epsilon \, \mathfrak{F}_1$ and $E_1 - E_1 E_2 \, \epsilon \, \mathfrak{F}_1$.[1]

DEFINITION 2. A collection $\mathfrak{B}_1$ of sets in a space Ω_1 is a *Borel field* if $\mathfrak{B}_1$ is a field and if $E_j \, \epsilon \, \mathfrak{B}_1$ $(j = 1, 2, \cdots)$ implies $\sum_{j=1}^{\infty} E_j \, \epsilon \, \mathfrak{B}_1$.

DEFINITION 3. A *probability measure* is an additive, non-negative set function $P_1(E)$ defined on a field $\mathfrak{F}_1$ in a space Ω_1, with $P_1(\Omega_1) = 1$, such that $P_1(\sum_{j=1}^{\infty} E_j) = \sum_{j=1}^{\infty} P_1(E_j)$ whenever $\{E_j\}$ is a sequence of disjunct sets belonging to $\mathfrak{F}_1$ whose sum is also in $\mathfrak{F}_1$.

DEFINITION 4. A space Ω_1 in which a probability measure P_1 has been defined on a Borel field $\mathfrak{B}_1$ is a *probability space*.

DEFINITION 5. A *measurable set* in a probability space Ω_1 is a set E such that $E \, \epsilon \, \mathfrak{B}_1$.

DEFINITION 6. Let Ω_1 be a probability space and Ω_1' a space on which there is given a Borel field $\mathfrak{B}_1'$ of measurable sets. Let $\phi(x)$ be a single-valued function whose domain is Ω_1 and whose range is in Ω_1' . $\phi(x)$ is a *measurable function* if the set E of points $x \, \epsilon \, \Omega_1$ for which $\phi(x)$ is in $E' \subseteq \Omega_1'$ is measurable whenever E' is.[2] If $\phi(x)$ is real valued, it is measurable if for every real number λ the

Received May 2, 1938.

[1] All the fields used in this paper will also satisfy the condition that if $E \, \epsilon \, \mathfrak{F}_1$, then $CE \, \epsilon \, \mathfrak{F}_1$, where CE is the complement in Ω_1 of the set E.

[2] The symbol $\{\phi(x) \, \epsilon \, E'\}$ will be used to denote E.

461

set $\{\phi(x) < \lambda\}$ is measurable. The Borel field of sets on the real line is taken, in this paper, as the collection of Borel sets.

DEFINITION 7. If $\phi(x)$ is a real-valued measurable function on Ω_1, the function $F(\lambda)$ of the real variable λ, $F(\lambda) = P_1\{\phi(x) < \lambda\}$, is its *distribution function*.

DEFINITION 8. Two measurable functions ϕ and ϕ' (defined on Ω_1 and Ω_1', respectively) whose ranges lie in the same space Ω_1'' have the *same distribution* if, for every $E'' \epsilon \mathfrak{B}_1''$, $P_1\{\phi \epsilon E''\} = P_1'\{\phi' \epsilon E''\}$. Thus, in particular, two real-valued measurable functions have the same distribution if and only if their distribution functions are identical.

DEFINITION 9. The class of all measurable functions, with ranges in some fixed space, but not necessarily all defined on the same space, such that every two have the same distribution is a *chance variable*. Any member of this class is a *representation* of the chance variable.

So far we have defined a single chance variable ϕ, in isolation from all other chance variables. If ϕ' is another chance variable (with range in the range space of ϕ), our definition enables us to answer the questions "What is the probability that $\phi \epsilon E$?" and "What is the probability that $\phi' \epsilon E'$?", but it does not give an answer to the question "What is the probability that both $\phi \epsilon E$ and $\phi' \epsilon E'$?" Since chance variables usually present themselves not singly but in sets and are connected with each other in rather special ways, we are led to the following considerations.

Associated with every probability space Ω_1 there is another space Ω defined as follows. Let Ω be the space of all infinite sequences $\omega = \{x_1, x_2, \cdots \}$, where $x_j \epsilon \Omega_1$ $(j = 1, 2, \cdots)$. Let $\mathfrak{B}$ be the smallest Borel field which contains every set determined by conditions of the form $x_j \epsilon E_j$, $E_j \epsilon \mathfrak{B}_1$ $(j = 1, \cdots, n)$. Until we define a probability measure on $\mathfrak{B}$, we may not consider Ω as a probability space. We make, however, the following definition.

DEFINITION 10. If a probability measure P is defined on the Borel field $\mathfrak{B}$ in Ω, the probability space Ω is a *stochastic process* associated with Ω_1.[3]

DEFINITION 11. Let $\alpha_1, \cdots, \alpha_n$ be any finite set of subscripts. The set E is a *cylinder set* over $x_{\alpha_1}, \cdots, x_{\alpha_n}$ if, whenever $\omega \epsilon E$, $\omega = \{x_1, x_2, \cdots \}$, then any point ω', obtained from ω by altering the coördinates $x_{\alpha_1}, \cdots, x_{\alpha_n}$ only, is also in E.

The collection of all measurable cylinder sets over $x_{\alpha_1}, \cdots, x_{\alpha_n}$ forms a Borel field $\mathfrak{B}_{\alpha_1, \ldots, \alpha_n} \subseteq \mathfrak{B}$.

DEFINITION 12. If two probability measures are defined on the Borel fields $\mathfrak{B}'$ and $\mathfrak{B}''$ respectively, $\mathfrak{B}' \subseteq \mathfrak{B}$, $\mathfrak{B}'' \subseteq \mathfrak{B}$, they are *coherent* if they assign the same values to sets common to $\mathfrak{B}'$ and $\mathfrak{B}''$.

In terms of the preceding three definitions we are now able to formulate a mathematical description of at least one of the ways in which chance variables occur in physical problems.

DEFINITION 13. A sequence, finite or infinite, of chance variables ϕ_n (with

---

[3] This is not the most general definition of stochastic process, but it is the one that is to be used exclusively in this paper.

ranges all in the same space Ω_1) is *stochastic* if the following conditions are satisfied.

(i) To every set of conditions of the form $\phi_j \,\epsilon\, E_j$, $E_j \,\epsilon\, \mathfrak{B}_1$ $(j = 1, \cdots, n)$ there corresponds a number $P_n(\phi_1 \,\epsilon\, E_1, \cdots, \phi_n \,\epsilon\, E_n)$.

(ii) A unique probability measure may be so defined on the Borel field $\mathfrak{B}_n = \mathfrak{B}_{1,\ldots,n}$ that the set $\{x_1 \,\epsilon\, E_1\} \cdots \{x_n \,\epsilon\, E_n\}$ has measure

$$P_n(\phi_1 \,\epsilon\, E_1, \cdots, \phi_n \,\epsilon\, E_n).$$

(iii) The probability measures on the Borel fields $\mathfrak{B}_j$ $(j = 1, 2, \cdots)$ are coherent, each with the others.

(iv) The function[4] $x_j(\omega)$ is a representation of ϕ_j $(j = 1, 2, \cdots)$.

The following is a fundamental theorem on the representation of stochastic sequences of chance variables. It was proved in the case where Ω_1 is the real line by Kolmogoroff[5] and in the general case by Doob.[6]

THEOREM 1. *Given a stochastic sequence $\{\phi_n\}$ of chance variables, a unique probability measure P may be so defined on $\mathfrak{B}$ that P is coherent with each P_n $(n = 1, 2, \cdots)$.*

Let Ω_1 be a probability space, and for each n let the chance variable ϕ_n, with domain and range on Ω_1, take the value x at the point $x \,\epsilon\, \Omega_1$. It is well known[7] that if to every set of conditions of the form $\phi_j \,\epsilon\, E_j$, $E_j \,\epsilon\, \mathfrak{B}_1$ $(j = 1, \cdots, n)$ we assign the number $P_1(\phi_1 \,\epsilon\, E_1) \cdots P_1(\phi_n \,\epsilon\, E_n)$ the sequence $\{\phi_k\}$ is stochastic.

THEOREM 2. *Given any probability space Ω_1, a unique probability measure P may be so defined on Ω that a set determined by the conditions $x_j \,\epsilon\, E_j$, $E_j \,\epsilon\, \mathfrak{B}_1$ $(j = 1, \cdots, n)$ has the measure $P_1(\phi_1 \,\epsilon\, E_1) \cdots P_1(\phi_n \,\epsilon\, E_n)$.*

DEFINITION 14. A *system* is a sequence $\{f_n\}$ of measurable functions on the stochastic process Ω satisfying the following conditions.

(i) $f_1(\omega) \equiv 0$ or else $f_1(\omega) \equiv 1$.

(ii) $f_n(\omega)$ $(n > 1)$ depends only on $x_1, \cdots, x_{n-1}$.

(iii) $f_n(\omega)$ $(n = 1, 2, \cdots)$ takes only the values 0 and 1.

(iv) $P(\lim\limits_{n \to \infty} \sup f_n(\omega) = 1) = 1$.[8]

---

[4] $x_j(\omega)$ is the function which at the point $\omega = \{x_1, x_2, \cdots\}$ takes the value x_j.

[5] X, p. 27. (See bibliography at the end of this paper.)

[6] VI, §1.

[7] Saks, XII, Chapter 3. The property referred to is the conclusion of Theorem 1 for a finite sequence of chance variables.

[8] A gambling system has been so defined by Doob, IV, p. 365. Essentially the same definition was suggested, independently, by Birnbaum and Schreier, I; Wald, XIII; and Huntemann, IX. This definition describes mathematically our intuitive idea of a gambling system. The player will bet on the outcome of the n-th play if $f_n = 1$, and will refrain from betting otherwise. Condition (ii) states that at each stage the player knows the results of the preceding trials only. Condition (iv) is merely a mathematical convenience which ensures that the probability is one that the player bet an infinite number of times.

3

With every system on Ω we associate a transformation T, defined almost everywhere on Ω and taking Ω into itself, as follows.

DEFINITION 15.   Let $a_n(\omega)$ be the lowest integer satisfying the equation $\sum_{j=1}^{a_n} f_j(\omega) = n$. Then, by condition (iv) in the definition of a system, with almost every ω there is associated an infinite sequence $\{a_n\}$ of subscripts. The *system transformation* T is defined by $T(x_1, x_2, \cdots) = \{x_1', x_2', \cdots\} = \{x_{a_1}, x_{a_2}, \cdots\}$.[9]

DEFINITION 16.   Let Ω_1 and Ω_1' be probability spaces and let T_1 be a transformation with domain Ω_1 and range in Ω_1'. T_1 is *measure preserving* if for every measurable set $E' \subseteq \Omega_1'$, the set $E = \{T_1(\omega) \, \epsilon \, E'\}$ is measurable and $P_1(E) = P_1'(E')$. We shall also write $T^{-1}(E')$ for E.

DEFINITION 17.   If the real-valued measurable function $\phi(x)$ defined on the probability space Ω_1 is summable,[10] $\int_{\Omega_1} \phi \, dP_1$ is the *expectation* of the chance variable represented by ϕ. Usually when we work with an integral on the whole space, we shall omit the range of integration in the symbol. We write $E(\phi)$ for the expectation of ϕ.

DEFINITION 18.   Let Ω be a stochastic process, n a positive integer, and Λ a measurable set, $\Lambda \subseteq \Omega$. The set function $P(\Lambda E)$, $E \, \epsilon \, \mathfrak{B}_{1,\ldots,n}$, is a probability measure on $\mathfrak{B}_{1,\ldots,n}$ that vanishes whenever $P(E) = 0$. Hence, by a well known theorem on absolutely continuous measures,[11] there exists a non-negative summable function on Ω, uniquely determined except for a set of measure zero and depending on the coördinates $x_1, \cdots, x_n$ only, say $P(x_1, \cdots, x_n; \Lambda)$, such that

$$P(\Lambda E) = \int_E P(x_1, \cdots, x_n; \Lambda) \, dP \text{ for every set } E \, \epsilon \, \mathfrak{B}_{1,\ldots,n}. \quad P(x_1, \cdots, x_n; \Lambda)$$

is the *conditional probability* of Λ for given $x_1, \cdots, x_n$.[12]

DEFINITION 19.   A stochastic process for which $P(x_1, \cdots, x_{n-1}; x_n \, \epsilon \, E)$ is almost everywhere independent of $x_1, \cdots, x_{n-1}$ for every $n = 1, 2, \cdots$, and every $E \, \epsilon \, \mathfrak{B}_1$ is *independent*. (A stochastic sequence of chance variables will be called independent if the corresponding stochastic process is.)

DEFINITION 20.   An independent stochastic process Ω for which the $x_n(\omega)$ all have the same distribution is *stationary*.

It is readily seen that if $\{\phi_n\}$ is a stochastic sequence of independent chance variables, $P(\phi_1 \, \epsilon \, E_1, \cdots, \phi_n \, \epsilon \, E_n) = P(\phi_1 \, \epsilon \, E_1) \cdots P(\phi_n \, \epsilon \, E_n)$ (for $n = 1, 2, \cdots$ and $E_j \, \epsilon \, \mathfrak{B}_1$ $(j = 1, \cdots, n))$, and conversely.

In terms of our definitions we can now state the following fundamental theorems.

THEOREM 3.   *If $Q(E)$ is an additive, non-negative set function defined on a field $\mathfrak{F}_1$ in a space Ω_1, with $Q(\Omega_1) = 1$, and if $Q(\sum_{j=1}^{\infty} E_j) = \sum_{j=1}^{\infty} Q(E_j)$ whenever*

---

[9] Theorem 1 shows that the transformed space Ω' is a stochastic process.

[10] For integration in abstract spaces see Saks, XII, Chapter 1.

[11] See Saks, XII, p. 36.

[12] Kolmogoroff, X, Chapter 5.

$\{E_n\}$ is a sequence of disjunct sets belonging to $\mathfrak{F}_1$ whose sum is also in $\mathfrak{F}_1$, then $P(E)$ may be so defined on the smallest Borel field $\mathfrak{B}_1$ including $\mathfrak{F}_1$ that it becomes a probability measure that coincides with Q on sets of $\mathfrak{F}_1$.[13]

THEOREM 4.   *Let Ω be an independent, stationary, stochastic process corresponding to a probability space Ω_1 and let E_1 be any measurable set, $E_1 \subseteq \Omega_1$. Then if $g(x)$ is the characteristic function of E_1 on Ω_1, we have*

$$\lim_{n \to \infty} \frac{1}{n} \sum_{j=1}^{n} g(x_j) = P_1(E_1)$$

almost everywhere on Ω.[14]

THEOREM 5.   *On an independent, stationary, stochastic process every system transformation is measure preserving.*[15]

2. **Application to "Kollektivs".**   Much work has been done in recent years on refining the definition of "Kollektiv" as formulated by von Mises.[16]   In this section we shall derive a consequence of Theorem 5 which includes some of the results concerning the existence of certain "admissible numbers" or "Kollektivs".[17]

DEFINITION 21.   Let $Q(E)$ be an additive, non-negative set function defined on a field $\mathfrak{F}_1$ in a space Ω_1, with $Q(\Omega_1) = 1$.   To every set $E \subseteq \Omega_1$ make correspond the numbers

$$Q^*(E) = \underset{\substack{E_1 \supseteq E \\ E_1 \,\epsilon\, \mathfrak{F}_1}}{\text{g.l.b. }} Q(E_1) \quad \text{and} \quad Q_*(E) = \underset{\substack{E_1 \subseteq E \\ E_1 \,\epsilon\, \mathfrak{F}_1}}{\text{l.u.b. }} Q(E_1).$$

Let $\mathfrak{J}_1$ be the collection of sets such that $Q^*(E) = Q_*(E)$.   It is readily verified that $\mathfrak{J}_1$ is a field, $\mathfrak{J}_1 \supseteq \mathfrak{F}_1$, and we define for every $E \,\epsilon\, \mathfrak{J}_1$, $Q(E) = Q^*(E) = Q_*(E)$. $\mathfrak{J}_1$ is the collection of *Jordan measurable sets* with respect to $\mathfrak{F}_1$.

THEOREM 6.   *Let P_1 be a probability measure defined on a denumerable field $\mathfrak{F}_1$ in a space Ω_1.   Let $\mathfrak{J}_1$ be the collection of Jordan measurable sets with respect to $\mathfrak{F}_1$, and $\mathfrak{B}_1$ the Borel extension of $\mathfrak{J}_1$.   Since P_1 may be defined on $\mathfrak{J}_1$ and then on $\mathfrak{B}_1$ coherently with its definition on $\mathfrak{F}_1$ so that it is a probability measure, Ω_1 becomes a probability space.   Let Ω be the independent, stationary, stochastic process associated with Ω_1 and let $T_1, T_2, \cdots$ be any sequence of system transformations on Ω.   Finally, write $T_n(x_1, x_2, \cdots) = \{x_1^n, x_2^n, \cdots\}$.   Then there is a set Z of measure zero on Ω such that*

$$\lim_{N \to \infty} \frac{1}{N} \sum_{j=1}^{N} g_E(x_j^n) = P_1(E)$$

[13] See, for example, Hahn, VII, p. 433.

[14] This theorem is known as the strong law of large numbers.   For a proof see VIII, p. 37, or V, p. 764.

[15] Doob (IV, p. 365) proved this for the case of real-valued chance variables.   The proof is the same as for this case.   This theorem (which we shall obtain later as a special case of Theorem 10 of this paper) is what corresponds in the classical theory to the von Mises "Regellosigkeit" principle.   See XI, p. 14.

[16] See von Mises, XI.

[17] See, for example, Copeland, II and III, and Wald, XIII.

for every $\omega \, \epsilon \, \Omega - Z$, *every* $E \, \epsilon \, \mathfrak{J}_1$ ($gE(x)$ *denoting the characteristic function of* E), *and all* $n = 1, 2, \cdots$.

   Proof.   Let E_1, E_2, $\cdots$ be the total collection of sets in $\mathfrak{J}_1$.   By Theorem 4

$$\lim_{N \to \infty} \sum_{j=1}^{N} g_{E_i}(x_j) = P_1(E_i)$$

except on a set A_i of measure zero ($i = 1, 2, \cdots$).   Write $A = A_1 + A_2 + \cdots$, and $C_n = T_n^{-1}(A)$.   Since A is of measure zero and each T_n is measure preserving (Theorem 5), C_n is also of measure zero.   Finally T_n may fail to be defined on a set D_n of measure zero.   Write $Z = A + (C_1 + C_2 + \cdots) + (D_1 + D_2 + \cdots)$; then $P(Z) = 0$.   Now take ω to be any point $\epsilon \, \Omega - Z$, E to be any set in $\mathfrak{J}_1$ and n to be any positive integer.   There exist two sequences of sets O_k and I_k such that $I_k \subseteq E \subseteq O_k$, $I_k \, \epsilon \, \mathfrak{J}_1$, $O_k \, \epsilon \, \mathfrak{J}_1$ ($k = 1, 2, \cdots$), and $\lim_{k \to \infty} P_1(I_k) = \lim_{k \to \infty} P_1(O_k) = P_1(E)$.   The sets I_k and O_k are among the sets E_1, E_2, $\cdots$, hence we have for all k

$$\lim_{N \to \infty} \frac{1}{N} \sum_{j=1}^{N} g_{I_k}(x_j) = P_1(I_k) \quad \text{and} \quad \lim_{N \to \infty} \frac{1}{N} \sum_{j=1}^{N} g_{O_k}(x_j) = P_1(O_k),$$

since ω was taken out of $Z \supseteq A_1 + A_2 + \cdots$.   Also $T_n(\omega)$ is not in $A_1 + A_2 + \cdots$, for otherwise ω would be contained in some C_n, and this contradicts the assumption that $\omega \, \epsilon \, \Omega - Z$.   Hence the last written relations hold when x_j is replaced by x_j^n; thus

$$\lim_{N \to \infty} \frac{1}{N} \sum_{j=1}^{N} g_{I_k}(x_j^n) = P_1(I_k) \quad \text{and} \quad \lim_{N \to \infty} \frac{1}{N} \sum_{j=1}^{N} g_{O_k}(x_j^n) = P_1(O_k).$$

Since $I_k \subseteq E \subseteq O_k$, we have $g_{I_k}(x) \leqq g_E(x) \leqq g_{O_k}(x)$ for all $x \, \epsilon \, \Omega_1$.   Hence

$$\frac{1}{N} \sum_{j=1}^{N} g_{I_k}(x_j^n) \leqq \frac{1}{N} \sum_{j=1}^{N} g_E(x_j^n) \leqq \frac{1}{N} \sum_{j=1}^{N} g_{O_k}(x_j^n);$$

whence

$$P(I_k) = \lim_{N \to \infty} \frac{1}{N} \sum_{j=1}^{N} g_{I_k}(x_j^n) \leqq \liminf_{N \to \infty} \frac{1}{N} \sum_{j=1}^{N} g_E(x_j^n)$$

$$\leqq \limsup_{N \to \infty} \frac{1}{N} \sum_{j=1}^{N} g_E(x_j^n) \leqq \lim_{N \to \infty} \frac{1}{N} \sum_{j=1}^{N} g_{O_k}(x_j^n) = P_1(O_k).$$

Since this is true for all k, we have

$$P_1(E) \leqq \liminf \leqq \limsup \leqq P_1(E),$$

so that

$$\liminf = \limsup = \lim_{N \to \infty} \frac{1}{N} \sum_{j=1}^{N} g_E(x_j^n) = P_1(E).[18]$$

---

[18] If we exclude trivial cases by insisting that P_1 have at least two different positive values, the set of points ω for which the conclusion of the theorem holds always has the power of the continuum.

3. **The von Mises definition of independence.** The axioms and definitions of the theory of probability as expounded by von Mises have been shown to correspond to theorems in the classical theory. Thus, for example, the various formulations of the law of large numbers correspond to the frequency definition of probability, and Theorem 5 of this paper corresponds to the "Regellosigkeit" principle. We shall now prove a theorem that expresses the fundamental idea in the von Mises definition of independence. There is a close analogy between this theorem and Theorem 5, and the method of proof here employed is similar to that of Doob.[19]

Let Ω be any stochastic process and Ω' an independent and stationary stochastic process. Let Ω^2 be the space of all sequences $\omega^2 = \{(x_1, y_1),$ $(x_2, y_2), \cdots \}$, where $\{y_1, y_2, \cdots \} \epsilon \Omega$ and $\{x_1, x_2, \cdots \} \epsilon \Omega'$. A unique probability measure is defined on Ω^2 by the conditions

$$P\{(x_1 \epsilon E_1) \cdots (x_n \epsilon E_n)(y_1 \epsilon F_1) \cdots (y_n \epsilon F_n)\}$$
$$= P(x_1 \epsilon E_1) \cdots P(x_n \epsilon E_n)P\{(y_1 \epsilon F_1) \cdots (y_n \epsilon F_n)\}.$$

THEOREM 7. *Let $E_1, E_2, \cdots$ be an infinite sequence of measurable sets in the range space of Ω such that the probability is one that an infinite number of the conditions $(y_n \epsilon E_n)$ are satisfied. Let $a_n(\omega^2)$ be the n-th subscript such that $y_{a_n} = E_{a_n}$. To every point ω^2 of Ω^2 make correspond the point $\omega' \epsilon \Omega'$, $\omega' = \{x_1', x_2', \cdots \}$, where $x_n' = x_{a_n}$. The transformation T so defined is defined almost everywhere on Ω^2 taking values in Ω' and is measure preserving in the sense that if Λ' is any measurable set in Ω' and $\Lambda^2 = T^{-1}(\Lambda')$ is the total set of points of Ω^2 whose images are in Λ', then Λ^2 is measurable and $P(\Lambda^2) = P(\Lambda').$*

Proof. It is sufficient to prove the theorem for sets Λ' of the form

(1) $$\Lambda' = \{x_1' \epsilon E_1', \cdots, x_n' \epsilon E_n'\},$$

where $E_1', \cdots, E_n'$ are measurable sets in the range space of Ω'. We proceed by induction.

Let $n = 1$; $\Lambda' = \{x_1' \epsilon E_1'\}$. We have, excepting always the set of measure zero where T may not be defined,

(2) $$\Lambda^2 = \{(y_1 \epsilon E_1)(x_1 \epsilon E_1')\} + \{(y_1 \epsilon CE_1)(y_2 \epsilon E_2)(x_2 \epsilon E_1')\} + \cdots .$$

Hence, since the sets in this sum are disjunct we have, by the definition of measure on Ω^2,

(3) $$P(\Lambda^2) = P(x_1 \epsilon E_1')P(y_1 \epsilon E_1) + P(x_2 \epsilon E_1')P\{(y_1 \epsilon CE_1)(y_2 \epsilon E_2)\} + \cdots .$$

The last written expression is the product of $P(x_n \epsilon E_1')$ by the probability that at least one of the conditions $(y_n \epsilon E_n)$ is satisfied, hence $P(\Lambda^2) = P(x_n \epsilon E_1') = P(\Lambda')$.

This proves the theorem for $n = 1$. Assume that it is true for $n - 1$. Write

[19] IV.

7

$\Lambda' = \{x_1' \epsilon E_1', \cdots, x_n' \epsilon E_n'\}$, $\Lambda^2 = \{T^{-1}(\Lambda')\}$. In order to prove that Λ^2 is measurable consider the functions $x_j'(\omega^2) = x_{a_j}(\omega^2)$.

$$\{x_{a_j}(\omega^2) \epsilon E\} = \{a_j = j\}\{x_j \epsilon E\} + \{a_j = j + 1\}\{x_{j+1} \epsilon E\} + \cdots.$$

The sets $\{x_m \epsilon E\}$ are measurable, if E is any measurable set in the range space of Ω'. The set $\{a_j = m\}$ is the set where the j-th y-coördinate of ω^2 which belongs to its E_i is y_m; this set is the sum of sets where y_m is in E_m and of the preceding y's precisely $j - 1$ are in their E_i. These summands, and hence their sum, and therefore, finally, the set $\{x_j' \epsilon E\}$ are measurable. Hence Λ, being the intersection of the n sets $\{x_j' \epsilon E_j'\}$, is measurable.

Denote by Λ_0 the set $\{x_j' \epsilon E_j' \ (j = 1, \cdots, n - 1)\}$. Then

$$(4) \qquad \Lambda^2 = \sum_{j=n}^{\infty} \{\omega^2 \epsilon \Lambda_0, a_n = j, x_j \epsilon E_n'\}.$$

Consider any summand in (4): $\{\omega^2 \epsilon \Lambda_0, a_n = j, x_j \epsilon E_n'\}$. It is readily verified that the set $\{\omega^2 \epsilon \Lambda_0, a_n = j\}$ is a cylinder set over the first $j - 1$ (x, y)-coördinates of ω^2. Hence

$$(5) \qquad P(\omega^2 \epsilon \Lambda_0, a_n = j, x_j \epsilon E_n') = P(\omega^2 \epsilon \Lambda_0, a_n = j)P(x_j \epsilon E_n').$$

From (4) and (5), the stationary character of Ω', and the disjunct nature of the summands in (4) we obtain

$$(6) \qquad P(\Lambda) = P(x_n \epsilon E_n') \sum_{j=n}^{\infty} P(\omega^2 \epsilon \Lambda_0, a_n = j).$$

The last written summation is the measure of the intersection of Λ_0 with the set where at least one of the conditions $(y_n \epsilon E_n)$ is satisfied for $n > j$: the measure of the latter set is one; hence

$$(7) \qquad P(\Lambda) = P(x_n \epsilon E_n') \cdot P(\Lambda_0).$$

The theorem follows immediately from the induction hypothesis.

4. **The invariance of expectation.** Theorem 5 asserts that under certain hypotheses all probability relations are invariant under a system transformation. In this section we make less restrictive hypotheses on the stochastic process to obtain the conclusion (Theorem 8) that the "fairness" of a gambling game of which the stochastic process is a mathematical description is invariant under a system—where the criterion for fairness is expressed, as usual, in terms of the vanishing of certain expectations.

DEFINITION 22. Let Ω be a stochastic process associated with a probability space consisting of real numbers; let n be a positive integer; and suppose that $x_j(\omega)$ is summable for $j = 1, 2, \cdots$. $Q(E) = \int_E x_n(\omega) \, dP$, $E \epsilon \mathfrak{B}_{1, \ldots, (n-1)}$, is a finite, completely additive set function on $\mathfrak{B}_{1, \ldots, (n-1)}$ that vanishes whenever $P(E) = 0$. Hence there exists a summable function on Ω, uniquely determined except for a set of measure zero and depending on the coördinates $x_1, \cdots, x_{n-1}$ only, say $E(x_1, \cdots, x_{n-1}; x_n)$, such that

$$\int_E x_n(\omega)\, dP = \int_E E(x_1, \cdots, x_{n-1}; x_n)\, dP$$

for every set $E \in \mathfrak{B}_{1,\cdots,(n-1)}$. $E(x_1, \cdots, x_{n-1}; x_n)$ is the *conditional expectation* of x_n for given $x_1, \cdots, x_{n-1}$.

THEOREM 8. *Let Ω be a real-valued stochastic process in which the functions $x_n(\omega)$ are uniformly bounded. Let T be a system transformation on Ω taking $\omega = \{x_1, x_2, \cdots\}$ into $\omega' = \{x_1', x_2', \cdots\}$. If $E(x_1, \cdots, x_{n-1}; x_n)$ vanishes for all n and almost all $x_1, \cdots, x_{n-1}$, then $E(x_1', \cdots, x_{n-1}'; x_n')$ vanishes for all n and almost all $x_1', \cdots, x_{n-1}'$.*

Proof. Our hypothesis is that

$$(8) \qquad \int_M E(x_1, \cdots, x_{n-1}; x_n)\, dP = \int_M x_n\, dP = 0$$

for all n and all measurable cylinder sets M over $x_1, \cdots, x_{n-1}$. We are to prove that

$$(9) \qquad \int_{M'} E(x_1', \cdots, x_{n-1}'; x_n')\, dP = \int_{M'} x_n'\, dP = 0$$

for all n and all measurable cylinder sets M' over $x_1', \cdots, x_{n-1}'$. It is sufficient to prove this result for sets M' of the form

$$M' = \{x_1' \in E_1, \cdots, x_{n-1}' \in E_{n-1}\},$$

where $E_1, \cdots, E_{n-1}$ are Borel sets on the real line.[20]

We start then with M' and derive an expression for $P(M'\{x_n' \in E_n\})$, where E_n is a Borel set on the real line. We have

$$\{x_m' \in E\} = \{a_m = m\}\{x_m \in E\} + \{a_m = m+1\}\{x_{m+1} \in E\} + \cdots$$

$$(10) \qquad\qquad\qquad = \sum_{j=0}^{\infty} \{a_m = m+j\}\{x_{m+j} \in E\}.$$

Hence

$$M'\{x_n' \in E_n\} = \{x_1' \in E_1\}\{x_2' \in E_2\} \cdots \{x_{n-1}' \in E_{n-1}\}\{x_n' \in E_n\}$$

$$= (\sum_{j_1=0}^{\infty} \{a_1 = j_1 + 1\}\{x_{j_1+1} \in E_1\}) \cdots (\sum_{j_n=0}^{\infty} \{a_n = j_n + n\}\{x_{j_n+n} \in E_n\})$$

$$(11) \qquad = \sum_{j_n=0}^{\infty} \cdots \sum_{j_1=0}^{\infty} \{a_1 = j_1 + 1\} \cdots \{a_n = j_n + n\}\{x_{j_1+1} \in E_1\}$$

$$\cdots \{x_{j_n+n} \in E_n\}.\text{[21]}$$

[20] Every measurable set over $x_1', \cdots, x_{n-1}'$ can be obtained from sets such as M' by at most a denumerable sequence of sum, product, and complement processes. If the integral vanishes for all sets such as M', it will vanish for sums of such sets, etc.

[21] $a_1(\omega), a_2(\omega), \cdots$ is the subscript sequence associated with T, as in Definition 15.

The last expression follows from the ordinary algebra of point sets.   We make certain easily verifiable remarks about the last written sum.

(12)   The sets being added are disjunct in pairs.

(13)   Since, for example, a_2 is always *greater than* a_1, the set $\{a_1 = j_1 + 1\}$ $\{a_2 = j_2 + 2\}$ is empty if $j_1 + 1 \geq j_2 + 2$; i.e., it is empty when $j_1 > j_2$.

In virtue of (13) we have from (11)

$$
(14) \quad M'\{x'_n \,\epsilon\, E_n\} = \sum_{j_n=0}^{\infty} \sum_{j_{n-1}=0}^{j_n} \cdots \sum_{j_1=0}^{j_2} \{a_1 = j_1 + 1\}
$$
$$
\cdots \{a_n = j_n + n\}\{x_{j_1+1} \,\epsilon\, E_1\} \cdots \{x_{j_n+n} \,\epsilon\, E_n\}.
$$

From this we have, in virtue of (12),

$$
(15) \quad P(M'\{x'_n \,\epsilon\, E_n\}) = \sum_{j_n=0}^{\infty} \cdots \sum_{j_1=0}^{j_2} P(\{a_1 = j_1 + 1\}
$$
$$
\cdots \{a_n = j_n + n\}\{x_{j_1+1} \,\epsilon\, E_1\} \cdots \{x_{j_n+n} \,\epsilon\, E_n\}).
$$

Let us write

$$
M_{j_1\cdots j_n} = \{a_1 = j_1 + 1\} \cdots \{a_n = j_n + n\}\{x_{j_1+1} \,\epsilon\, E_1\}\{x_{j_{n-1}+n-1} \,\epsilon\, E_{n-1}\}.
$$

It is immediately verifiable that the set $M_{j_1\cdots j_n}$ is a cylinder set over the coördinates $x_1, \cdots, x_{j_n+n-1}$.[22]   Write also

$$
M_{j_n} = \sum_{j_{n-1}=0}^{j_n} \cdots \sum_{j_1=0}^{j_2} M_{j_1\cdots j_n}.
$$

Since we already saw that the summands are disjunct sets, we have

$$
(16) \quad P(M_{j_n}) = \sum_{j_{n-1}=0}^{j_n} \cdots \sum_{j_1=0}^{j_2} P(M_{j_1\cdots j_n}).
$$

Since the x_n were assumed to be uniformly bounded, the x'_n will be uniformly bounded, and therefore summable.

$$
(17) \quad \int_{M'} x'_n \, dP = \lim_{\delta\to 0} \sum_{r=-\infty}^{\infty} r\delta P(M'\{r\delta \leq x'_n < (r + 1)\delta\}).
$$

Then, from (15)

$$
(18) \quad \int_{M'} x'_n \, dP = \lim_{\delta\to 0} \sum_{r=-\infty}^{\infty} r\delta \sum_{j_n=0}^{\infty} P(M_{j_n}\{r\delta \leq x_{j_n+n} < (r + 1)\delta\}).
$$

We may now define the function $y(\omega)$ to be equal to x_{j_n+n} on M_{j_n} ($j_n = 0, 1, 2, \cdots$), and to be zero elsewhere.   (The sets M_{j_n} are disjunct sets.   The function $y(\omega)$ really depends on n, but since n is being held fixed in this discussion we do not indicate this dependence.)   Then

---

[22] See (ii), Definition 14.

$$\int_{M'} x_n' \, dP = \lim_{\delta \to 0} \sum_{r=-\infty}^{\infty} r\delta \sum_{j_n=0}^{\infty} P(M_{j_n}\{r\delta \leqq y < (r+1)\delta\})$$

$$= \lim_{\delta \to 0} \sum_{r=-\infty}^{\infty} r\delta P \left(\sum_{j_n=0}^{\infty} M_{j_n}\{r\delta \leqq y < (r+1)\delta\}\right)$$

$$= \int_{\sum_{j_n=0}^{\infty} M_{j_n}} y \, dP = \sum_{j_n=0}^{\infty} \int_{M_{j_n}} y \, dP = \sum_{j_n=0}^{\infty} \int_{M_{j_n}} x_{j_n+n} \, dP$$

(19)
$$= \sum_{j_n=0}^{\infty} \lim_{\delta \to 0} \sum_{r=-\infty}^{\infty} r\delta P(M_{j_n}\{r\delta \leqq x_{j_n+n} < (r+1)\delta\})$$

$$= \sum_{j_n=0}^{\infty} \lim_{\delta \to 0} \sum_{r=-\infty}^{\infty} r\delta \sum_{j_{n-1}=0}^{j_n} \cdots \sum_{j_1=0}^{j_2} P(M_{j_1\cdots j_n}\{r\delta \leqq x_{j_n+n} < (r+1)\delta\})$$

$$= \sum_{j_n=0}^{\infty} \sum_{j_{n-1}=0}^{j_n} \cdots \sum_{j_1=0}^{j_2} \lim_{\delta \to 0} \sum_{r=-\infty}^{\infty} r\delta P(M_{j_1\cdots j_n}\{r\delta \leqq x_{j_n+n} < (r+1)\delta\})$$

$$= \sum_{j_n=0}^{\infty} \sum_{j_{n-1}=0}^{j_n} \cdots \sum_{j_1=0}^{j_2} \int_{M_{j_1}} x_{j_n+n} \, dP = 0.$$

This concludes the proof of Theorem 8.

We remark on the analogy between this theorem and Theorem 5. Here we proved, essentially, that if $E(x_1, \cdots, x_{n-1} ; x_n)$ has a constant value independent of n and $x_1, \cdots, x_{n-1}$, then $E(x_1', \cdots, x_{n-1}' ; x_n')$ will also have that constant value independent of n and $x_1', \cdots, x_{n-1}'$. Theorem 5, on the other hand, may be phrased as follows. If $P(x_1, \cdots, x_{n-1} ; x_n \, \epsilon \, E)$ has a constant value depending only on E, but not on n or $x_1, \cdots, x_{n-1}$, then $P(x_1', \cdots, x_{n-1}' ; x_{n-1}' \, \epsilon \, E)$ will also have that constant value dependent only on E, but not on n or $x_1', \cdots, x_{n-1}'$.

5. The invariance of asymptotic independence.

DEFINITION 23. The stochastic process Ω is *uniformly asymptotically independent* if there exists a probability measure F on its range space such that $P(x_1, \cdots, x_{n-1} ; x_n \, \epsilon \, E)$ converges uniformly in ω (but not necessarily in E) almost everywhere to $F(E)$, where E is any measurable set.

The first purpose of this section is to prove the following theorem.

THEOREM 9. *The property of uniform asymptotic independence is invariant under every system transformation.*

Instead of proving this theorem we shall state and prove the following slightly stronger one.

THEOREM 10. *If for some $\epsilon > 0$, measurable set E, and positive integer n_0,* $|P(x_1, \cdots, x_{n-1} ; x_n \, \epsilon \, E) - F(E)| < \epsilon$ *for all $n \geqq n_0$ almost everywhere, then* $|P(x_1', \cdots, x_{n-1}' ; x_n' \, \epsilon \, E) - F(E)| < \epsilon$ *for all $n \geqq n_0$ almost everywhere (where the x_n' are obtained from the x_n by a system transformation).*

The important difference between this theorem and the preceding one is that here we definitely state that the ϵ and n_0 after the system transformation are the same as before.

Proof. Let I'_{n-1} be any measurable cylinder set over $x'_1, \cdots, x'_{n-1}$. Then

$$\left| \int_{I'_{n-1}} P(x'_1, \cdots, x'_{n-1}; x'_n \epsilon E) - F(E)\, dP \right| = |P(I'_{n-1}\{x'_n \epsilon E\}) - P(I'_{n-1})F(E)|$$

$$= \left| \sum_{j=0}^{\infty} P(I'_{n-1}\{a_n = n+j\}\{x_{n+j} \epsilon E\}) - F(E)P(I'_{n-1}) \right|$$

(20)
$$\leqq \sum_{j=0}^{\infty} \int_{I'_{n-1}\{a_n=n+j\}} |P(x_1, \cdots, x_{n+j-1}; x_{n+j} \epsilon E) - F(E)|\, dP$$

$$\leqq \sum_{j=0}^{\infty} \int_{I'_{n-1}\{a_n=n+j\}} \epsilon\, dP = \int_{I'_{n-1}} \epsilon\, dP.$$

Since this is true uniformly in I'_{n-1}, we have $|P(x'_1, \cdots, x'_{n-1};$ $x'_n \epsilon E) - F(E)| < \epsilon$ for $n \geqq n_0$ almost everywhere. This concludes the proof of Theorem 10, and therefore of Theorem 9.

The reason for stating Theorem 10 in its present form is that in this form Theorem 5 is easily seen to follow from it. For, according to the hypotheses of Theorem 5, the hypotheses of Theorem 10 are satisfied with $n_0 = 1$ and every $\epsilon > 0$. The conclusion of Theorem 10 then assures us that after any system transformation the conditions are still satisfied, with $n_0 = 1$ and every $\epsilon > 0$. This implies that the transformed process is independent, stationary, and has the same distributions as the original stochastic process.

Besides the situation just mentioned, there are many other important examples of uniformly asymptotically independent stochastic processes. It is clear, however, that uniform asymptotic independence is a very strong condition. There may be stochastic processes which from a practical point of view are independent in the long run, without being uniformly asymptotically independent in the sense of Definition 23. In order to investigate such processes we make the following definitions.

DEFINITION 24. The stochastic process Ω is *asymptotically independent in probability* if there exists a probability measure F on its range space such that $P(x_1, \cdots, x_{n-1}; x_n \epsilon E)$ converges in probability to $F(E)$, where E is any measurable set. (That is: to every positive number ϵ and measurable set E there corresponds a positive integer n_0 such that $P\{| P(x_1, \cdots, x_{n-1};$ $x_n \epsilon E) - F(E) | > \epsilon\} < \epsilon$ for $n > n_0$.)

DEFINITION 25. The stochastic process Ω is ϵ-*multiplicative* if there exists a probability measure F on its range space such that to every positive number ϵ there corresponds a positive integer n_0 such that

$$\lceil P(\Lambda\{x_n \epsilon E\}) - P(\Lambda)F(E) | < \epsilon \qquad \text{for } n > n_0,$$

where E is a preassigned measurable set, uniformly in the measurable cylinder set Λ over $x_1, \cdots, x_{n-1}$.

Before investigating the invariance under system transformations of asymptotic independence in probability, we need the following auxiliary theorem.

THEOREM 11. *A necessary and sufficient condition that a stochastic process be asymptotically independent in probability is that it be ϵ-multiplicative.*

Proof. We use the notation of Definitions 24 and 25. We first prove that if the process is ϵ-multiplicative, then it is asymptotically independent in probability. Suppose that the conclusion of the theorem is false: then there exists a positive number ϵ such that

$$(21) \qquad P\{|\ P(x_1, \cdots, x_{n-1}\ ; x_n\ \epsilon\ E) - F(E)\ | > \epsilon\} \geqq \epsilon$$

for an infinite number of values of n. This implies that either there is an infinite number of values of n for which the difference is $> \epsilon$ and positive on a set of measure $\geqq \epsilon$, or else that there is an infinite number of values of n for which the difference is negative and $< - \epsilon$ on a set of measure $\geqq \epsilon$. In either case there is an infinite number of values of n corresponding to which we may find a measurable cylinder set Λ_n over $x_1, \cdots, x_{n-1}$ such that $P(\Lambda_n) \geqq \epsilon$ and such that the difference in (21) is of one sign on all Λ_n and is greater in absolute value than ϵ. For all these values of n we have

$$|\ P(\Lambda_n\{x_n\ \epsilon\ E\}) - P(\Lambda_n)F(E)\ | = \left|\int_{\Lambda_n} P(x_1, \cdots, x_{n-1}; x_n\ \epsilon\ E) - F(E)\ dP\right|$$

$$= \int_{\Lambda_n} |\ P(x_1, \cdots, x_{n-1}; x_n\ \epsilon\ E) - F(E)\ |\ dP \geqq \epsilon P(\Lambda_n) \geqq \epsilon^2.$$

Since this contradicts the assumption of ϵ-multiplicativity, the sufficiency of the condition is proved.

Assume now that the process is asymptotically independent in probability. Then

$$|\ P(\Lambda\{x_n\ \epsilon\ E\}) - P(\Lambda)F(E)\ | = \left|\int_{\Lambda} P(x_1, \cdots, x_{n-1}; x_n\ \epsilon\ E) - F(E)dP\right|$$

$$\leqq \int_{\Lambda} |\ P(x_1, \cdots, x_{n-1}; x_n\ \epsilon\ E) - F(E)\ |dP$$

$$= \int_{\Lambda^\epsilon} + \int_{\Lambda_\epsilon} |\ \cdots\ |dP,$$

where Λ_ϵ is the part of Λ on which the integrand is $\leqq \epsilon$, and $\Lambda^\epsilon = \Lambda - \Lambda_\epsilon$. For n sufficiently large we have, using asymptotic independence in probability,

$$|\ P(\Lambda\{x_n\ \epsilon\ E\}) - P(\Lambda)F(E)\ | \leqq \epsilon\ P(\Lambda_\epsilon) + 2P(\Lambda^\epsilon) \leqq \epsilon + 2\epsilon$$

(since $|\ P(x_1, \cdots, x_{n-1}\ ; x_n\ \epsilon\ E) - F(E)| \leqq 2$). This concludes the proof of Theorem 11.

We are now able to begin the examination of the behavior of asymptotic independence in probability under system transformations.

13

THEOREM 12.   *Let Ω be a stochastic process and let F be a probability measure on its range space such that $P(x_1, \cdots, x_{n-1}; x_n \, \epsilon \, E)$ converges almost everywhere to $F(E)$, where E is any measurable set.   Then, if T is any system transformation on Ω, the transformed process $T(\Omega)$ is asymptotically independent in probability.*

We note that the conditions put on Ω and those that $T(\Omega)$ is proved to satisfy are not the same.   The process Ω is asymptotically independent in probability, and a little more: the conditional probability distributions are assumed not merely to converge in probability but to converge almost everywhere; but the process $T(\Omega)$ is merely asserted to be asymptotically independent in probability.   That this lack of symmetry is in a certain sense in the nature of things will be shown later.

Proof.   According to Theorem 11 it is sufficient to prove that $T(\Omega)$ is ϵ-multiplicative; that is, it is sufficient to show that, for every ϵ, $|\, P(\Lambda'\{x_n' \, \epsilon \, E\}) - P(\Lambda')F(E)\,| < \epsilon$, for n sufficiently large, uniformly in the measurable cylinder set Λ' over $x_1', \cdots, x_{n-1}'$.   We have, using our usual notation for system transformations,

$$(22) \qquad P(\Lambda'\{x_n' \, \epsilon \, E\}) = \sum_{j=0}^{\infty} P(\Lambda'\{a_n = n + j\}\{x_{n+j} \, \epsilon \, E\}).$$

Hence

$$|\, P(\Lambda'\{x_n' \, \epsilon \, E\}) - P(\Lambda')F(E)\,|$$

$$= \left| \sum_{j=0}^{\infty} P(\Lambda'\{a_n = n + j\}\{x_{n+j} \, \epsilon \, E\}) - P(\Lambda'\{a_n = n + j\})F(E) \right|$$

$$(23) \qquad = \left| \sum_{j=0}^{\infty} \int_{\Lambda'\{a_n = n+j\}} P(x_1, \cdots, x_{n+j-1}; x_{n+j} \, \epsilon \, E) - F(E) \, dP \right|$$

$$\leqq \sum_{j=0}^{\infty} \int_{\Lambda'\{a_n = n+j\}} |\, P(x_1, \cdots, x_{n+j-1}; x_{n+j} \, \epsilon \, E) - F(E)\,|\, dP.$$

Since $P(x_1, \cdots, x_{n-1}; x_n \, \epsilon \, E)$ converges to $F(E)$, it converges, by Egoroff's theorem,[23] uniformly on a set D, with $P(D)$ arbitrarily close to one.   Hence for given $\epsilon > 0$ we may select D so that $P(CD) = 1 - P(D) < \epsilon$.   On D we have

$$(24) \qquad |\, P(x_1, \cdots, x_{n+j-1}; x_{n+j} \, \epsilon \, E) - F(E)\,| < \epsilon$$

for n sufficiently large and $j = 0, 1, 2, \cdots$.   Hence, from (23)

$$|\, P(\Lambda'\{x_n' \, \epsilon \, E\}) - P(\Lambda')F(E)\,|$$

$$\leqq \sum_{j=0}^{\infty} \int_{\Lambda'\{a_n=n+j\}D} + \int_{\Lambda'\{a_n=n+j\}CD} |\, P(x_1, \cdots, x_{n+j-1}; x_{n+j} \, \epsilon \, E) - F(E)\,|\, dP$$

$$(25) \qquad \leqq \sum_{j=0}^{\infty} \int_{\Lambda'\{a_n=n+j\}D} \epsilon \, dP + \sum_{j=0}^{\infty} \int_{\Lambda'\{a_n=n+j\}CD} 2 \, dP$$

$$= \epsilon P(\Lambda'D) + 2 P(\Lambda' \cdot CD) \leqq \epsilon + 2\epsilon,$$

as was to be proved.

[23] Saks, XII, p. 18.

THEOREM 13. *Under the hypotheses of Theorem 12, to every positive integer k, measurable sets $E_1, \cdots, E_k$, and positive number ϵ there corresponds a positive integer n_0 such that*

$$| P(\Lambda'\{x'_{n+1} \epsilon E_1\} \cdots \{x'_{n+k} \epsilon E_k\}) - F(E_1) \cdots F(E_k)P(\Lambda') | < \epsilon$$

for all $n > n_0$, uniformly in the measurable cylinder set Λ' over $x'_1, \cdots, x'_n$.

This slightly more general theorem follows readily from Theorem 12 by mathematical induction.

In order to show that without some extra hypothesis (such as convergence in Theorem 12) we may not assert asymptotic independence in probability about $T(\Omega)$, we construct an example of a stochastic process which is asymptotically independent in probability, but which loses this property under a suitable system transformation.

Take $0 < q < p < 1$. Let Ω be a stochastic process in which each x_n takes only the values 0 and 1, and in which the following conditions are satisfied. For $2^m \leqq n \leqq 2^{m+1} - 1$, $P(x_1, \cdots, x_{n-1} ; x_n = 0)$ depends only on the coordinates $x_1, \cdots, x_m$ $(m = 0, 1, 2, \cdots)$. For each positive integer m arrange the 2^m possible sets of values of $(x_1, \cdots, x_m)$ in some order: say, for definiteness, that these sets are ordered according to the magnitude of the dyadic fractions $.x_1 x_2 \cdots x_m$. Let $P(x_1 = 0) = p$. Let $P(x_1, \cdots, x_{2m-1} ; x_{2m} = 0)$ be p when $(x_1, \cdots, x_m)$ is the first set in this ordering and let $P(x_1, \cdots, x_{2m-1} ; x_{2m} = 0)$ be q otherwise. Let $P(x_1, \cdots, x_{2m} ; x_{2m+1} = 0)$ be p when $(x_1, \cdots, x_m)$ is the second set and q otherwise; and so on, for each n between 2^m and $2^{m+1} - 1$, and every $m = 0, 1, 2, \cdots$. According to a theorem of Doob's[24] these conditions are consistent. That is, if we define $P(x_1, \cdots, x_{n-1} ; x_n = 1)$ to be $1 - P(x_1, \cdots, x_{n-1} ; x_n = 0)$, then there is a probability measure on Ω which has precisely these functions for its conditional probability functions.

We have to show that this process is asymptotically independent in probability. Let ϵ be any positive number, $0 < \epsilon < p - q$. The probability measure of the set where $| P(x_1, \cdots, x_{n-1} ; x_n = 0) - q | > \epsilon$ is the measure of the set where $P(x_1, \cdots, x_{n-1} ; x_n = 0) = p$. For $2^m \leqq n \leqq 2^{m+1} - 1$, this set is of the form $\{x_1 = x_1^0, \cdots, x_m = x_m^0\}$. We are to prove then that the measure of such a set approaches zero with n^{-1}. But

$$P(x_1 = x_1^0, \cdots, x_m = x_m^0) = P(x_1 = x_1^0, \cdots, x_{m-1} = x_{m-1}^0)$$
$$\cdot P(x_1^0, \cdots, x_{m-1}^0 ; x_m = x_m^0).$$

The last-written conditional probability is not greater than $r = \max \{p, q, (1 - p), (1 - q)\}$; whence $P(x_1 = x_1^0, \cdots, x_m = x_m^0) \leqq r^m$. Hence $P(x_1, \cdots, x_{n-1} ; x_n = 0)$ converges in measure to q.

We now define a system transformation T by defining the sequence of choice functions $\{f_n\}$.[25] Let $f_n(x_1, \cdots, x_{n-1})$ be 1 or 0 according as $P(x_1, \cdots, x_{n-1} ; x_n = 0) = p$ or q. It is readily verified that this sequence of functions satisfies all the conditions of Definition 14.

[24] VI, §4.
[25] Definition 14.

For $2^m \leq n \leq 2^{m+1} - 1$, f_n depends on the coördinates $x_1, \cdots, x_m$ only ($m = 0, 1, 2, \cdots$). Thus $f_1 \equiv 1$; $f_2 = 1$ if $x_1 = 0$ and $f_2 = 0$ otherwise; $f_3 = 0$ if $x_1 = 0$ and $f_3 = 1$ otherwise. Hence we see that a_2 takes only the values 2 and 3 and takes these values on the sets $\{P(x_1; x_2 = 0) = p\}$ and $\{P(x_1, x_2; x_3 = 0) = p\}$, respectively. It is similarly verified that for all n, $2^{n-1} \leq a_n \leq 2^n - 1$, and that the set where a_n takes one of its values, say $\{a_n = k\}$, coincides with the set $\{P(x_1, \cdots, x_{k-1}; x_k = 0) = p\}$.

It is now easy to see that $T(\Omega)$ is not asymptotically independent in probability. For if it were, $P(x_n' = 0)$ would certainly converge to q. But

$$P(x_n' = 0) - q = \sum_{j=0}^{\infty} P(\{a_n = n + j\}\{x_{n+j} = 0\}) - P(a_n = n + j)q$$

$$= \sum_{j=0}^{\infty} \int_{\{a_n = n+j\}} P(x_1, \cdots, x_{n+j-1}; x_{n+j} = 0) - q \, dP$$

$$= \sum_{j=2^{n-1}}^{2^n-1} \int_{\{a_n = j\}} P(x_1, \cdots, x_{j-1}; x_j = 0) - q \, dP$$

$$= \int_{\{a_n = 2^{n-1}\}} (p - q) \, dP + \cdots + \int_{\{a_n = 2^n - 1\}} (p - q) \, dP = p - q.$$

6. **Asymptotic expectation theorems.** In this section we shall prove the theorems that stand in the same relation to the theorems of the preceding section as Theorem 8 stands to Theorem 5.

THEOREM 14. *If Ω is a real-valued stochastic process such that the functions $x_n(\omega)$ are uniformly bounded and such that $| E(x_1, \cdots, x_{n-1}; x_n) | < \epsilon$ for all $n \geq n_0$ almost everywhere, then $| E(x_1', \cdots, x_{n-1}'; x_n') | < \epsilon$ for all $n \geq n_0$ almost everywhere (where the x_n' are obtained from the x_n by a system transformation).*

Proof. Let I_{n-1}' be any measurable cylinder set over $x_1', \cdots, x_{n-1}'$. Then

$$\left| \int_{I_{n-1}'} E(x_1', \cdots, x_{n-1}'; x_n') \, dP \right| = \left| \int_{I_{n-1}'} x_n' \, dP \right|$$

$$(26) \qquad = \left| \lim_{\delta \to 0} \sum_{r=-\infty}^{\infty} r\delta P(I_{n-1}'\{r\delta \leq x_n' < (r+1)\delta\}) \right|$$

$$= \left| \lim_{\delta \to 0} \sum_{r=-\infty}^{\infty} r\delta \sum_{j=0}^{\infty} P(I_{n-1}'\{a_n = n + j\}\{r\delta \leq x_{n+j} < (r+1)\delta\}) \right|.$$

We now define the auxiliary function y to be x_{n+j} on $I_{n-1}'\{a_n = n + j\}$ and zero elsewhere. Then

$$\left| \int_{I_{n-1}'} E(x_1', \cdots, x_{n-1}'; x_n') \, dP \right|$$

$$= \left| \lim_{\delta \to 0} \sum_{r=-\infty}^{\infty} r\delta \sum_{j=0}^{\infty} P(I_{n-1}'\{a_n = n + j\}\{r\delta \leq y < (r+1)\delta\}) \right|$$

16

$$= \left| \int_{\sum\limits_{j=0}^{\infty} I'_{n-1}\{a_n=n+j\}} y \, dP \right| = \left| \sum_{j=0}^{\infty} \int_{I'_{n-1}\{a_n=n+j\}} x_{n+j} \, dP \right|$$

$$\leqq \sum_{j=0}^{\infty} \int_{I'_{n-1}\{a_n=n+j\}} | E(x_1, \cdots, x_{n+j-1} ; x_{n+j}) | \, dP$$

$$\leqq \sum_{j=0}^{\infty} \int_{I'_{n-1}\{a_n=n+j\}} \epsilon \, dP = \int_{I'_{n-1}} \epsilon \, dP \quad \text{for} \quad n \geqq n_0.$$

Since this is true uniformly in I'_{n-1}, we have $| E(x'_1, \cdots, x'_{n-1} ; x'_n \epsilon E) | < \epsilon$, for all $n \geqq n_0$, almost everywhere. This concludes the proof of Theorem 14.

We note that the proof of this theorem remains unaltered if we remove the absolute value signs from its statement. This shows, for example, that if the conditional expectations are all negative, they will remain negative after any system transformation.

We note also that this theorem stands in the same relation to Theorem 8 as Theorem 10 to Theorem 5.

THEOREM 15. *If Ω is a real-valued stochastic process such that the functions $x_n(\omega)$ are uniformly bounded and such that $E(x_1, \cdots, x_{n-1} ; x_n)$ converges to zero almost everywhere as $n \to \infty$, then $E(x'_1, \cdots, x'_{n-1} ; x'_n)$ converges to zero in probability (where the x'_n are obtained from the x_n by a system transformation).*

Proof. A necessary and sufficient condition that $E(x'_1, \cdots, x'_{n-1} ; x'_n)$ converge to zero in probability is that to every positive number ϵ there correspond a positive integer n_0, such that for $n > n_0 \left| \int_{\Lambda'} x'_n \, dP \right| < \epsilon$ whatever the measurable set Λ' over $x'_1, \cdots, x'_{n-1}$ may be. (By definition $\int_{\Lambda'} E(x'_1, \cdots, x'_{n-1} ; x'_n) dP = \int_{\Lambda'} x'_n \, dP$.) We have now

$$(27) \quad \left| \int_{\Lambda'} x'_n \, dP \right| = \left| \lim_{\delta \to 0} \sum_{r=-\infty}^{\infty} r\delta P(\Lambda'\{r\delta \leqq x'_n < (r+1)\delta\}) \right|$$

$$= \left| \lim_{\delta \to 0} \sum_{r=-\infty}^{\infty} r\delta \sum_{j=0}^{\infty} P(\Lambda'\{a_n = n+j\}\{r\delta \leqq x_{n+j} < (r+1)\delta\}) \right|.$$

We define the function y to be x_{n+j} on $\Lambda'\{a_n = n+j\}$ and zero elsewhere. Then

$$(28) \quad \left| \int_{\Lambda'} x'_n \, dP \right| = \left| \lim_{\delta \to 0} \sum_{r=-\infty}^{\infty} r\delta \sum_{j=0}^{\infty} P(\Lambda'\{a_n = n+j\}\{r\delta \leqq y < (r+1)\delta\}) \right|$$

$$= \left| \int_{\sum\limits_{j=0}^{\infty} \Lambda'\{a_n=n+j\}} y \, dP \right| = \left| \sum_{j=0}^{\infty} \int_{\Lambda'\{a_n=n+j\}} x_{n+j} \, dP \right|$$

$$\leqq \sum_{j=0}^{\infty} \int_{\Lambda'\{a_n=n+j\}} | E(x_1, \cdots, x_{n+j-1} ; x_{n+j}) | \, dP.$$

By hypothesis $E(x_1, \cdots, x_{n-1} ; x_n)$ converges to zero almost everywhere. We apply Egoroff's theorem. Given any positive number ϵ, there exists a set D such that $P(CD) = 1 - P(D) < \epsilon$ and such that on D we have

$$(29) \qquad | E(x_1, \cdots, x_{n+j-1} ; x_{n+j}) | < \epsilon$$

for n sufficiently large and $j = 0, 1, 2, \cdots$. Then

$$(30) \quad \left| \int_{\Lambda'} x_n' \, dP \right| \leqq \sum_{j=0}^{\infty} \int_{\Lambda'\{a_n=n+j\}D} + \int_{\Lambda'\{a_n=n+j\}CD} | E(x_1, \cdots, x_{n+j-1} ; x_{n+j}) | \, dP$$

$$\leqq \sum_{j=0}^{\infty} \int_{\Lambda'\{a_n=n+j\}} \epsilon \, dP + \int_{\Lambda'\{a_n=n+j\}CD} K \, dP,$$

where K is the common upper bound of the functions x_n. (If $| x_n | < K$, $| E(x_1, \cdots, x_{n-1} ; x_n) | \leqq K$.) Hence, finally,

$$(31) \qquad \left| \int_{\Lambda'} x_n' \, dP \right| \leqq \epsilon P(\Lambda'D) + KP(\Lambda' \cdot CD) \leqq \epsilon + K\epsilon .$$

This concludes the proof of Theorem 15.

We note that the example of the preceding section shows that the convergence in probability of $E(x_1, \cdots, x_{n-1} ; x_n)$ is not invariant under every system transformation.

BIBLIOGRAPHY

I. Z. W. BIRNBAUM AND J. SCHREIER, *Eine Bemerkung zum starken Gesetz der Grossen Zahlen*, Studia Math., vol. 4(1933), pp. 85–89.

II. A. H. COPELAND, *Admissible numbers in the theory of geometrical probability*, Am. Jour. of Math., vol. 53(1931), pp. 153–162.

III. A. H. COPELAND, *Admissible numbers in the theory of probability*, Am. Jour. of Math., vol. 50(1928), pp. 535–552.

IV. J. L. DOOB, *Note on probability*, Annals of Math., vol. 37(1936), pp. 363–367.

V. J. L. DOOB, *Probability and statistics*, Trans. Amer. Math. Soc., vol. 36(1934), pp. 759–775.

VI. J. L. DOOB, *Stochastic processes with an integral-valued parameter*, Trans. Amer. Math. Soc., vol. 44(1938), pp. 87–150.

VII. H. HAHN, *Über die Multiplikation total-additiver Mengenfunktionen*, Annali della R. Sc. Norm. Sup. di Pisa, vol. 2(1933), pp. 429–452.

VIII. E. HOPF, *On causality, statistics, and probability*, Jour. of Math. and Phys., vol. 13 (1934), pp. 51–102.

IX. H. HUNTEMANN, *Über den mathematischen Kern des Prinzips vom ausgeschlossenen Spielsystem und eine darauf gegründete Wahrscheinlichkeitstheorie*, Deutsche Mathematik, vol. 2(1937), pp. 593–622.

X. A. KOLMOGOROFF, *Grundbegriffe der Wahrscheinlichkeitsrechnung*, Berlin, 1933.

XI. R. VON MISES, *Wahrscheinlichkeitsrechnung*, Wien, 1931.

XII. S. SAKS, *Theory of the Integral*, Warsaw, 1937.

XIII. A. WALD, *Die Widerspruchfreiheit der Kollektivbegriffes der Wahrscheinlichkeitsrechnung*, Ergebnisse eines math. Kolloq., vol. 8(1936), pp. 38–72.

UNIVERSITY OF ILLINOIS.

Reprinted from the Proceedings of the NATIONAL ACADEMY OF SCIENCES,
Vol. 28, No. 6, pp. 254–258. June, 1942

ON MONOTHETIC GROUPS

By Paul R. Halmos and H. Samelson

THE INSTITUTE FOR ADVANCED STUDY.

Communicated April 17, 1942

§ 1. A topological group G is called *monothetic* (following van Dantzig[1]) if there exists a cyclic subgroup H which is dense in G, i.e., the closure of H is G. A generating element of such a cyclic subgroup is called a *generator* of G.

All groups considered in the sequel are abelian; monothetic groups are evidently abelian.

The elements of finite order of a discrete group K form a subgroup, the *torsion group* $T(K)$ of K.

We call a group G *separable* if there exists a countable subset which is dense in G.

We use the theory of character groups. We denote by C the value group for the characters, the group of real numbers mod 1 with the usual topology; let $\overline{C}$ be the same group, but with the discrete topology. We denote by h the natural mapping of $\overline{C}$ on C; it is continuous and an algebraic (but not of course topological) isomorphism.

The character group of a group G is denoted by G^*. The *annihilator* of a subset H of G is the set of those characters of G which map every element of H into the zero element of C; it is a (closed) subgroup of G^*; we denote it by $A(H)$. We recall that the character group of a compact (discrete) group is discrete (compact); for both types of groups we have the duality theorem, which says $G^{**} = G$, and more generally, $A(A(H)) = H$ for every closed subgroup H.

§ 2. We restrict our considerations to locally compact groups (as is customary in the theory of abelian groups). It is known that a locally compact monothetic group is either compact or discrete.[2] The discrete case being trivial we consider only the compact case. We prove

THEOREM I. *A compact group G is monothetic if and only if its character group G^* is isomorphic to a subgroup of $\overline{C}$.*

Proof. (a) Suppose G is monothetic; let d be a generator of G. Let $f = f(x)$ ($x \in G$, $f(x) \in C$) be an arbitrary character of G; the mapping $f \to h^{-1} (f(d))$ is obviously an isomorphism of G^* into $\overline{C}$, because every character f is completely determined by its value $f(d)$ for the generator d of G.

(b) Suppose G^* is a subgroup of $\overline{C}$. The mapping h of $\overline{C}$ into C induces a homomorphic mapping of G^* into C, i.e., a character of G^*; call this character d. Since $G = (G^*)^*$, d may be considered as an element of G. Let D be the subgroup of G generated by d, i.e., the closure of the cyclic group generated by d. The character d maps only the zero element of G^* into the zero of C; this means that the annihilator of d, and hence also that of D, contains only the zero of G^*. But this means obviously that D equals G. This proves Theorem I.

(c) From the considerations in (a) and (b) it is clear that an element d of G is a generator of G if and only if its annihilator contains only the zero element of G^*.

(d) Suppose again G^* is a subgroup of $\overline{C}$. It can be shown easily by elementary methods, without using the duality theorem, that the cyclic group generated by the character d of G^*, defined in (b), is dense in the character group G^{**} of G^*, by direct consideration of the neighborhoods of an arbitrary element of G^{**}.

§ 3.   THEOREM II. *A discrete group H is isomorphic to a subgroup of $\overline{C}$ if and only if its power (cardinal number) is $\leqq \mathfrak{c}$ ($=$ the power of the continuum) and its torsion group $T(H)$ is isomorphic to a subgroup of $\overline{C}$ and so of $T(\overline{C})$.*

Theorem II is a consequence of the following

LEMMA. *Let H be a discrete group of power $\leqq \mathfrak{c}$; every isomorphic mapping f_1 of $T(H)$ into $\overline{C}$ can be extended to an isomorphic mapping f of H into $\overline{C}$.*

Proof. The complement of $T(\overline{C})$ contains $\mathfrak{c}$ linearly independent elements (elements of a Hamel basis for the real numbers, reduced mod 1); let $\lambda_1, \lambda_2, \ldots$ be a well ordering of these elements. We well order also the elements of the complement of $T(H)$: $x_1, x_2, \ldots$. We consider the groups H_α generated by the elements of $T(H)$ and the x_β with $\beta < \alpha$, and construct isomorphic mappings f_α of H_α into $\overline{C}$ such that f_α is an extension of f_β for $\beta < \alpha$. We start with the given f_1. Suppose f_β is constructed for $\beta < \alpha$.

(a) If α is a limit ordinal then H_α is the union of the H_β with $\beta < \alpha$. Every $z \in H_\alpha$ is contained in some H_β, $\beta < \alpha$; we define $f_\alpha(z) = f_\beta(z)$. It is clear that this f_α has the desired properties.

(b) If α is not a limit ordinal then H_α is generated by $H_{\alpha-1}$ and $x_{\alpha-1}$.

(b') If no multiple of $x_{\alpha-1}$ belongs to $H_{\alpha-1}$, then H_α is the direct sum of $H_{\alpha-1}$ and the cyclic group generated by $x_{\alpha-1}$; we extend $f_{\alpha-1}$ by putting $f_\alpha(x_{\alpha-1}) = \lambda_{\alpha-1}$. This gives an isomorphic extension, because under our constructions λ_γ and the elements of the groups $f_\beta(\dot{H}_\beta)$ with $\beta \leq \gamma$ are always linearly independent.

(b'')   If a multiple of $x_{\alpha-1}$ belongs to $H_{\alpha-1}$, let n be the smallest positive integer for which $nx_{\alpha-1} \epsilon H_{\alpha-1}$. Every element z of H_α can then be written in a unique manner as $y + mx_{\alpha-1}$ with $y \epsilon H_{\alpha-1}$ and $0 \leq m < n$. Let λ be an element of $\overline{C}$ with $n\lambda = f_{\alpha-1}(nx_{\alpha-1})$. Put $f_\alpha(z) = f_{\alpha-1}(y) + m\lambda$. It is easily verified that f_α is a homomorphic mapping of H_α, and that it is an extension of $f_{\alpha-1}$. To prove that it is an isomorphism, suppose $f_\alpha(z) = 0$. We have then $nz = 0$, because $nz \epsilon H_{\alpha-1}$, $f_{\alpha-1}(nz) = nf_\alpha(z) = 0$, and $f_{\alpha-1}$ is isomorphic.   So we have $z \epsilon T(H)$, and $f_1(z) = f_\alpha(z) = 0$; but f_1 is isomorphic, and so $z = 0$.

The desired extension f of f_1 is now given by $f(x) = f_1(x)$ for $x \epsilon T(H)$ and $f(x_\alpha) = f_{\alpha+1}(x_\alpha)$ for the x_α which form the complement of $T(H)$.

§ 4.   We come now to a theorem which is the "dual" of Theorem II.

THEOREM II*.   *Let G be a compact group and let G_1 be its component of the identity; G is monothetic if and only if it is separable and the totally disconnected factor group G/G_1 is monothetic.*

For the proof we note that the (obviously necessary) separability guarantees that the power condition of Theorem II is fulfilled for $H = G^*$. Therefore, by Theorem II G^* is isomorphic to a subgroup of $\overline{C}$ if and only if $T(G^*)$ is.   But $T(G^*)$ is, as is well known, the character group of G/G_1. Applying now Theorem I to G and G/G_1 we obtain Theorem II*.

From this follows immediately the

COROLLARY.   *Every compact connected separable (abelian) group is monothetic.*

A discrete group is called *locally cyclic* if every subgroup which is generated by a finite number of elements can be generated by a single element. It is easy to see that a group without elements of infinite order is locally cyclic if and only if it is isomorphic to a subgroup of $T(\overline{C})$. We may accordingly restate Theorem II* as

THEOREM II$'$ *A compact group G is monothetic if and only if it is separable and the torsion group $T(G^*)$ of its character group is locally cyclic.*

Let Z be a compact totally disconnected monothetic group.   Since the character group of a totally disconnected group has no elements of infinite order it follows from Theorem I that Z^* is (isomorphic to) a subgroup of $T(\overline{C})$.   Consequently Z^* is the direct sum of groups Z^*_p, p running over the prime numbers, where each Z^*_p is either the zero group or cyclic of order p^n for some n or isomorphic to the group T_p of all elements of $T(\overline{C})$ the order of which is a power of p.   It follows that Z is the direct sum of groups Z_p, p running over the primes, where Z_p is either the zero group or cyclic of order p^n for some n or the p-adic group, the p-adic group being the character group of the group T_p just mentioned. Conversely, every such direct sum is a compact totally disconnected monothetic group.

§ 5. We prove now a theorem on the Haar measure of the set of generators of a group which is a partial strengthening of the corollary of § 4.

THEOREM III. *The set of generators of a compact connected (abelian) group satisfying the second countability axiom has Haar measure 1.*[3]

Proof. The character group G^* is countable; let $0, x_1, x_2, \ldots$ be its elements. Let G_i be the set of those elements of G which are mapped by the character x_i of G into the zero of C. Each G_i is a closed proper subgroup of G, and so of measure 0 (because, G being connected, G_i has an infinite number of disjoint cosets of equal measure); the union of the G_i has therefore measure 0 too. Using now remark (c) of §2 we see that the set of generators of G is identical with the complement of the union of the G_i, and so of measure 1.

Theorem III is not necessarily true for a group which does not fulfil the second countability axiom. An example is the toral group of dimension c (the direct sum of c copies of the group C); the set of its generators has inner measure 0 and outer measure 1. To see this, we call a subset of that group a c-set if it is (in an obvious sense) a cylinder set over a countable number of coördinates. It has been shown to us by S. Kakutani that the inner measure of a subset S of our group is equal to the supremum of the measures of measurable c-sets contained in S. Now it is clear that an element of the group is a generator if and only if its coördinates are linearly independent elements of $\overline{C}$. Therefore every c-set contains elements which are not generators; and this means that the inner measure of the set of generators is 0. On the other hand it is easily seen that every c-set of positive measure contains a generator (using Theorem III for the special case, G = direct sum of countably many copies of C); hence the complement of the set of generators is of inner measure 0, or the outer measure of the set of generators is 1.

The situation for totally disconnected monothetic groups is this. The set of generators of a group which is cyclic of order p^n, or p-adic, has measure $1 - 1/p$. From this and the structure of an arbitrary totally disconnected monothetic group Z (§ 4) one concludes easily that the measure of the set of generators of Z is equal to $\Pi (1 - 1/p)$, extended over those primes p for which $Z_p \neq 0$, and hence that this measure may take any value between 0 and 1.

[1] Van Dantzig, D., "Zur topologischen Algebra," *Mathematische Annalen*, **107**, 591 (1933)

[2] Weil, André, "L'integration dans les groupes topologiques et ses applications," Paris, p. 97 (1938).

[3] This theorem was stated by Schreier, J., and Ulam, S., "Sur le nombre des générateurs d'un groupe topologique compact et connexe," *Fund. Math.*, **24**, 304 (1935).

ANNALS OF MATHEMATICS
Vol. 43, No. 2, April, 1942

OPERATOR METHODS IN CLASSICAL MECHANICS, II

By Paul R. Halmos and John von Neumann

(Received December 23, 1941)

Introduction

The purpose of this paper is two-fold: to map all measure spaces for which this is possible on the unit interval, and to apply such mapping theorems to the study of ergodic measure preserving transformations with a pure point spectrum.

"Mappings" between two measure spaces may be interpreted in two ways, as set mappings and as point mappings, and accordingly we give below two sets of necessary and sufficient conditions for the existence of a mapping from a given space to the interval. The first of these, the set mapping or algebraic isomorphism theorem, seems to be known, and although it has never been explicitly stated in the literature there are many proofs of special cases of it on record. We give an explicit proof of it and use a construction of the proof in proving the second, point mapping or geometric isomorphism, theorem. This second theorem depends on the new concept of normal measure space: a seemingly artificial concept which is, however, useful for two reasons. First, it is purely measure theoretic (and not topological), in character, and hence is applicable to the measure spaces usually discussed in probability theory; second it is hereditary under all the usual operations on measure spaces (such as the formation of direct products, decomposition into direct sums, etc.).

Using the concepts and results of the mapping theorems just described, and of the Pontrjagin duality theorem concerning compact and discrete abelian groups, we are able to show that every ergodic measure preserving transformation with a pure point spectrum is isomorphic to a rotation on a compact abelian group. This is a "normal form" theorem for a certain class of measure preserving transformations and can be used to answer many questions, such as the existence of square roots, commutative transformations, etc., concerning such transformations.

Although this paper is a continuation of an earlier work of one of us[1] it is to a large extent independent of this earlier work. The proofs of the main theorems mentioned above are logically complete here; only in some of the applications, as for example in discussing the relation between point mappings and set mappings, do we make use of the results of (I).

1. General measure spaces; the algebraic isomorphism theorem

Let X be any set, and $\mathfrak{X}$ any Borel field of subsets of X; let m be a non negative, contably additive, finite measure defined on $\mathfrak{X}$. The system $\{X, \mathfrak{X}, m\}$, which we shall usually denote by X, or, if necessary to indicate its

---

[1] See John von Neuman, *Zur Operatorenmethode in der klassischen Mechanik*, Annals of Mathematics, vol. 33, (1932), pp. 587–642. In the sequel we shall refer to this paper as (I).

dependence on $\mathfrak{X}$ and m, by $X(\mathfrak{X}, m)$ is called a *measure space*. Sets $E \epsilon \mathfrak{X}$ are called *measurable*; we shall use also the usual terminology of the Lebesgue theory in describing functions as measurable, integrable, etc.   A measure space is *complete* if every subset of a measurable set of measure zero is itself measurable (and has, of course, measure zero).   Since it is always possible to extend the definition of m to a Borel field $\mathfrak{X}' \supset \mathfrak{X}$ so that $X(\mathfrak{X}', m)$ is complete, we shall lose no generality, and gain somewhat in simplicity, by assuming completeness.

In any measure space X we shall write $\mathfrak{B} = \mathfrak{B}(\mathfrak{X})$ for the Boolean algebra of measurable sets modulo sets of measure zero.   We shall make use of the notations of set theory, ($\subset$, $+$, etc.) in $\mathfrak{B}$, and of the fact that we may consider m as defined on $\mathfrak{B}$.

We discuss now the concept of separability in measure spaces.   A Borel field $\mathfrak{X}$ (or a measure space $X(\mathfrak{X}, m)$) is *strictly separable* if it contains a countable collection of sets such that the smallest Borel field containing all of them, (the Borel field *spanned* by them), is $\mathfrak{X}$ itself.   Two sub Borel fields, $\mathfrak{A}$ and $\mathfrak{B}$, of the Borel field $\mathfrak{X}$ of measurable sets in a measure space $X(\mathfrak{X}, m)$ are *equivalent* if to every set E in either one of them there corresponds a set F in the other such that the symmetric difference $(E - F) + (F - E)$ has measure zero.   A measure space is *separable* if there exists a strictly separable Borel field $\mathfrak{A}$ contained in and equivalent to $\mathfrak{X}$.[2]   A concept, which lies logically between separability and strict separability, more useful than either of these, is proper separability.   A measure space $X(\mathfrak{X}, m)$ is *properly separable* if there exists a strictly separable Borel field $\mathfrak{A} \subset \mathfrak{X}$, such that to every $E \epsilon \mathfrak{X}$ there corresponds an $F \epsilon \mathfrak{A}$ with $E \subset F$ and $m(F - E) = 0$.[3]   We observe that this definition is self dual: by applying the condition to $X - E$ we readily obtain a set $F \epsilon \mathfrak{A}$ with $F \subset E$ and $m(E - F) = 0$.   We shall make use of the fact that if X is separable (or properly separable) and $\mathfrak{A}$ is the strictly separable Borel field described in the definitions above then $\mathfrak{B}(\mathfrak{X}) = \mathfrak{B}(\mathfrak{A})$.   In the case of (properly) separable measure spaces it will be necessary to indicate in the notation the strictly separable Borel field used; we shall write $X = X(\mathfrak{X}, \mathfrak{A}, m)$.   We shall call sets of $\mathfrak{A}$ *Borel sets*, and functions measurable ($\mathfrak{A}$) *Baire functions*.   (A real valued function $f(x)$ is measurable ($\mathfrak{A}$) if the inverse image under f of every real Borel set S, i.e. the set $\{x \mid f(x) \epsilon S\}$, belongs to $\mathfrak{A}$.)

---

[2] This is not the usual form in which this definition is given.   Cf., for example, J. L. Dobb, *One—parameter families of transformation*, Duke Mathematical Journal, vol. 4, (1938), p. 753.   That our definition is, however, equivalent to the usual one is proved by Paul R. Halmos, *The decomposition of measures*, Duke Mathematical Journal, vol. 8, (1941), p. 387.   We observe X is separable if and only if the Boolean algebra $\mathfrak{B}(\mathfrak{X})$ has a countable number of generators.

[3] The concept of proper separability, first introduced by W. Ambrose and S. Kakutani, *Structure and continuity of measurable flows*, Duke Mathematical Journal, vol. 9, (1942), pp. 25–42, is fundamental in measure theory.   Although it is possible to give examples of separable but not properly separable measure spaces, these examples are all of a more or less pathological kind.   One such example is the unit interval, with the Borel field of all sets of Lebesgue measure zero and their complements in the role of $\mathfrak{X}$.

A measurable set E in the measure space $X(\mathfrak{X}, m)$ is *indecomposable* if it contains no proper measurable subsets other than the empty set; an element $E \in \mathfrak{B}(\mathfrak{X})$ is an *atom* if it contains no proper subelements, other than 0, in $\mathfrak{B}(\mathfrak{X})$. A measure space is *non atomic* if $\mathfrak{B}(\mathfrak{X})$ has no atoms: in other words if every measurable set of positive measure contains measurable subsets of smaller positive measure. From the point of view of a study of the structure of measure spaces indecomposable sets and atoms are uninteresting: we shall generally assume that the former consist of exactly one point and the latter are absent. More specifically our assumption will be described in the following terms.

A countable sequence, $A_1, A_2, \cdots$, of subsets of X is a *separating sequence* if to every pair of points, $x \neq y$, we may find an integer n with $x \in A_n$, $y \in X - A_n$. If there exists in X a separating sequence of measurable sets, an indecomposable set contains exactly one point. We shall now show that the assumption of the existence of a separating sequence of measurable sets has a similar effect on atoms. Let E be a set of positive measure which contains no measurable subsets of smaller positive measure. It follows that for each n one of the two sets, EA_n, and $E(X - A_n)$ has measure zero and the other one has measure $m(E)$. By a slight change of notation we may assume $m(EA_n) = m(E)$ for $n = 1, 2, \cdots$. If we write $\prod_{n=1}^{\infty} A_n = A$, then we have $m(EA) = m(E)$; since, however, A can contain at most one point, this implies that for some point $x \in E$ we have $m(E - x) = 0$. In other words the existence of a measurable separating sequence implies that the weight of an atom is concentrated at one point; if, for example, we assume that the measure of a point is always zero, we may infer that the space is non atomic. Since in a measure space, which has by definition finite measure, there can be at most a countable set of points of positive measure, and since their measure theoretic structure is clear, we shall generally assume non-atomicity explicitly.

If $X_1(\mathfrak{X}_1, m_1)$ and $X_2(\mathfrak{X}_2, m_2)$ are measure spaces, a *set isomorphism* between X_1 and X_2 is a measure preserving isomorphism between the Boolean algebras $\mathfrak{B}(\mathfrak{X}_1)$ and $\mathfrak{B}(\mathfrak{X}_2)$. More specifically a set isomorphism is a one to one mapping T from $\mathfrak{B}(\mathfrak{X}_1)$ on $\mathfrak{B}(\mathfrak{X}_2)$ which is such that

$$T(X_1 - E) = X_2 - TE,$$

$$T(\textstyle\sum_{n=1}^{\infty} E_n) = \sum_{n=1}^{\infty} TE_n,$$

$$m_1(E) = m_2(TE).$$

If such a mapping T exists, X_1 and X_2 are *set isomorphic*.

After one more comment on notation we shall be ready to state and prove our first result. Since the unit interval plays a fundamental role in our investigations and is used as a yardstick with which to compare other measure spaces, we find it convenient to introduce a special notation for it. We shall denote the unit interval by $\tilde{X}$, the collection of Lebesgue and Borel measurable sets by

$\mathfrak{X}$ and $\tilde{\mathfrak{a}}$ respectively, and Lebesgue measure by $\tilde{m}$. In our terminology $\tilde{X} = \tilde{X}(\mathfrak{X}, \tilde{\mathfrak{a}}, \tilde{m})$ is a properly separable measure space.[4]

THEOREM 1. *A necessary and sufficient condition that a measure space of total measure one be set isomorphic to the unit interval is that it be separable and non-atomic.*

PROOF. Since the unit interval is separable and non-atomic and since these properties are evidently invariant under set isomorphisms, the necessity of our conditions is clear. To prove their sufficiency, let $X(\mathfrak{X}, \mathfrak{a}, m)$ be the given measure space, $m(X) = 1$, and let $A_1, A_2, \cdots$ be a countable sequence of Borel sets which span $\mathfrak{a}$. We may assume (by adding a superfluous set to the $\{A_n\}$ if necessary) that $\sum_{n=1}^{\infty} A_n = X$. Then we may make correspond to every rational number r, $0 \leq r \leq 1$, a set B_r such that

(i) $\{A_n\}$ and $\{B_r\}$ span the same field;

(ii) $r < s$ implies $B_r \subset B_s$;

(iii) $\prod_{r>s} B_r = B_s$;

(iv) $\prod_r B_r = 0$; $\sum_r B_r = X$.[5]

We now define, for every real number a, $0 \leq a \leq 1$, a set B_a by $B_a = \prod_{r>a} B_r$. It is clear that this definition of B_a is consistent with its previous definition in case a is rational, and that the family of sets $\{B_a\}$ satisfies the conditions (ii), (iii), (iv), (where in (iii) and (iv) we extend the products and sums over an arbitrary countable set of real numbers r for which inf $r = s$ in (iii), inf $r = 0$ and sup $r = 1$, respectively, in (iv)). Moreover, condition (i) implies that $B_a \epsilon \mathfrak{a}$ for all a and that the Borel field spanned by the B_a is $\mathfrak{a}$ itself.

Given now the family B_a we may find a (uniquely determined) function $f(x)$, defined for $x \epsilon X$, $0 \leq f(x) \leq 1$, for which $\{x \mid f(x) \leq a\} = B_a$; we may, for example, define

$$(1) \qquad\qquad f(x) = \inf \{a \mid x \epsilon B_a\}.$$

The class of all sets of the form

$$(2) \qquad\qquad f^{-1}(\tilde{E}) = \{x \mid f(x) \epsilon \tilde{E}\},$$

where $\tilde{E}$ is an arbitrary Borel set in the unit interval, is a Borel field contained in $\mathfrak{a}$; since it contains all B_a , and therefore all A_n , it coincides with $\mathfrak{a}$.

Let $F(a) = m\{x \mid f(x) \leq a\} = m(B_a)$ be the distribution function of $f(x)$: $F(a)$ is monotone non-decreasing from 0 to 1 as a ranges between 0 and 1, and is continuous from the right. (This much is always true, of an arbitrary distribution function.) In our special case we assert that $F(a)$ is continuous. For

---

[4] In the sequel we shall sometimes use the notation $\tilde{X}(\mathfrak{X}, \tilde{\mathfrak{a}}, \tilde{m})$ for the perimeter of the unit circle in the complex plane: it is clear that this space has the same measure theoretic structure as the unit interval. We shall always make it clear whether the symbol $\tilde{X}$ has its real or its complex meaning.

[5] Cf. (I), p. 602; see also J. L. Doob, *Stochastic processes with an integral valued parameter*, Transactions of the American Mathematical Society, vol. 44, (1938), p. 91.

if $a = a_0$ is a discontinuity of $F(a)$, then $\{x \mid f(x) = a_0\}$ is a set of positive measure which therefore, (non-atomicity), has Borel subsets of smaller positive measure. Such a subset cannot be put in the form $f^{-1}(\tilde{E})$, contrary to what we have already proved.

For any $\tilde{x}$, $0 \leq \tilde{x} \leq 1$, we define $\tilde{f}(\tilde{x}) = \inf \{a \mid F(a) \geq \tilde{x}\}$. It is well known (and easily verified) that $\tilde{f}(\tilde{x})$ is a strictly monotone increasing (not necessarily continuous) function of $\tilde{x}$, which increases from 0 to 1 as $\tilde{x}$ does, and which is continuous on the left. Moreover the distribution function of $\tilde{f}(\tilde{x})$ is again $F(a)$.

For any Borel set $\tilde{E} \subset \tilde{X}$, consider the set $\tilde{f}^{-1}(\tilde{E})$: we assert that the collection of all sets of this form, (which clearly forms a Borel field), coincides with $\tilde{\mathfrak{A}}$. This is true since the increasing character of $\tilde{f}(\tilde{x})$ implies that every interval $(0, \tilde{x})$ has the form $\tilde{f}^{-1}(\tilde{E})$, where $\tilde{E}$ can even be chosen as an interval.

Suppose that it ever happens that $f^{-1}(\tilde{E}_1) = f^{-1}(\tilde{E}_2)$. (We shall now make use of the fact that for an arbitrary Baire function $g(x)$, $0 \leq g(x) \leq 1$, the correspondence $\tilde{E} \to g^{-1}(\tilde{E}) = \{x \mid g(x) \, \epsilon \, \tilde{E}\}$, is a homomorphism of $\tilde{\mathfrak{A}}$ into $\mathfrak{A}$, i.e. that $g^{-1}(\tilde{X} - \tilde{E}) = X - g^{-1}(\tilde{E})$, and $g^{-1}(\tilde{E}_1 + \tilde{E}_2 + \cdots) = g^{-1}(\tilde{E}_1) + g^{-1}(\tilde{E}_2) + \cdots$). If we write $\tilde{E}' = (\tilde{E}_1 - \tilde{E}_2) + (\tilde{E}_2 - \tilde{E}_1)$ for the symmetric difference between $\tilde{E}_1$ and $\tilde{E}_2$, then it follows from the equality of the distributions of $f(x)$ and $\tilde{f}(\tilde{x})$, that $m\{f^{-1}(\tilde{E}')\} = \tilde{m}\{\tilde{f}^{-1}(\tilde{E}')\} = 0$. Conversely, of course, $\tilde{f}^{-1}(\tilde{E}_1) = f^{-1}(\tilde{E}_2)$ implies the same result.

Consequently the correspondence $f^{-1}(\tilde{E}) \rightleftarrows \tilde{f}^{-1}(\tilde{E})$ is one to one, not necessarily between $\mathfrak{A}$ and $\tilde{\mathfrak{A}}$, but certainly between $\mathfrak{B} = \mathfrak{B}(\mathfrak{X}) = \mathfrak{B}(\mathfrak{A})$ and $\tilde{\mathfrak{B}} = \mathfrak{B}(\tilde{\mathfrak{X}}) = \mathfrak{B}(\tilde{\mathfrak{A}})$. It is clear that this correspondence preserves measure, and the homomorphic nature of the mappings $\tilde{E} \to f^{-1}(\tilde{E})$ and $\tilde{E} \to \tilde{f}^{-1}(\tilde{E})$ shows that it is also an algebraic isomorphism.

This concludes the proof of Theorem 1.

2. Normal spaces; the geometric isomorphism theorem

If $X_1(\mathfrak{X}_1, m_1)$ and $X_2(\mathfrak{X}_2, m_2)$ are measure spaces, a *point isomorphism* between X_1 and X_2 is a one to one mapping from almost all of X_1 on almost all of X_2 such that $E_1 \, \epsilon \, \mathfrak{X}_1$ if and only if $E_2 = TE_1 \, \epsilon \, \mathfrak{X}_2$, and then $m_1(E_1) = m_2(E_2)$. If such a mapping T exists, X_1 and X_2 are *point isomorphic*. Our problem in this section is to find necessary and sufficient conditions in order that a measure space be point isomorphic to the unit interval. The fundamental concept in this connection is that of a normal space.

DEFINITION 1. *A measure space is* proper *if it is complete, properly separable, and non-atomic, and if it contains a separating sequence of Borel sets.*

DEFINITION 2. *A proper measure space is* normal *if to each real valued univalent Baire function $f(x)$ there corresponds a set X_0 of measure zero such that the range, $f(X - X_0)$, is a Borel set.*

The following lemmas concerning proper and normal spaces will be useful in the sequel.

LEMMA 1. *On every proper measure space $X(\mathfrak{X}, \mathfrak{A}, m)$ there exist real valued bounded univalent Baire functions.*

PROOF. Since X is certainly separable and non-atomic the construction of the proof of Theorem 1 applies. We assert that the real valued bounded Baire function $f(x)$ defined by (1) is univalent. For if the set $\{x \mid f(x) = a\}$ contained more than one point, then the intersection of this set with a Borel set separating two of its points could not be expressed in the form $\{x \mid f(x) \in \bar{E}\}$. Since, however, the proof of Theorem 1 establishes that every Borel set has this form, $f(x)$ must be univalent.

LEMMA 2. *If $X(\mathfrak{X}, \mathfrak{a}, m)$ is a proper measure space with the property that the condition of Definition 2 is satisfied by every bounded function then X is normal.*

PROOF. Let $f(x)$ be any univalent Baire function, and let $G(y)$ be any continuous function which maps the infinite interval, $-\infty < y < +\infty$, in a one to one way on a finite interval. Then $g(x) = G(f(x))$ is a Baire function which is univalent and bounded, hence, by hypothesis, there is a set X_0 of measure zero such that $g(X - X_0)$ is a Borel set. The image of this Borel set under the one to one continuous mapping $G^{-1}(y)$ is the range $f(X - X_0)$ which is therefore also a Borel set.

LEMMA 3. *If $X(\mathfrak{X}, \mathfrak{a}, m)$ is a normal space, $B \subset X$ is a Borel set, and $f(x)$ is a real valued univalent Baire function, then there is a set $B_0 \subset B$ of measure zero such that $f(B - B_0)$ is a Borel set. B_0 can even be chosen in the form BX_0', where X_0' is a Borel set of measure zero, depending on f but not on B.*

PROOF. We shall carry out the proof in three steps, first establishing the existence of a suitable B_0 corresponding to a fixed B, then showing that B_0 may even be chosen as a Borel set, and, finally, proving on the basis of our separability hypotheses, that we may choose B_0 in the form described in the statement of the lemma.

(i) We observe that the first statement asserts, essentially, that a Borel set in a normal space is itself a normal space. Accordingly, using Lemma 2, we may assume that $f(x)$ is bounded. Let $f'(x)$ be a bounded univalent Baire function on X, (Lemma 1); by appropriate linear transformations of $f(x)$ and of $f'(x)$ we can secure

$$0 \leqq f(x) \leqq 1 < f'(x)$$

throughout X. Then the function $f^*(x)$, defined to be equal to $f(x)$ on B and to $f'(x)$ on $B' = X - B$ is a univalent Baire function on X, hence for a suitable set X_0 of measure zero, $f^*(X - X_0)$ is a Borel set. The intersection of this Borel set with the closed interval (0, 1) is also a Borel set: this intersection is, however, precisely $f(B - B_0)$, where $B_0 = BX_0$.

(ii) Let B_1 be a Borel set of measure zero, $B_1 \supset B_0$. Applying the result of (i) to $X - B_1$ we may find a set $B_2 \supset B_1$ of measure zero such that $f(X - B_2)$ is a Borel set. We proceed similarly by induction, choosing $B_3 \supset B_2$ to be a Borel set of measure zero, choosing $B_4 \supset B_3$ so that $f(X - B_4)$ is a Borel set, and so on. We have $B_0 \subset B_1 \subset B_2 \subset B_3 \subset \cdots$; all B_n are of measure zero; or n odd B_n is a Borel set; for n even $f(X - B_n)$ is a Borel set. We write $B_0^* = \sum_{n=0}^{\infty} B_n$. Then B_0^* has measure zero, and, because of the monotone character of the sequence $\{B_n\}$, $B_0^* = \sum_{n=0}^{\infty} B_{2n+1}$, so that B_0^* is a Borel set. Similarly $X - B_0^* = X - \sum_{n=0}^{\infty} B_{2n} = \prod_{n=0}^{\infty} (X - B_{2n})$, so that $f(X - B_0^*)$ is a Borel set, and also $B(X - B_0^*) = (B - B_0)(X - B_0^*)$, so that $f(B(X - B_0^*)) = f(B - B_0)f(X - B_0^*)$ is a Borel set. We may accordingly change notation and denote by B_0 the intersection of B and B_0^* : this new B_0 is a Borel set of measure zero with the property that $f(B - B_0)$ is a Borel set.

(iii) Let A_1, A_2, $\cdots$ be a sequence which spans $\mathfrak{A}$, and apply the result of (ii) to find, for each n, a Borel set $A_n^0 \subset A_n$, of measure zero, such that $f(A_n - A_n^0)$ is a Borel set. We write $A^0 = \sum_{n=1}^{\infty} A_n^0$, and we apply (ii) once more, this time to $X - A^0$, to find a Borel set $X_0' \supset A^0$, of measure zero, such that $f(X - X_0')$ is a Borel set. Let us write $A_n' = A_n - A_n'$, and let $\mathfrak{A}'$ be the Borel field $(\subset \mathfrak{A})$ spanned by the A_n'. Then we have $(X - X_0')A_n = (X - X_0')A_n'$ for all n, and we see, moreover, that to every Borel set B, (i.e. to every set $B \epsilon \mathfrak{A}$), there corresponds a set $B' \epsilon \mathfrak{A}'$ such that $(X - X_0')B = (X - X_0')B'$. Since $f(A_n')$ is a Borel set, and since the collection of sets A for which $f(A)$ is a Borel set is clearly a Borel field, (because f is univalent), it follows that for every $B' \epsilon \mathfrak{A}'$, $f(B')$ is a Borel set. Consequently for every $B \epsilon \mathfrak{A}$

$$f(B - BX_0') = f(B(X - X_0')) = f(B'(X - X_0')) = f(B')f(X - X_0'),$$

so that $f(B - BX_0')$ is a Borel set, and the proof of the lemma is complete.

LEMMA 4. *If* $X(\mathfrak{X}, \mathfrak{A}, m)$ *is a proper measure space, and if for a single real valued univalent Baire function* $g(x)$ *we can find a set* X_0 *of measure zero such that* $g(B - BX_0)$ *is a Borel set whenever* B *is, then* X *is normal and, moreover, this same set* X_0 *will satisfy the condition of definition 2 for any real valued univalent Baire function* $f(x)$.

PROOF. We write $Y = g(X - X_0)$; for every $y_0 \epsilon Y$, $y_0 = g(x_0)$, we define $F(y_0) = f(x_0)$. $F(y)$ is then a real valued univalent function of the real variable $y \epsilon Y$. Since

(3) $\{y \mid F(y) < a\} = g[\{x \mid f(x) < a\}(X - X_0)],$

and since the right member is a Borel set by hypothesis, $F(y)$ is a Baire function. Since $f(x) = F(g(x))$, we have $f(X - X_0) = F(Y)$, and therefore $f(X - X_0)$ is a Borel set.[6]

An important class of measure spaces is the class of *m-spaces*. An *m-space* is a complete measure space $X(\mathfrak{X}, m)$ on which a metric is defined so that, topologically, it is a complete separable space, and which satisfies the following two conditions:

(i) the measure of an open set is positive;

(ii) for every measurable set E, $m(E) = \inf \{m(O) \mid E \subset O, O \text{ open}\}$. With the Borel field $\mathfrak{A}$ of Borel sets (in the usual topological sense of the word) $X = X(\mathfrak{X}, \mathfrak{A}, m)$ becomes a proper measure space; it is a known result of topology

---

[6] See F. Hausdorff, *Mengenlehre*, Berlin, 1935, p. 266.

that it is even normal in our sense of the word, and that the exceptional set X_0 of measure zero may even be chosen as the empty set.[7]

We shall use m-spaces later; at present we mention them only as examples of normal spaces. The following theorem, the main theorem of the present section, applies to m-spaces, (since they are normal), and shows that, measure theoretically, they are isomorphic to the unit interval.

THEOREM 2. *A necessary and sufficient condition that a measure space of total measure one be point isomorphic to the unit interval is that it be normal.*

PROOF. The necessity of our condition is obvious: the unit interval is normal and normality is invariant under point isomorphism. Before giving a proof of sufficiency we remark on the hypotheses. Since the various conditions in the definition of a *proper* space are logically independent, they are obviously indispensable for a sufficiency proof. It is possible that the condition of *normality* could be replaced by a weaker one, but examples seem to indicate that it is the best way of expressing that the space is "measurable in itself."

For the proof of sufficiency we use the notations of the proof of Theorem 1; in particular we use the functions $f(x)$ and $\tilde{f}(\tilde{x})$ that we defined there. We denote by D and $\tilde{D}$ the ranges of $f(x)$ and $\tilde{f}(\tilde{x})$ respectively. By omitting from X a set of measure zero we may, by normality, assume that D is a Borel set; $\tilde{D}$ is also a Borel set. (We observe that the omission of a set of measure zero does not change the distribution of f and hence does not change $\tilde{f}$ at all). Form the set $R = (D - \tilde{D}) + (\tilde{D} - D)$. Since $f^{-1}(\tilde{D} - D) = f^{-1}(\tilde{D}) - f^{-1}(D)$ lies entirely in the complement of $f^{-1}(D)$, and since this complement is empty, $f^{-1}(\tilde{D} - D)$ is empty. Since $\tilde{f}^{-1}(\tilde{D} - D)$ has the same measure as $f^{-1}(\tilde{D} - D)$, this proves that the measure of $\tilde{f}^{-1}(\tilde{D} - D)$ is zero. Similarly we can prove that the measure of both $f^{-1}(D - \tilde{D})$ and $\tilde{f}^{-1}(D - \tilde{D})$ is zero, (and, in fact, the latter is empty). Hence if we omit from both X and $\tilde{X}$ a Borel set, namely $f^{-1}(R)$ and $\tilde{f}^{-1}(\tilde{R})$ respectively, of measure zero, on the remainder f and $\tilde{f}$ are univalent Baire functions with identical (Borel measurable) ranges.

If to every $x \, \epsilon \, X$ (after the omission, as described, of a set of measure zero), we make correspond the point $\tilde{f}^{-1}(f(x)) \, \epsilon \, \tilde{X}$, the correspondence is one to one. Moreover if B is any Borel set in X, and $B' = f(B)$, then B' is a Borel set and $f^{-1}(B') = B$. Consequently, considered as an element of the Boolean algebra $\mathfrak{B}(\mathfrak{X})$, the correspondent, under the set mapping described in the proof of theorem 1, of B is $\tilde{f}^{-1}(B') = \tilde{B} = \tilde{f}^{-1}(f(B))$, so that the point mapping just described induces precisely the same set isomorphism between $\mathfrak{B}$ and $\tilde{\mathfrak{B}}$. It follows that this point correspondence is measure preserving. This concludes the proof of Theorem 2.

3. The relation between set transformations and point transformations

If T is a measure preserving transformation (i.e. a point isomorphism) of a measure space $X(\mathfrak{X}, m)$ on itself, then T induces a set mapping (of $\mathfrak{B} = \mathfrak{B}(\mathfrak{X})$

---

[7] See Hausdorff, op. cit., p. 269.

on itself) by making correspond to every set $E \, \epsilon \, \mathfrak{X}$ the set $TE \, \epsilon \, \mathfrak{X}$. It is known that in an m-space the converse is true: every set isomorphism is induced in this way by a point isomorphism.[8] Motivated by this we give the following definition.

DEFINITION 3. *A measure space* $X(\mathfrak{X}, m)$ *has* sufficiently many measure preserving transformations *if every set isomorphism of* $\mathfrak{B}$ *on itself is induced by a point isomorphism of* X *on itself.*

It follows from Theorem 2 that every normal space has sufficiently many measure preserving transformations. In between the two concepts (normal spaces and spaces with sufficiently many measure preserving transformations) there is, however, room for a pathological occurrence which we shall describe in this section. We begin by proving some auxiliary results.

LEMMA 5. *If two point mappings, on a measure space* X *which contains a separating sequence* E_1, E_2, $\cdots$ *of measurable sets, induce the same set mapping on* $\mathfrak{B}$ *then they differ on at most a set of measure zero.*

PROOF. It is sufficient to consider the case where one of the transformations is the identity. If then TE_n and E_n differ only on a set of measure zero, for $n = 1, 2, \cdots$, it follows that all $T^k E_n$ differ from each other only on sets of measure zero. Hence the invariant set

$$F_n = \sum_{k=-\infty}^{\infty} T^k E_n - \prod_{k=-\infty}^{\infty} T^k E_n$$

has measure zero. We form the invariant set X' by omitting from X the set $\sum_{n=1}^{\infty} F_n$ of measure zero. If now $x \neq Tx$, then some E_n contains one but not both of x and Tx, and therefore x is contained in one but not both of E_n and $T^{-1} E_n$. Consequently $x \, \epsilon \, F_n$, so that $x \notin X'$.

LEMMA 6. *Let* $X(\mathfrak{X}, m)$ *be a measure space and let* $X' \subset X$ *be any (not necessarily measurable) subset of* X. *Let* $\mathfrak{X}'$ *be the collection of all sets of the form* $E' = X'E$, *with* $E \, \epsilon \, \mathfrak{X}$; *for every* $E' \, \epsilon \, \mathfrak{X}'$, $E' = X'E$, *define* $m'(E') = m(E)$. *With these definitions* m' *is uniquely determined (so that* $X'(\mathfrak{X}', m')$ *is a measure space) if and only if the outer measure of* X' *in* X *is equal to the measure of* X.[9]

LEMMA 7. *If* $\{\phi_n(x)\}$, $n = 1, 2, \cdots$, *is a complete orthonormal set of functions in* $L_2(X)$, *where* $X(\mathfrak{X}, m)$ *is a measure space which contains a separating sequence,* E_1, E_2, $\cdots$, *of measurable sets, then there is a set* $N \, \epsilon \, \mathfrak{X}$ *of measure zero such that* $x, y \notin N$ *and* $\phi_n(x) = \phi_n(y)$ *for* $n = 1, 2, \cdots$, *implies* $x = y$.

---

[8] See John von Neumann, *Einige Sätze über messbare Abbildungen*, Annals of Mathematics, vol. 33, (1932), p. 582. In definition 5, p. 576, all descriptive properties of the transformation (such for example as $M_1 + M_2 \to M_1' + M_2'$) should be modified by the phrase "neglecting sets of measure zero."

[9] The outer measure of E_0, $m^*(E_0)$, is defined by $m^*(E_0) = \inf \{m(E) \mid E_0 \subset E \, \epsilon \, \mathfrak{X}\}$. Similarly we may define the inner measure, $m_*(E_0) = \sup \{m(E) \mid E_0 \supset E \, \epsilon \, \mathfrak{X}\}$. If X is complete then E_0 is measurable (i.e. $E_0 \, \epsilon \, \mathfrak{X}$) if and only if $m_*(E_0) = m^*(E_0) = m(E_0)$. In case X is properly separable it is sufficient to take the supremum and infimum over Borel sets E. For the proof of Lemma 6, see J. L. Doob, *Stochastic processes depending on a continuous parameter*, Transactions of the American Mathematical Society, vol. 42, (1937), pp. 109–110.

PROOF. Let $\psi_m(x)$ be the characteristic function of E_m; we have

(4) $$\psi_m(x) = \sum_{n=1}^{\infty} a_{nm} \phi_n(x),$$

in the sense of convergence in the mean (or order two). Consequently, for each m, a subsequence of the partial sums of the series in (4) converges to $\psi_m(x)$ almost everywhere; for each m we choose a fixed subsequence with this property and we let N be the union of all the sets of measure zero at which these subsequences do *not* converge to $\psi_m(x)$. If x, $y \notin N$ and $\phi_n(x) = \phi_n(y)$ for all n, then it follows that $\psi_m(x) = \psi_m(y)$ for all m, whence (using the fact that E_1, E_2, $\cdots$ is a separating sequence) $x = y$.

LEMMA 8. *Let $\tilde{X}(\tilde{\mathfrak{X}}, \tilde{\mathfrak{A}}, \tilde{m})$ be the perimeter of the unit circle in the complex plane, and let $\mu(\tilde{E})$ be any measure (i.e. a countably additive, non-negative set function with $\mu(\tilde{X}) = 1$) defined for $\tilde{E} \in \tilde{\mathfrak{A}}$. If for a single number λ, with $|\lambda| = 1$ and $(\arg \lambda)/2\pi$ irrational, μ is invariant under rotation through $\arg \lambda$, i.e. $\mu(\lambda\tilde{E}) = \mu(\tilde{E})$ for every $\tilde{E} \in \tilde{\mathfrak{A}}$, then $\mu(\tilde{E}) = \tilde{m}(\tilde{E})$.*

PROOF. Let $\tilde{A}_1$ and $\tilde{A}_2$ be any two closed intervals (arcs) of the same length in $\tilde{X}$. Since the sequence $\{\lambda^n\}$ of powers of λ is everywhere dense in X, we may find a sequence $\{n_j\}$ of positive integers, so that

(5) $$\lim_{j \to \infty} \lambda^{n_j} \tilde{A}_1 = \tilde{A}_2,^{[10]}$$

and consequently

(6) $$\lim_{j \to \infty} \mu(\lambda^{n_j} \tilde{A}_1) = \mu(\tilde{A}_2).^{[11]}$$

Since $\mu(\lambda^{n_j} \tilde{A}_1) = \mu(\tilde{A}_2)$, we have proved that $\mu(\tilde{A}_1) = \mu(\tilde{A}_2)$. Thus $\mu(\tilde{A})$ is a function of the arc length of $\tilde{A}$, i.e. of $\tilde{m}(\tilde{A})$. This numerical function is clearly monotone and additive, hence proportional to $\tilde{m}(\tilde{A})$. Considering $\tilde{A} = \tilde{X}$ shows that the factor of proportionality is 1. Thus $\mu(\tilde{E})$ and $\tilde{m}(\tilde{E})$ agree for arcs, and therefore for all Borel sets.

As an immediate consequence of this lemma we observe that if for any Borel set $\tilde{E}_0$ we have $\tilde{E}_0 = \lambda\tilde{E}_0$, then $\tilde{m}(\tilde{E}_0) = 0$ or else $\tilde{m}(\tilde{E}_0) = 1$, for otherwise

$$\mu(\tilde{E}) = \tilde{m}(\tilde{E}\tilde{E}_0)/\tilde{m}(\tilde{E}_0)$$

would contradict what we just proved.

After these preliminaries we are now ready to introduce the pathological concept we mentioned at the beginning of this section.

DEFINITION 4. *A (not necessarily measurable) subset E of a measure space X is absolutely invariant if for every measure preserving transformation T of X on itself, the symmetric difference $(E - TE) + (TE - E)$ is measurable and has measure zero.*

LEMMA 9. *If E is measurable and $m(E) = 0$ or $m(E) = m(X)$ then E is absolutely invariant. Conversely if X is separable and non-atomic and $E \subset X$*

---

[10] See S. Saks, *Theory of the integral*, Warszawa, 1937, p. 5.

[11] See Saks, *op. cit.*, p. 8.

is measurable and absolutely invariant, then $m(E) = 0$ *or* $m(E) = m(X)$; *if* X *is not measurable and absolutely invariant then* $m_*(E) = 0$, $m^*(E) = m(X)$.

PROOF. The first statement is obvious. To prove the remaining statements we observe that if T is a measure preserving transformation and if A is a set (almost) invariant under T, in the sense that $(A - TA) + (TA - A)$ is measurable and has measure zero, then any measurable cover, A^*, and any measurable kernel, A_*, of A are also (almost) invariant under T.[12] For $A \subset A^*$ implies $TA \subset TA^*$; since TA and A are almost equal, and T is measure preserving, TA^* is a measurable cover of TA, and therefore $TA^* + (A - TA)$ is a measurable cover of A. It follows (since any two measurable covers of A are almost equal) that TA^* is almost equal to A^*, as was to be proved. A similar argument applies to measurable kernals.

It follows from the preceding paragraph that if E is absolutely invariant then so are E_* and E^*. If we knew that a measurable absolutely invariant set must have measure zero or $m(X)$, we could conclude that for a non-measurable absolutely invariant E, $m_*(E) = 0$ and $m^*(E) = m(X)$. In the case where X is the perimeter of the unit circle, there are many examples of measure preserving transformations whose measurable invariant sets all have measure zero or $m(X)$: in fact the rotations described in Lemma 8 are such. If a set is invariant under all measure preserving transformations it is *á fortiori* invariant under these and hence if it is measurable it will have measure zero or $m(X)$. The general case is, however, reduced to the case of the circle by Theorem 1.

To show that the concept of absolute invariance is not vacuous we shall now show that non-measurable absolutely invariant sets exist. In the existence proof we make free use of the continuum hypothesis and well ordering.

LEMMA 10. *If* $X = X(\mathfrak{X}, \mathfrak{A}, m)$ *is a proper measure space of total measure one, there exists an absolutely invariant set* $E \subset X$ *with* $m_*(E) = 0$, $m^*(E) = 1$.

PROOF. Since on a separable measure space there are at most $\mathfrak{c}$ ($=$ the power of the continuum) set transformations (since a set transformation is completely determined by its behavior on a countable collection of sets, and the set of all functions from a set of power $\aleph_0$ to a set of power $\mathfrak{c}$ has power $\mathfrak{c}$), it follows from lemma 5 that we may find a set of at most $\mathfrak{c}$ measure preserving transformations of X on itself with the property that every measure preserving transformation differs on at most a set of measure zero from one of the given set. Let this set be well ordered, so that to each ordinal $\alpha < \Omega$ ($=$ the first uncountable ordinal) there corresponds a measure preserving transformation T_α. We may similarly enumerate the collection of all Borel sets of positive measure: let these be denoted by E_α, $\alpha < \Omega$.

For any $x \, \epsilon \, X$ and any $\alpha < \Omega$ we write

$$C_\alpha(x) = \left\{ \prod_{i=1}^{k} T_{\alpha_i}^{n_i} x \mid \alpha_i \leqq \alpha, k = 1, 2, \cdots; n_i = 0, \pm 1, \pm 2, \cdots \right\}.$$

---

[12] A^* [or A_*] is a measurable cover [or kernel] of A if it is measurable, if $A \subset A^*$ [or $A_* \subset A$], and if $m^*(A) = m(A^*)$ [or $m_*(A) = m(A_*)$]. If A_1^* and A_2^* are measurable covers of A then $(A_1^* - A_2^*) + (A_2^* - A_1^*)$ has measure zero.

$C_\alpha(x)$ is the smallest set containing x and invariant under T_β for all $\beta \leq \alpha$. Further relevant properties of $C_\alpha(x)$ are the following. $C_\alpha(x)$ is a countable set; for $\alpha \leq \beta$, $C_\alpha(x) \subset C_\beta(x)$; if $y \notin C_\alpha(x)$, then $C_\alpha(y)$ and $C_\alpha(x)$ are disjoint.

By transfinite induction we now define points x_α and y_α. x_1 is chosen in E_1 ; y_1 is chosen in E_1 but not in $C_1(x_1)$. Since $C_1(x_1)$ is countable and E_1 (being a Borel set of positive measure) is not, the choice of y_1 is possible. If x_α and y_α are defined for all $\alpha < \beta$, we define x_β as follows. Since the set

$$\sum\nolimits_{\alpha < \beta} \{C_\beta(x_\alpha) + C_\beta(y_\alpha)\}$$

is countable, we may choose $x_\beta \in E_\beta$ so that x_β is not in this set. After this is done we may add $C_\beta(x_\beta)$ to this set and choose y_β so that $y_\beta \in E_\beta$, but y_β is not in the enlarged set.

Concerning the points x_α and y_α we now assert: for any α and β, $\alpha \neq \beta$, $C_\alpha(x_\alpha)$ and $C_\beta(y_\beta)$ are disjoint. If $\alpha \leq \beta$, then we know, by definition, that $y_\beta \notin C_\beta(x_\alpha)$ so that $C_\beta(y_\beta)$ and $C_\beta(x_\alpha)$ are disjoint—á fortiori $C_\beta(y_\beta)$ and $C_\alpha(x_\alpha)$ are disjoint. If $\alpha > \beta$, then again x_α is not in $C_\alpha(y_\beta)$ so that $C_\alpha(x_\alpha)$ and $C_\alpha(y_\beta)$ are disjoint, and therefore so also are $C_\alpha(x_\alpha)$ and $C_\beta(y_\beta)$.

We write

$$A = \sum\nolimits_{\alpha < \Omega} C_\alpha(x_\alpha);$$
$$B = \sum\nolimits_{\beta < \Omega} C_\beta(y_\beta);$$

it follows that A and B are disjoint. Since A contains x_α and B contains y_β, both A and B have at least one point in common with every Borel set of positive measure; consequently $X - A$ and $X - B$ cannot contain any such sets. It follows that both A and B have outer measure one (since their complements have inner measure zero), and since each is contained in the complement of the other, they both have inner measure zero.

It is now easy to see that A is (almost) invariant under every measure preserving transformation T. Given T we may find $\beta < \Omega$, such that T and T_β differ on at most a set of measure zero. Also we have

$$T_\beta A = \sum\nolimits_{\alpha < \Omega} T_\beta C_\alpha(x_\alpha).$$

Since for $\alpha \geq \beta$, $C_\alpha(x_\alpha)$ is invariant under T_β, A and $T_\beta A$ can differ at most on the countable set $\sum_{\alpha < \beta} T_\beta C_\alpha(x_\alpha)$. Since $T_\beta A$ and TA differ on at most a set of measure zero, we have proved that A and TA differ on at most a set of measure zero. We may choose either A or B for the E of Lemma 10.

The following two lemmas establish the connection between absolute invariance and the property of having sufficiently many measure preserving transformations.

LEMMA 11. *Let $X(\mathfrak{X}, m)$ be a measure space of total measure one with sufficiently many measure preserving transformations, and let $X' \subset X$ be any subset of X with $m^*(X') = 1$. If X' is absolutely invariant, then the measure space $X'(\mathfrak{X}', m')$ (defined in Lemma 6) has sufficiently many measure preserving transformations.*

PROOF. The correspondence $E \rightleftarrows E' = X'E$ is a set isomorphism between $\mathfrak{B} = \mathfrak{B}(\mathfrak{X})$ and $\mathfrak{B}' = \mathfrak{B}(\mathfrak{X}')$. Through this isomorphism any set mapping of X' on itself (i.e. any set isomorphism of $\mathfrak{B}'$ on itself) induces a set mapping of X on itself. Since X has, by hypothesis, sufficiently many measure preserving transformations, it follows that to any set mapping T' on X' there corresponds a measure preserving transformation T of X on itself, such that T induces the same set mapping of X as T'. Since X' is absolutely invariant, $(X' - TX') + (TX' - X')$ has measure zero; let N' be the smallest set invariant under T which contains this set of measure zero. We may redefine T on N' to be the identity; the resulting T leaves X' strictly invariant and may therefore be considered as a measure preserving transformation of X' on itself. It is clear that this measure preserving transformation induces the set isomorphism T' on X' and that, therefore, X' has sufficiently many measure preserving transformations.

LEMMA 12. *Let $X(\mathfrak{X}, m)$ be a measure space of total measure one which has a separating sequence of measurable sets, and let $X' \subset X$ be any subset of X with $m^*(X') = 1$. If the measure space $X'(\mathfrak{X}', m')$ (defined in Lemma 6) has sufficiently many measure preserving transformations then X' is an absolutely invariant subset of X.*

PROOF. We use the notation introduced in the proof of Lemma 11. Let T be any measure preserving transformation on X; through the correspondence $E \rightleftarrows E' = X'E$, T induces a set mapping T' on X'. Since X' has sufficiently many measure preserving transformations, the set mapping T' of X' is induced by some measure preserving transformation, say S, of X' on itself. We shall prove that for almost every point $x \epsilon X'$, $Sx = Tx$.

For any set $E \epsilon \mathfrak{X}$ we know that $S^{-1}E' = S^{-1}(X'E)$ and $X' \cdot T^{-1}E$ differ on at most a set of measure zero (since S and T induce the same set mapping on X'): we denote this set of measure zero by N_E, and we write N for the union of all N_E, where we allow E to run through a separating sequence. Let x be any point in $X' - N$; we assert that $Sx = Tx$. If this were not true, we could find a set E, belonging to the separating sequence used above, such that $Sx \epsilon E$ and $Tx \notin E$. Since $x \epsilon X'$, $Sx \epsilon X'$, and therefore $x \epsilon S^{-1}(X'E)$; since $Tx \notin E$, á fortiori $x \notin X' \cdot T^{-1}E$. It follows that $x \epsilon N_E \subset N$; since this contradicts the choice of x, we must have $Sx = Tx$.

We have proved that T leaves almost every point of X' in X': in other words X' is almost invariant under T. Since T was arbitrary, it follows that X' is absolutely invariant.

We conclude this section with an isomorphism theorem that makes clear the structure of measure spaces with sufficiently many measure preserving transformations.

THEOREM 3. *A necessary and sufficient condition that a proper measure space of total measure one have sufficiently many measure preserving transformations is that it be point isomorphic to an absolutely invariant subset of the unit interval.*

PROOF. Since the property of possessing sufficiently many measure preserving transformations is invariant under point isomorphism, and since, by Lemma 11, an absolutely invariant set has this property, sufficiency is clear.

To prove necessity we first observe that the given measure space, $X(\mathscr{X}, \mathcal{C}, m)$ is set isomorphic with $\tilde{X}(\tilde{\mathscr{X}}, \tilde{\mathcal{C}}, \tilde{m})$ in virtue of Theorem 1. (It will be most convenient in this proof to think of $\tilde{X}$ as the perimeter of the unit circle in the complex plane.) Consider on $\tilde{X}$ the measure preserving transformation $\tilde{x} \to \lambda \tilde{x}$, where $\lambda \in \tilde{X}$ is a fixed number with $(\arg \lambda)/2\pi$ irrational. The set isomorphism between $\tilde{X}$ and X makes correspond to this transformation on $\tilde{X}$ a certain measure preserving transformation T on X. A set isomorphism may also be considered as a mapping of the characteristic functions of X on the characteristic functions of X: this mapping may be extended to all $L_2(\tilde{X})$ and thus generates an isomorphism between $L_2(\tilde{X})$ and $L_2(X)$. Let $\phi(x)$ be the correspondent on X of the function $\tilde{\phi}(\tilde{x}) \equiv \tilde{x}$ on $\tilde{X}$; the function $\phi(x)$ has the following properties:

(i) $$|\phi(x)| \equiv 1;$$

(ii) $$\phi(Tx) \equiv \lambda\phi(x);$$

(iii) $\{\phi^n(x)\} = \{(\phi(x))^n\}$, $n = 0, \pm1, \pm2, \cdots$, is a complete orthonormal set in $L_2(X)$.

(To be precise: since $\phi(x)$ is determined only up to a set of measure zero, properties (i) and (ii) need to be true only almost everywhere. It is clear, however, that by changing ϕ on a set of measure zero we may assume that (i) and (ii) are always true. We may also assume, and we find it convenient to do so, that $\phi(x)$ is a Baire function.)

We apply Lemma 7 to $\{\phi^n(x)\}$ to obtain a set N of measure zero with the property described there. By increasing N, if necessary, we may assume that N is invariant under T. We now omit the points of N from X: we shall show that the remainder (henceforth to be denoted by X again) is in one to one measure preserving correspondence with an absolutely invariant subset of $\tilde{X}$.

The function $x' = \phi(x)$ defines a mapping from X to $\tilde{X}$; we know that this mapping is Borel measurable (i.e. that the inverse image of a set in $\tilde{\mathcal{C}}$ lies in $\mathcal{C}$), and we assert furthermore that it is univalent. For if we had $\phi(x) = \phi(y)$, then we should also have $\phi^n(x) = \phi^n(y)$ for all n, and this possibility is precisely what we eliminated when we threw away the set N.

The transformation T is carried by the mapping ϕ into some transformation T' of the range $\phi(X) = X' \subset \tilde{X}$ into itself; since

$$T'x' = \phi(T\phi^{-1}(x')) = \lambda x',$$

we see that X' is invariant under the rotation $\tilde{x} \to \lambda \tilde{x}$.

For every Borel set $\tilde{E} \subset \tilde{X}$ (i.e. $\tilde{E} \in \tilde{\mathcal{C}}$) we define $\mu(\tilde{E}) = m(\phi^{-1}(\tilde{E}))$. Since $\phi(T\phi^{-1}(\tilde{E})) = X' \cdot \lambda\tilde{E}$, we have $T\phi^{-1}(\tilde{E}) = \phi^{-1}(X' \cdot \lambda\tilde{E}) = \phi^{-1}(\lambda\tilde{E})$. Since T is measure preserving it follows that

$$\mu(\tilde{E}) = m(\phi^{-1}(\tilde{E})) = m(T\phi^{-1}(\tilde{E})) = m(\phi^{-1}(\lambda\tilde{E})) = \mu(\lambda\tilde{E}).$$

Hence, by Lemma 8, $\mu(\tilde{E}) = \tilde{m}(\tilde{E})$.

Suppose, finally, that $\tilde{E}_1$ and $\tilde{E}_2$ are Borel subsets of $\tilde{X}$ for which $X'\tilde{E}_1 = X'\tilde{E}_2$. Write $\tilde{E} = (\tilde{E}_1 - \tilde{E}_2) + (\tilde{E}_2 - \tilde{E}_1)$: it follows that $X'\tilde{E}$ is empty, so that $\phi^{-1}(\tilde{E})$ is empty and $\mu(\tilde{E}) = m(\phi^{-1}(\tilde{E})) = \tilde{m}(\tilde{E}) = 0$. This implies that $\tilde{m}(\tilde{E}_1) = \tilde{m}(\tilde{E}_2)$; it follows from Lemma 6 that $\tilde{m}^*(X') = 1$.

To sum up: we have proved that X is point isomorphic with a possibly nonmeasurable subset X' of $\tilde{X}$, with $\tilde{m}^*(X') = 1$; since X has sufficiently many measure preserving transformations, so does X'. Lemma 12 now applies: X' is absolutely invariant and the theorem is proved.

4. Application of the geometric isomorphism theorem to measure preserving transformations

In this section we shall have occasion to use certain facts about measure preserving transformations and the Pontrjagin duality theory: we describe briefly the parts of these theories that we need. Throughout the remainder of our work we consider only normal spaces of total measure one.

Two measure preserving transformations T_1 and T_2, defined, say, on X_1 and X_2, are (point —) *isomorphic*[13] if there is a point isomorphism T from X_1 to X_2 with the property that TT_1T^{-1} is almost everywhere equal to T_2. With every measure preserving transformation T we associate a unitary transformation U defined on $L_2(X)$ by $Uf(x) = f(Tx)$. A measure preserving transformation T has *pure point spectrum* if U has; in other words if there exists a complete orthonormal sequence, $\{f_n(x)\}$ of functions in $L_2(X)$ and a sequence $\Lambda = \{\lambda_n\}$ of complex numbers (of absolute value one) such that $f_n(Tx) = \lambda_n f_n(x)$ almost everywhere, for $n = 1, 2, \cdots$. T is *ergodic* if $f(Tx) = f(x)$ almost everywhere, with $f \in L_2(X)$, is equivalent to $f(x) = $ constant almost everywhere. The *spectrum*, Λ, of an ergodic measure preserving transformation with pure point spectrum is a subgroup of the multiplicative group of complex numbers of absolute value one. The numbers $\lambda_n \in \Lambda$ are, moreover, a complete set of invariants of T, in the sense that if two measure preserving transformations with pure point spectrum have the same set Λ of eigenvalues with the same multiplicities then they are isomorphic.[14]

Concerning groups we shall need the following. A compact abelian separable topological group, X, as an m-space, in the sense that we may define on it an invariant metric $d(x, y)$ and (unique) invariant Haar measure $m(E)$ in such a way that it becomes an m-space.[15] Let Λ' be the character group of X; i.e. Λ' is the set of all complex valued continuous functions $f(x)$ with $|f(x)| \equiv 1$

---

[13] Since this is the only kind of isomorphism for measure preserving transformations that we shall use, we shall in the sequel omit the qualifying 'point —'.

[14] All these statements are proved in (*I*) for flows: it is easy, however, to make the translation from the one parametric case to the discrete case.

[15] Invariance means that for all points x, y, and a, and all measurable sets E, we have $d(x, y) = d(ax, ay)$ and $m(aE) = m(E)$. We find it convenient to write all groups multiplicatively, even though they are abelian.

and $f(xy) = f(x)f(y)$.   Then Λ' is countable, and the functions $f(x)$ ϵ Λ' form a complete orthonormal set in $L_2(X)$.   Conversely let Λ be any countable abelian group, and let X be its character group; i.e. X is the set of all complex valued functions $x(\lambda)$, defined on Λ, with $|\,x(\lambda)\,| \equiv 1$ and $x(\lambda\mu) = x(\lambda)x(\mu)$. X may be so topologized that it becomes a compact separable (and, of course, abelian) group.   If to every λ ϵ Λ we make correspond the function $f(x)$ on X, defined by $f(x) = x(\lambda)$ then this correspondence is an isomorphism between Λ and the entire character group Λ' of X.[16]

The fact that Haar measure is invariant means that the *rotation* $x \rightarrow ax$, where a is any fixed element of the group, is a measure preserving transformation.   The point of introducing the seemingly irrelevant compact groups into the study of measure preserving transformations is that such rotations are normal forms for a large class of transformations.

THEOREM 4.   *An ergodic measure preserving transformation with pure point spectrum on a normal space is isomorphic to a rotation on a compact separable abelian group.*

PROOF.   Let Λ be the spectrum of the given measure preserving transformation; let X be the character group of Λ, and Λ' that of X.   If for every λ ϵ Λ we define $a(\lambda) \equiv \lambda$, then $a = a(\lambda)$ is in X.   For every x ϵ X we define $Tx = ax$; we assert that T has pure point spectrum and that its spectrum is simple and precisely equal to Λ.   It has pure point spectrum because the characters $f(x)$ ϵ Λ' form a complete orthonormal system on $L_2(X)$, and every such f is an eigenfunction of T belonging to the eigenvalue $f(a)$, $f(ax) = f(a)f(x)$.   This shows, moreover, that the spectrum of T, including multiplicities, is obtained by forming the numbers $f(a)$ for all f ϵ Λ'.   Since to each f there corresponds (through the isomorphism described above) an element λ ϵ Λ for which $f(x) = x(\lambda)$ for all x, we see that we may equally well form the numbers $a(\lambda)$, i.e. λ, for all λ ϵ Λ.   Hence T is ergodic and it follows, from the previously quoted result of (I), that the given transformation and T are isomorphic.

Since in this proof we used only the group Λ of eigenvalues and not the actual transformation we have also the following corollary.

COROLLARY 1.   *Every countable group of complex numbers of absolute value one is the spectrum of an ergodic measure preserving transformation with pure point spectrum.*

Theorem 4 also enables us to characterize the set of all transformations which commute with a given ergodic transformation with pure point spectrum.   The solution of this problem for general measure preserving transformations is probably very difficult.

COROLLARY 2.   *If $x \rightarrow ax = Tx$ is an ergodic rotation on a compact abelian group X and if S is any measure preserving transformation on X for which $ST = TS$ then S is also a rotation.*

---

[16] For the proof of all these statements see L. Pontrjagin, *Topological groups*, Princeton, 1939, Chapter V.

PROOF. We have $S(ax) = aS(x)$, so that if we write $b(x) = Sx \cdot x^{-1}$, then

$$b(Tx) = S(ax)(ax)^{-1} = Sx \cdot x^{-1} = b(x).$$

In other words $b(x)$ is invariant under T; since T is ergodic $b(x) = b =$ constant[17] and $Sx = bx$, as was to be proved.

We shall call a measure preserving transformation R an *involution* if $R^2 = I$ (= the identity), and we shall call an involution a *factor* of a given transformation T if $S = RT$ is also an involution (so that $T = SR$).

COROLLARY 3. *If $x \to ax = Tx$ is any rotation on a compact abelian group X, then T may be factored, $T = SR$, $S^2 = R^2 = I$; if T is ergodic every factor R of T is a reflection, $Rx = bx^{-1}$.*

PROOF. Clearly if $Rx = bx^{-1}$ then R is an involution; also $Sx = RTx = R(ax) = ba^{-1} \cdot x^{-1}$ is an involution. Conversely if T is ergodic and if $T = SR$, $S^2 = R^2 = I$, then $TRT = SR \cdot R \cdot SR = R$, so that $aR(ax) = Rx$. It follows as in the proof of Corollary 2 that $b(x) = x \cdot R(x)$ is invariant under T, (i.e. $b(ax) = ax \cdot R(ax) = x \cdot R(x) = b(x)$), so that $b(x) = b =$ constant, and $Rx = bx^{-1}$.

COROLLARY 4. *Any ergodic measure preserving transformation T with pure point spectrum is isomorphic to its own inverse, $T^{-1} = RTR^{-1}$, where R may even be chosen as an involution.*

PROOF. From Corollary 3 we know that $T = SR$, $S^2 = R^2 = I$; since $T^{-1} = R^{-1}S^{-1} = RS$, we have $T^{-1} = R \cdot SR \cdot R = R \cdot T \cdot R^{-1}$.

There seems to be some reason for the conjecture that the results of Corollaries 3 and 4 are valid for an arbitrary measure preserving transformation.

We have seen that every rotation is a measure preserving transformation with pure point spectrum; the question arises as to when a rotation is ergodic. The following theorem asserts that for rotations ergodicity (i.e. metric transitivity) is equivalent to regional transitivity.[18]

THEOREM 5. *If a is a fixed element of the compact abelian group X, the rotation $x \to ax$ is ergodic if and only if the sequence $\{a^n\}$ is everywhere dense in X.*

PROOF. If $x \to ax$ is ergodic then the iterates of some point, say x_0, are everywhere dense.[19] Since the transformation $x \to x \cdot x_0^{-1}$ is a homeomorphism, it carries the sequence $\{a^n x_0\}$ of iterates of x_0 into a dense sequence; but $a^n x_0 x_0^{-1} = a^n$.

Suppose, conversely, that $\{a^n\}$ is everywhere dense. We have already seen that any rotation has every function f in the character group Λ' of X for an eigenfunction, and that the functions of Λ' are a complete orthonormal set in $L_2(X)$. Since eigenfunctions belonging to different eigenvalues are orthogonal, every function invariant under the rotation $x \to ax$ must be a linear combina-

---

[17] The definition of ergodicity says that numerically valued invariant functions are constant. It is easy to verify that this implies the same result for functions (such as $b(x)$) whose values are in the group X.

[18] For a discussion of the various kinds of transitivity see G. A. Hedlund, *The dynamics of geodesic flows*, Bulletin of the American Mathematical Society, vol. 45, (1939), p. 243.

[19] See Eberhard Hopf, *Ergodentheorie*, Berlin, 1937, p. 29.

tion of the invariant functions of the set Λ': if the only invariant function in Λ' is $f(x) \equiv 1$, the rotation is ergodic. Suppose then that $f(ax) = f(x)$ for some $f \in \Lambda'$. It follows (taking x to be the unit element of X, $x = 1$) that $f(a^n) = f(1) = 1$ for all n; since $\{a^n\}$ is dense and f is continuous it follows that $f(x) = 1$.

To introduce the final result of this paper we observe that Theorem 4, and the existence of an invariant metric on any compact separable group, imply that every ergodic measure preserving transformation with pure point spectrum is isomorphic to an isometric transformation on an m-space. Conversely:

THEOREM 6. *If T is an ergodic measure preserving transformation on an m-space $X(\mathfrak{X}, m)$ such that to every $\epsilon > 0$ there corresponds a $\delta = \delta(\epsilon) > 0$ in such a way that $d(x, y) < \delta$ implies $d(T^n x, T^n y) < \epsilon$, $n = 0, \pm 1, \pm 2, \cdots$, (in other words if the family $\{T^n\}$ of transformations is equicontinuous), then T has pure point spectrum: in fact it is possible to introduce into X a multiplication so that it becomes (with the original topology of X) a compact separable abelian group and T becomes a rotation.*

We comment first of all on the hypothesis. Since an isometric transformation clearly has the described equicontinuity property, on the face of it our hypothesis is weaker than isometry. But if our hypothesis is satisfied we may introduce into X a new metric, $d'(x, y)$, defined by

$$d'(x, y) = \sup \{\min(1, d(T^n x, T^n y)) \mid n = 0, \pm 1, \pm 2, \cdots\};$$

it is easy to verify that d and d' induce the same topology on X, and that $d'(Tx, Ty) = d'(x, y)$. We may (and do) therefore assume that T is isometric in the first place.

We shall make the proof of Theorem 6 depend on the following two lemmas which have an interest of their own.

LEMMA 13. *If on an m-space X there exists an ergodic and isometric measure preserving transformation then X is compact.*

PROOF. Let T be an ergodic and isometric transformation; since X is complete we have to show only that it is totally bounded. If it is not, then there is an $\epsilon > 0$ and an infinite sequence of points $x_1, x_2, \cdots$ in X such that the open spheres S_n of radius ϵ with center at x_n are pairwise disjoint. Let x_0 be any point of X whose iterates $\{T^k x_0\}$ are everywhere dense in X, and choose for each $n = 1, 2, \cdots$ an integer $k = k(n)$ such that $d(x_n, T^k x_0) < \epsilon/2$. If we denote by S_0 the open sphere of radius $\epsilon/2$ with center at x_0, then for each n, $T^{k(n)} S_0 \subset S_n$, so that $m(S_n) \geq m(S_0) > 0$. Since a measure space has, by definition, finite measure, there cannot exist an infinite sequence of pairwise disjoint sets whose measure is bounded away from zero; it follows that X is totally bounded and therefore compact.

LEMMA 14. *Let X be any compact group (not necessarily separable or abelian) and let $m(E)$ be any finite measure, defined (at least) for all Borel sets of X, such that the measure of an open set is positive and that the measure of any measurable set is the lower bound of the measure of open sets containing it. Then the set X_0 of all $x \in X$ for which $m(xE) = m(E)$ for all measurable sets E is a closed subgroup of X.*

PROOF. Since $x \, \epsilon \, X_0$ and $y \, \epsilon \, X_0$ implies

$$m(xy^{-1}E) \; = \; m(y^{-1}E) \; = \; m(y(y^{-1}E)) \; = \; m(E),$$

X_0, and consequently its closure $\bar{X}_0$, is a subgroup; we shall prove $\bar{X}_0 \subset X_0$.

Take $x \, \epsilon \, \bar{X}_0$, and let E be any closed (and hence compact) subset of X. Let O be any open set, $O \supset xE$, and let N be a neighborhood of 1 (= the unit element of X) such that for $a \, \epsilon \, N$, $axE \subset O$. Then Nx is a neighborhood of x, so that the intersection of Nx and X_0 is not empty; say $y = ax$, $a \, \epsilon \, N$, $y \, \epsilon \, X_0$. Then

$$m(E) \; = \; m(yE) \; = \; m(axE),$$

and since $axE \subset O$, $m(E) \leqq m(O)$. In other words $xE \subset O$ implies that $m(E) \leqq m(O)$: our condition on m implies that $m(E) \leqq m(xE)$. Applying this result to the compact set xE and the point $x^{-1} \, \epsilon \, \bar{X}_0$ (in place of E and x) we obtain $m(xE) \leqq m(E)$, so that $m(xE) = m(E)$ for all closed sets E. It follows that $m(xE) = m(E)$ for all measurable sets E, as was to be proved.

PROOF OF THEOREM 6. Let x_0 be any point in X for which $\{T^n x_0\}$ is everywhere dense; write $x_n = T^n x_0$ for $n = \pm 1, \pm 2, \cdots$. For $x = x_n$ and $y = x_m$ we define $p(x, y) = x_{n+m}$, and $r(x) = x_{-n}$. If $x' = x_{n'}$, $x'' = x_{n''}$, $y' = x_{m'}$, $y'' = x_{m''}$, then

$$\begin{aligned}
d(p(x', y'), p(x'', y'')) &= d(x_{n'+m'}, x_{n''+m''}) \\
&\leqq d(x_{n'+m'}, x_{n'+m''}) + d(x_{n'+m''}, x_{n''+m''}) \\
&= d(x_{m'}, x_{m''}) + d(x_{n'}, x_{n''}) \\
&= d(y', y'') + d(x', x'');
\end{aligned}$$

in other words $p(x, y)$ is uniformly continuous throughout its domain of definition; similarly since we have

$$d(r(x), r(y)) \; = \; d(x_{-n}, x_{-m}) \; = \; d(x_{-n+n+m}, x_{-m+n+m}) \; = \; d(y, x),$$

$r(x)$ is uniformly continuous throughout its domain. The domain of $p(x, y)$ is an everywhere dense subset of the product space of X with itself, and the domain of $r(x)$ is an everywhere dense subset of X, consequently they each have a unique continuous extension, to all the product space and all X respectively.

The rest of the proof is now easy. We define, for every x and y in X, $xy = p(x, y)$ and $x^{-1} = r(x)$; it is clear that with these definitions X becomes an abelian topological group. We may write, for any $x = x_n$ and an arbitrary y, $p'(x, y) = T^n y$; then $p'(x, y)$ is a continuous extension of our original $p(x, y)$ and therefore (because of the uniqueness of extension) $T^n y = x_n y$. (For $n = 1$, we obtain, in particular, $Ty = x_1 y$ for all y. The originally chosen element x_0 is now the unit element of the group.) If E is any measurable set then $T^n E = x_n E$ has the same measure as E, so that measure is preserved by an everywhere dense set of x's; since, by Lemma 13, X is compact, Lemma 14 implies that for all x and all measurable sets E, $m(xE) = m(E)$. The uniqueness of Haar measure implies that m is the Haar measure of the group X; this completes the proof that T is a rotation, and hence has pure point spectrum.

INSTITUTE FOR ADVANCED STUDY

Reprinted from the
BULLETIN OF THE AMERICAN MATHEMATICAL SOCIETY
Vol. 49, No. 8, pp. 619-624, Aug. 1943

ON AUTOMORPHISMS OF COMPACT GROUPS

PAUL R. HALMOS

1. Introduction and definitions. Let G be a compact abelian group and α a continuous automorphism of G. We write G multiplicatively and use, accordingly, the exponent notation for automorphisms. Thus the image under α of the element $x \in G$ will be denoted by x^α; similarly we shall write for (complex valued) functions $f(x)$, $f^\alpha(x)$ $= f(x^\alpha)$.[1]

If m is Haar measure[2] in G (normalized so that $m(G) = 1$) we consider the set function $m'(E) = m(E^\alpha)$. (E^α is the set of all x^α, $x \in E$.) Since m' is a measure on G possessing all defining properties of m it follows from the uniqueness of Haar measure[3] that $m'(E) = m(E)$ for every measurable set E. In other words α is a measure preserving transformation of G; the purpose of this note is to investigate a few simple properties of α from the point of view of measure theory.

We shall make use of the Pontrjagin duality theory,[4] and, in particular, we shall need the fact that the group of automorphisms of G is essentially the same as that of the character group G^*. More precisely: if to any $\phi = \phi(x) \in G^*$ we make correspond $\phi^\alpha = \phi^\alpha(x) = \phi(x^\alpha)$, then $\phi^\alpha \in G^*$, and the correspondence $\phi \rightarrow \phi^\alpha$ is an automorphism of G^*. The duality theory also enables us to reverse this argument: every automorphism of G^* is induced in this way by a continuous automorphism of G.

We recall some standard definitions from ergodic theory. A measure preserving transformation α (not necessarily an automorphism) is *ergodic* if the only (complex valued, measurable) solutions f of the equation $f^\alpha = f$ are constant almost everywhere. The transformation α is *mixing* if the only (complex valued, measurable) solutions f of the equation $f^\alpha = \lambda f$, for any constant λ, are constant almost everywhere.[5] (It is true, though irrelevant, that for $\lambda \neq 1$ even a constant fails to be a solution unless it is zero.) It is well known that the mapping

---

Presented to the Society, December 27, 1942; received by the editors November 23, 1942.

[1] This notation dovetails, as usual, with ordinary exponentiation in G; thus $x^{3\alpha^2} = (x^3)^{\alpha^2} = (x^{\alpha^2})^3$, and so on.

[2] For a general discussion of measure theory in topological groups see A. Weil, *L'intégration dans les groupes topologiques et ses applications*, Paris, 1938.

[3] Weil, op. cit., pp. 36–38.

[4] Weil, op. cit., chap. 6.

[5] See E. Hopf, *Ergodentheorie*, Berlin, 1937, chap. 3, for a discussion of the fact that these definitions are equivalent to the ones more commonly given.

$f \rightarrow f^\alpha$ induced by a measure preserving transformation α on the space of functions over G is a unitary transformation of the Hilbert space $L_2(G)$.[6] Two measure preserving transformations α and β are of the *same spectral type* if there is a unitary transformation ω of $L_2(G)$ (ω need not be induced by a transformation of G) for which $f^{\omega\alpha\omega^{-1}} = f^\beta$ for all $f \in L_2(G)$. Given any point $x \in G$ (or function f on G) the set of all x^{α^n} (or f^{α^n}), $n = 0, \pm 1, \pm 2, \cdots$ is the *orbit* of x (or f). If α is an automorphism the orbit of the identity consists of the identity only; if this is the only finite orbit we shall say, for the sake of brevity, that α has no finite orbits.

In terms of the definitions of the preceding paragraph we can state our results quite concisely. In Theorem 1 we obtain a simple characterization of ergodicity and mixing in terms of the orbits of α in G^*. Theorem 2 is a statement concerning abstract groups and, prima facie, has nothing to do with measure theory; together with Theorem 1 and the duality theory it yields, however, a complete description of the spectral type of ergodic automorphisms. In the concluding section we state some unsolved problems and emphasize the importance of group automorphisms as a source of many new and simple examples of transformations with properties that were once considered difficult to obtain.[7]

2. **Ergodic and mixing automorphisms.** We prove the following theorem:

THEOREM 1. *A continuous automorphism α of a compact abelian group G is ergodic (or mixing) if and only if the induced automorphism on the character group G^* has no finite orbits.*

We call attention to the somewhat surprising fact that for continuous automorphisms ergodicity is equivalent to the apparently stronger mixing condition. We shall see later that much more than this is true: if α is ergodic then it automatically has the strongest of the whole known hierarchy of mixing properties.

The similarity of our definitions of ergodicity and mixing to each other enables us to prove both parts of the theorem simultaneously. Suppose that $f \in L_2(G)$ and λ, $|\lambda| = 1$, are such that $f^\alpha = \lambda f$.[8] We may expand f in a Fourier series in the characters $\phi \in G^*$, $f(x) = \sum_\phi a(\phi)\phi(x)$.[9] Concerning this series we must make two comments. First, even if G^*

---

[6] Hopf, op. cit., p. 9.

[7] Hopf, op. cit., p. 42.

[8] The fact that, considered as a transformation of L_2, α is a unitary operator implies that if there is any proper value λ at all then it must be of modulus one.

[9] Weil, op. cit., p. 76.

is uncountable, at most a countable number of ϕ's fail to be orthogonal to f, so that at most a countable number of a's are different from zero. Second, the series need converge only in the sense of L_2 (mean square convergence); the known fact that a sub-sequence of its partial sums converges almost everywhere is sufficient to justify the simple formal steps that follow. Replacing x by x^α we obtain

$$\sum_\phi \lambda a(\phi)\phi(x) = \lambda f(x) = f^\alpha(x) = f(x^\alpha)$$

$$= \sum_\phi a(\phi)\phi(x^\alpha) = \sum_\phi a(\phi)\phi^\alpha(x) = \sum_\phi a(\phi^{\alpha-1})\phi(x).$$

Hence (using the orthogonality of the characters) we may equate coefficients and obtain $\lambda a(\phi) = a(\phi^{\alpha-1})$, or $|a(\phi)| = |a(\phi^{\alpha-1})|$. Since ϕ is arbitrary it follows that all coefficients $a(\phi)$ corresponding to ϕ's in the same orbit are equal in modulus. Since $\sum_\phi |a(\phi)|^2 < \infty$ it follows that all a's corresponding to ϕ's in the same infinite orbit vanish. This settles the *only if* part of our theorem: a non-constant f can exist only if α (considered as an automorphism of G^*) has finite orbits.

The converse is easier. Let $\phi \in G^*$ ($\phi \neq 1$) have a finite orbit; suppose, for definiteness, that n is the least positive integer for which $\phi^{\alpha^n} = \phi$. It follows that for the function $f = \phi + \phi^\alpha + \cdots + \phi^{\alpha^{n-1}}$ we have $f^\alpha = f$. The orthogonality and, a fortiori, linear independence of the ϕ's show that f is not constant. Since this shows that the existence of a finite orbit implies non-ergodicity, the proof of Theorem 1 is complete.

THEOREM 2. *Let α be any automorphism of the discrete abelian group H; if α has no finite orbits then it has an infinite number of orbits.*[10]

Case I. We assume first that there is in H an element ϕ_0 of finite order. By raising ϕ_0 to a suitable power we may assume that the order of ϕ_0 is a prime p. Write $\phi_n = \phi_0^{\alpha^n}$, $n = 0, \pm 1, \pm 2, \cdots$; we shall prove that the ϕ_n are independent mod p. (It is clear that the order of each ϕ_n is p.) Suppose, on the contrary, that $\phi_{i_1}^{r_1} \cdots \phi_{i_k}^{r_k} = 1$; it is merely a notational change to write $i_1 = 1, \cdots, i_k = k$, and we may, of course, assume that r_1 and r_k are not congruent to zero mod p. We have then

$$(1) \qquad \phi_1 = (\phi_2^{-r_2} \cdots \phi_k^{-r_k})^{r_1^{-1}}, \qquad \phi_k = (\phi_1^{-r_1} \cdots \phi_{k-1}^{-r_{k-}})^{r_k^{-1}}.$$

(The exponents r_1^{-1}, r_k^{-1} make sense since we may interpret them in the modular field $GF(p)$.) Consider now the finite subgroup H_0 of H

---

[10] The author's thanks are due to R. Baer, R. H. Fox, and H. Samelson for many valuable discussions of this theorem and its proof.

generated by $\phi_1, \cdots, \phi_k$. It follows from the definition of the ϕ_n and the relations (1) that H_0 is invariant under both α and α^{-1} and contains consequently the entire orbit of ϕ_0. Since this contradicts the assumed nonexistence of finite orbits, the ϕ_n must indeed be independent. The desired conclusion follows at once: the elements $\psi_n = \phi_1 \cdots \phi_n$, $n = 1, 2, \cdots$, must all lie in different orbits.

Case II. If ϕ_0 is an element of infinite order we write $\phi_i = \phi_0^{\alpha^i}$ as before and we ask for which (positive or negative) integers i it is true that ϕ_i is a positive rational power of ϕ_0 (that is, a positive integral power of some root of ϕ_0). If $\phi_i = \phi_0^r$ then $(\phi_{-i})^r = (\phi_0^r)^{\alpha^{-i}} = (\phi_i)^{\alpha^{-i}} = \phi_0$, so that ϕ_{-i} is an rth root of ϕ_0. If also $\phi_j = \phi_0^s$ then $\phi_{i+j} = (\phi_j)^{\alpha^i} = \zeta_0^{s\alpha^i} = \phi_0^{\alpha^i s} = \phi_0^{rs}$. In other words the set of i's under consideration is an additive group of integers; let i_0 be a generator of this group, $\phi_{i_0} = \zeta_0^{r_0}$. Then for any integer n

$$\phi_{i_0 n} = (\cdots ((\phi_0)^{r_0})^{r_0} \cdots)^{r_0} = \phi_0^{r_0^n};$$

in other words the set of possible r's that may occur as exponents in a relation $\phi_i = \phi_0^r$ consists of all (positive, negative, or zero) integral powers of r_0. Hence the set of powers of ϕ_0 which are in the same orbit as ϕ_0 consists only of powers of r_0, and consequently there is a power of ϕ_0 which does not lie in this orbit. We may choose this power of ϕ_0 as a new starting element (of infinite order) and repeat the above argument ad infinitum. This completes the proof of Theorem 2.

We are now prepared to describe the spectral type of ergodic automorphisms. Let S be a complete orthonormal set in an abstract, not necessarily separable, Hilbert space and denote by ψ any particular element of S. Arrange all remaining elements of S as an infinite matrix in such a way that each row contains a countably infinite number of elements. Use as row index any set of suitable power and as column index the set of all (positive, negative, or zero) integers. A unique unitary operator σ is defined on Hilbert space by the requirement that it send ψ into itself and $\phi_{i,j}$ into $\phi_{i,j+1}$ for $j = 0, \pm 1, \pm 2, \cdots$, and all i. The spectral type of σ depends only on the number of rows; if we agree to use $\aleph$ for this cardinal number we may write $\sigma = \sigma(\aleph)$. We summarize our result in terms of these σ's.

THEOREM 3. *If α is a continuous ergodic automorphism of a compact abelian group G and $\aleph$ is the (necessarily infinite) cardinal number of G^* then α has the spectral type of the unitary operator $\sigma(\aleph)$.*

For the proof we need remember only that the characters of G form a complete orthonormal set of elements of $L_2(G)$. The principal character plays the role of ψ and the orbits of the other characters may be written as the rows of the matrix mentioned above. Theorem 2 shows that there must be an infinite number of rows, and the well known fact that for any infinite cardinal $\aleph \cdot \aleph_0 = \aleph$ [11] shows that the number of rows is the same as the total number of elements in G^*.[12]

3. **Examples and questions.** Let H be any compact abelian group and let G be the direct product of H with itself a countable number of times. We write G as the set of all sequences $\{x_n \mid n = 0, \pm 1, \pm 2, \cdots \}$, and we define a continuous ergodic automorphism α of G by the relations

$$\{ x_n \}^\alpha = \{ x_n' \}, \quad x_n' = x_{n+1}, \quad n = 0, \pm 1, \pm 2, \cdots .$$

Transformations isomorphic to such α's (not only in the spectral but even in the stronger, measure theoretic, sense) were among the first known examples of ergodic transformations.

Examples apparently very simple from the algebraic point of view, but very difficult to handle geometrically, are furnished by the solenoids. Consider for instance the multiplicative group of all real numbers of the form e^r, where r is a dyadic rational number. The operation of squaring is an automorphism of this group (with no finite orbits), and hence an ergodic automorphism of its compact character group.

More in the classical spirit than either of the last two examples are the (continuous) automorphisms of the n-dimensional toral group. In order to retain our multiplicative notation we write the torus as n-tuples $(x_1, \cdots, x_n)$ of complex numbers of modulus one; thus the product of $(x_1, \cdots, x_n)$ by $(y_1, \cdots, y_n)$ is $(x_1 y_1, \cdots, x_n y_n)$ and the identity element is $(1, \cdots, 1)$. It is well known that the automorphism group of the torus is the unimodular group, in the following sense. Given any $n \times n$ matrix $\alpha = (a_{ij})$ whose elements are integers and whose determinant is ± 1, we consider the mapping $(x_i) \rightarrow (\prod_j x_j^{a_{ij}})$: this mapping is the most general continuous automorphism of the torus. The condition of ergodicity—that is, the condition that, considered as an automorphism of the character group, α have no finite orbits—is equivalent in classical terms to the requirement that no root of unity should be a proper value of α. This remark enables us to write down any desired number of quite different looking analytic ergodic (and hence mixing) measure preserving transformations on the finite dimensional torus.

---

[11] See F. Hausdorff, *Mengenlehre*, Berlin, 1935, p. 71.

[12] The first explicit discussion of the spectral form of a measure preserving transformation of type $\sigma(\aleph_0)$ was carried out by J. L. Doob and R. A. Leibler, *On the spectral analysis of a certain transformation*, Amer. J. Math. vol. 65 (1943) pp. 263–272.

This last example and the particular structure of the automorphisms it describes suggests a new, purely measure theoretic, invariant of measure preserving transformations. The Hamilton-Cayley equation says, in our notation, that if p is the characteristic polynomial of α then for every x in the torus $x^{p(\alpha)} = 1$. It follows that for every character f of the torus we have, similarly, $f^{p(\alpha)} = 1$. The existence or nonexistence of f's thus *annihilated* by certain polynomials in α (with, of course, integer coefficients) and, if they exist, their algebraic and measure theoretic structure, furnishes the invariant to which we referred. To illustrate the possible application of these invariants we mention the following: if it could be proved that the characters are (except for trivial changes on a set of measure zero) the only measurable functions of constant modulus one which are *annihilated* by $p(\alpha)$, it would follow rather easily that two ergodic automorphisms of the torus are measure theoretically isomorphic if and only if they correspond to conjugate elements in the unimodular group. We could thus obtain the first examples of measure theoretically distinct transformations of the spectral type of $\sigma(\aleph_0)$,[13] and the usual proper value theory would then point the way to further, more delicate, invariants.

In conclusion we mention an unsolved problem of purely technical interest, but one whose solution may throw some light on the deeper problems raised above. The question is simply: do there exist measure preserving transformations (on spaces of finite measure) of the spectral type of $\sigma(\aleph)$, where $\aleph$ is a *finite* cardinal?

SYRACUSE UNIVERSITY

---

[13] The first examples of measure theoretically different but spectrally isomorphic transformations are due to J. von Neumann. These examples (not yet published) are, however, not mixing transformations.

Reprinted from the
TRANSACTIONS OF THE AMERICAN MATHEMATICAL SOCIETY
Vol. 55, No. 1, pp. 1-18, Jan. 1944

APPROXIMATION THEORIES FOR MEASURE PRE-SERVING TRANSFORMATIONS

BY

PAUL R. HALMOS

1. Introduction. The purpose of this paper is to introduce approximative notions into the theory of measure preserving transformations in the hope that they will turn out to be useful tools in investigating some of the outstanding problems of ergodic theory.

By a measure preserving transformation one means, ordinarily, a one-to-one mapping of a measure space on itself which is such that both it and its inverse preserve the measurability and the numerical measure of measurable sets. Since, however, it is customary in measure theory (and, from the point of view of purity of method, highly desirable) to identify two transformations, or functions, or sets, which agree except for a set of measure zero, the treatment below discusses not measure spaces but rather their Boolean algebras, consisting of the measurable sets modulo sets of measure zero. It is known that such "measure algebras" are simply characterizable in algebraic terms and that from this point of view measure preserving transformations are nothing other than automorphisms of the underlying algebra. A consistent adherence to this point of view led me to some simple but illuminating facts (Theorem 1) which exhibit the reason why some common measure theoretic devices work.

Three notions of approximation—that is, three topologies—are introduced into the set G of all measure preserving transformations. For the reader familiar with von Neumann's fundamental work on Hilbert space two of these are best described as the analogues (and, indeed, relativizations) of the "strong neighborhood" and the "uniform" topologies for bounded operators. The third (called the "metric" topology below) is defined in terms of a distance function, the distance between two transformations being the measure of the set of points where they differ.

The results proved are of three types. The first type is purely technical: I investigate the relations of the various topologies to each other and to the group structure of G. It is not surprising that G turns out to be a topological group with more or less decent properties in all three cases (Theorems 2 and 7). The interesting and nontrivial fact along these lines is that the uniform and the metric topologies are the same (Theorem 10).

The second type of result is the one I consider most important: it asserts that arbitrary measure preserving transformations may be approximated by

Presented to the Society, April 24, 1943; received by the editors March 10, 1943.

1

transformations with comparatively simple properties (for example, by transformations of finite period). These results (Theorems 3, 4, and 8) are the analogues of various theorems asserting the density of step functions in sundry function spaces and will, I hope, find similar use.

The third and final type of result, generally an easy consequence of the approximation theorems just mentioned, is motivated by the oldstanding conjecture that "in general a measure preserving transformation is ergodic." I adopt the usual, by now classical, interpretation of "in general" (that is, "except for a set of first category") and show that the adage is true in the neighborhood topology[1] (Theorem 6) and false in the metric topology (Theorem 9). I hope in the near future to be able to apply these same methods to prove analogous theorems about the much more difficult class of mixing transformations.

2. The Boolean algebra. Let B be a Boolean σ-algebra[2] in which the null and unit elements, the Boolean operations of union and intersection, the inclusion relation, and the complement of an element a are denoted by the familiar symbols 0, 1, $\cup$, $\cap$, $\subset$, and a', respectively. It is well known that if addition and multiplication are defined in B by the formulas

$$a + b = (a \cap b') \cup (a' \cap b), \qquad ab = a \cap b,$$

then B becomes an algebraic ring[3]. In this ring multiplication is commutative ($ab = ba$), every element is idempotent ($a^2 = a$), and addition is modulo 2 ($a + a = 0$). It follows that the relations

$$a + b = a - b, \qquad a' = 1 - a,$$
$$a \cup b = a + b + ab, \qquad a - b = a' - b'$$

are identities in B; in the sequel their right and left sides will be used interchangeably, the choice in any particular case being guided by convenience and intuitive content alone.

Suppose moreover that there is defined on B a numerically valued, positive, countably additive, finite measure: the measure of an element a is to be denoted by $|a|$[4]. The existence of such a measure has a profound effect on

---

   (1) The first theorem of this type is due to J. C. Oxtoby and S. M. Ulam, *Measure-preserving homeomorphisms and metrical transitivity*, Ann. of Math. (2) vol. 42 (1941) p. 880. Their topology is, however, very different from mine and depends on the topological and metric (as opposed to purely measure theoretic) structure of the underlying space.

   (2) See Garrett Birkhoff, *Lattice theory*, Amer. Math. Soc. Colloquium Publications, vol. 25, New York, 1940, pp. 29 and 88.

   (3) See M. H. Stone, *The theory of representations for Boolean algebras*, Trans. Amer. Math. Soc. vol. 40 (1936) p. 43.

   (4) In other words: $|a| = 0$ is equivalent to $a = 0$, $a_i a_j = 0$ for $i \neq j$, $i, j = 1, 2, \cdots$ implies $|\cup_i a_i| = \sum_i |a_i|$, and $|1| < \infty$. Boolean algebras satisfying all these conditions are called *measure algebras*. Cf. Dorothy Maharam, *On homogeneous measure algebras*, Proc. Nat. Acad. Sci. U.S.A. vol. 28 (1942) p. 108.

the algebraic structure of B. Possibly the most surprising result is that the presence of the measure implies that B is complete in the sense of lattice theory. If, in other words, E is an arbitrary, not necessarily countable, subset of B then there exists an element $a \in B$ such that $x \subset b$ for all $x \in E$ is equivalent to $a \subset b$([5]). This fact will be used quite frequently in the work that follows: the element a so defined will be denoted by $a = \sup E$.

The most usual form in which such suprema are encountered is the following. If E is any subset of B denote by E^* the set of all those elements of B every nonzero subelement of which contains a nonzero subelement belonging to E. (In other words, E^* is the set of all elements arbitrarily small parts of which belong to E.) Similarly, denote by E_* the set of all those elements of B *every* nonzero subelement of which belongs to E. Then the lattice completeness of B implies the following.

THEOREM 1. *E^* is always a principal ideal (that is, there is an element $e^* \in B$ such that $a \in E^*$ if and only if $a \subset e^*$); E_* is a principal ideal if and only if for all subsets A of E_*, containing a nonzero element, $\sup A \in E$. The generator $e^* = \sup E^*$ of E^* and the element $\tilde{e}_* = \sup (E')_*$ (which is the generator of $(E')_*$ in case $(E')_*$ is a principal ideal) are each other's complements*([6]).

Proof. Observe first that $0 \in E^*$ and $0 \in E_*$. This paradoxical fact is due to the wording of the definitions of E^* and E_*: since 0 contains no nonzero subelements it vacuously satisfies the conditions of both definitions.

To prove that E^* is a principal ideal observe that if $a \in E^*$ and $b \subset a$ then $b \in E^*$. In consequence of this fact it is sufficient to prove that $e^* = \sup E^*$ belongs to E^*. Suppose therefore that $a \subset e^*$, $a \neq 0$. Then (from the definition of e^*) there must be an element $b \in E^*$ for which $ab \neq 0$. Since $ab \subset b$ and $b \in E^*$ there is an element $c \subset ab$, $c \neq 0$, for which $c \in E$. Since $c \subset ab \subset a$, this shows that every nonzero $a \subset e^*$ contains at least one nonzero c in E.

The sufficiency proof for E_* is similar to the above: once more $a \in E_*$ and $b \subset a$ implies $b \in E_*$, so that it is sufficient to prove that $e_* = \sup E_*$ belongs to E_*. Take $a \subset e_*$, $a \neq 0$; then $a = \sup \{b : b \subset a, b \in E_*\}$([7]), whence, by the assumed condition, $a \in E$ and therefore $e_* \in E_*$. The necessity of the condition is even easier to see: if E_* is a principal ideal and $A \subset E_*$ then $\sup A \in E_*$, and consequently, since every nonzero element of E_* belongs to E, $\sup A \in E$.

Finally, if $a \in E^*$ and $b \in (E')_*$ then $ab = 0$. For otherwise $ab \subset b$ would imply that all nonzero subelements of ab are in E' and $ab \subset a$ would imply that some nonzero subelement of ab is in E. Hence e^* and $\tilde{e}_*$ are disjoint; to prove that their union is 1, suppose that $a \neq 0$ is disjoint from $\tilde{e}_*$. Then no nonzero subelement of a can be in $(E')_*$. Hence every nonzero subelement

---

([5]) See Birkhoff, op. cit., p. 100.

([6]) The Boolean algebra symbols, E', $E \cup F$, and so on, are used for the algebra of *subsets* of B, as well as for the abstract algebra of *elements* of B.

([7]) The symbol $\{x : \cdots \}$ stands, as usual, for "the set of all x's for which $\cdots$."

of a contains a nonzero subelement in E; that is, $a \in E^*$ or $a \subset e^*$. This completes the proof of Theorem 1.

COROLLARY. *If, for every subset A of E, sup A belongs to E then E_* is a principal ideal.*

This follows immediately from Theorem 1 and the remark (already used above) that every nonzero element of E_* belongs to E.

The presence of a measure serves also to introduce a natural metric topology into B: the distance between a and b may be defined by $|a-b|$. It is easy to verify that the distance axioms are satisfied; a nontrivial conclusion is that B thereby becomes a *complete* metric space[8]. The following inequalities will be useful later:

(1)
$$\Big| \, |a| - |b| \, \Big| \leq |a - b|,$$

(2)
$$\big| (a_1 + b_1) - (a_2 + b_2) \big| \leq |a_1 - a_2| + |b_1 - b_2|,$$

(3)
$$\big| a_1 b_1 - a_2 b_2 \big| \leq |a_1 - a_2| + |b_1 - b_2|,$$

(4)
$$\big| (a_1 \cup b_1) - (a_2 \cup b_2) \big| \leq |a_1 - a_2| + |b_1 - b_2|.$$

The first of these follows from the relations $a = ab + ab'$ and $b = ab + a'b$, (2) is a consequence of the additivity of measure, (3) follows from (2) and the fact that $|ab| \leq |a|$, and (4) follows from (3) by taking complements. It follows, of course, from these inequalities of Lipschitz type that each of the functions $|a|$, $a+b$, ab, and $a \cup b$ is a (uniformly) continuous function of all its arguments.

Finally the following three normalizing assumptions about B will be made throughout, unless the contrary is explicitly stated.

(*) $|1| = 1$.

(**) B is non-atomic (that is, for every $a \in B$, $a \neq 0$, there is an element $b \in B$ such that $b \neq 0$, $b \neq a$, $b \subset a$).

(***) As a metric space B is separable.

Most of the interesting results of this paper remain true, though not interesting, in case (**) is not assumed. The cardinal number restriction (***) is merely a matter of convenience. One of its main purposes is to concretize the object of study (that is, the algebra B). It is known that the axioms of a measure algebra together with (*), (**), and (***) are categorical: any system satisfying them is isomorphic to the measure algebra of all measurable sets modulo sets of measure zero of, say, the unit interval[9]. This representation is frequently useful and will be exploited below.

---

[8] The proof of this fact is very similar to the proof of the Riesz-Fischer theorem and can, in fact, be made to follow from it.

[9] See for example Paul R. Halmos and John von Neumann, *Operator methods in classical mechanics.* II, Ann. of Math. (2) vol. 43 (1942) p. 335.

3. **The neighborhood topology.** An *automorphism* of B is a one-to-one mapping T of B onto itself such that

$$| Ta | = | a |, \qquad Ta' = (Ta)',$$
$$T(\cup_i a_i) = \cup_i Ta_i, \qquad T(\cap_i a_i) = \cap_i Ta_i (^{10}).$$

Under the operation of composition the set G of all automorphisms of B is a group.

Let S be any element of G and take $a \in B$, $\epsilon > 0$. Writing

$$N(S) = N(S; a, \epsilon) = \{ T : | Sa - Ta | < \epsilon \},$$

a unique topology is defined in G by the requirement that the collection of all sets of the form $N(S)$ be a subbase for the open sets[11]. I shall call this the *neighborhood topology* of G[12]; sets of the form $N(S)$ are *subbasic neighborhoods* of S, and finite intersections of subbasic neighborhoods are *basic neighborhoods* of S.

THEOREM 2. *In the neighborhood topology G is a complete topological group satisfying the first countability axiom*[13].

Proof. (1) If S_1, $S_2 \in G$ and $S_1 \neq S_2$ then there is an $a \in B$ such that $\epsilon = | S_1 a - S_2 a | > 0$. Consequently $N(S_1) = \{ T : | S_1 a - Ta | < \epsilon \}$ is a neighborhood of S_1 which does not contain S_2, so that G is a T_0-space.

(2) Consider any S_0, $T_0 \in G$ and any subbasic neighborhood N_0 of their "quotient" $S_0 T_0^{-1}$,

$$N_0 = N(S_0 T_0^{-1}; a, \epsilon) = \{ R : | S_0 T_0^{-1} a - Ra | < \epsilon \}.$$

Write $b = T_0^{-1} a$ (so that $a = T_0 b$) and consider any $S \in N(S_0; b, \epsilon/2)$ and $T \in N(T_0; b, \epsilon/2)$. Then

$$| S_0 T_0^{-1} a - ST^{-1} a | \leqq | S_0 T_0^{-1} a - ST_0^{-1} a | + | ST_0^{-1} a - ST^{-1} a |$$
$$= | S_0 b - Sb | + | b - T^{-1} T_0 b |$$
$$= | S_0 b - Sb | + | Tb - T_0 b | < \epsilon.$$

Consequently ST^{-1} is a continuous function of both its arguments.

(¹⁰) It is known that the conditions $| Ta | = | a |$ and $T0 = 0$ are sufficient to imply that T is an automorphism. This fact is a member of a class of theorems of which a much better known specimen is the assertion that an isometry of a finite dimensional unitary space, which leaves the origin fixed, is a unitary transformation, that is, an automorphism of the space. Such theorems greatly reduce the labor of verifying that a given transformation is an automorphism.

(¹¹) See Solomon Lefschetz, *Algebraic topology*, Amer. Math. Soc. Colloquium Publications, vol. 27, New York, 1942, p. 6.

(¹²) This topology is the analogue of the "strong" topology of operators on Hilbert space and may in fact be obtained from the strong topology by relativization.

(¹³) For the notion of completeness for topological groups see André Weil, *Sur les espaces à structure uniforme et sur la topologie générale*, Paris, 1938, p. 29.

(3) Since G has already been shown to be a topological group it is sufficient to investigate first countability in the neighborhood of the identity. Let a_1, a_2, $\cdots$ be a countable dense set in B; write $N_{ij} = N(I;\ a_i,\ 1/j)$, $i,\ j = 1,\ 2,\ \cdots$; then the N_{ij} are a subbase at I (the identity). For given any $N = N(I;\ a,\ \epsilon)$, choose i so that $|a - a_i| < \epsilon/3$ and choose j so that $1/j < \epsilon/3$. Then, if $T \in N_{ij}$,

$$|a - Ta| \leq |a - a_i| + |a_i - Ta_i| + |Ta_i - Ta| < \epsilon,$$

so that $N_{ij} \subset N$; q.e.d.

(4) In a topological group satisfying the first countability axiom completeness is equivalent to sequential completeness. In order to prove that every convergent sequence has a limit, I first remark that $\{T_n\}$ is convergent if and only if the sequence $\{T_n a\}$ of elements of B is convergent for every $a \in B$. Consequently, if $\{T_n\}$ is convergent a unique T is defined by $Ta = \lim_n T_n a$. It is easy to see that $Ta' = (Ta)'$ and $T(a_1 \cup a_2 \cup \cdots \cup a_n) = Ta_1 \cup Ta_2 \cup \cdots \cup Ta_n$; the continuity of $|a|$ implies then that T is measure preserving and therefore that $T \in G$. This concludes the proof of Theorem 2.

In order to state one of the principal results of this paper it is convenient (though not necessary) to assume for a moment that the elements of B are measurable sets in the unit interval (or, more precisely, residue classes of such sets modulo sets of measure zero). Call an interval $(k/2^n,\ [k+1]/2^n)$, $k = 0,\ 1,\ \cdots,\ n-1$; $n = 0,\ 1,\ 2,\ \cdots$, a *dyadic interval of rank n*, and a union of such intervals a *dyadic set of rank n*. A *permutation P* (or, more precisely, a dyadic permutation of rank n) is a one-to-one transformation of the interval which maps each dyadic interval of rank n into itself or into another one by an ordinary translation. The reason for introducing the interval was to avoid an abstract and complicated description of this comparatively simple class (the dyadic sets) of elements of B and the related class (the dyadic permutations) of elements of G. In what follows only one property of dyadic sets will be used (in addition, that is, to their simple algebraic structure) and that is that they form a (countable) dense set in B.

THEOREM 3. *In the neighborhood topology permutations are dense in G*[14].

Proof. (1) The idea of the proof is simple. It is to be proved that given any n elements $a_1,\ \cdots,\ a_n$ of B and any automorphism $T \in G$ there is a

---

[14] This theorem has the rank of a "folk theorem," that is, one that most measure theorists conjecture and prove in one form or another. It has a satisfying intuitive content: it says that, in the limit, every measure preserving transformation is obtained by cutting up the space (with an ordinary pair of Euclidean scissors) into a finite number of pieces and then merely permuting the pieces. The first precise formulation of this result (not the one above) I heard from John von Neumann in November, 1940. His formulation appears below in Theorem 8. The first published version (different from both von Neumann's and mine) is due to Oxtoby and Ulam, op. cit. p. 919.

permutation P which affects the a_i approximately the same way as T. To do this consider the atoms of the (finite) Boolean algebra generated by $a_1, \cdots, a_n, Ta_1, \cdots, Ta_n$. Approximate each of these atoms (and consequently all unions of such atoms) by dyadic sets and then permute the dyadic sets as necessary. The somewhat cumbersome details follow.

(2) Let $b_1, \cdots, b_m$ be a decomposition of 1 (that is, $b_ib_j=0$ for $i \neq j$ and $b_1 \cup \cdots \cup b_m = 1$), and let δ be any positive number. Let $\bar{b}_1, \cdots, \bar{b}_m$ be dyadic elements such that $|b_i - \bar{b}_i| < \delta$, $i = 1, \cdots, m$. Write

$$\tilde{b}_i = \bar{b}_1' \cdots \bar{b}_{i-1}' \bar{b}_i \bar{b}_{i+1}' \cdots \bar{b}_m',$$

and observe that the disjointness of the b_i implies that

$$b_i = b_1' \cdots b_{i-1}' b_i b_{i+1}' \cdots b_m'.$$

Consequently

$$|b_i - \tilde{b}_i| \leq |b_1' - \bar{b}_1'| + \cdots + |b_{i-1}' - \bar{b}_{i-1}'| + |b_i - \bar{b}_i|$$
$$+ |b_{i+1}' - \bar{b}_{i+1}'| + \cdots + |b_m' - \bar{b}_m'| < m\delta$$

(since $|a-b| = |a'-b'|$). The $\tilde{b}_i$, being finite intersections of dyadic elements, are themselves dyadic; moreover they are pairwise disjoint. Hence $\tilde{b} = \tilde{b}_1 \cup \cdots \cup \tilde{b}_m$ is dyadic. Also, since $b_1 \cup \cdots \cup b_m = 1$, it follows that

$$|1 - \tilde{b}| \leq \sum_i |b_i - \tilde{b}_i| < m^2\delta.$$

Split the complement $1 - \tilde{b}$ of the dyadic element $\tilde{b}$ into m equal dyadic pieces and add one of these pieces to each $\tilde{b}_i$: denote the elements so obtained by b_i^*. Then $|b_i^* - \tilde{b}_i| < m^2\delta/m = m\delta$, so that

$$|b_i - b_i^*| \leq |b_i - \tilde{b}_i| + |\tilde{b}_i - b_i^*| < 2m\delta.$$

Finally let $(i_1, i_2, \cdots, i_k)$ be any subset of $(1, 2, \cdots, m)$. Then

$$|(b_{i_1} \cup \cdots \cup b_{i_k}) - (b_{i_1}^* \cup \cdots \cup b_{i_k}^*)| \leq \sum_j |b_{i_j} - b_{i_j}^*| < 2km\delta \leq 2m^2\delta.$$

To sum up: corresponding to each decomposition $\{b_i\}$ of 1 there is a dyadic decomposition $\{b_i^*\}$ of 1 such that every union of the b's is arbitrarily close to the corresponding union of the b^*'s.

(3) Suppose now that $b_1, \cdots, b_m$ and $c_1, \cdots, c_m$ are two decompositions of 1 with the property $|b_i| = |c_i|$, $i = 1, \cdots, m$. Then there exist arbitrarily close dyadic decompositions $\{b_i^*\}$ and $\{c_i^*\}$ with $|b_i^*| = |c_i^*|$. To see this apply (2) to the b's and c's separately and then alter the c^*'s (say) by subtracting from the fat ones and adding to the thin ones until the desired result is achieved. The hypothesis $|b_i| = |c_i|$ insures that the approximating property of the c_i^* will not be lost during the alterations.

(4) Consider now any automorphism $S \in G$, and any (basic) neighborhood N of S,

$$N(S) = \{T : \left| Sa_i - Ta_i \right| < \epsilon, i = 1, \cdots, n\}.$$

Let $b_1, \cdots, b_m$ be the atoms of the Boolean algebra generated by $a_1, \cdots, a_n$, and write $c_i = Sb_i, i = 1, \cdots, m$. According to (3) there are dyadic decompositions $\{b_i^*\}$ and $\{c_i^*\}$ such that every union of b's (or of c's) differs by less than $\epsilon/2$ from the corresponding union of the b^*'s (or of the c^*'s) and such that $\left| b_i^* \right| = \left| c_i^* \right|$. Construct a dyadic decomposition of 1 so fine that each b_i^* and each c_i^* is a union of intervals of this decomposition. Since the b_i^* (as well as the c_i^*) are pairwise disjoint and since (having the same measure) corresponding ones contain the same number of these small dyadic intervals, it is clear that there exists a permutation P for which $Pb_i^* = c_i^*$.

(5) Each $a_j, j = 1, \cdots, n$, is a union of certain b_i's: denote by a_j^* the corresponding union of the b_i^*'s. Then $\left| Sa_j - Pa_j \right| \leq \left| Sa_j - Pa_j^* \right| + \left| Pa_j^* - Pa_j \right|$. Since Sa_j is a union of c's and Pa_j^* is the corresponding union of c^*'s, each of the two terms of the right side of the last written inequality is dominated by $\epsilon/2$, so that $P \in N$; q.e.d.

COROLLARY. *In the neighborhood topology G satisfies the second countability axiom.*

For in any topological group (uniform structure) the first countability axiom and the existence of a countable dense set are together equivalent to the second countability axiom. The set of permutations is countable.

The next theorem is to be used in proving Theorem 6 below; it is however of a certain interest of its own as a sharpening of Theorem 3.

THEOREM 4. *In the neighborhood topology cyclic permutations are dense in G.*

Proof. Since it is already known that permutations are dense it is sufficient to prove that if P is any permutation and

$$N = \{T : \left| Pa_i - Ta_i \right| < \epsilon, i = 1, \cdots, n\}$$

is any neighborhood of it then there is a cyclic permutation $Q \in N$. It is, moreover, no loss of generality to assume that the a_i are dyadic since the neighborhoods for which this is true form a base for open sets. Since in this case the Pa_i are also dyadic, there exists a decomposition of 1 into dyadic intervals b_j of sufficiently high constant rank such that all a_i and all Pa_i are unions of the b_j. Write P in the ordinary cyclic notation as a permutation on the b's indicating all cycles (even those of length one):

$$P : (b_1^{(1)} \cdots b_{n_1}^{(1)})(b_1^{(2)} \cdots b_{n_2}^{(2)}) \cdots (b_1^{(p)} \cdots b_{n_p}^{(p)}).$$

(It is true, but irrelevant, that $n_1 + n_2 + \cdots + n_p$ is a power of two.) Choose

m so large that $1/2^m < \epsilon/2p$, and consider the decomposition of 1 into the dyadic intervals of rank m. All $b_i^{(j)}$ contain the same number, say q, of these small dyadic intervals:

$$b_i^{(j)} = {}^{1}c_i^{(j)} \cup {}^{2}c_i^{(j)} \cup \cdots \cup {}^{q}c_i^{(j)}.$$

A unique (cyclic) permutation Q is defined by the conditions

$$
\begin{aligned}
{}^{k}c_i^{(j)} &\to {}^{k}c_{i+1}^{(j)}, & \text{except for } i = n_j; \\
{}^{k}c_{n_j}^{(j)} &\to {}^{k+1}c_1^{(j)}, & \text{except for } k = q; \\
{}^{q}c_{n_j}^{(j)} &\to {}^{1}c_1^{(j+1)}, & \text{except for } j = p; \\
{}^{q}c_{n_p}^{(p)} &\to {}^{1}c_1^{(1)}. &
\end{aligned}
$$

It follows that for $i \neq n_j$, $Qb_i^{(j)} = Pb_i^{(j)}$, and $|Qb_{n_j}^{(j)} - Pb_{n_j}^{(j)}| = 2/2^m < \epsilon/p$. Since any union of b's contains at most p of the $b_{n_j}^{(j)}$, and since a_i is such a union for all $i = 1, \cdots, n$, it follows finally that $|Qa_i - Pa_i| < p \cdot \epsilon/p = \epsilon$, so that $Q \in N$, q.e.d.

The proof above establishes also the following corollary.

COROLLARY. *If P is any permutation and $N = N(P)$ any dyadic neighborhood, $N(P) = \{ T: |Pa_i - Ta_i| < \epsilon, i = 1, \cdots, n \}$, then there is a positive integer m and a cyclic permutation Q of rank m such that $Q \in N$ and such that each a_i is a union of the dyadic intervals of rank m.*

To state the next theorem it is necessary to recall the definition of an ergodic automorphism. $T \in G$ is *ergodic* if $Ta = a$ implies $a = 0$ or 1. The notion of an ergodic automorphism is the weakest formulation, which is simultaneously useful and precise, of the intuitive concept of a transformation which is a thorough shuffling.

THEOREM 5. *In the neighborhood topology ergodic automorphisms are dense in G.*

Proof. In view of Theorem 3 and the corollary to Theorem 4 it will be sufficient to prove the following statement: given any cyclic permutation Q of rank m there is an ergodic transformation T which agrees with Q on all dyadic elements of rank m. To prove this let $c_0, c_1, \cdots, c_{r-1}$ be the dyadic intervals of rank m $(r = 2^m)$ with the notation so chosen that $Qc_i = c_{i+1}$ for $i = 0, 1, \cdots, r-1 \pmod r$. Let S be an ergodic automorphism of the Boolean algebra of all subelements of c_0[15], and define, for any $a \in b$, $a = ac_0 \cup ac_1 \cup \cdots \cup ac_{r-1}$,

---

[15] That such an S exists is well known: it follows, for example, from the possibility of representing B on the unit interval, as described above.

$$Ta = QS(ac_0) \cup Q(ac_1) \cup Q(ac_2) \cup \cdots \cup Q(ac_{r-1}).$$

The verification that T is an automorphism is quite mechanical; it is necessary only to prove that T is ergodic. Suppose therefore that $Ta = a$. This implies that $(Ta)c_i = ac_i$ for $i = 0, 1, \cdots, r-1$; since $(Ta)c_i = T^i(ac_0)$ it follows that $ac_i = T^i(ac_0)$ for all i (mod r). Using, however, the definition of T,

$$T(ac_0) = QS(ac_0), \; T^2(ac_0) = Q^2S(ac_0), \cdots, T^r(ac_0) = Q^rS(ac_0) = S(ac_0).$$

Consequently $ac_0 = T^r(ac_0) = S(ac_0)$, whence, by the ergodicity of S, $ac_0 = 0$ or $ac_0 = c_0$. Since $ac_i = T^i(ac_0)$ it follows that either $ac_i = 0$ for all i or else $ac_i = c_i$ for all i. In other words $a = 0$ or 1; q.e.d.

THEOREM 6. *In the neighborhood topology the set E of ergodic automorphisms is a residual G_δ.*

Proof. Since E is already known to be dense and since a dense G_δ is residual, it will be sufficient to prove that E is a G_δ. To prove this I shall make use of the following fact[16]: an automorphism $T \in G$ is ergodic if and only if, for every a and $b \in B$, $\lim_n n^{-1} \sum_{k=0}^{n-1} |a \cdot T^k b| = |a| \cdot |b|$. Let $a_1, a_2, \cdots$ be a dense sequence in B; write

$$E(i, j, m, n) = \left\{ T : \left| n^{-1} \sum_{k=0}^{n-1} |a_i \cdot T^k a_j| - |a_i| \cdot |a_j| \right| < m^{-1} \right\}$$

and

$$F = \cap_i \cap_j \cap_m \cup_n E(i, j, m, n).$$

It follows trivially that $E \subset F$. Suppose therefore that $T \in G$ is not ergodic; it is to be shown that T does not belong to F. Since T is not ergodic there is an element $b \in B$ for which $Tb = b$ and $|b| \cdot |b'| > 0$. Write $\delta = |b| |b'|/8$, choose i so that $|a_i - b'| < \delta$, choose j so that $|a_j - b| < \delta$, and choose m so that $1/m < 4\delta$. Then

$$\left| |a_i \cdot T^k a_j| - |b' \cdot T^k b| \right| \leqq |a_i \cdot T^k a_j - b' \cdot T^k b| \leqq |a_i - b'| + |a_j - b| < 2\delta,$$

for all k and therefore

$$\left| n^{-1} \sum_{k=0}^{n-1} |a_i \cdot T^k a_j| - n^{-1} \sum_{k=0}^{n-1} |b' \cdot T^k b| \right| < 2\delta$$

for all n. Also

[16] The first use of this result for a very similar purpose is due to Oxtoby and Ulam, op. cit. p. 904. For a proof of the equivalence of this limiting condition to ergodicity see Eberhard Hopf, *Ergodentheorie*, Berlin, 1937, p. 30. The condition has a very natural intuitive interpretation. Two elements a and b of the measure algebra B are called *independent* if $|ab| = |a| |b|$. (This definition is motivated by probability theory.) Hence the condition says that an element moving under the influence of an ergodic automorphism tends, on the average, to become independent of every fixed element.

$$\left|\, |b'|\,|b| - |a_i|\,|a_j|\,\right| < |b'| \cdot \left|\,|b| - |a_j|\,\right| + |a_j| \cdot \left|\,|b'| - |a_i|\,\right| < 2\delta,$$

whence it follows (by the familiar trick of adding and subtracting both $n^{-1}\sum_{k=0}^{n-1}|a_i \cdot T^k a_j|$ and $|a_i|\,|a_j|$) that

$$\left|\, n^{-1}\sum_{k=0}^{n-1} |\,b' \cdot T^k b\,| - |b'|\,|b|\,\right| < \left|\, n^{-1}\sum_{k=0}^{n-1} |\,a_i \cdot T^k a_j\,| - |a_i|\,|a_j|\,\right| + 4\delta.$$

Since $Tb = B$ implies $T^k b = b$ for all k, the left side of this inequality is equal to $|b'|\,|b|$ and consequently to 8δ. Hence, finally,

$$\left|\, n^{-1}\sum_{k=0}^{n-1} |\,a_i \cdot T^k a_j\,| - |a_i|\,|a_j|\,\right| > 4\delta > m^{-1}$$

for all n: this proves that T is not in F, q.e.d.

Two comments are in order concerning the interpretation of Theorem 6.

I. Category results are usually stated for *complete* metric spaces only, whereas the topology of G was defined in terms of neighborhoods. It is easy to see, however, that the ordinary proof of Baire's theorem goes through with only verbal changes in any uniform space satisfying the first countability axiom. More than this is true: such a uniform space is always metrizable[17]. For topological groups a still stronger result is true: the first countability axiom implies the existence of a metric invariant under all left translations[18]. It is easy to exhibit such a metric for the group G discussed above. Let $a_1, a_2, \cdots$ be a dense sequence in B; the distance function $d(S, T) = \sum_{n=1}^{\infty}(1/2^n)|Sa_i - Ta_i|$ has all the desired properties. Since, however, this rather artificial construction did not seem to throw any light on the structure of G it was omitted from the main body of the discussion.

The ideal situation would be the existence of a metric for G invariant simultaneously under right and left translations. The following (unfortunately quite involved and indirect) argument will show that this is not the case. I shall need to make use of van Dantzig's necessary and sufficient condition for the (invariant) metrizability of a separable topological group[19] and of the following fact concerning compact groups. If Γ is any compact separable topological group, denote by T_α, for $\alpha \in \Gamma$, the (Haar) measure preserving transformation defined by $T_\alpha \gamma = \alpha \gamma$. T_α may be looked upon as an automorphism of the Boolean algebra B of all (Haar) measurable sets modulo sets of measure zero; the correspondence $\alpha \to T_\alpha$ is a homeomorphic isomor-

---

[17] See André Weil, op. cit. p. 16.

[18] See Shizuo Kakutani, *Über die Metrisation der topologischen Gruppen*, Proceedings of the National Academy of Japan vol. 12 (1936) p. 82.

[19] The condition is that $T_n \to I$ should imply $S_n T_n S_n^{-1} \to I$ for all sequences $\{S_n\}$. See D. van Dantzig, *Zur topologischen algebra. I. Komplettirungstheorie*, Math. Ann. vol. 107 (1932) p. 16.

phism between all Γ and a part of G [20] if G is thought of as topologized by the neighborhood topology. If S is any continuous automorphism of the compact group Γ then S is a (Haar) measure preserving transformation [21] (so that S is also an automorphism of B, $S \in G$); a trivial computation shows that $ST_\alpha S^{-1} = T_{S(\alpha)}$. Putting all this together, the impossibility of invariant metrizability will be proved if I can exhibit a topological group Γ, a sequence of continuous automorphisms S_n of Γ, and a sequence α_n of elements of Γ, such that α_n converges to 1, whereas $S_n(\alpha_n)$ does not converge to 1. (For then T_{α_n} converges to I in G whereas $S_n T_{\alpha_n} S_n^{-1} = T_{S_n(\alpha_n)}$ does not.) Such an example is easy to construct. Let Γ be the two-dimensional toral group (that is, the set of all pairs (ξ, η), $0 \leqq \xi$, $\eta \leqq 1$; the group operation is coordinatewise addition modulo 1). Let S_n be the automorphism defined by the unimodular matrix

$$\begin{pmatrix} n^2 & n+1 \\ n-1 & 0 \end{pmatrix},$$

that is, $S_n(\xi, \eta) = (n^2\xi + (n+1)\eta, (n-1)\xi)$, and $\alpha_n = (0, 1/2(n+1))$. Clearly $\alpha_n \to (0, 0)$ (the identity of Γ) and $S_n(\alpha_n) = (1/2, 0)$.

II. Baire's theorem, as is well known, is often used to give existence proofs. Theorem 6, however, is certainly not presented in the spirit of an existence theorem; in fact the existence of an ergodic automorphism was used in the course of the proof. The purpose of the theorem is rather to give further support (along the lines of the work of Oxtoby and Ulam [22]) to the familiar conjecture that "in general a measure preserving transformation is ergodic." There is, however, no implication between Theorem 6 and the corresponding result of Oxtoby and Ulam: they define a stronger topology and I consider a wider class of transformations.

4. **The metric topology.** There are several other interesting and natural topologies on the group G of automorphisms; in this section I shall discuss one of these. Since this topology is defined by a metric I shall refer to it as the *metric topology* of G, in distinction to the previously discussed neighborhood topology. Using the notations and results of Theorem 1 it is not difficult to define the metric. Given S and T in G, write $E = E(S, T) = \{a : Sa \neq Ta\}$, then the distance between S and T is given by $d(S, T) = |\sup E^*(S, T)|$. In words: the distance between S and T is the measure of the largest element on arbitrarily small subelements of which S and T are different.

([20]) This result is usually stated in terms of the "strong" topology of operators on Hilbert space; the translation from the usual proof to one applicable in the present case is, however, quite trivial. See footnote 13 above, and André Weil, *L'integration dans les groupes topologiques et ses applications*, Paris, 1938, p. 141.

([21]) See Paul R. Halmos, *On automorphisms of compact groups*, Bull. Amer. Math. Soc. vol. 49 (1943) p. 619.

([22]) See Oxtoby and Ulam, op. cit. p. 876.

The lack of immediate intuitive content of this definition is a typical example of the one disadvantage of the Boolean algebra formulation as compared with the more customary "point" formulation. If S and T had been measure preserving transformations on a measure space, I could have defined the distance between S and T to be the measure of the set of points where they differ. Not having any points to talk about I had to adopt the above circumlocution. This process is not as bad, however, as it seems, since in proving "almost everywhere" statements points always have to be ignored and the only effective tools are the algebraic properties of sets of positive measure.

THEOREM 7. *In the metric topology G is a complete topological group and, in fact, the metric $d(S, T)$ is invariant under both left and right translation.*

Proof. (1) The first thing to prove is that $d(S, T)$ satisfies the distance axioms. It is clear that $d(S, T) \geq 0$ and $E(S, T) = E(T, S)$ implies that $d(S, T) = d(T, S)$. If $d(S, T) = 0$ then $E^*(S, T)$ contains only the element 0. Hence every nonzero $a \in B$ contains a nonzero b no nonzero subelement of which is in E. In other words, every $a \neq 0$ has a nonzero subelement belonging to $(E')_*$. Since E' satisfies the hypotheses of the corollary to Theorem 1, $(E')_*$ is a principal ideal; write $e = \sup (E')_*$. Then for every $a \neq 0$, $ae \neq 0$: this means that $e = 1$, that is, that every a is in E'. Hence E is empty and $S = T$. To prove, finally, the triangle inequality, take $R, S, T \in G$ and write $A = E(R, S)$, $B = E(S, T)$, $C = E(R, T)$. Clearly $(C')_* \supset (A')_* \cap (B')_*$. Writing $a = \sup (A')_*$, $b = \sup (B')_*$, $c = \sup (C')_*$, it follows that $c \supset ab$, so that $c' \subset a' \cup b'$. From the second half of Theorem 1, $a' = \sup A^*$, $b' = \sup B^*$, $c' = \sup C^*$, whence $d(R, T) \leq d(R, S) + d(S, T)$.

(2) Since $E(S, T) = E(RS, RT)$, it is clear that $d(S, T) = d(RS, RT)$. Moreover $S^{-1}a = T^{-1}a$ ($= b$, say) if and only if $Sb = Tb$ ($= a$); consequently $E'(S^{-1}, T^{-1})$ is the transform under S (or under T) of $E'(S, T)$. It follows that $d(S^{-1}, T^{-1}) = d(S, T)$. Since in any group a metric which is invariant under two of the three obvious group operations (right translation, left translation, taking inverses) is also invariant under the third, d has the stated invariance properties.

(3) If $\{T_i\}$ is a Cauchy sequence in G, write $E_{ij} = \{a : T_i a \neq T_j a\}$; then $d(T_i, T_j) = |\sup E_{ij}^*| \to 0$. Choose m_k so that for $i, j > m_k$, $|\sup E_{ij}^*| < 1/2^k$; write $n_1 = m_1$, $n_k = \max (n_{k-1} + 1, m_k)$. Define also

$$e_k = \sup (E_{n_k n_{k+1}})^*, \quad d_k = e_k \cup e_{k+1} \cup e_{k+2} \cup \cdots, \quad c_k = d_k'.$$

Since $c_k \subset e_m'$ for $m \geq k$, $a \subset c_k$ implies $T_{n_k} a = T_{n_{k+1}} a = \cdots$. Since $|d_k| \leq |e_k| + |e_{k+1}| + \cdots < 1/2^{k-1}$, $c_k \to 1$, clearly $c_1 \subset c_2 \subset c_3 \subset \cdots$. Consequently a unique automorphism T is defined by the conditions $Ta = T_{n_k} a$ for $a \subset c_k$; clearly $d(T, T_{n_k}) \leq |d_k| < 1/2^{k-1}$. The Cauchy property of the $\{T_i\}$ implies, as usual, that $T_n \to T$. This completes the proof of Theorem 7.

G is not separable in the metric topology. For, representing B on the unit interval as usual, let T_α $(0 \leq \alpha < 1)$ be translation by α (mod 1). Clearly $d(T_\alpha, T_\beta) = 1$ if $a \neq \beta$ [23].

The remainder of this section is dedicated to deriving the analogues of the approximation theorem (Theorem 3) and the category theorem (Theorem 6) of the previous section. For this purpose it is necessary to go into some detail concerning "local" properties of automorphisms: this is done in the lemmas below.

For any $T \in G$ write $E_n = E_n(T) = \{a : T^n a = a\}$. According to the corollary to Theorem 1, $(E_n)_*$ is a principal ideal; write $e_n = \sup (E_n)_*$. It will be convenient to use the following terminology: if $a \subset e_n$, T *has period* n *in* a (in particular if $a \subset e_1$, T *is the identity in* a); if T has some finite period in a, T *is periodic in* a; if T is periodic in 1, T *is periodic*; if T is not periodic in any $b \subset a$, $b \neq 0$, T *is nowhere periodic in* a; if T is nowhere periodic in 1, T *is nowhere periodic*.

LEMMA 1. *If nowhere in* a $(a \neq 0)$ *does* T *have a period smaller than* n, $n = 1, 2, \cdots$, *then there is at least one* $b_n \subset a$, $b_n \neq 0$, *such that* $b_n T^i b_n = 0$ *for* $0 < i \leq n - 1$.

Proof. For $n = 1$ the lemma is vacuously true; I proceed by induction. If, for $n \geq 2$, b_{n-1} has already been found then, since $b_{n-1} \subset a$, T does not have the period $n - 1$ in b_{n-1}. Consequently there is a nonzero subelement c of b_{n-1} for which $c \neq T^{n-1} c$. It follows (using the fact that T preserves measure) that $c \cdot T^{n-1} c' \neq 0$; write $b_n = c \cdot T^{n-1} c'$. Since $b_n \subset T^{n-1} c'$ and $T^{n-1} b_n \subset T^{n-1} c$, $b_n \cdot T^{n-1} b_n = 0$; the fact that $b_n \subset b_{n-1}$ implies that $b_n \cdot T^i b_n = 0$ for $0 < i \leq n - 2$.

LEMMA 2. *If* T *has nowhere a period smaller than* n, $n = 2, 3, \cdots$, *then there is at least one* b, $b \neq 0$, *such that* b, Tb, $\cdots$, $T^{n-1} b$ *are pairwise disjoint and* $1/n \geq |b| \geq 1/(2n - 1)$.

Proof. (1) This lemma is a sharpening of Lemma 1 (at least for the case $a = 1$) in that seemingly more is required of b than of b_n, and in that b is quantitatively estimated. Two-thirds of this improvement is trivial. First: if $b \cdot T^i b = 0$ for $i = 1, \cdots, n - 1$ then b, Tb, $\cdots$, $T^{n-1} b$ are pairwise disjoint since for any $i < j$, $T^i b \cdot T^j b = T^i(b \cdot T^{j-i} b) = 0$. Also: if b, Tb, $\cdots$, $T^{n-1} b$ are pairwise disjoint then $b \cup Tb \cup \cdots \cup T^{n-1} b \subset 1$ shows that $n|b| \leq 1$. It is only the lower inequality that is not trivial.

(2) Consider the set $E = \{a : a \cdot T^i a = 0, i = 1, \cdots, n - 1\}$. Lemma 1 shows that E is not empty. With the ordering $(\subset)$ of B, E is a partially ordered set; let $A \subset E$ be a linearly ordered subset (that is, for $a_1, a_2 \in A$ either $a_1 \subset a_2$ or $a_2 \subset a_1$). Write $a^* = \sup A$; then $T^i a^* = \sup T^i A$; I assert that $a^* \in E$. For, if $a_1 \in A$ and $T^i a_2 \in T^i A$, then either $a_1 \subset a_2$ and $a_2 T^i a_2 = 0$ or $a_2 \subset a_1$ and $a_1 T^i a_1 = 0$ implies $a_1 \cdot T^i a_2 = 0$, so that $\sup A$ and $\sup T^i A$ are indeed disjoint

---

[23] This example, as well as many other properties of the metric topology, is reminiscent of the "uniform" topology of operators on Hilbert space; the connection between them will become clearer in the next section.

for $i=1, \cdots, n-1$. These considerations prove that the hypothesis, and therefore the conclusion, of Zorn's lemma[24] is valid; E contains at least one maximal element b.

(3) Using this maximal $b \in E$ write $a = b \cup Tb \cup \cdots \cup T^{n-1}b$. Then for every nonzero $c \subset a'$, $b \cdot T^i c \neq 0$ for at least one $i=1, \cdots, n-1$. For if this were not true, in other words if $0 \neq c \subset a'$, $b \cdot T^i c = 0$ for $i=1, \cdots, n-1$, then by Lemma 1 there is a nonzero subelement $d \subset c$ for which $d \cdot T^i d = 0$, $i=1, \cdots, n-1$. Hence $(b \cup d) \cdot T^i(b \cup d) = b \cdot T^i b \cup d \cdot T^i b \cup b \cdot T^i d \cup d \cdot T^i d = 0$ for $i=1, \cdots, n-1$; but this is impossible since it contradicts the maximality of b. In other words: for every $c \subset a'$, $c(T^{-1}b \cup T^{-2}b \cup \cdots \cup T^{-n+1}b) \neq 0$; consequently $a' \subset T^{-1}b \cup \cdots \cup T^{-n+1}b$ and $|a'| \leq (n-1)|b|$. Since $1 = b \cup Tb \cup \cdots \cup T^{n-1}b \cup a'$, $1 = n|b| + |a'| \leq (2n-1)|b|$, q.e.d.

LEMMA 3. *If T is nowhere periodic and ϵ is positive, then there exists an automorphism $S \in G$, an element $a \in B$, and a positive integer n such that*

(i) $Sa = a$,

(ii) *S has the period n in a,*

(iii) $|a| > 1/2$,

(iv) $d(S, T) \leq \epsilon$.

Proof. Choose n so that $2/n \leq \epsilon$, and choose b, in accordance with Lemma 2, so that $b \cdot T^i b = 0$ for $i=1, \cdots, n-1$, and $|b| \geq 1/(2n-1)$. Write $a = b \cup Tb \cup \cdots \cup T^{n-1}b$, as above; then $|a| \geq n/(2n-1) > 1/2$. For $d \subset b \cup Tb \cup \cdots \cup T^{n-2}b$ define $Sd = Td$; for $d \subset T^{n-1}b$ define $Sd = T^{-n+1}d$. These requirements determine a unique automorphism S of the algebra of subelements of a; it is clear that (i), (ii), and (iii) are satisfied.

Consider $e = a' \cdot T^{-1}a'$ and $Te = Ta' \cdot a'$. Then $a' = e \cup a' \cdot T^{-1}a$ and $a' = Te \cup a' \cdot Ta$, so that $|a' \cdot T^{-1}a| = |a' \cdot Ta|$. Consequently there is an automorphism $R \in G$ for which $R(a' \cdot T^{-1}a) = a' \cdot Ta$. For $d \subset e$ define $Sd = Td$ and for $d \subset a' \cdot T^{-1}a$ define $Sd = Rd$; these requirements (together with the ones already described in the preceding paragraph) determine a unique automorphism $S \in G$. Since $S = T$ on all subelements of $b \cup Tb \cup \cdots \cup T^{n-2}b \cup e$, $d(S, T) \leq |b| + |a' \cdot T^{-1}a|$. Since $a' \cdot T^{-1}a \subset T^{-1}b$, $d(S, T) \leq 2|b| \leq 2/n \leq \epsilon$, q.e.d.

THEOREM 8. *In the metric topology periodic automorphisms are dense in G.*[25]

Proof. The main idea of this proof is to show that Lemma 3 remains valid even if the hypothesis of nowhere periodicity is removed. Assuming for a moment that this has already been accomplished, I shall show how the theorem follows from it. Given T and ϵ, apply this sharpened form of Lemma 3 to T and $\epsilon/2$, obtaining an automorphism $S_1 \in G$, an element $a_1 \in B$, and an in-

---

[24] See Lefschetz, op. cit. p. 5.

[25] My thanks are due to R. H. Fox for several valuable discussions of this theorem and its proof. Among other things he discovered the original version (which I subsequently modified slightly) of the statement and proof of Lemma 3.

teger $n_1 > 0$, such that $S_1 a_1 = a_1$, S_1 has the period n_1 in a_1, $|a_1| > 1/2$, and $d(S_1, T) \leq \epsilon/2$. Next apply the same result to $\epsilon/4$ and S_1 considered as an automorphism of the algebra of subelements of a_1' only, obtaining S_2, $a_2 \subset a_1'$, and n_2 such that $S_2 a_2 = a_2$, S_2 has the period n_2 in a_2, $|a_2| \geq (1/2)|a_1'|$ and $d(S_1, S_2) \leq \epsilon/4$. (Since the algebra of subelements of a_1' is not normalized, the form of the third condition had to be modified slightly.) Repeating this process sufficiently often, say k times, a stage is reached where $|a_1' \cdots a_k'| < \epsilon$. A unique automorphism S is defined by the conditions $Sd = S_i d$ for $d \subset a_i$, $i = 1, \cdots, k$, $Sd = d$ for $d \subset a_1' a_2' \cdots a_k'$. For this S, $d(S, T) < \epsilon/2 + \epsilon/4 + \cdots + \epsilon/2^k + \epsilon < 2\epsilon$, and S has everywhere the period $n_1 n_2 \cdots n_k$.

In order to obtain the required sharpening of Lemma 3, consider once more the sets $E_n = \{a : T^n a = a\}$, and the elements $e_n = \sup (E_n)_*$. Since $TE_n = E_n$, it follows that $Te_n = e_n$ and consequently, writing $e = e_1 \cup e_2 \cup e_3 \cup \cdots$, that $Te = e$. Given $\epsilon < |e|/2$ choose n so that $|e_1 \cup \cdots \cup e_n| > |e| - \epsilon/2$; define T_0 by $T_0 d = Td$ for $d \subset e_1 \cup \cdots \cup e_n$ and for $d \subset e'$, and $T_0 d = d$ for $d \subset e \cdot e_1' \cdots e_n'$. Clearly $d(T_0, T) \leq \epsilon/2$. If $|e_1 \cup \cdots \cup e_n| > 1/2$ the desired result is already achieved (with S replaced by T_0 and n replaced by $n!$); if $|e_1 \cup \cdots \cup e_n| < 1/2$, a single application of Lemma 3 to $\epsilon/2$ and T_0 considered on the subelements of e' only, in the way already described above, concludes the proof.

The "size" of the set of ergodic automorphisms is easily determined by means of Theorem 8.

THEOREM 9. *In the metric topology the set of ergodic automorphisms is nowhere dense in G.*

Proof. Let K be any sphere in G; it follows from Theorem 8 that K contains a periodic automorphism S of, say, period n. Choose $\epsilon > 0$ so that $\epsilon < 1/n$ and so that the sphere K_1 of radius ϵ about S is contained in K. The nowhere denseness follows from the fact that no T in K_1 is ergodic. To prove this choose $T \in K_1$ and consider $E = E(S, T) = \{a : Sa \neq Ta\}$ and $e = \sup E^*$. Then $d(S, T) = |e| < \epsilon < 1/n$; if $|e| = 0$, $S = T$ and T is obviously not ergodic. If $|e| > 0$, write $a = e \cup Se \cup S^2 e \cup \cdots \cup S^{n-1} e$. Then $Sa = a$ and $0 < |a| < 1$. Since $e \subset a$, S and T agree on all subelements of a', so that $Ta' = Sa' = a'$ and $0 < |a'| < 1$. In other words, every $T \in K_1$ has a nontrivial invariant element and is therefore nonergodic, q.e.d.

5. **The uniform topology.** The metric discussed in the previous section is not perhaps one that most people would consider "natural." The topology it defines is, for one thing, very strong (that is, has many open sets); so strong in fact that it is practically discrete. It is very hard for two automorphisms to be close in this topology: the assertion that any particular set (such as the set of periodic automorphisms) is dense reveals a deep structural property of automorphisms. By the same token, however, it is very easy for any particular set (such as the set of ergodic automorphisms) to be nowhere dense and

any assertion concerning topological smallness (for example, being nowhere dense or having first category) is essentially a property of the topology only. The metric topology is, however, of considerable interest because, as I shall presently show, it coincides with one of the most natural topologies of G and is much easier to handle analytically.

This natural topology is also defined by a metric; by analogy with the space of bounded operators on Hilbert space I shall call it the *uniform topology*. The distance between two elements S and T of G is defined by

$$\delta(S, T) = \sup \{ | Sa - Ta | : a \in B \}.$$

The elementary properties of δ are so easy to see that it is not worthwhile to state them in a formal theorem. That δ satisfies the distance axioms is a well known fact in general metric spaces. Since $| RSa - RTa | = | Sa - Ta |$, it is clear that δ is invariant under left translation in G. Also, as a runs through B, so does $T^{-1}a$, and

$$| S^{-1}a - T^{-1}a | = | S(S^{-1}a - T^{-1}a) | = | a - S(T^{-1}a) | = | T(T^{-1}a) - S(T^{-1}a) |;$$

it follows that $\delta(S^{-1}, T^{-1}) = \delta(S, T)$, and consequently that δ is invariant under all the group operations. The interesting fact concerning δ, namely the coincidence of the uniform and the metric topologies of G, is a consequence of the following theorem.

THEOREM 10. $(2/3)d(S, T) \leqq \delta(S, T) \leqq d(S, T)$.

Proof. Since both d and δ are invariant under the group operations it is sufficient to prove the theorem for the case $S = I$. To prove the upper inequality, suppose that $d(I, T) = \alpha$; then there is an element $a \in B$ of measure α such that $Ta = a$ and for all $b \subset a'$, $Tb = b$. Hence, for any b,

$$| Tb - b | \leqq | T(ab) - (ab) | + | T(a'b) - (a'b) | = | T(ab) - (ab) |$$
$$= | a(Tb - b) | \leqq | a | = \alpha = d(I, T).$$

It follows that $\delta(I, T) = \sup \{ | Tb - b | : b \in B \} \leqq d(I, T)$. This inequality is rather natural. It shows that every δ-sphere contains the corresponding d-sphere of the same radius; in other words that the metric topology is stronger (has more open sets) than the uniform topology. This is in accordance with my earlier remarks to the effect that no reasonable topology ought to be stronger than the metric topology.

To prove the lower inequality it is necessary to examine in a little more detail the way in which an arbitrary automorphism T is made up of periodic pieces. Consider, as in the proof of Theorem 8, the elements

$$e_k = \sup (\{a : T^k a = a\}_*), \qquad k = 1, 2, \cdots.$$

Write $\bar{e}_1 = e_1$, $\bar{e}_k = e_k e'_{k-1} e'_{k-2} \cdots e'_2 e'_1$, and $\bar{e}_0 = e'_1 e'_2 e'_3 \cdots = \bar{e}'_1 \bar{e}'_2 \bar{e}'_3 \cdots$.

(Observe that, according to Theorem 1, $|\bar{e}_1'| = d(I, T)$.) Since $Te_k = e_k$ for $k = 1, 2, \cdots$, it follows that $T\bar{e}_k = \bar{e}_k$ for $k = 0, 1, 2, \cdots$; clearly $1 = \bar{e}_0 \cup \bar{e}_1 \cup \bar{e}_2 \cup \cdots$. T is nowhere periodic in $\bar{e}_0$, T has period k in $\bar{e}_k$ for $k = 1, 2, \cdots$, and, in $\bar{e}_k$, T has nowhere a period smaller than k. Hence Lemma 1 may be applied to each nonzero $\bar{e}_k$, with a, n replaced by $\bar{e}_k$, k for $k = 1, 2, \cdots$, and a, n replaced by $\bar{e}_0$, 2 for $k = 0$. The conclusion is that for each k there is an element $b_k \subset \bar{e}_k$ ($b_k \neq 0$ if $\bar{e}_k \neq 0$) such that $b_k \cdot T^i b_k = 0$ for $0 < i \leq k - 1$, $k = 1, 2, \cdots$, and $b_0 \cdot T b_0 = 0$ for $k = 0$. Lemma 2, applied to the (unnormalized) algebra of all subelements of $\bar{e}_0$, shows that b_0 may be chosen so that $|b_0| \geq (1/3)|\bar{e}_0|$. (In case $\bar{e}_0 = 0$ this is, of course, trivial.) For larger values of k the corresponding estimate for b_k is not good enough for my purpose; I shall show, in fact, that for $k = 1, 2, \cdots$ b_k can be chosen so that $|b_k| = |\bar{e}_k|/k$. To see this it is necessary only to apply Zorn's lemma, exactly as in step (2) of the proof of Lemma 2. This procedure shows that the b_k may be assumed maximal. If now $|b_k| < |\bar{e}_k|/k$, then Lemma 1 may be applied to $\bar{e}_k(b_k' \cdot T b_k' \cdots T^{k-1} b_k')$, leading to a contradiction of the maximality of b_k.

Now write $a_0 = b_0$, $a_1 = b_1$ ($= \bar{e}_1$); $a_{2k} = \cup_{i=0}^{k-1} T^{2i} b_{2k}$, $a_{2k+1} = \cup_{i=0}^{k-1} T^{2i} b_{2k+1}$, $k = 1, 2, \cdots$. Then, for all $k \neq 1$, $a_k \cdot T a_k = 0$. Moreover for $k = 1, 2, \cdots$,

$$|a_{2k}| = k|b_{2k}| = (k/2k)|\bar{e}_{2k}| = |\bar{e}_{2k}|/2,$$
$$|a_{2k+1}| = k|b_{2k+1}| = (k/(2k+1))|\bar{e}_{2k+1}| \geq |\bar{e}_{2k+1}|/3;$$

consequently for all $k = 0, 1, 2, \cdots$, $|a_k| \geq |\bar{e}_k|/3$.

The desired inequality between d and δ is a consequence of these last written estimates. Writing $a = a_0 \cup a_2 \cup a_3 \cup a_4 \cup \cdots$, it follows that $a \cdot Ta = 0$ and $|a| = |a_0| + |a_2| + |a_3| + \cdots \geq (|\bar{e}_0| + |\bar{e}_2| + |\bar{e}_3| + \cdots)/3 \geq |\bar{e}_1'|/3$. Consequently $\delta(I, T) \geq |Ta - a| \geq 2|\bar{e}_1'|/3 = 2d(I, T)/3$, q.e.d.

The translation of the unit interval by 1/3, mod 1, shows that both bounds in Theorem 10 are best possible.

In order for S and T to be close in the uniform (and hence in the metric) topology $|Sa - Ta|$ has to be uniformly small, whereas closeness in the neighborhood topology requires only that $|Sa - Ta|$ be small for a finite number of a. Clearly then the uniform topology is no weaker than the neighborhood topology: that they do not coincide, and consequently that the uniform topology is really stronger, follows from the already demonstrated topological differences between them (such as separability, density of ergodic automorphisms, invariant metrizability, and so on).

SYRACUSE UNIVERSITY,
SYRACUSE, N. Y.

ANNALS OF MATHEMATICS
Vol. 45, No. 4, October, 1944

IN GENERAL A MEASURE PRESERVING TRANSFORMATION IS MIXING

By Paul R. Halmos

(Received November 13, 1943)

In this note I continue the study of topological properties of the group of measure preserving transformations begun in an earlier paper.[1] Using the methods and results of that paper I present the first proof of the old standing conjecture stated in the title.[2] "In general" means of course that the exceptional set is of the first category in one of the usual "natural" topologies (the strong neighborhood topology) for measure preserving transformations. The principal new and quite surprising fact used in the proof is that for any almost nowhere periodic measure preserving transformation T (and *a fortiori* for any mixing T) the set of all conjugates of T (i.e. the set of all STS^{-1}) is everywhere dense. It is this possibility of a dense conjugate class in a comparatively well behaved topological group (a rather natural generalization of the finite symmetric groups) that is contrary to naive intuition.

Let G be the group of all measure preserving transformations of the unit interval.[3] For any $S \epsilon G$, measurable set a, and positive number ϵ, write

$$N(S) = N(S; a, \epsilon) = \{T: |Sa - Ta| < \epsilon\}.^4$$

A unique topology (called the neighborhood topology) is defined in G by the requirement that the collection of all sets of the form $N(S)$ be a subbase for open sets.[5] (With this topology G becomes a complete topological group satisfying the second countability axiom.[6] A permutation (or more precisely a dyadic permutation of rank m) is a measure preserving transformation which maps each dyadic interval of rank m onto itself or onto another one by a translation. (A dyadic interval of rank m is an interval of the form $(k/2^m, (k + 1)/2^m)$, $k = 0, 1, \cdots, 2^m - 1$, and a dyadic set of rank m is a union of such dyadic intervals.) A dyadic neighborhood is a set of the form $\{T: |Pa_i - Ta_i| < \epsilon, i = 1, \cdots, n\}$ where P is a permutation and the a_i are dyadic sets. The dyadic

---

[1] *Approximation theories for measure preserving transformations*, Transactions of the A. M. S., vol. 55 (1944), pp. 1–18, In the sequel I shall refer to this paper as (A).

[2] See Eberhard Hopf, *Ergodentheorie*, Berlin, 1937, p. 41. Cf. also (A), §1. The word 'mixing' is used in this paper in the weak sense of Hopf.

[3] It is known that the apparently great restriction to the unit interval is in fact very slight: measure theoretically most interesting spaces are isomorphic to each other and hence to the unit interval. See Paul R. Halmos and John von Neumann, *Operator methods in classical mechanics, II*, Annals of Math., vol. 43 (1942), Theorems 1 and 2.

[4] The symbol $\{- : -\}$ denotes the set of all those objects named before the colon which satisfy the condition stated after it. $|-|$ is used to denote the measure of a measurable set and $a - b$ standards for the symmetric difference of two sets, $a - b = ab' \cup a'b$, where a' is the complement of a.

[5] See Solomon Lefschetz, *Algebraic Topology*, New York, 1942, p. 6.

[6] See (A), Theorems 2 and 3.

786

neighborhoods form a base for the open sets of G. A measure preserving transformation T is almost nowhere periodic (or for brevity non periodic) if the set of those ξ, $0 \leq \xi \leq 1$, for which there is a positive integer n such that $T^n \xi = \xi$, has measure zero. Finally I define the symbols $d(S, T)$ and $\delta(S, T)$ by the formulae

$$d(S, T) = | \{\xi : S\xi \neq T\xi\} |, \quad \delta(S, T) = \sup \{ | Sa - Ta | \}.$$

In terms of these definitions it is now possible to state the following four results (proved in (A)) which will be needed in the sequel.

(1) *For every non periodic T and positive integer n there is a measurable set b such that $b, Tb, \cdots, T^{n-1}b$ are pairwise disjoint and $1/n \geq |b| \geq 1/(2n - 1)$.*

(2) *Every dyadic neighborhood contains cyclic permutations of arbitrarily large ranks.*

(3) *If T has (almost everywhere) exactly the period m then there exists a measurable set b such that $b, Tb, \cdots, T^{m-1}b$ are pairwise disjoint and $|b| = 1/m$.* (See (A), p. 18).

(4) *Both $d(S, T)$ and $\delta(S, T)$ are group invariant metrics for G (giving rise, however, to a topology different from the neighborhood topology) and, for all S and T, $\delta(S, T) \leq d(S, T)$.*

LEMMA 1. *For every non periodic T and positive integer n there exists a measure preserving transformation S and a measurable set a such that (i) $Sa = a$, (ii) S has exactly the period n in a, (iii) $|a| > 1/2$, (iv) $d(S, T) \leq 2/n$, and (v) S is nowhere periodic in a'.*[7]

PROOF: Choose b in accordance with (1) so that $b, Tb, \cdots, T^{n-1}b$ are pairwise disjoint and $|b| \geq 1/(2n - 1)$. Write $a = b \cup Tb \cup \cdots \cup T^{n-1}b$; then $|a| \geq n/(2n - 1) > 1/2$. For $\xi \epsilon b \cup Tb \cup \cdots \cup T^{n-2}b$ define $S\xi = T\xi$; for $\xi \epsilon T^{n-1}b$ define $S\xi = T^{-n+1}\xi$. It follows that (i), (ii), and (iii) are satisfied: it remains to extend the definition of S in such a way as to satisfy (iv) and (v).

Consider $e = a' \cdot T^{-1}a'$ and $Te = a' \cdot Ta'$. Since $a' = e \cup (a' \cdot T^{-1}a)$ and $a' = Te \cup (a' \cdot Ta)$ it follows that $x = a' \cdot T^{-1}a$ and $y = a' \cdot Ta$ have the same measure. For $\xi \epsilon e$ define $S\xi = T\xi$; it remains now to define S on x so that $Sx = y$. Before doing this I remark that (iv) is already achieved. This follows from the fact that

$$\{\xi : S\xi \neq T\xi\} \subset (T^{n-1}b) \cup (a' \cdot T^{-1}a),$$

whence (using the fact that T is measure preserving and that $a' \cdot T^{-1}a \subset T^{-1}b$)

$$d(S, T) \leq 2 |b| \leq 2/n.$$

[7] This result is a strengthening of Lemma 3 of (A) and the first two paragraphs of the proof here given are a slight modification of the discussion in (A). A modification in this, as well as in several other results and definitions cited from (A), is necessary because the language and notation of (A) are considerably more algebraic than those of the present paper.

It may be assumed without loss of generality that x and y are disjoint, for otherwise S may be defined in xy to be any non periodic transformation and thus the problem reduces to finding a suitable transformation from xy' to $x'y$. This assumption makes it possible to draw a schematic sketch of all steps considered so far.

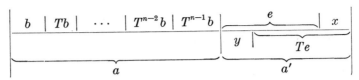

Let c be the set of those points $\xi \,\epsilon\, e$ for which $T^n\xi \,\epsilon\, e$ for all $n = 1, 2, \cdots$. Since $Tc \subset c$, the fact that T is measure preserving implies that (after possibly discarding a set of measure zero from c) $Tc = c$. It follows (since $c \subset e$) that $c \subset Te$. In other words almost every point of y gets out of e (and hence into x) in a finite number of steps. Write

$$y_k = \{\xi : \xi \,\epsilon\, y, \, T^i\xi \,\epsilon\, e \text{ for } i < k, \, T^k\xi \,\epsilon\, x\},$$

and define for $\xi \,\epsilon\, y_k$, $Q\xi = T^k\xi$. Let P be any non periodic transformation of y into itself and define for $\xi \,\epsilon\, x$, $S\xi = PQ^{-1}\xi$. S is now everywhere defined and it remains to prove that it is non periodic.

If $\xi \,\epsilon\, Te$ is such that $T^n\xi \,\epsilon\, Te$ for all $n > 0$ then (by an argument similar to that in the preceding paragraph) except for a set of ξ's of measure zero ξ and all its T transforms must belong to e also and consequently $S^n\xi = T^n\xi$. In this case the non periodicity of T ensures that of S. Hence it is sufficient to show that no $\xi \,\epsilon\, y$ has a finite orbit under S. To prove this take $\xi \,\epsilon\, y_k$. Then since $T^i\xi \,\epsilon\, e$ for $i = 0, \cdots, k - 1$, $S^i\xi = T^i\xi$. Also since $T^i\xi \,\epsilon\, Te$ for $i = 1, \cdots, k$, $T\xi, \cdots, T^k\xi$ do not belong to y. Hence ξ is not only not periodic of some period $\leq k$ but it does not even return to y in k steps. At the $(k + 1)^{\text{th}}$ step (since $S^k\xi = T^k\xi = Q\xi \,\epsilon\, x$), $S^{k+1}\xi = PQ^{-1}T^k\xi$, whence, using the definition of Q, $S^{k+1}\xi = P\xi$. In other words: the first time that a point of y returns to y it is transformed by P. Hence the S-orbit of ξ contains the (infinite) P-orbit of ξ and cannot therefore be finite. This completes the proof of Lemma 1.

LEMMA 2. *For every non periodic T and positive integer n there exists a measure preserving transformation S which has (almost) everywhere the period n and for which $d(S, T) \leq 4/n$.*

PROOF: The proof of this lemma is an application of Lemma 1 and consists of iterating that former result an infinite number of times. More precisely: after one application of Lemma 1 one obtains a set a_1 and a transformation S_1 with properties there described. The second time Lemma 1 is applied to S_1 on a_1' to find a set $a_2 \subset a_1'$ and a transformation S_2 such that S_2 has the period n on a_2, $|a_2| \geq (1/2)|a_1'|$, and $d(S_1, S_2) \leq (2/n)|a_1'|$. Proceeding in this way by induction one obtains a sequence of disjoint measurable sets $a_1, a_2, \cdots$ and a sequence of measure preserving transformations $S_1, S_2, \cdots$ such that

$d(S_i, S_{i+1}) \leq (2/n) \mid (a_1 \cup \cdots \cup a_i)' \mid, \mid a_{i+1} \mid \geq \mid (a_1 \cup \cdots \cup a_i)' \mid /2, \, S_i a_i = a_i$, and S_i has period n in a_i. It follows also (by induction) that $\mid a_1 \cup \cdots \cup a_i \mid > 1/2 + \cdots + 1/2^i = 1 - 1/2^i$. Hence the union of all the a's is (except possibly for a set of measure zero) the entire unit interval and therefore a measure preserving transformation S of the entire interval may be defined by writing, for $\xi \, \epsilon \, a_i$, $S\xi = S_i \xi$. S will have almost everywhere the period n and, moreover,

$$d(T, S) \leq d(T, S_1) + d(S_1, S_2) + d(S_2, S_3) + \cdots$$

$$\leq (2/n)(1 + 1/2 + 1/2^2 + 1/2^3 + \cdots) = 4/n.$$

With these preliminary results out of the way I am now in a position to state and prove the first main theorem of this paper.

THEOREM 1. *The conjugate class of any non periodic measure preserving transformation $T_0 \, \epsilon \, G$ is (in the neighborhood topology) everywhere dense in G.*

PROOF: Let $N = N(P) = \{T : \mid Pa_i - Ta_i \mid < \epsilon, i = 1, \cdots, n\}$ be any dyadic neighborhood of some permutation P: it is to be proved that for some $R \, \epsilon \, G$, $RT_0R^{-1} \, \epsilon \, N$. Write $M = \{T : \mid Pa_i - Ta_i \mid < \epsilon/2, i = 1, \cdots, n\}$ and apply to M the result (2) quoted from (A). It follows that M contains a cyclic permutation Q of rank m greater than the ranks of all a_i and such that $1/2^{m-3} < \epsilon$. Then Lemma 2 may be applied to T_0 to obtain an S which has (almost) everywhere the period 2^m and for which $d(S, T_0) \leq 4/2^m < \epsilon/2$.

Observe next that Q and S are conjugate in G. For denote by $q_0, \cdots, q_{r-1}$ ($r = 2^m$) the dyadic intervals of rank m, so arranged that $Qq_i = q_{i+1}$ ($i \bmod r$) and denote by s_0 a set of measure $1/2^m$ which is disjoint from $S^i s_0$ for $i = 1, \cdots, r - 1$. (The existence of such an s_0 is guaranteed by (3)). Write $s_i = S^i s_0$, $i = 1, \cdots, r - 1$, and let R be any (measure preserving) transformation taking s_0 into q_0. To extend the definition of R write, for any $\xi \, \epsilon \, s_i$, $R\xi = Q^i R S^{-i} \xi$. Schematically:

It is quite easy to see that $Q = RSR^{-1}$.

The proof can now be quickly finished. Since $d(S, T_0) < \epsilon/2$ and since (by (4)) d is invariant under the group operations,

$$d(Q, RT_0R^{-1}) = d(RSR^{-1}, RT_0R^{-1}) = d(S, T_0) < \epsilon/2.$$

Using the definition of δ and the inequality stated in (4) it follows that

$$\mid Pa_i - RT_0R^{-1}a_i \mid \leq \mid Pa_i - Qa_i \mid + \mid Qa_i - RT_0R^{-1}a_i \mid$$

$$\leq \mid Pa_i - Qa_i \mid + \delta(Q, RT_0R^{-1})$$

$$< \epsilon/2 + d(Q, RT_0R^{-1}) < \epsilon.$$

Consequently $RT_0R^{-1} \, \epsilon \, N(P)$; q.e.d.

71

To establish the residual character of any set M of measure preserving transformations it is sufficient, in view of the theorem just proved, to show that (i) M contains a non periodic transformation, (ii) M is self conjugate, and (iii) M is a G_δ.[8] In the sequel I shall show that the set M of mixing transformations has these three properties.

Recall first the definition of mixing. A measure preserving transformation is mixing if there exists a set I_0 of positive integers of density zero such that for any two measurable sets a and b

$$\lim_{n \notin I_0} | T^n a \cdot b | = | a | \cdot | b |.[9]$$

The properties (i) and (ii) of M are easy consequences of this definition. (i) Not only does M contain non periodic transformations but in fact every mixing T is non periodic.[10] For if the set of those ξ whose period is say m had positive measure then (3) (applied to T considered on this set only) would yield a set c of positive measure with the properties $T^m c = c$, $T^j c \cdot c = 0$ for $0 < j < m$. Writing $a = b = c$, it follows that $| T^n a \cdot b | = | c |$ or 0 according as $n \equiv 0$ (mod m) or not. Since this sort of behavior clearly excludes the possibility of the limiting property of mixing transformations it follows that a T which is periodic on a set of positive measure cannot be mixing. (ii) If T is mixing and $S = RTR^{-1}$ then S is mixing. For

$$| S^n a \cdot b | = | (RTR^{-1})^n a \cdot b | = | RT^n R^{-1} a \cdot b |$$
$$= | R(T^n R^{-1} a \cdot R^{-1} b) | = | T^n (R^{-1} a) \cdot (R^{-1} b) |;$$

the limiting property of T (with a, b replaced by $R^{-1}a$, $R^{-1}b$) together with the fact that R is measure preserving yields the desired conclusion.

To prove (iii), i.e. to prove that M is a G_δ, it is necessary to use the known Hilbert space characterization of mixing. A measure preserving transformation T may be regarded as a unitary operator on the complex Hilbert space L_2, by defining $Tf(\xi) = f(T\xi)$. (The unitary character of T follows from the fact that measure and, consequently, integral are invariant.) Mixing in Hilbert space terms amounts to the existence of a set I_0 of positive integers of density zero such that for any two functions $f, g \in L_2$,

$$\lim_{n \notin I_0} (T^n f, g) = (f, 1)(1, g)$$

(where (f, g) is the inner product $\int_0^1 f(\xi)\overline{g(\xi)}\, d\xi$ and 1 stands for the function identically equal to 1, so that $(f, 1) = \int_0^1 f(\xi)\, d\xi$ and $(1, g) = \int_0^1 \overline{g(\xi)}\, d\xi$.)

---

[8] Cf. the proof of Theorem 6, (A).

[9] See Hopf, op. cit., p. 36. For all concepts and results quoted in this paragraph and the next the reader is referred to Chapter III of Hopf's book.

[10] To see that this statement is really stronger than the original formulation of (i) it is necessary to make the non trivial observation that there exists at least one mixing transformation. See for example Paul R. Halmos, *On automorphisms of compact groups*, Bulletin of the A. M. S., vol. 49 (1943), pp. 619–624.

From the point of view of the preceding paragraph the set G of all measure preserving transformations becomes a subset of the set U of all unitary operators. On this latter set von Neumann's "strong neighborhood topology" may be defined: a subbasic neighborhood of $T_0 \, \epsilon \, U$ is of the form

$$N(T_0) = N(T_0 \, ; f, \, \epsilon) = \{T: \| \, T_0 f - Tf \, \| < \epsilon\},$$

where $f \, \epsilon \, L_2$, $\epsilon > 0$, and $\| \, f \, \|$ stands, as usual, for $(f, f)^{1/2}$. I shall need to make use of the fact that this topology when specialized to the subset G of U coincides with the neighborhood topology defined earlier. (This fact is proved by showing that a topology equivalent to the "strong neighborhood topology" of von Neumann is obtained if f is restricted to be the characteristic function of a measurable set.[11]) Two further facts will be used in the proof that follows: first that for fixed f, $g \, \epsilon \, L_2$ the function $\varphi(T) = T(f, g)$ is a continuous function of T (in the neighborhood topology) and, last but far from least, that T is mixing if and only if it has no non trivial proper values, i.e. $Tf = \lambda f$ ($f \neq 0$) implies that f is (almost everywhere) equal to the constant $\lambda = 1$. (This last fact is known as the mixing theorem.[12])

The remainder of the proof (that M is a G_δ) is somewhat analogous to the proof of Theorem 6 of (A). Let f_1, f_2, $\cdots$ be a dense sequence in L_2 ; write

$$E(i, j, m, n) = \{T: | \, (T^n f_i, f_j) - (f_i, 1)(1, f_j) \, | < 1/m\}$$

and

$$F = \cap_i \cap_j \cap_m \cup_n E(i, \ j, \ m, \ n).$$

The fact that (Tf, g) is a continuous function of T implies that $E(i, j, m, n)$ is open and therefore that F is a G_δ. The limiting property of mixing T's shows that $M \subset F$: I shall now show that, conversely, if T is not mixing then T is not in F.

If T is not mixing then, by the mixing theorem, there is an $f \neq 0$ and a complex number λ of modulus one (since T is unitary) such that $Tf = \lambda f$. Without loss of generality it may be assumed that f is orthogonal to the trivial proper function 1, $(f, 1) = 0$, and that f is normalized, $\| \, f \, \| = 1$. Choose now an i for which $\| \, f - f_i \, \| < .1$ and choose $j = i$, $m = 2$. I shall show that T does not belong to F. Observe first that $\| \, f_i \, \| \leq \| \, f \, \| + \| \, f_i - f \, \| \leq 1.1$. It follows that

$$1 = | \, (T^n f, f) - (f, 1)(1, f) \, |$$

$$\leq | \, (T^n f, f) - (T^n f, f_i) \, | + | \, (T^n f, f_i) - (T^n f_i, f_i) \, | + | \, (T^n f_i, f_i) - (f_i, 1)(1, f_i) \, |$$

$$+ | \, (f_i, 1)(1, f_i) - (f_i, 1)(1, f) \, | + | \, (f_i, 1)(1, f) - (f, 1)(1, f) \, |$$

---

[11] See John von Neumann, *Zur Algebra der Funktionaloperationen und Theorie der normalen Operatoren*, Math. Ann., vol. 102 (1929), p. 386.

[12] Hopf, op. cit., p. 37.

$$\leqq \| f - f_i \| + \| f - f_i \| \, \| f_i \| + \| f_i \| \, \| f_i - f \| + \| f_i - f \|$$
$$+ \, | \, (T^n f_i, \, f_i) \, - \, (f_i, \, 1)(1, \, f_i) \, |$$
$$\leqq .1 + .11 + .11 + .1 + | \, (T^n f_i, \, f_i) \, - \, (f_i, \, 1)(1, \, f_i) \, |$$
$$< .5 + | \, (T^n f_i, \, f_i) \, - \, (f_i, \, 1)(1, \, f_i) \, | \, .$$

In other words

$$| \, (T^n f_i, f_i) - (f_i, 1)(1, f_i) \, | > .5 = 1/m$$

for all n.   Hence:

THEOREM 2.   *In the neighborhood topology of G the set M of mixing transformations is a residual G_δ.*

SYRACUSE UNIVERSITY

ANNALS OF MATHEMATICS
Vol. 48, No. 3, July, 1947

INVARIANT MEASURES

By Paul R. Halmos

(Received April 10, 1946)

1. Introduction

Given a measurable transformation T on a measure space X with measure m, under what conditions does there exist a measure m^*, suitably related to m and X, and invariant under T? The purpose of this paper is to discuss this question and incidentally to investigate systematically its relation to various possible conditions on T (such as the preservation of sets of measure zero, incompressibility, etc.). The main contributions of sections 2 and 3 are (a) to free the related results of Hopf from their unnecessary non singularity restrictions (Theorems 1 and 2, particularly (12)) and (b) to present the first known example of an incompressible (and even non singular) transformation which preserves no suitable finite measure (Theorem 3, particularly (18) and (19)). Section 4 treats the extent to which a finite invariant measure m^* is uniquely determined by the transformation (Theorem 4). Roughly speaking the result is that m^* is uniquely determined by its values on the measurable invariant sets and that these values may in fact be prescribed arbitrarily. In section 5 a simple proof is given of a slightly extended version of Poincaré's recurrence theorem, and the conclusion is applied to derive the main result (Theorem 5) consisting of a necessary and sufficient condition for the existence of a not necessarily finite invariant measure. Section 6 discusses the main unsolved problem of the theory (does an invariant measure always exist?) and presents a fundamental structure theorem (Theorem 6) as a possible first step toward its solution. The structure or factorization theorem asserts essentially that every transformation is the product of a purely dissipative transformation by a measure preserving transformation. The result of this theorem is used to present some likely looking candidates for the role of an example to prove the conjecture that even an infinite invariant measure need not exist.

2. Definitions and preliminary results

A *measure space* is a set X on a certain collection of subsets of which a non negative and countably additive measure m is defined. It is assumed that the specified collection of subsets (called the *measurable* sets of X) is closed under complementation and the formation of countable unions and intersections. A measure μ, that is a non negative and countably additive set function defined for all measurable sets, is *finite* if $\mu(X) < \infty$; it is *σ-finite* if X is the union of a countable collection $\{X_n\}$ of measurable sets for which $\mu(X_n) < \infty$. If the given basic measure m is finite or σ-finite it is often convenient to refer to the

735

measure space X itself as finite or σ-finite respectively.[1] *Throughout this paper it is assumed that the measure space X under consideration is σ-finite.* A measure μ is *stronger* than m if $\mu(E) = 0$ implies $m(E) = 0$, (and in this case m is *weaker* than μ); μ is *equivalent* to m if it is simultaneously stronger and weaker than m, i.e. if μ and m vanish for the same sets. If E and F are measurable sets then F is a proper subset of E if $F \subseteq E$ and $m(E - F) > 0$.

The principal objects of the present study are transformations of a measure space. A one to one transformation T of X onto itself is *measurable* if for every integer n and measurable set E the set $T^n E$ is also measurable. A measure μ is *invariant* (under T) if for every measurable set E, $\mu(TE) = \mu(E)$. Two measurable sets E and F are *k-equivalent* (under T), $1 \leq k \leq \infty$, if both may be decomposed into k pairwise disjoint measurable sets, $E = \bigcup_{i=1}^{k} E_i$, $F = \bigcup_{i=1}^{k} F_i$, so that for each i an integer n_i exists with $T^{n_i} E_i = F_i$. A measurable set E is *compressible* if $TE \subseteq E$ and $m(E - TE) > 0$, i.e. if it is 1-equivalent to a proper subset of itself, with the exponent $n_1 = 1$. A measurable set E is *unbounded* if it is ∞-equivalent to a proper subset of itself (and *bounded* otherwise). If compressible (or unbounded) measurable sets exist, the transformation T will be called compressible (or unbounded); otherwise T is *incompressible* (or *bounded*). The transformation T is $(+)$ *non singular* if $m(E) = 0$ implies $m(TE) = 0$, $(-)$ *non singular* if T^{-1} is $(+)$ non singular, *non singular* if it is both $(+)$ and $(-)$ non singular, and *singular* otherwise. It is sometimes convenient to use abbreviations for the names of the concepts introduced. Specifically, $Z(+)$, $Z(-)$, and Z will denote $(+)$, $(-)$, and unqualified non singularity respectively; C_* and C^* will denote the conditions that T is incompressible and bounded respectively; and M_*, M^*, and M will denote the conditions that there exists a finite invariant measure weaker than m but not identically zero, stronger than m, and equivalent to m, respectively.

THEOREM 1. *The following three conditions on the transformation T are equivalent to each other.*

(1) *T is compressible.*

(2) *For some positive integer k and measurable set E, E is k-equivalent to a proper subset of itself.*

(3) *For some positive integer k, X is k-equivalent to a proper subset.*

PROOF. $(1) \to (2)$. This is trivial from the definition of compressibility, with $k = 1$.

$(2) \to (3)$. Suppose that $E = \bigcup_{i=1}^{k} E_i$, $F = \bigcup_{i=1}^{k} F_i$, $F_i = T^{n_i} E_i$, $F \subseteq E$, $m(E - F) > 0$, where the $\{E_i\}$ and the $\{F_i\}$ are each pairwise disjoint sequences. Write $E_{k+1} = F_{k+1} = E'$ ($=$ the complement of E), and $n_{k+1} = 0$; it follows that X is $(k + 1)$-equivalent to the proper subset $(E - F)'$.

---

[1] The use of the notational device σ has become increasingly popular in recent years, and I employ it several times in this paper in its customary sense. That is, if P is a property of sets, then σ-P is another property which is possessed by a set X if and only if it is the union of a countable collection of sets with property P. Thus, for example, the F_σ sets of a topological space could be called σ-closed.

$(3) \rightarrow (1)$. This implication is non trivial. Suppose that $X = \bigcup_{i=1}^{k} E_i = \bigcup_{i=0}^{k} F_i$ are pairwise disjoint decompositions with $F_i = T^{n_i}E_i$ for $i = 1, \cdots, k$ and $m(F_0) > 0$. Consider the sequence $F_0, TF_0, T^2F_0, \cdots$; suppose first that there is a positive integer n_0 such that for $n > n_0$, $m(T^nF_0) = 0$, and take for n_0 the least positive integer with this property. Write $E = \bigcup_{n \geq n_0} T^nF_0$; then $m(E) \geq m(T^{n_0}F_0) > 0$, and $m(TE) = m(\bigcup_{n > n_0} T^nF_0) = 0$. Clearly $TE \subsetneq E$ and $m(E - TE) = m(E) - m(TE) = m(E) > 0$, so that in this case the set E, and accordingly the transformation T, is compressible.

It remains to consider the case in which there exist arbitrarily large values of n for which $m(T^nF_0) > 0$. In this case a value of n exists for which $m(T^nF_0) > 0$ and $n + n_i \geq 1$ for $i = 1, \cdots, k$. Then $X = \bigcup_{i=0}^{k} T^nF_i$ is a decomposition of X with $m(T^nF_0) > 0$ and $T^nF_i = T^{n+n_i}E_i$, $i = 1, \cdots, k$. Hence (replacing F_i by T^nF_i) it is permissible to assume that $n_i \geq 1$. There is no loss of generality, and a small gain in elegance, in assuming also that $n_i = i$, $i = 1, \cdots, k$; this can be achieved by inserting empty sets whenever necessary in both the E and the F decompositions, and uniting two E_i (and the corresponding F_i) if the two associated exponents are equal.

Consider now the array,

E_1	E_2	E_3	E_4	$\cdots$	E_k	
	TE_2	TE_3	TE_4	$\cdots$	TE_k	I
		T^2E_3	T^2E_4	$\cdots$	T^2E_k	
				$\cdots$		
					$T^{k-1}E_k$	
$TE_1 = F_1$	$T^2E_2 = F_2$	$T^3E_3 = F_3$	$T^4E_4 = F_4$	$\cdots$	$T^kE_k = F_k$	II

Call all sets above the line II upper sets, all sets below the line I lower sets, and all those between the two lines middle sets. Let A_j, $j = 0, 1, 2, \cdots$, be the set of points which belong to at least j upper sets. (Since $X = \bigcup_{i=1}^{k} E_i$, it is clear that $A_0 = A_1 = X$; it is clear also that A_j is empty for sufficiently large j. It will not, however, be necessary to make use of these facts below.) Then TA_j is the set of points which belong to at least j lower sets. If a point is in TA_j then (since it can belong to at most one F_i) it belongs to at least $j - 1$ middle sets. Since, however, it also has to belong to some E_i, it belongs to at least j upper sets. In other words $TA_j \subsetneq A_j$.

Consider any point of $A_j - TA_j$. Such a point cannot belong to any F_i, $i = 1, \cdots, k$, and belongs therefore to F_0. For, since it is not in TA_j, it belongs to fewer than j lower sets, and (if it did belong to some F_i, $i = 1, \cdots, k$) it would therefore belong to fewer than $j - 1$ middle sets. But its presence in A_j, together with its presence in one and only one E_i, implies that it has to belong to at least $j - 1$ middle sets. Hence $A_j - TA_j \subsetneq F_0$, and it is, moreover, clear from this analysis that $A_j - TA_j$ is the set of those points of F_0 which belong to exactly $j - 1$ middle sets, $j = 1, 2, \cdots$. It follows that the sets $A_j - TA_j$ are pairwise disjoint sets whose union is F_0, and consequently that the measure of at least one of them is positive. The corresponding A_j, and hence the transformation T, is compressible.

COROLLARY 1. *Every power of an incompressible transformation is incompressible.*

PROOF. The assertion that E is compressible under T^n is equivalent to the assertion that E is 1-equivalent, under T, to a proper subset, and hence, by Theorem 1, to the compressibility of T.

COROLLARY 2. *The following two conditions on the transformation T are equivalent to each other.*

(4) *T is unbounded.*

(5) *X is unbounded under T.*

PROOF. The implication (5) → (4) is trivial; (4) → (5) is proved exactly the same way as (2) → (3) in Theorem 1.

Theorem 1, together with its corollaries and its proof, appears in substantially the same form in the work of E. Hopf.[2] My presentation is a slightly simplified and extended version—extended in that it dispenses with Hopf's hypothesis concerning the preservation of sets of measure zero.

3. Finite measures

The purpose of the next three theorems is to present a complete survey of the relations among the various conditions of non singularity, compressibility, and preservation of a *finite* measure.

THEOREM 2. *The following implications are valid.*

(6) $Z \leftrightarrow Z(+) \wedge Z(-)$;

(7) $M \rightarrow M_* \wedge M^*$;

(8) $C^* \rightarrow C_*$;

(9) $M \rightarrow Z$, *i.e. the existence of an equivalent finite invariant measure implies non singularity*;

(10) $C_* \wedge Z(\pm) \rightarrow Z$, *i.e. if T is incompressible and either $(+)$ or $(-)$ non singular, then it is non singular*;

(11) $M^* \wedge Z(\pm) \rightarrow M$, *i.e. if a finite invariant measure stronger than m exists for a transformation which is either $(+)$ or $(-)$ non singular then a finite invariant measure equivalent to m also exists*;

(12) $M^* \leftrightarrow C^*$, *i.e. a finite invariant measure stronger than m exists if and only if the transformation is bounded.*

PROOF. Since the possibility that m is identically zero was excluded in the description of the condition M_*, and there only, (7) is not true in this trivial case. Otherwise the relations (6), (7), and (8) are immediate consequences of the definitions.

(9) If $\bar{m}$ is a finite invariant measure equivalent to m then

$$m(E) = 0 \leftrightarrow \bar{m}(E) = 0 \leftrightarrow \bar{m}(TE) = 0 \leftrightarrow m(TE) = 0,$$

so that Z is satisfied.

---

[2] See [5], Theorem 5, p. 391.

(10) Suppose first that the incompressible transformation T satisfies $Z(+)$ and that $m(E) = 0$; it is to be proved that $m(T^{-1})E = 0$. Write $A = \bigcup_{n=0}^{\infty} T^n E$) then $Z(+)$ implies that $m(A) = 0$. Also

$$m(T^{-1}E) = m(A \cdot T^{-1}E) + m(A' \cdot T^{-1}E) = m(A' \cdot T^{-1}E);$$

hence it is sufficient to show that $B = A' \cdot T^{-1}E = T^{-1}E \cdot \bigcap_{n=0}^{\infty} T^n E'$ has measure zero.

For $m = 1, 2, \cdots, T^m B = T^{m-1}E \cdot \bigcap_{n=0}^{\infty} T^{n+m}E' = T^{m-1}E \cdot \bigcap_{n=m}^{\infty} T^n E' \subsetneq T^{m-1}E$, whereas $B \subseteq T^{m-1}E'$. In other words B is disjoint from all its positive iterates. Write $C = \bigcup_{n=0}^{\infty} T^n B$, then $TC = \bigcup_{n=1}^{\infty} T^n B \subsetneq C$ and $C - TC = B$. The incompressibility of T implies the desired conclusion.

To prove the second half of (10), i.e. that $C_* \wedge Z(-)$ implies Z, observe that (by Corollary 1 to Theorem 1) if T satisfies C_* and $Z(-)$ then T^{-1} satisfies C_* and $Z(+)$. The first half of (10), just proved, implies that T^{-1} satisfies Z, and this is clearly equivalent to T satisfying Z.

(11) Suppose first that T satisfies M^* and $Z(+)$ and that, accordingly, there exists a finite invariant measure m^* stronger than m. The first step in the proof of (11) is to show that T satisfies $Z(-)$, i.e. that if $m(E) = 0$ then $m(T^{-1}E) = 0$. Write $F = \bigcup_{n=0}^{\infty} T^n E$; then by $Z(+)$, $m(F) = 0$. Also $T^{-1}F = \bigcup_{n=-1}^{\infty} T^n E = T^{-1}E \cup F$, or $T^{-1}F' = T^{-1}E' \cdot F'$. Hence

$$m^*(F') = m^*(T^{-1}F') = m^*(T^{-1}E' \cdot F')$$

or $m^*(T^{-1}E \cdot F') = 0$. It follows from the fact that m^* is stronger than m that $m(T^{-1}E \cdot F') = 0$. Hence $m(T^{-1}E) = m(T^{-1}E \cdot F') + m(T^{-1}E \cdot F) = 0$.

Next, apply the Radon-Nikodym theorem[3] to find a non negative measurable function $f(x)$ for which, identically in E,

$$m(E) = \int_E f(x) \, dm^*(x).$$

(The possible infinity of m causes no difficulty in the application of the Radon-Nikodym theorem, since it is obviously valid for subsets of a fixed set of finite measure, and since the whole space is, by hypothesis, a countable union of such sets.) Write $A = \{x : f(x) = 0\}$; then $m(A) = 0$ and, since Z was just proved to be satisfied, $m(T^n A) = 0$ for $n = 0, \pm 1, \pm 2, \cdots$. Hence for $B = \bigcup_{n=-\infty}^{\infty} T^n A$, $TB = B$ and $m(B) = 0$. I define $\bar{m}(E) = m^*(E \cdot B')$ for all E. It is clear that $\bar{m}$ is finite and invariant. If $m(E) = 0$ then

$$0 = \int_E f(x) \, dm^*(x) \geqq \int_{EB'} f(x) \, dm^*(x).$$

But from $\{x : f(x) = 0\} = A \subseteq B$ it follows that $\{x : f(x) > 0\} \supseteq B' \supseteq EB'$, so that $m^*(EB') = \bar{m}(E)$ must be zero. Conversely, if $\bar{m}(E) = 0$ then $m^*(EB') = 0$, whence $m(EB') = 0$; but $m(EB') = m(E) - m(EB) = m(E)$, so that $m(E) = 0$.

The second half of (11) is proved by the observation that if T satisfies M^*

---

[3] See [8], Theorem 14.11, p. 36.

and $Z(-)$ then T^{-1} satisfies M^* and $Z(+)$ and therefore, by the half just proved, T^{-1} satisfies M.

(12) If M^* is satisfied and if X is ∞-equivalent to a subset E then the invariance of the measure m^* shows that $m^*(E) = m^*(X)$ and therefore $m^*(E') = 0$. The relation between m^* and m implies then that $m(E') = 0$, i.e. that X is bounded and C^* is satisfied.

The implication $C^* \to M^*$ is the deepest part of this theorem. By a very intricate set theoretical analysis Hopf has shown that $Z \wedge C^* \to M$, and while it would be highly desirable to give a direct, simple, and intuitive proof of (12), for the present I shall base the completion of the proof of Theorem 2 on Hopf's result.[4]

There is no loss of generality in assuming that the measure m is finite. (Cf. remark I at the end of this section.) Let $\{c_n : n = 0, \pm1, \pm2, \cdots\}$ be a sequence of positive constants with $\sum_{-\infty}^{\infty} c_n = 1$, and write $m_1(E) = \sum_{-\infty}^{\infty} c_n m(T^n E)$. Clearly $m_1(E) \geqq 0$; if $E_1, E_2, \cdots$ is a countable sequence of pairwise disjoint measurable sets and $E = \bigcup_{i=1}^{\infty} E_i$, then

$$m_1(E) = \sum_n c_n m(T^n E) = \sum_n c_n m(\bigcup_i T^n E_i) = \sum_n c_n \sum_i m(T^n E_i)$$
$$= \sum_i \sum_n c_n m(T^n E_i) = \sum_i m_1(E_i),$$

so that m_1 is a measure. (The interchange of the order of summations is justified by the fact that all the terms are non negative.) The positiveness of c_0 implies that the measure m_1 is stronger than m; and since, finally, an application of a power of T to the sequence $\{T^n E\}$ merely displaces the terms by a finite number of steps, it follows that $m_1(E) = 0$ is equivalent to $m_1(TE) = 0$. It is easy to see that a set E is bounded with respect to the measure m_1 if and only if every $T^n E$ is bounded with respect to m. It follows that the boundedness of T with respect to m implies its boundedness with respect to m_1. Consequently Hopf's result may be applied to yield a measure $\bar{m}_1$, finite, invariant, and equivalent to m_1, and therefore stronger than m, which is exactly what was to be proved.

The force of the statement of Theorem 2 is considerably strengthened by the assertion that it is the best possible theorem of its type. The phrase "best possible" is used here in its technical sense, meaning that except for the implications explicitly asserted in Theorem 2 and those which follow trivially from the transitivity of the implication relation, no others hold among the conditions M, M^*, M_*, Z, $Z(+)$, $Z(-)$, C^*, and C_*. By a small number of elementary applications of Theorem 2, this very inclusive statement may be precisely formulated as follows.

THEOREM 3. *For each of the list of conditions written below the set of transformations satisfying it is not empty, and, moreover, this statement remains true if in each condition in which M'_* ($=$ the negation of M_*) appears it is replaced by M_*.*

---

[4] See [5], Theorem 4, p. 387.

(13) M;

(14) $M'_* \wedge M^* \wedge Z(\pm)'$;

(15) $M'_* \wedge Z \wedge C'_*$;

(16) $M'_* \wedge Z(\mp) \wedge Z(\pm)' \wedge C'_*$;

(17) $M'_* \wedge Z(+)' \wedge Z(-)' \wedge C'_*$;

(18) $M'_* \wedge M^{*'} \wedge Z \wedge C_*$, *i.e. there exists an incompressible and non singular transformation which preserves no finite measure either weaker or stronger than m*;

(19) $M'_* \wedge M^{*'} \wedge Z(+)' \wedge Z(-)' \wedge C_*$.

PROOF. (13) The consideration of any measure preserving transformation T on a space X of finite measure (e.g. the identity on a space consisting of a single point of measure one) establishes the possibility of (13).

(14) Let X be a space of two points, $X = \{x, y\}$, with $m(x) = 1$, $m(y) = 0$, and define $Tx = y$, $Ty = x$. The measure m^*, defined by $m^*(x) = m^*(y) = 1$, shows that M^* is satisfied; it is clear, however, that $Z(+)$, $Z(-)$, and M_* are not.

(15) Let X be the set of integers; for every set $E \subsetneq X$ define $m(E)$ to be the number of points in E; write $Tx = x + 1$. It is clear that Z is satisfied. The set $E_+ = \{x: x > 0\}$, and therefore the transformation T, is compressible; the only finite, invariant measure is identically zero.

(16) Let X and T be as in (15) but, instead of the measure m, use the measure $m_1(E) = m(E \cdot E'_+)$. Just as in (15), $Z(+)$ is valid, while M_* and C_* are not; the set E_+ (for which $m_1(E_+) = 0$, $m_1(T^{-1}E_+) = 1$) shows that $Z(-)$ fails. For the transformation T^{-1} the roles of $Z(+)$ and $Z(-)$ in this statement are interchanged.

(17) Let X and T be as in (15) and (16); let E_2 be the set of even integers and use, instead of m or m_1, the measure $m_2(E) = m(E \cdot E'_2)$. Then $m_2(E_2) = 0$ but $m_2(TE_2)$ and $m_2(T^{-1}E_2)$ are both equal to $m_2(E'_2) = \infty$, so that both $Z(+)$ and $Z(-)$ fail.

The proofs of (18) and (19) require considerably deeper methods and constitute, in fact, the first solution of a problem explicitly raised in the literature several times—the first time by E. Hopf in 1932.[5]

(18) To construct an example for (18), let X be the real line, with the notions of Lebesque measurability and measure defined as usual, and let T be a one to one measure preserving and *irreducible* transformation of X onto itself. The term "irreducible" is intended to mean that the only measurable sets invariant under T are sets of measure zero, or the complements of sets of measure zero. (Irreducible transformations have been called *metrically transitive*. Because of its useful algebraic connotations, the shorter and more intuitive term is, in my opinion, preferable to the older one.) It is far from obvious that such a transformation exists. It is convenient, however, to complete the proof of Theorem 3 under the assumption of the existence of an irreducible transformation, thus slightly postponing the presentation of an example.

---

[5] See [5], p. 391.

Since T is measure preserving, Z is satisfied; to prove that C_* is satisfied suppose that $TE \subseteq E$ and $m(E - TE) > 0$. For $n \geq 1$, $T^n(E - TE) \subseteq T^n E \subseteq TE$, whereas $E - TE \subseteq TE'$; in other words $E - TE$ is disjoint from all its positive iterates. It follows easily that any two different (not necessarily positive) iterates of $E - TE$ are disjoint from each other. Hence if $F \subseteq E - TE$ is such that $0 < m(F) < m(E - TE)$, then the set $F^* = \bigcup_{n=-\infty}^{\infty} T^n F$ is invariant under T, but neither it nor its complement has measure zero. Since this contradicts the assumed irreducibility of T, it follows that T is indeed incompressible.

Suppose now that m_* is a finite invariant measure weaker than m. The Radon-Nikodym theorem implies the existence of a non negative integrable function $f(x)$ for which, identically in E, $m_*(E) = \int_E f(x)\; dm(x)$. It follows that $m_*(TE) = \int_E f(Tx)\; dm(x)$; the invariance of m_* and the uniqueness assertion of the Radon-Nikodym theorem yield the conclusion that $f(Tx) = f(x)$ almost everywhere. By changing f on a set of measure zero (without changing its relation to m and m_*) it is permissible to assume that $f(Tx) = f(x)$ for all x. But then for every real constant k, the set $\{x: f(x) < k\}$ is invariant under T—the irreducibility of T implies immediately that $f(x)$ is identically equal to some constant k. Since the only integrable constant function is identically zero, it follows that $m_*(E)$ is necessarily identically zero, and that therefore T cannot satisfy M_*. The implication $M^* \wedge Z \rightarrow M_*$ (a trivial consequence of Theorem 2) show that T must also fail to satisfy M^*.

(19) Let X and T be as in (18), let E_0 be any measurable set with $0 < m(E_0) < \infty$, and use, instead of m, the measure $m_0(E) = m(E \cdot E_0')$. Since m_0 is weaker than m, C_* is still true and M^* and M_* are still false. The irreducibility of T shows that both TE_0 and $T^{-1}E_0$ intersect E_0' in a set for which m is positive; in other words if the measure m_0 is used neither $Z(+)$ nor $Z(-)$ is satisfied.

The assertion of the theorem concerning the replacement of M_*' by M_* is implied by the following statement. If T is a measurable transformation on a measure space X, and if T does not satisfy M_*, then it is always possible to construct a measurable transformation T_0 on a measure space X_0, so that T_0 does satisfy M_*, and so that in addition T_0 satisfies exactly those among the remaining conditions M, M^*, Z, $Z(+)$, $Z(-)$, C^*, and C_*, which T satisfies. To construct X_0 it is necessary merely to adjoin to X a single point x_0, define $m_0(E)$, for $E \subseteq X_0$, to be $m(E) + 1$ or $m(E)$ according as E does or does not contain x_0, and to define $T_0 x$ to be Tx or x_0 according as $x \neq x_0$ or $x = x_0$. The measure m_*, whose value $m_*(E)$ is 1 or 0 according as E does or does not contain x_0, shows that T_0 satisfies M_*; a routine verification establishes the remaining parts of the assertion.

In order to clarify and summarize Theorems 2 and 3, I subjoin a Venn diagram, illustrating the implication relations among the conditions investigated. Each of the conditions M, M^*, M_*, $Z(+)$, $Z(-)$, and C_* is represented by the triangle in the two base angles of which its name appears. Since $C^* = M^*$, and $Z = Z(+) \wedge Z(-)$, C^* and Z are omitted from the diagram. The logical product of two conditions corresponds to the intersection of the two corresponding regions; the fact that one condition implies another is represented in the diagram by the region corresponding to the former being contained in the region corresponding to the latter.

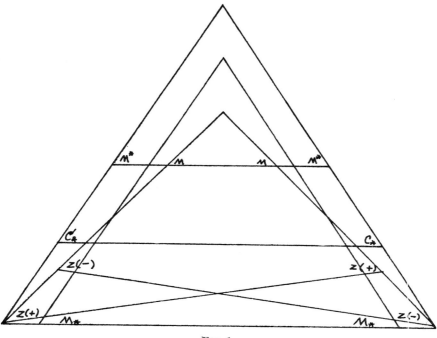

Fig. 1.

It remains now to fill the gap left in the proof of Theorem 3, by establishing the existence of an irreducible transformation on the real line.[6] For this purpose let J_n be the interval

$$J = \{x: (\tfrac{1}{2})^{n+1} \leqq x < (\tfrac{1}{2})^n\}, \; n = 0, 1, 2, \cdots;$$

let the measure space X (measure theoretically obviously equivalent to the

---

[6] I could not find any example in the literature. Hopf ([4], pp. 65–67) discusses a transformation in the plane which very likely has all the properties needed for the present purpose, but he gives no proof of its irreducibility. The first example I saw was communicated to me in a letter by J. C. Oxtoby, dated October 26, 1945. Oxtoby remarks also that the methods of his joint paper with Ulam [7], specifically Lemma 5, p. 884, of that paper, serve to construct an irreducible measure preserving homeomorphism of n dimensional Euclidean space onto itself for every $n \geqq 2$. The method of construction used above has been discussed recently by S. Kakutani [6].

real line) be the collection of intervals defined in the (x, y) plane for $n = 0$, $1, 2, \cdots$, by $x \in J_n$, $y = 0, 1, \cdots, 2^{n+1} - 1$. Write X_0 for the base unit interval, $X_0 = \{(x, y): (x, y) \in X, y = 0\}$, and let T_0 be any one to one, measure preserving, and irreducible ($=$ metrically transitive, ergodic) transformation of the finite measure space X_0 onto itself. (For instance T_0 could be translation modulo 1 by an irrational constant.) Write for $n = 0, 1, 2, \cdots$,

$$T(x, y) = (x, y + 1) \text{ for } x \in J_n, \quad y = 0, \cdots, 2^{n+1} - 2;$$

$$T(x, y) = T_0(x, 0) \text{ for } x \in J_n, y = 2^{n+1} - 1.$$

It is clear that T is a one to one, measurable, and measure preserving transformation of X onto itself.

Suppose now that a measurable set E is invariant under T. It follows from the definition of T that $(x, y) \in E$ if and only if $(x, 0) \in E$, and that for $x \in J_n$, $(x, 0) \in E$ implies

$$T_0(x, 0) = T^{2^{n+1}}(x, 0) \in E.$$

Consequently E consists of all those points (x, y) for which $(x, 0)$ belongs to $E_0 = E \cdot X_0$, where E_0 is invariant under T_0. The irreducibility of T_0 implies that either $m(E_0) = 0$ or else $m(X_0 - E_0) = 0$, whence either $m(E) = 0$ or else $m(X - E) = 0$. Hence T is irreducible.

The proof of Theorem 3 is now complete: I adjoin two relevant remarks.

I. The possible infinity of the original measure m is *not* the cause of the existence of the examples used to prove Theorem 3. Infinite measures were employed in some of the examples, but for convenience only. Since, by hypothesis, all the measure spaces X considered here are σ-finite, $X = \bigcup_{n=1}^{\infty} X_n$, $m(X_n) < \infty$, $n = 1, 2, \cdots$, it is always possible, without destroying the validity or invalidity of any of the conditions under discussion to introduce a finite measure equivalent to m. For this purpose it is sufficient, for instance, to produce a suitable sequence $\{c_n\}$ of convergence factors and to define the new measure of a set E by the sum

$$\sum_{n=1}^{\infty} c_n m(E \cdot X_n).$$

II. It is often customary to consider only measures m which are *complete* in the sense that every subset of a set of measure zero is measurable (and has, of course, measure zero). The presence of this assumption does not invalidate (it merely simplifies slightly) the results above. If in fact it is also assumed (and this is the situation in most non pathological measure spaces) that every set of positive measure contains a non measurable set, then the completeness of the measure (together with the measurability of the transformation) serves merely automatically to ensure the validity of the condition Z.

4. Uniqueness

Since Theorems 2 and 3 describe completely the conditions for the existence of finite invariant measure of various kinds, it becomes of interest to raise the question of the extent of uniqueness. The detailed answer is given in the following theorem and its corollary.

THEOREM 4. (i) *If $M_*[M^*]$ is satisfied then among all finite invariant measures weaker [stronger] than m there is a (not necessarily unique) strongest [weakest], denoted by m_*^0 [m_0^*, equivalent to m on the invariant sets]. (ii) If μ is any finite measure defined for the invariant sets and there weaker than m_*^0 [m_0^*, and stronger than m], then there exists a unique finite invariant measure μ_* [μ^*] weaker [stronger] than m, defined for all measurable sets, and equal to μ on the invariant sets.*

COROLLARY. *If M is satisfied then m_*^0 and m_0^* are both equivalent to a finite invariant measure $\bar{m}_0$ equivalent to m. If μ is any finite measure defined for the invariant sets and there equivalent to m, then there exists a unique finite invariant measure $\bar{\mu}$, equivalent to m, defined for all measurable sets, and equal to μ on the invariant sets.*

PROOF. (i) There is no loss of generality in assuming that the measure m is finite. (Cf. remark I at the end of section 3.) Suppose then that M_* is satisfied and let m_* be any finite invariant measure weaker than m. Write $\mathfrak{E}(m_*)$ for the class of measurable sets E for which the two conditions $F \subseteq E$ and $m_*(F) = 0$ together imply that $m(F) = 0$. In other words, $\mathfrak{E}(m_*)$ is the class of sets on subsets of which m_* is stronger than, and therefore equivalent to m. (An easy application of the Radon-Nikodym theorem, similar to the one made in the next paragraph, shows that $\mathfrak{E}(m_*)$ is not empty.) Write $\mathfrak{E}$ for the union of all $\mathfrak{E}(m_*)$, the union being extended over all finite invariant m_* weaker than m, and define $\alpha = \sup_{E \in \mathfrak{E}} m(E)$. Select a sequence $\{E_n\}$ of sets in $\mathfrak{E}$ for which $m(E_n) \to \alpha$, and write $\bar{E} = \bigcup_n E_n$. Then $m(\bar{E}) \geqq \alpha$. Since $E_n \in \mathfrak{E}$, it follows that $E_n \in \mathfrak{E}(m_*^n)$, where m_*^n is some finite invariant measure weaker than m. Let $\{c_n\}$ be a sequence of positive constants for which the series $\sum_n c_n m_*^n(X)$ converges; then $m_*^0(E) = \sum_n c_n m_*^n(E)$ is a finite invariant measure weaker than m.

Let F be any subset of $\bar{E}$ with $m_*^0(F) = 0$. Then $m_*^n(F) = 0$ for all n, whence $m_*^n(F \cdot E_n)$ and therefore $m(F \cdot E_n)$ both vanish for all n. It follows that $m(F) = 0$, so that $\bar{E} \in \mathfrak{E}(m_*^0)$, and therefore, incidentally, $m(\bar{E}) = \alpha$. On the other hand let m_* be any finite invariant measure weaker than m. By the Radon-Nikodym theorem, m_* is an integral, $m_*(E) = \int_E f(x) \, dm(x)$, with a non negative integrand $f(x)$. Suppose that the set E of x's in $\bar{E}'$ for which $f(x) > 0$ had positive measure. Then for any $F \subseteq E$, $m_*(F) = 0$ would imply $m(F) = 0$, i.e. $E \in \mathfrak{E}(m_*)$. It would follow easily that $E \cup \bar{E} \in \mathfrak{E}(m_* + m_*^0) \subsetneq \mathfrak{E}$, and $m(E \cup \bar{E}) > \alpha$. Since, by the definition of α, this is impossible, $m(E)$ must vanish, and therefore $m_*(\bar{E}') = 0$. Consequently m_* is weaker than m_*^0, and the part of (i) concerning m_*^0 is proved.

The proof of the part of (i) concerning m_0^* is somewhat different. Suppose that M_* is satisfied, and let m_1^* be any finite, invariant measure stronger than m. Write $\mathfrak{E}$ for the class of measurable *invariant* sets E for which $m(E) = 0$ and $m_1^*(E) > 0$, and define $\alpha = \sup_{E \in \mathfrak{E}} m_1^*(E)$ ($= 0$ in case $\mathfrak{E}$ is empty). If $\mathfrak{E}$ is not empty, select a sequence $\{E_n\}$ of sets in $\mathfrak{E}$ for which $m^*(E_n) \to \alpha$, and write

$\bar{E} = \bigcup_n E_n$; if $\mathfrak{E}$ is empty let $\bar{E}$ be the empty set. In either case it is very easy to see that $m(\bar{E}) = 0$ and $m_1^*(\bar{E}) = \alpha$; define $m_0^*(E) = m_1^*(E \cdot \bar{E}')$. It follows (using the invariance of $\bar{E}$) that m_0^* is a finite invariant measure. If $m_0^*(E) = 0$, then $m_0^*(E \cdot \bar{E}') = 0$ whence $m(E \cdot \bar{E}') = 0$, so that $m(E) \leqq m(E \cdot \bar{E}') + m(\bar{E}) = 0$. Hence m_0^* is indeed stronger than m. If E is invariant and $m(E) = 0$, then $m_0^*(E)$ has to vanish also, for otherwise $E \cup \bar{E}$ would have the contradictory properties $m(E \cup \bar{E}) = 0$, $m_1^*(E \cup \bar{E}) > \alpha$. In other words m_0^* is weaker than m, and therefore equivalent to m, on the invariant sets. If, finally, m^* is any finite invariant measure stronger than m then the chain of implications, $m^*(E) = 0 \leftrightarrow m^*(\bigcup_{-\infty}^{\infty} T^n E) = 0 \rightarrow m(\bigcup_{-\infty}^{\infty} T^n E) = 0 \leftrightarrow m_0^*(\bigcup_{-\infty}^{\infty} T^n E) = 0 \leftrightarrow m_0^*(E) = 0$, shows that m_0^* is weaker than m^*. This concludes the proof of (i).

(ii) If μ is any finite measure defined for the invariant sets and there weaker than m_0^* and stronger than m, then, by the Radon-Nikodym theorem there exists a non negative function $\omega(x)$, measurable with respect to the collection of invariant sets, such that identically for all invariant sets E, $\mu(E) = \int_E \omega(x) \, dm_0^*(x)$. The fact that $\omega(x)$ is measurable with respect to the invariant sets implies that $\omega(x)$ is invariant under T, $\omega(x) = \omega(Tx)$; it follows that the measure $\mu^*(E) = \int_E \omega(x) \, dm^*(x)$, defined for all measurable sets E, is a finite invariant measure. It is clear that μ^* is weaker than m_0^* and equal to μ on the invariant sets. The relations, $\mu^*(E) = 0 \leftrightarrow \mu^*(\bigcup_{-\infty}^{\infty} T^n E) = 0 \leftrightarrow \mu(\bigcup_{-\infty}^{\infty} T^n E) = 0 \rightarrow m(\bigcup_{-\infty}^{\infty} T^n E) = 0 \rightarrow m(E) = 0$, show that μ^* is also stronger than m. The assertion concerning the existence of μ_* is proved by exactly the same argument.

To prove, finally, the uniqueness statement, I shall prove that if m_1 and m_2 are any two finite invariant measures which agree on the invariant sets then they are identical. For this purpose observe first that $m_1(E) = 0 \leftrightarrow m_1(\bigcup_{-\infty}^{\infty} T^n E) = 0 \leftrightarrow m_2(\bigcup_{-\infty}^{\infty} T^n E) = 0 \leftrightarrow m_2(E) = 0$, so that m_1 and m_2 are equivalent. It follows that

$$m_1(E) = \int_E f(x) \, dm_2(x)$$

and, consequently, that

$$m_1(TE) = \int_{TE} f(x) \, dm_2(x) = \int_E f(Tx) \, dm_2(x).$$

The invariance of m_1 implies therefore that $f(Tx) = f(x)$ almost everywhere; by an argument employed once before (in the proof of (18), Theorem 3) it is permissible to assume that f is invariant under T. Since m_1 and m_2 agree on the invariant sets, it follows that $\int_E f(x) \, dm_2(x) = \int_E 1 \cdot dm_2(x)$ for all invariant sets E. The uniqueness assertion of the Radon-Nikodym theorem (applied to the collection of invariant sets only) implies that $f(x) = 1$ almost everywhere, whence it follows that m_1 and m_2 are identical.

To prove the corollary it is sufficient to remark that a finite invariant measure $\bar{m}_0$ equivalent to m is simultaneously stronger and weaker than both m_*^0 and m_0^* and is therefore equivalent to both. The remaining assertions of the corollary are immediate consequences of the theorem.

5. Infinite measures

Several of the examples in section 3 establish the possibility that a transformation which preserves an infinite (but σ-finite) measure need not preserve any finite measure. Accordingly there arises the problem of characterizing those transformations which preserve a σ-finite measure stronger than m. It might appear that for the sake of completeness one should introduce and discuss systematically conditions similar to M_*, M^*, and M, but different from these in that the requirement of finiteness is replaced by σ-finiteness. It is easy to see, however, that almost all the theory remains unchanged; the only significant new question is indeed the one concerning the existence of a σ-finite invariant measure stronger than m. The main purpose of this section is to present (in Theorem 5 below) the answer to this question. It is worth while to remark first that the requirement of σ-finiteness, imposed on the desired invariant measure, is not an artificial one: without it the answer is trivial and useless. For if $m^*(E)$ is defined as the number of points in E, then $m^*(E)$ is clearly an invariant measure stronger than m; but it is in general so very infinite that no useful consequences can follow from its existence.

For the statement of an important auxiliary result I need the notion of a recurrent set: E is *recurrent* if $E \subseteq \bigcap_{n=1}^\infty \bigcup_{m=n}^\infty T^{-m}E$, i.e. if every point of E returns to E an infinite number of times under positive iterations of T.

LEMMA. *If T is incompressible then every set E becomes recurrent upon the omission of a suitable set of measure zero.*

PROOF. For any set E, write $F = \Phi(E) = \bigcap_{p=1}^\infty \bigcup_{q=p}^\infty T^{-q}E$. Then for $m \geq 1$, $T^{-m}F = \bigcap_{p=m+1}^\infty \bigcup_{q=p}^\infty T^{-q}E \supseteq F$. (In fact it is easy to show that $TF = F$.) It follows that $\Phi(EF) \supseteq \bigcap_{n=1}^\infty \bigcup_{m=n}^\infty T^{-m}E \cdot F = F \supseteq EF$, so that EF is recurrent. On the other hand write $G = \bigcup_{n=0}^\infty T^{-n}E$; then $T^{-p}G = \bigcup_{q=p}^\infty T^{-q}E \subseteq G$, whence it follows, by the incompressibility of T^{-p}, that $0 = m(G \cdot T^{-p}G') = m(\bigcup_{n=0}^\infty T^{-n}E \cdot \bigcap_{q=p}^\infty T^{-q}E') \geq m(E \cdot \bigcap_{q=p}^\infty T^{-q}E')$. Since this is true for all $p = 1, 2, \cdots$, it follows that $m(E \cdot F') = m(E \cdot \bigcup_{p=1}^\infty \bigcap_{q=p}^\infty T^{-q}E') = 0$, which proves the lemma.[7]

THEOREM 5. *A necessary and sufficient condition that there exist a σ-finite invariant measure m^* stronger than m is that T be σ-bounded.*[8]

PROOF. The necessity of the condition is clear. If m^* exists then every

---

[7] A special case of this result is known as the Poincaré recurrence theorem.

[8] In accordance with the notational convention discussed in footnote (1), a set E is σ-bounded if it is the union of countably many bounded sets; if X is σ-bounded, the transformation T is called σ-bounded.

set of finite m^* measure is bounded (cf. the beginning of the proof of part (12) of Theorem 2) so that the σ-finiteness of m^* implies the σ-boundedness of T.

To prove the sufficiency of the condition, observe first (as was remarked several times before) that it may be assumed that m is finite. Introduce, as before, the measure $m_1(E) = \sum_{-\infty}^{\infty} c_n m(T^n E)$, (with $c_n > 0$, $\sum_{-\infty}^{\infty} c_n = 1$); then, with respect to m_1, T is still σ-bounded. To prove this, let E be any measurable set, and let $\mathfrak{F}$ be the class of all measurable sets $F \subseteq E$ for which $m(F) = 0$ but $m_1(F) > 0$. Write $\alpha = \sup_{F \in \mathfrak{F}} m_1(F)$ and choose a sequence $\{F_n\}$ of sets in $\mathfrak{F}$ with $m_1(F_n) \to \alpha$. If $\bar{F} = \bigcup_{n=1}^{\infty} F_n$ then $\bar{F} \subseteq E$, $m(\bar{F}) = 0$, and $m_1(\bar{F}) = \alpha$. If, moreover, $F \subseteq E - \bar{F}$ and $m(F) = 0$ then $m_1(F)$ also has to vanish, for otherwise the maximality of α would be contradicted. Hence within the set $E - \bar{F}$ (of the same measure as E) m and m_1 are equivalent, and consequently the σ-boundedness of $E - \bar{F}$ with respect to m implies its σ-boundedness with respect to m_1. Since T is non singular with respect to m_1, and since a non singular transformation sends bounded sets into bounded sets, it follows that the invariant set $\bigcup_{-\infty}^{\infty} T^n(E - \bar{F})$ is σ-bounded with respect to m_1. It follows by the familiar exhaustion procedure (and the σ-finiteness of X) that X is σ-bounded with respect to m_1. Hence there is no loss of generality in assuming that T is non singular in the first place.

It is easy to see also that given any measurable transformation T, the space X can be decomposed into two disjoint invariant measurable sets on one of which the transformation is incompressible and on the other of which it is dissipative.[9] (A measurable set E is a *wandering set* if all its iterated images are disjoint from one another; T is *dissipative* if there exists a wandering set E whose images cover the space.) Hence the dissipative and the incompressible parts of T may be treated separately. The existence of m^* for a non singular dissipative transformation is trivial: one has merely to define $m^*(E)$ to be equal to $m(E)$ for all measurable subsets of a wandering set, and then to extend $m^*(E)$ to all iterates of this wandering set by mapping it under T. Hence it is sufficient to discuss non singular incompressible transformations only.

It is convenient to make two more simplifying assumptions. First: if E is any measurable bounded set of positive measure, it is sufficient to define m^* for the measurable subsets of the invariant set $E^* = \bigcup_{-\infty}^{\infty} T^n E$. For if an m^* satisfying the requirements can be found in this case, then since X is assumed to be σ-bounded) at most countably many applications of this result will serve to define m^* over all X. Second: since it follows from the preceding lemma that upon the omission of a set of measure zero E becomes recurrent, it may be assumed that E itself is recurrent.

The recurrence of E implies that $E \subseteq \bigcup_{n=1}^{\infty} T^{-n} E$. A transformation S may be defined on E as follows. Write $E_n = E \cdot T^{-n} E \cdot \bigcap_{0 < i < n} T^{-i} E'$. Clearly the E_n form a sequence of pairwise disjoint measurable sets whose union is E; for $x \in E_n$, define $Sx = T^n x$. More simply: Sx is the first member of the se-

---

[9] See [4], Theorem 13.1, p. 46. Hopf states this result for measure preserving transformations only, but the proof covers the general case.

quence Tx, T^2x, T^3x, $\cdots$ that belongs to E. It is easy to verify that S is a one to one measurable transformation of E onto itself. If, moreover, E were ∞-equivalent under S to a proper subset of itself then, since for each x, S is equal to a (positive) power of T, E would be ∞-equivalent to a proper subset of itself under T. Since E is by hypothesis bounded under T, it follows that S is bounded. Consequently (by parts (11) and (12) of Theorem 2) there exists a finite measure μ, defined for all measurable subsets of E, eqivalent to m, and invariant under S.

It is now possible to define $m^*(F)$ for every measurable set F. Given F, consider all decompositions of F into countably many pairwise disjoint measurable sets, $F = \bigcup_{i=1}^{\infty} F_i$, for which integers n_i can be found with $T^{n_i}F_i \subseteq E$, and write $m^*(F)$ for the upper bound of all sums $\sum_{i=1}^{\infty} \mu(T^{n_i}F_i)$.[10] The assumption $E^* = X$ implies that such a decomposition of F can always be found[11]; a straightforward varification shows that m^* is an invariant measure[12] which is (as a consequence of the non singularity of T) stronger than m. The only new and non trivial detail is the proof that m^* is σ-finite; for this purpose it will be sufficient to prove that $m^*(E) = \mu(E) < \infty$. The proof of this fact is broken up into three steps presented below.

(i) If $F \subseteq E$ and $T^kF \subseteq E$ for some $k = 1, 2, \cdots$, then $\mu(F) = \mu(T^kF)$. This statement may be proved by induction, as follows. For $k = 1$, $F \subseteq E$, and $TF \subseteq E$, whence $F \subseteq T^{-1}E$, so that, by the definition of S, $SF = TF$. It follows that $\mu(TF) = \mu(SF) = \mu(F)$. If the statement has been proved for $1, \cdots, k$, and if $F \subseteq E$ and $T^{k+1}F \subseteq E$, then write $F = FE_1 \cup FE_2 \cup \cdots$, where the E_n are the subsets of E described in the definition of S. The fact that $T^{k+1}F \subseteq E$ implies that this apparently infinite union stops in reality at the $(k+1)^{\text{th}}$ step: $F = FE_1 \cup \cdots \cup FE_{k+1}$. From the definition of S, $S(FE_i) = T^i(FE_i)$. For $1 \leq i \leq k$, the relations $FE_i \subseteq E$ and $T^i(FE_i) \subseteq E$, together with the induction hypothesis, imply that $\mu(FE_i) = \mu(T^i(FE_i))$. Since, also for $1 \leq i \leq k$, $T^i(FE_i) \subseteq E$ and $T^{k+1-i}(T^i(FE_i)) = T^{k+1}(FE_i) = E$, and since $1 \leq k + 1 - i \leq k$, another application of the induction hypothesis yields $\mu(T^{k+1}(FE_i)) = \mu(FE_i)$. For $i = k + 1$, the definition of S shows that $S(FE_{k+1}) = T^{k+1}(FE_{k+1})$; this completes the proof of statement (i).

(ii) If $F \subseteq E$ and $T^kF \subseteq E$ for some $k = 0, \pm 1, \pm 2, \cdots$, then $\mu(F) = \mu(T^kF)$. For $k = 0$ the statement is trivial, and for $k = 1, 2, \cdots$ it is proved by (i). If k is negative then write $G = T^kF$, $T^{-k}G = F$, and apply (i) to G.

(iii) If $E = \bigcup_{i=1}^{\infty} F_i$ is a decomposition of E into pairwise disjoint measurable sets and if for each i, n_i is an integer for which $T^{n_i}F_i \subseteq E$, then $\sum_{i=1}^{\infty} \mu(T^{n_i}F_i) = \sum_{i=1}^{\infty} \mu(F_i) = \mu(E)$. It follows that $m^*(E) = \mu(E)$, and the proof of Theorem 5 is complete.

---

[10] This construction was introduced by G. D. Birkhoff and P. A. Smith, [1], p. 361, and systematically exploited by Hopf, [5], p. 377 et seq.

[11] See [5], p. 376.

[12] See [5], p. 377. Hopf carries through the details of the verification for the lower bound of the sums, but his proof extends almost without change to the upper bound.

6. The unsolved problem

Although the condition of Theorem 5 sheds considerable light on the problem of the existence of (possibly infinite) invariant measures, and although it puts into clear evidence the relation of this problem to the analogous problem for finite measures, it is not sufficiently tractable to help decide whether or not an arbitrary measurable transformation preserves some σ-finite measure m^* stronger than m. This question is the fundamental unsolved problem of the theory. I conjecture that the answer is no, that in other words there exists a measurable transformation which preserves no σ-finite measure stronger than m, but I am unable to produce an example. The purpose of this section is to present two reformulations of the fundamental problem in the hope that one of them may ultimatly lead to its solution, and also to prove a structure theorem for all non pathological measurable transformations which should simplify considerably the construction of a counter example, and which might by the way be considered to have an interest of its own.

Suppose that the measurable transformation T is such that every invariant set of positive measure contains a bounded set of positive measure. In this case it is easy to see that the σ-finiteness of m implies that X (and therefore T) is σ-bounded, and consequently a σ-finite invariant measure exists. If, accordingly, a transformation exists which preserves no σ-finite measure, then there must exist an invariant set of positive measure whose bounded subsets all have measure zero. It might as well be assumed that the invariant set of positive measure is the whole space X, so that the problem becomes that of searching for a *completely unbounded* transformation, i.e. a transformation which possesses no bounded sets of positive measure.

It has been remarked several times in the preceding sections that there is no loss of generality in assuming that the given transformation T is non singular: in the remainder of this section this will be assumed. It follows then from the Radon-Nikodym theorem that a positive measurable function $\omega(x)$ can be found such that for every measurable set E, $m(TE) = \int_E \omega(x) \, dm(x)$.[13] (If T is measure preserving $\omega(x)$ is identically 1.) More generally, for every integer $n = 0, \pm 1, \pm 2, \cdots$, there is a positive measurable function $\omega_n(x)$ such that for every measurable set E, $m(T^n E) = \int_E \omega_n(x) \, dm(x)$. Clearly $\omega_1(x) = \omega(x)$. For $n = i + j$ the formula for $m(T^n E)$ becomes $m(T^{i+j} E) = \int_{T^i E} \omega_i(x) \, dm(x) = \int_E \omega_i(T^j x) \, dm(T^j x) = \int_E \omega_i(T^j x) \omega_j(x) \, dm(x)$ whence follows, from the uniqueness part of the Radon-Nikodym theorem, the recursion relation $\omega_{i+j}(x) = \omega_i(T^j x) \omega_j(x)$ satisfied by almost every x. It is easy (but for the present purpose not particularly useful) to deduce from this relation an expression for $\omega_i(x)$ in terms of $\omega(x)$ and its iterates $\omega(T^j x)$.

---

[13] The function $\omega(x)$ might well be called the Jacobian of the transformation T: it plays the same role for T as the Jacobian of a differentiable transformation of Euclidean space.

Now suppose for a moment that a σ-finite invariant measure m^* stronger than m does exist; since T is non singular it may be assumed that m^* is equivalent to m. It follows that there exists a positive measurable function $f(x)$ such that for every measurable set E, $m(E) = \int_E f(x)\, dm^*(x)$. Hence $m(T^nE) =$

$$\int_{T^nE} f(x)\, dm^*(x) = \int_E f(T^nx)\, dm^*(x) \quad \text{and also} \quad m(T^nE) = \int_E \omega_n(x)\, dm(x) =$$

$\int_E \omega_n(x)f(x)\, dm^*(x)$, so that

$$f(T^nx) = \omega_n(x)f(x)$$

almost everywhere. Suppose conversely that the functional equation $f(Tx) = \omega(x)f(x)$ has a positive measurable solution f, and write $m^*(E) = \int_E (1/f(x))\, dm(x)$. Clearly m^* is a σ-finite measure equivalent to m; since, moreover,

$$m^*(TE) = \int_{TE} (1/f(x))\, dm(x) = \int_E (1/f(Tx))\, dm(Tx)$$

$$= \int_E (1/f(Tx))\omega(x)\, dm(x) = \int_E (1/f(x))\, dm(x) = m^*(E),$$

it follows also that m^* is invariant. In other words an invariant measure exists if and only if the functional equation has a solution.

Concerning this functional equation certain comments should be made. First of all it is clear that a positive (but not necessarily measurable) solution always exists: define $f(x)$ arbitrarily for one point out of each orbit of T, and use the functional equation to define f at the iterated images of x. If T is measure preserving (in which case, to be sure, there is no point in considering the functional equation) the equation becomes $f(Tx) = f(x)$, i.e. the equation of an invariant function. and any positive constant is an acceptable solution. In general the product of an invariant function and a solution of the functional equation is another solution, and moreover every solution may be obtained from a particuar one by multiplying it by a suitable invariant function. Hence if, in particular, T happens to be irreducible, then the solution (if any) is unique up to a constant factor.

From the recursion formula $\omega_{n+1}(x) = \omega(x)\omega_n(Tx)$ it follows that if the sequence $\{\omega_n(x)\}$ had a non zero finite limit $g(x)$ almost everywhere (possibly after transforming it by a suitable summability method) then $1/g(x)$ would be a solution of the functional equation. This seemingly helpful comment is however most probably useless. To see this, consider the case of an irreducible transformation T for which an infinite (but of course σ-finite) invariant measure m^* does exist; then $\omega_n(x) = f(T^nx)/f(x)$. An application of the Birkhoff ergodic theorem shows (if $f(x)$ is integrable with respect to m^*) that the sequence $\{\omega_n(x)\}$ does indeed converge almost everywhere $(C, 1)$, but the limit function is necessarily identically zero.

The existence of completely unbounded transformations or solutions of the functional equation: these are the two reformulations of the main problem. I turn now to the structure theorem mentioned at the beginning of this section.

From the point of view of searching for an invariant measure there are two particularly easy classes of transformations. One of these is the class of measure preserving transformations—for these there is no problem. The other is the class of *simple* transformations, where a transformation T is called simple if there exists a measurable set E whose iterated images are all disjoint from one another and which is such that on the complement of $\bigcup_{-\infty}^{\infty} T^n E$ the transformation is the identity. (In other words, T is simple if its incompressible or non dissipative part is the identity.)

It has been remarked already (see comment I at the end of section 3) that there is no loss of generality in assuming that the originally given measure m is finite. Hence only pathological cases (atomic and non separable measures) are neglected if it is assumed that the measure space X is the unit interval.[14] After these preliminary remarks and definitions it is now possible to state and evaluate the range of application of the following structure theorem.

THEOREM 6. *Every one to one measurable and non singular transformation T of the unit interval onto itself is a product, $T = PS$, where P is measure preserving and S is simple.*

PROOF. For every x, $0 \leq x \leq 1$, write K_x for the closed interval, $K_x = \{t : 0 \leq t \leq x\}$, and define $S(x) = m(TK_x)$. It is clear that S is a monotone non decreasing function of x with $S(0) = 0$ and $S(1) = 1$. If for $0 \leq a \leq b \leq 1$ it happens that $S(a) = S(b)$, then $0 = S(b) - S(a) = m(TK_b) - m(TK_a) = m(T(K_b - K_a))$, whence $m(K_b - K_a) = 0$, so that $b = a$, and it follows that $S(x)$ is strictly increasing.

If $1 \geq x_1 \geq x_2 \geq \cdots$ is a sequence converging to $x_0 \geq 0$, then $\bigcap_{n=1}^{\infty} TK_{x_n} = TK_{x_0}$ (and, of course, $TK_{x_1} \supseteq TK_{x_2} \supseteq \cdots$), so that $S(x_0) = m(TK_{x_0}) = \lim_{n \to \infty} m(TK_{x_n}) = \lim_{n \to \infty} S(x_n)$: in other words S is continuous on the right. If $0 \leq x_1 \leq x_2 \leq \cdots$ is a sequence converging to $x_0 \leq 1$ then $K_{x_1} \subseteq K_{x_2} \subseteq \cdots \subseteq K_{x_0}$ and, moreover, $m(K_{x_0} - \bigcup_{n=1}^{\infty} K_{x_n}) = 0$. It follows that

$$0 = m(TK_{x_0} - \bigcup_{n=1}^{\infty} TK_{x_n}) = m(TK_{x_0}) - \lim_{n \to \infty} m(TK_{x_n})$$

$$= S(x_0) - \lim_{n \to \infty} S(x_n),$$

so that S is continuous on the left. Consequently S defines a one to one transformation of the unit interval on itself: I shall show next that this transformation is measurable and non singular.

Observe first that $m(SK_x) = S(x) = m(TK_x)$. It follows easily that for every interval J, with or without either of its endpoints, $m(SJ) = m(TJ)$, and consequently that for every set E, $\bar{m}(SE) = \bar{m}(TE)$, where $\bar{m}$ denotes Lebesque outer measure. Now let E be any measurable set and A an arbitrary set. It follows that

----

[14] See [3], Theorem 1, p. 335.

$$\bar{m}(A \cdot SE) \ + \ \bar{m}(A \cdot SE') \ = \ \bar{m}(S(S^{-1}A \cdot E)) \ + \ \bar{m}(S(S^{-1}A \cdot E'))$$

$$= \ \bar{m}(T(S^{-1}A \cdot E)) + \bar{m}(T(S^{-1}A \cdot E')) \ = \ \bar{m}(TS^{-1}A \cdot TE) + \bar{m}(TS^{-1}A \cdot TE')$$

$$= \ \bar{m}(TS^{-1}A) \ \text{(since } TE \text{ is measurable}^{15}) \ = \ \bar{m}(SS^{-1}A) \ = \ \bar{m}(A),$$

so that SE is measurable and, of course, $m(SE) = m(TE)$.

If $m(E) = 0$ then $S^{-1}E$ has to be measurable and to have measure zero, since T is non singular and $0 = \bar{m}(E) = \bar{m}(SS^{-1}E) = \bar{m}(TS^{-1}E)$. The continuity of S implies that if E is a Borel set then so is $S^{-1}E$; since every measurable set is the union of a Borel set and a set of measure zero, it follows that for every measurable set E, $S^{-1}E$ is also measurable. This concludes the proof that S is measurable and non singular.

Write $P = TS^{-1}$; then for any measurable set E, $m(PE) = m(TS^{-1}E) = m(SS^{-1}E) = m(E)$, so that P is measure preserving and $T = PS$; it remains only to prove that S is simple.

The unit interval is the union of the three sets $\{x: S(x) = x\}$, $\{x: S(x) < x\}$, and $\{x: S(x) > x\}$, and each of the latter two, being open, is the union of countably many open intervals whose end points are in the first set. If $S(a) < a$ then the iterated images under S of the interval $S(a) < x \leqq a$ are all disjoint from one another and exhaust one of the open intervals of the set $\{x: S(x) < x\}$. For every open interval of each of the sets $\{x: S(x) < x\}$ and $\{x: S(x) > x\}$ a generating subinterval such as $S(a) < x \leqq a$ can be constructed; it is clear that the union of these generating subintervals is a wandering set whose iterated images cover all the dissipative part of the interval. This completes the proof of Theorem 6.

The formula $\int_E dS(x) = m(SE)$, (verified immediately for intervals and therefore valid for every measurable set) together with the non singularity of S shows that $S(x)$ is an absolutely continuous function of x.[16] It is clear that the derivative $\omega(x) = S'(x)$ is also the Jacobian of the transformation S, i.e. that it satisfies the relation $m(SE) = \int_E \omega(x) \, dx$. Since, moreover, P is measure preserving, the relations $m(TE) = m(PSE) = m(SE)$ show that $\omega(x)$ is also the Jacobian of T. The Jacobian of every non singular transformation of the unit interval is a positive integrable function $\omega(x)$ for which $\int_0^1 \omega(x) \, dx = 1$; conversely every such function is the Jacobian of some transformation. (For, given ω, write $S(x) = \int_0^x \omega(t) \, dt$: it is clear that S is a simple transformation whose Jacobian is ω.)

The preceding survey over all non singular transformations and their Jacobians makes it easy to propose likely looking candidates for the role of a completely

---

[15] This is an application of the Caratheodory definition of measurability, [2], p. 246.
[16] Compare also [8], Theorem 6.7, p. 227.

unbounded transformation. Consider for instance the simple (in fact dissipative) transformations S_1 and S_2 defined by $S_1 x = x^2$ and $S_2(x) = x/2$ for $0 \leq x \leq \frac{1}{2}$, $S_2(x) = (3x - 1)/2$ for $\frac{1}{2} \leq x \leq 1$, and let P be a translation of the interval modulo 1 by an irrational constant, or perhaps a mixing transformation. The proposed candidates (probably as simple as any that can be found) are those of the form PS_1 or PS_2.

UNIVERSITY OF CHICAGO

BIBLIOGRAPHY

1. G. D. BIRKHOFF AND P. A. SMITH, *Structure analysis of surface transformations*, Jour. de Math. pures et appliquées, 9. sér., vol. 7 (1928), pp. 345–379.
2. C. CARATHEODORY, *Vorlesungen über reelle Funktionen*, Leipzig, 1918.
3. P. R. HALMOS AND J. V. NEUMANN, *Operator methods in classical mechanics*, II, Annals of Mathematics, vol. 43 (1942), pp. 332–350.
4. E. HOPF, *Ergodentheorie*, Berlin, 1937.
5. E. HOPF, *Theory of measure and invariant integrals*, Trans., A. M. S., vol. 34 (1932), pp. 373–393.
6. S. KAKUTANI, *Induced measure preserving transformations*, Proc. Imp. Acad. Tokyo, vol. 19 (1943), pp. 635–641.
7. J. C. OXTOBY AND S. M. ULAM, *Measure preserving hemeomorphisms and metrical transitivity*, Annals of Math., vol. 42 (1941), pp. 874–920.
8. S. SAKS, *Theory of the integral*, Warszawa, 1937.

Reprinted from THE ANNALS OF MATHEMATICAL STATISTICS
Vol. XX, No. 2, June, 1949

APPLICATION OF THE RADON-NIKODYM THEOREM TO THE THEORY OF SUFFICIENT STATISTICS[1]

BY PAUL R. HALMOS[2] AND L. J. SAVAGE

University of Chicago

Summary. The body of this paper is written in terms of very general and abstract ideas which have been popular in pure mathematical work on the theory of probability for the last two or three decades. It seems to us that these ideas, so fruitful in pure mathematics, have something to contribute to mathematical statistics also, and this paper is an attempt to illustrate the sort of contribution we have in mind. The purpose of generality here is not to solve immediate practical problems, but rather to capture the logical essence of an important concept (sufficient statistic), and in particular to disentangle that concept from such ideas as Euclidean space, dimensionality, partial differentiation, and the distinction between continuous and discrete distributions, which seem to us extraneous.

In accordance with these principles the center of the stage is occupied by a completely abstract sample space—that is a set X of objects x, to be thought of as possible outcomes of an experimental program, distributed according to an unknown one of a certain set of probability measures. Perhaps the most familiar concrete example in statistics is the one in which X is n dimensional Cartesian space, the points of which represent n independent observations of a normally distributed random variable with unknown parameters, and in which the probability measures considered are those induced by the various common normal distributions of the individual observations.

A statistic is defined, as usual, to be a function T of the outcome, whose values, however, are not necessarily real numbers but may themselves be abstract entities. Thus, in the concrete example, the entire set of n observations, or, less trivially, the sequence of all sample moments about the origin are statistics with values in an n dimensional and in an infinite dimensional space respectively. Another illuminating and very general example of a statistic may be obtained as follows. Suppose that the outcomes of two not necessarily statistically independent programs are thought of as one united outcome—then the outcome T of the first program alone is a statistic relative to the united program. A technical measure theoretic result, known as the Radon-Nikodym theorem, is important in the study of statistics such as T. It is, for example, essential to the very definition of the basic concept of conditional probability of a subset E of X given a value y of T.

The statistic T is called sufficient for the given set $\mathfrak{M}$ of probability measures

---

[1] This paper was the basis of a lecture delivered upon invitation of the Institute at the meeting in Chicago on December 30, 1947.

[2] Fellow of the John Simon Guggenheim Memorial Foundation.

225

if (somewhat loosely speaking) the conditional probability of a subset E of X given a value y of T is the same for every probability measure in $\mathfrak{M}$. It is, for instance, well known that the sample mean and variance together form a sufficient statistic for the measures described in the concrete example.

The theory of sufficiency is in an especially satisfactory state for the case in which the set $\mathfrak{M}$ of probability measures satisfies a certain condition described by the technical term *dominated*. A set $\mathfrak{M}$ of probability measures is called dominated if each measure in the set may be expressed as the indefinite integral of a density function with respect to a fixed measure which is not itself necessarily in the set. It is easy to verify that both classical extremes, commonly referred to as the discrete and continuous cases, are dominated.

One possible formulation of the principal result concerning sufficiency for dominated sets is a direct generalization to the abstract case of the well known Fisher-Neyman result: T is sufficient if and only if the densities can be written as products of two factors, the first of which depends on the outcome through T only and the second of which is independent of the unknown measure. Another way of phrasing this result is to say that T is sufficient if and only if the likelihood ratio of every pair of measures in $\mathfrak{M}$ depends on the outcome through T only. The latter formulation makes sense even in the not necessarily dominated case but unfortunately it is not true in that case. The situation can be patched up somewhat by introducing a weaker notion called pairwise sufficiency.

In ordinary statistical parlance one often speaks of a statistic sufficient for some of several parameters. The abstract results mentioned above can undoubtedly be extended to treat this concept.

1. Basic definitions and notations. A measurable space (X, S) is a set X and a σ-algebra S of subsets of X.[3] If (X, S) and (Y, T) are measurable spaces and if T is a transformation from X into Y (or, in other words, if T is a function with domain X and range in Y), then T is *measurable* if, for every F in T, $T^{-1}(F) \in S$. If Y is a Borel set in a finite dimensional Euclidean space, then we shall always understand that T is the class of all Borel subsets of Y, and the measurability of a function f from X to Y will be expressed by the notation $f(\epsilon) S$.

Throughout most of what follows it will be assumed that (X, S) and (Y, T) are fixed measurable spaces and that T is a measurable transformation (also called a *statistic*) from X *onto* Y. A helpful example to keep in mind is the Cartesian plane in the role of X, its horizontal coordinate axis in the role of Y, and perpendicular projection from X onto Y in the role of T.

The following notations will be used. If g is a point function on Y (with arbitrary range), then gT is the point function on X defined by $gT(x) = g(T(x))$. If μ is a set function (with arbitrary range) on S, then μT^{-1} is the set function

---

[3] A σ-algebra is a non empty class S of sets, closed under the formation of complements and countable unions. If (X, S) is a measurable space, the sets of S will be called the measurable sets of X.

on T defined by $\mu T^{-1}(F) = \mu(T^{-1}(F))$. The class of all sets of the form $T^{-1}(F)$, with $F \in T$, will be denoted by $T^{-1}(T)$; the characteristic function of a set A (in any space) will be denoted by χ_A.

LEMMA 1. *If g is any function on Y and A is any set in the range of g, then*

$$\{x: gT(x) \in A\} = T^{-1}(\{y: g(y) \in A\});$$

hence, in particular, $\chi_{T^{-1}(F)} = \chi_F T$ for every subset F of Y.[4]

PROOF. The following statements are mutually equivalent: (a) $x_0 \in \{x: gT(x) \in A\}$, (b) $g(T(x_0)) \in A$, (c) if $y_0 = T(x_0)$, then $g(y_0) \in A$, and (d) $T(x_0) \in \{y: g(y) \in A\}$. The equivalence of the first and last ones of these statements is exactly the assertion of the lemma.

We shall have frequent occasion to deal with functions on X which are induced by measurable functions on Y; the following result is a useful and direct structural characterization of such functions.

LEMMA 2. *If f is a real valued function on X, then a necessary and sufficient condition that there exist a measurable function g on Y such that $f = gT$ is that $f (\epsilon) T^{-1}(T)$; if such a function g exists, then it is unique.*[5]

PROOF. The necessity of the condition is clear. To prove sufficiency, suppose that $f (\epsilon) T^{-1}(T)$, $y_0 \in Y$, and write $X_0 = T^{-1}(\{y_0\})$. Suppose $x_0 \in X_0$ and write $E = \{x: f(x) = f(x_0)\}$. Since $f (\epsilon) T^{-1}(T)$, there is a set F in T such that $E = T^{-1}(F)$. Since $x_0 \in E$, it follows that $y_0 \in F$ and therefore that

$$X_0 = T^{-1}(\{y_0\}) \subset T^{-1}(F) = E.$$

In other words f is a constant on X_0 and consequently the equation $g(y_0) = f(x_0)$ unambiguously defines a function g on Y. The facts that $f = gT$ and that g is measurable are clear; the uniqueness of g follows from the fact that T maps X onto Y.

2. Measures and their derivatives.

A *measure* is a real valued, non negative, finite (and therefore bounded), countably additive function on the measurable sets of a measurable space.[6] An integral whose domain of integration is not indicated is always to be extended over the whole space. If the symbol $[\mu]$, pronounced "modulo μ", follows an assertion concerning the points x of X, it is to be understood that the set E of those points for which the assertion is not true is such that $E \in S$ and $\mu(E) = 0$. Thus, for instance, if f and g are functions (with arbitrary range) on X, then $f = g [\mu]$ means that

---

[4] The symbol $\{ - : - \}$ stands for the set of all those objects named before the colon which satisfy the condition stated after it.

[5] The notation $f (\epsilon) T^{-1}(T)$ means of course that f is a measurable function not only on the measurable space (X, S) but also on the measurable space $(X, T^{-1}(T))$. The restriction to real valued functions is inessential and is made only in order to avoid the introduction of more notation.

[6] Although most of the measures occurring in the applications of our theory are *probability measures* (i.e. measures whose value for the whole space is 1), the consideration of probability measures only is, in many of the proofs in the sequel, both unnecessary and insufficient.

$\mu(\{x: f(x) \neq g(x)\}) = 0$. Similarly, if f is a real valued function on X, then f (ϵ) $T^{-1}(T)$ [μ] means that there exists a real valued function g on X such that g (ϵ) $T^{-1}(T)$ and $f = g$ [μ].

If μ and ν are two measures on S, ν is *absolutely continuous* with respect to μ, in symbols $\nu \ll \mu$, if $\nu(E) = 0$ for every measurable set E for which $\mu(E) = 0$. The measures μ and ν are *equivalent*, in symbols $\mu \equiv \nu$, if simultaneously $\mu \ll \nu$ and $\nu \ll \mu$.[7] One of the most useful results concerning absolute continuity is the Radon-Nikodym theorem, which may be stated as follows.[8]

A necessary and sufficient condition that $\nu \ll \mu$ is that there exist a non negative function f on X such that

$$\nu(E) = \int_E f(x) \, d\mu(x)$$

for every E in S. The function f is unique in the sense that if also

$$\nu(E) = \int_E g(x) \, d\mu(x)$$

for every E in S, then $f = g$ [μ]. If $\nu(E) \leq \mu(E)$ for every E in S, then $0 \leq f(x) \leq 1$ [μ].

It is customary and suggestive to write $f = d\nu/d\mu$. Since $d\nu/d\mu$ is determined only to within a set for which μ vanishes, it follows that in a relation of the form

$$\frac{d\nu}{d\mu} \ (\epsilon) \ T^{-1}(T) \ [\mu]$$

the symbol [μ] is superfluous and may be omitted.

For typographical and heuristic reasons it is convenient sometimes to write the relation $f = d\nu/d\mu$ in the form $d\nu = f d\mu$; all the properties of Radon-Nikodym derivatives which are suggested by the well known differential formalism correspond to true theorems. Some of the ones that we shall make use of are trivial (e.g. $d\nu_1 = f_1 d\mu$ and $d\nu_2 = f_2 d\mu$ imply $d(\nu_1 + \nu_2) = (f_1 + f_2) d\mu$), while others are well known facts in integration theory (e.g. (i) $d\lambda = f d\nu$ and $d\nu = g d\mu$ imply $d\lambda = fg d\mu$, and (ii) $d\nu = f d\mu$ and $d\mu = g d\nu$ imply $fg = 1$ [μ]).

We conclude this section with a simple but useful result concerning the transformations of integrals.

LEMMA 3. *If g is a real valued function on Y and μ is a measure on S, then*

$$\int_F g(y) \, d\mu T^{-1}(y) = \int_{T^{-1}(F)} gT(x) \, d\mu(x)$$

for every F in T, in the sense that if either integral exists, then so does the other and the two are equal.

---

[7] It is clear that the relation of equivalence is reflexive, symmetric, and transitive, and hence deserves its name.

[8] For a proof of the Radon-Nikodym theorem and similar facts concerning the measure and integration theory which we employ, see S. Saks, *Theory of the Integral*, Warszawa—Lwów, 1937.

PROOF. Replacing g by $g\chi_F$ we see that it is sufficient to consider the case $F = Y$. The proof for this case follows from the observation that every approximating sum

$$\Sigma_i \, g(y_i)\mu T^{-1}(F_i)$$

of $\int g d\mu T^{-1}$ is also an approximating sum

$$\Sigma_i \, g T(x_i)\mu(E_i)$$

of $\int g T d\mu$, and conversely.[9]

3. Conditional probabilities and expectations.

LEMMA 4. *If μ and ν are measures on S such that $\nu \ll \mu$, then $\nu T^{-1} \ll \mu T^{-1}$.*

PROOF. If $F \in T$ and $0 = \mu T^{-1}(F) = \mu(T^{-1}(F))$, then

$$0 = \nu(T^{-1}(F)) = \nu T^{-1}(F).\text{[10]}$$

Lemma 4 is the basis of the definition of a concept of great importance in probability theory. If μ is a measure on S and f is a non negative integrable function on X, then the measure ν defined by $d\nu = f d\mu$ is absolutely continuous with respect to μ. It follows from Lemma 4 that νT^{-1} is absolutely continuous with respect to μT^{-1}; we write $d\nu T^{-1} = g d\mu T^{-1}$. The function value $g(y)$ is known as the *conditional expectation* of f given y (or given that $T(x) = y$); we shall denote it by $e_\mu(f \mid y)$. If $f = \chi_E$ is the characteristic function of a set E in S, then $e_\mu(f \mid y)$ is known as the *conditional probability* of E given y; we shall denote it by $p_\mu(E \mid y)$.[11]

The abstract nature of these definitions makes an intuitive justification of them desirable. Observe that since $\nu T^{-1}(F) = \nu(T^{-1}(F)) = \int_{T^{-1}(F)} f(x) \, d\mu(x)$, the defining equation of $e_\mu(f \mid y)$, written out in full detail, takes the form

$$\int_{T^{-1}(F)} f(x) \, d\mu(x) = \int_F e_\mu(f \mid y) \, d\mu T^{-1}(y), \qquad F \in T.$$

---

[9] It is of interest to observe that either side of the equation in Lemma 3 may be obtained from the other by the formal substitution $y = T(x)$. A special case of this lemma is the celebrated and often misunderstood assertion that the expectation of a random variable is equal to the first moment of its distribution function.

[10] That the converse of Lemma 4 is not true is shown by the following example. Let X be the unit square, let Y be the unit interval, and let T be the perpendicular projection from X onto Y. Let μ be ordinary (Borel-Lebesgue) measure and let ν be linear measure on the intersection of X with, say, the horizontal line whose ordinate is $\frac{1}{2}$. Clearly ν is not absolutely continuous with respect to μ, but $\nu T^{-1} = \mu T^{-1}$.

[11] Definitions in this form were first proposed by A. Kolmogoroff, *Grundbegriffe der Wahrscheinlichkeitsrechnung*, Berlin, 1933. With a slight amount of additional trouble, conditional expectation could be defined for more general functions, but only the non negative case will occur in our applications.

If $f = \chi_E$, then this equation becomes the defining equation of $p_\mu(E \mid y)$:

$$\mu(E \cap T^{-1}(F)) = \int_F p_\mu(E \mid y)\, d\mu T^{-1}(y), \qquad F \epsilon T.^{12}$$

The customary definition of "the conditional probability of E given that $T(x) \epsilon F$" is $\mu(E \cap T^{-1}(F))/\mu(T^{-1}(F))$, (assuming that the denominator does not vanish). Since $\mu(T^{-1}(F)) = \mu T^{-1}(F)$, we have

$$\frac{\mu(E \cap T^{-1}(F))}{\mu(T^{-1}(F))} = \frac{1}{\mu T^{-1}(F)} \int_F p_\mu(E \mid y)\, d\mu T^{-1}(y).$$

It is now formally plausible that if "F shrinks to a point y," then the left side of the last written equation should tend to the conditional probability of E given y and the right side should tend to the integrand $p_\mu(E \mid y)$. The use of the Radon-Nikodym differentiation theorem is a rigorous substitute for this rather shaky difference quotient approach.

Since $p_\mu(E \mid y)$ is determined, for each E, only to within a set for which μT^{-1} vanishes, it would be too optimistic to expect that, for each y, it behaves, regarded as a function of E, like a measure. It is, however, easy to prove that

(i) $\qquad p_\mu(X \mid y) = 1 \; [\mu T^{-1}]$,

(ii) $\qquad 0 \leq p_\mu(E \mid y) \leq 1 \; [\mu T^{-1}]$,

(iii) $\qquad$ if $\{E_n\}$ is a disjoint sequence of measurable sets, then $p_\mu(\cup_{n=1}^\infty E_n \mid y) = \sum_{n=1}^\infty p_\mu(E_n \mid y) \; [\mu T^{-1}].^{13}$

The exceptional sets of measure zero depend in general on E in (ii) and on the particular sequence $\{E_n\}$ in (iii). It is interesting to observe that, despite the fact that μ need not be a probability measure, p_μ turns out always to have the normalization property (i). It is natural to ask whether or not the indeterminacy of $p_\mu(E \mid y)$ may be resolved, for each E, in such a way that the resulting function is a measure for each y, except possibly for a fixed set of y's on which μT^{-1} vanishes. Doob[14] has shown that this is the case when X is the real line; in the general case such a resolution is impossible. Fortunately, however, conditional probabilities are sufficiently tractable for most practical and theoretical purposes, and the requirement that they should behave like probability measures in the strict sense described above is almost never needed.

---

[12] We observe that it is not sufficient to require this for $F = Y$ only, i.e. to require $\mu(E) = \int p_\mu(E \mid y)\, d\mu T^{-1}(y)$. This special equation is satisfied by many functions which do not deserve the name conditional probability; e.g. it is satisfied by $p_\mu(E \mid y) =$ constant $= \mu(E)/\mu T^{-1}(Y)$.

[13] See J. L. Doob, "Stochastic processes with an integral-valued parameter," *Am. Math. Soc. Trans.*, Vol. 44 (1938), pp. 95–98.

[14] See Doob, *loc. cit.* Doob asserts the theorem in much greater generality, but his proof is incorrect. The error in the proof and a counterexample to the general theorem were communicated to us by J. Dieudonné in a letter dated September 4, 1947. Doob's proof is valid for more general spaces than the real line (e.g. for finite dimensional Euclidean spaces and for compact metric spaces). The details of Dieudonné's counterexample will appear in a forthcoming book (entitled *Measure theory*) by Halmos.

We conclude this section with two easy but useful results which might also serve as illustrations of the method of finding conditional probabilities and expectations in certain special cases.

LEMMA 5. *If μ is a measure on S, if g is a non negative function on Y, integrable with respect to μT^{-1}, and if ν is the measure on S defined by $d\nu = gT d\mu$, then $d\nu T^{-1} = g d\mu T^{-1}$, or, equivalently, $e_\mu(gT \mid y) = g(y) \; [\mu T^{-1}]$.*

PROOF. From $\nu(E) = \displaystyle\int_E gT(x) \, d\mu(x)$ and Lemma 3 it follows that

$$\nu T^{-1}(F) = \nu(T^{-1}(F)) = \int_F g(y) \, d\mu T^{-1}(y).$$

LEMMA 6. *If μ is a measure on S, if f and g are non negative functions on X and Y respectively, and if f, gT, and $f \cdot gT$ are all integrable with respect to μ, then*

$$e_\mu(f \cdot gT \mid y) = e_\mu(f \mid y) \cdot g(y) \; [\mu T^{-1}].$$

Hence, in particular, if $F \in T$, then

$$p_\mu(E \cap T^{-1}(F) \mid y) = p_\mu(E \mid y) \chi_F(y) \; [\mu T^{-1}]$$

for every E in S.

PROOF. If $d\nu = f d\mu$, then, by definition of e_μ, $\nu T^{-1}(F) = \displaystyle\int_F e_\mu(f \mid y) \, d\mu T^{-1}(y)$. Applications of Lemmas 3 and 5 yield

$$\int_F e_\mu(f \mid y) g(y) \, d\mu T^{-1}(y) = \int_F g(y) \, d\nu T^{-1}(y) = \int_{T^{-1}(F)} gT(x) \, d\nu(x)$$

$$= \int_{T^{-1}(F)} f(x) gT(x) \, d\mu(x) = \int_F e_\mu(f \cdot gT \mid y) \, d\mu T^{-1}(y),$$

and therefore the desired conclusion follows from the uniqueness assertion of the Radon-Nikodym theorem.

4. Dominated sets of measures. In many statistical situations it is necessary to consider simultaneously several measures on the same σ-algebra. The concept of absolute continuity is easily extended to sets of measures. If $\mathfrak{M}$ and $\mathfrak{N}$ are two sets of measures on S and if, for every set E in S, the vanishing of $\mu(E)$ for every μ in $\mathfrak{M}$ implies the vanishing of $\nu(E)$ for every ν in $\mathfrak{N}$, then we shall call $\mathfrak{N}$ absolutely continuous with respect to $\mathfrak{M}$ and write $\mathfrak{N} \ll \mathfrak{M}$. If $\mathfrak{N} \ll \mathfrak{M}$ and $\mathfrak{M} \ll \mathfrak{N}$, the sets $\mathfrak{M}$ and $\mathfrak{N}$ are called equivalent and we write $\mathfrak{M} \equiv \mathfrak{N}$. If, in particular, $\mathfrak{M}$ contains exactly one measure μ, $\mathfrak{M} = \{\mu\}$, the abbreviated notations $\mathfrak{N} \ll \mu$, $\mu \ll \mathfrak{N}$, and $\mu \equiv \mathfrak{N}$, will be employed for $\mathfrak{N} \ll \mathfrak{M}$, $\mathfrak{M} \ll \mathfrak{N}$, and $\mathfrak{M} \equiv \mathfrak{N}$, respectively.

A set $\mathfrak{M}$ of measures on S will be called *dominated* if there exists a measure λ on S (not necessarily in $\mathfrak{M}$) such that $\mathfrak{M} \ll \lambda$. In applications there frequently occur sets of measures which are dominated in a sense apparently weaker than the one just defined—weaker in that the measure λ, which may for instance be

Lebesgue measure on the Borel sets of a finite dimensional Euclidean space, is not necessarily finite. It is easy to see, however, that whenever λ has the property (possessed by Lebesgue measure) that the space X is the union of countably many sets of finite measure, then a finite measure equivalent to λ exists and the two possible definitions of domination coincide.

The following result on dominated sets of measures may be found to have some interest of its own and will be applied in the sequel.

LEMMA 7. *Every dominated set of measures has an equivalent countable subset.*

PROOF. Let $\mathfrak{M}$ be a dominated set of measures on S, $\mathfrak{M} \ll \lambda$; for any μ in $\mathfrak{M}$ write $f_\mu = d\mu/d\lambda$ and $K_\mu = \{x : f_\mu(x) > 0\}$. We define (for the purposes of this proof only) a *kernel* as a set K in S such that, for some measure μ in $\mathfrak{M}$, $K \subset K_\mu$ and $\mu(K) > 0$; we define a *chain* as a disjoint union of kernels. Since $\lambda(K) > 0$ for every kernel K, it follows from the finiteness of λ that every chain is a *countable* disjoint union of kernels. It follows also from these definitions that if C is a measurable subset of a chain, such that $\mu(C) > 0$ for at least one measure μ in $\mathfrak{M}$, then C is a chain, and that a disjoint union of chains is a chain. The last two remarks imply, through the usual process of disjointing any countable union, that a countable (but not necessarily disjoint) union of chains is a chain.

Let $\{C_j\}$ be a sequence of chains such that, as $j \to \infty$, $\lambda(C_j)$ approaches the supremum of the values of λ on chains. If $C = \bigcup_{j=1}^{\infty} C_j$, then C is a chain for which $\lambda(C)$ is maximal. The definition of a chain yields the existence of a sequence $\{K_i\}$ of kernels such that $C = \bigcup_{i=1}^{\infty} K_i$, and the definition of a kernel yields the existence, for each $i = 1, 2, \cdots$, of a measure μ_i in $\mathfrak{M}$ such that $K_i \subset K_{\mu_i}$ and $\mu_i(K_i) > 0$. We write $\mathfrak{N} = \{\mu_1, \mu_2, \cdots\}$; since $\mathfrak{N} \subset \mathfrak{M}$, the relation $\mathfrak{N} \ll \mathfrak{M}$ is trivial. We shall prove that $\mathfrak{M} \ll \mathfrak{N}$.

Suppose that $E \in S$, $\mu_i(E) = 0$ for $i = 1, 2, \cdots$, and let μ be any measure in $\mathfrak{M}$. It is to be proved that $\mu(E) = 0$. Since $\mu(E - K_\mu) = 0$, there is no loss of generality in assuming that $E \subset K_\mu$. If $\mu(E - C) > 0$, then $\lambda(E - C) > 0$ and therefore (since $E - C$ is a kernel) $E \cup C$ is a chain with $\lambda(E \cup C) > \lambda(C)$. Since this is impossible, it follows that $\mu(E - C) = 0$. Since $0 = \mu_i(E) = \mu_i(E \cap K_i) = \int_{E \cap K_i} f_{\mu_i} \, d\lambda$ and since $K_i \subset K_{\mu_i}$, it follows that $\lambda(E \cap K_i) = 0$. We conclude that $\lambda(E \cap C) = \sum_{i=1}^{\infty} \lambda(E \cap K_i) = 0$ and therefore $\mu(E \cap C) = 0$. Since $\mu(E) = \mu(E - C) + \mu(E \cap C)$, the proof of the lemma is complete.

5. Sufficient statistics for dominated sets. The statistic T is sufficient for a set $\mathfrak{M}$ of measures on S if, for every E in S, there exists a measurable function $p = p(E \mid y)$ on Y, such that

$$p_\mu(E \mid y) = p(E \mid y) \, [\mu T^{-1}]$$

for every μ in $\mathfrak{M}$.[15] In other words, T is sufficient for $\mathfrak{M}$ if there exists a condi-

---

[15] The original definition of sufficiency was given by R. A. Fisher, "On the mathematical foundations of theoretical statistics," *Roy. Soc. Phil. Trans.*, Series A, Vol. 222 (1922), pp. 309–368.

tional probability function common to every μ in $\mathfrak{M}$, or, crudely speaking, if the conditional distribution induced by T is independent of μ.

THEOREM 1. *A necessary and sufficient condition that the statistic T be sufficient for a dominated set $\mathfrak{M}$ of measures on S is that there exist a measure λ on S such that $\mathfrak{M} \equiv \lambda$ and such that $d\mu/d\lambda$ (ϵ) $T^{-1}(T)$ for every μ in $\mathfrak{M}$.*

Proof of necessity. Let $\mathfrak{N} = \{\mu_1, \mu_2, \cdots\}$ be a countable subset equivalent to $\mathfrak{M}$ (Lemma 7), and write λ for the measure on S defined by

$$\lambda(E) = \sum_{i=1}^{\infty} a_i\mu_i(E),$$

where $a_i = 1/2^i\mu_i(X)$, $i = 1, 2, \cdots$. Clearly $\mathfrak{M} \equiv \lambda$.

If p is a conditional probability function common to every μ in $\mathfrak{M}$, then, for every F in $\boldsymbol{T}$,

$$\lambda(E \cap T^{-1}(F)) = \sum_{i=1}^{\infty} a_i\mu_i(E \cap T^{-1}(F))$$

$$= \sum_{i=1}^{\infty} a_i \int_F p(E \mid y) \, d\mu_i T^{-1}(y) = \int_F p(E \mid y) \, d\lambda T^{-1}(y),$$

i.e. p serves also as a conditional probability for λ.

Take any fixed μ in $\mathfrak{M}$, write $d\mu/d\lambda = f$, and $e_\lambda(f \mid y) = g(y)$; then $d\mu T^{-1} = g d\lambda T^{-1}$, and we have, for every E in $\boldsymbol{S}$,

$$\int_E f(x) \, d\lambda(x) = \mu(E) = \int p(E \mid y) \, d\mu T^{-1}(y)$$

$$= \int p(E \mid y)g(y) \, d\lambda T^{-1}(y) = \int e_\lambda(\chi_E \mid y)e_\lambda(gT \mid y) \, d\lambda T^{-1}(y)$$

$$= \int e_\lambda(\chi_E \cdot gT \mid y) \, d\lambda T^{-1}(y) = \int \chi_E(x)gT(x) \, d\lambda(x) = \int_E gT(x) \, d\lambda(x).$$

The desired result, $f(x) = gT(x)$ [λ], follows from a comparison of the first and last terms in the last written chain of equations.

Proof of Sufficiency. We shall prove that p_λ is a conditional probability function common to every μ in $\mathfrak{M}$. Take any fixed E in $\boldsymbol{S}$ and μ in $\mathfrak{M}$ and write $d\mu/d\lambda = gT$. If the measure ν is defined by $d\nu = \chi_E d\mu$, then $d\nu T^{-1} = p_\mu d\mu T^{-1}$, where $p_\mu = p_\mu(E \mid y)$. The hypothesis $d\mu = gTd\lambda$ implies that $d\mu T^{-1} = gd\lambda T^{-1}$ and hence that

$$d\nu T^{-1} = p_\mu \cdot gd\lambda T^{-1}.$$

On the other hand $d\nu = \chi_E d\mu = \chi_E \cdot gTd\lambda$, so that

$$d\nu T^{-1} = e_\lambda d\lambda T^{-1},$$

where $e_\lambda = e_\lambda(\chi_E \cdot gT \mid y) = p_\lambda(E \mid y)g(y)$. It follows from a comparison of the two expressions for $d\nu T^{-1}$ that

$$p_\mu(E \mid y)g(y) = p_\lambda(E \mid y)g(y) \quad [\lambda T^{-1}].$$

Since the relation $d\mu T^{-1} = gd\lambda T^{-1}$ clearly implies that $g(y) \neq 0 \ [\mu T^{-1}]$ (i.e. that $\mu T^{-1}(\{y: g(y) = 0\}) = 0$), it follows, finally that

$$p_\mu(E \mid y) = p_\lambda(E \mid y) \ [\mu T^{-1}].$$

6. Special criteria for sufficiency. Theorem 1 may be recast in a form more akin in spirit to previous investigations of the concept of sufficiency.[16]

COROLLARY 1. *A necessary and sufficient condition that the statistic T be sufficient for a dominated set $\mathfrak{M}$ ($\ll \lambda_0$) of measures on S is that, for every μ in $\mathfrak{M}$, $f_\mu = d\mu/d\lambda_0$ be factorable in the form $f_\mu = g_\mu \cdot t$, where $0 \leq g_\mu$ (ϵ) $T^{-1}(T)$, $0 \leq t$, t and $g_\mu \cdot t$ are integrable with respect to λ_0, and t vanishes $[\lambda_0]$ on each set in S for which every μ in $\mathfrak{M}$ vanishes.*

In more customary statistical language the condition asserts essentially that "each density is factorable into a function of the statistic alone and a function independent of the parameter."

PROOF. If T is sufficient for $\mathfrak{M}$, then there exists a measure λ with the properties described in Theorem 1. It follows that

$$f_\mu = \frac{d\mu}{d\lambda_0} = \frac{d\mu}{d\lambda}\frac{d\lambda}{d\lambda_0}$$

and we may write $g_\mu = d\mu/d\lambda$ and $t = d\lambda/d\lambda_0$. The only assertion that is not immediately obvious is the one concerning the vanishing of t. To prove it, suppose that $\mu(E) = 0$ for every μ in $\mathfrak{M}$; the fact that then

$$0 = \lambda(E) = \int_E t(x) \, d\lambda_0(x)$$

implies the desired conclusion.

If, conversely, $f_\mu = g_\mu \cdot t$, then we may write $d\lambda = td\lambda_0$. The relation $\mathfrak{M} \equiv \lambda$ follows from the statement concerning the vanishing of t, and the relation $d\mu/d\lambda$ (ϵ) $T^{-1}(T)$ is implied by the equation $d\mu = g_\mu \cdot td\lambda_0 = g_\mu d\lambda$.

For the statement of the next consequence of Theorem 1 it is convenient to call a set $\mathfrak{M}$ of measures on S *homogeneous* if $\mu \equiv \nu$ for every μ and ν in $\mathfrak{M}$.

COROLLARY 2. *A necessary and sufficient condition that the statistic T be sufficient for a homogeneous set $\mathfrak{M}$ of measures on S is that, for every μ and ν in $\mathfrak{M}$, $d\nu/d\mu$ (ϵ) $T^{-1}(T)$.*

PROOF. Since a homogeneous set is dominated (by any one of its elements), Theorem 1 is applicable. If T is sufficient for $\mathfrak{M}$ and if λ has the properties described in Theorem 1, then $d\nu/d\mu = (d\nu/d\lambda)/(d\mu/d\lambda)$. The converse follows, through Theorem 1, by letting λ be any measure in $\mathfrak{M}$.

We shall say that the statistic T is *pairwise sufficient* for a set $\mathfrak{M}$ of measures

---

[16] See J. Neyman, "Su un teorema concernente le cosiddette statistiche sufficienti," *Inst. Ital. Atti. Giorn.*, Vol. 6 (1935), pp. 320–334. In this paper Neyman is somewhat restricted by his use of classical analytical methods, but he points out the possibility and desirability of extending his results to a much more general domain. For a recent presentation of the theory and further references to the literature cf. H. Cramér, *Mathematical Methods of Statistics*, Princeton, 1946.

on S if it is sufficient for every pair $\{\mu, \nu\}$ of measures in $\mathfrak{M}$. In other words, T is pairwise sufficient for $\mathfrak{M}$ if, for every E in S and μ and ν in $\mathfrak{M}$, there exists a measurable function $p_{\mu\nu}(E \mid y)$ on Y such that

$$p_\mu(E \mid y) = p_{\mu\nu}(E \mid y) \; [\mu T^{-1}] \quad \text{and} \quad p_\nu(E \mid y) = p_{\mu\nu}(E \mid y) \; [\nu T^{-1}].$$

Since pairwise sufficiency is (at least apparently) weaker than sufficiency, it is not surprising that there is a simple criterion for it even in the case of quite arbitrary (not necessarily homogeneous or dominated) sets of measures.

COROLLARY 3. *A necessary and sufficient condition that T be pairwise sufficient for a set $\mathfrak{M}$ of measures on S is that, for any two measures μ and ν in $\mathfrak{M}$, $d\mu/d(\mu + \nu)$ (ϵ) $T^{-1}(T)$.*

PROOF. If T is sufficient for μ and ν, then there exists a measure $\lambda \equiv \mu + \nu$ such that $d\mu/d\lambda$ (ϵ) $T^{-1}(T)$ and $d\nu/d\lambda$ (ϵ) $T^{-1}(T)$. It follows that

$$\frac{d\mu}{d(\mu + \nu)} = \frac{d\mu}{d\lambda} \bigg/ \frac{d(\mu + \nu)}{d\lambda} = \frac{d\mu}{d\lambda} \bigg/ \left(\frac{d\mu}{d\lambda} + \frac{d\nu}{d\lambda} \right).$$

The sufficiency of the condition follows immediately by applying Theorem 1 to the two-element set $\{\mu, \nu\}$.

7. Pairwise sufficiency and likelihood ratios. It is sometimes convenient to express the result of Corollary 3 in slightly different language. If λ is a measure on S and if f and g are real valued measurable functions on X such that $\lambda(\{x: f(x) = g(x) = 0\}) = 0$, we shall say that the pair (f, g) is *admissible* $[\lambda]$. (Intuitively an admissible pair (f, g) is to be thought of as a ratio f/g, which, however, may not be formed directly at the points x for which $g(x) = 0$.) Two admissible pairs (f_1, g_1) and (f_2, g_2) will be called *equivalent* $[\lambda]$, in symbols $(f_1, g_1) \equiv (f_2, g_2). [\lambda]$, if there exists a real valued measurable function t on X such that $t(x) \neq 0$ $[\lambda]$ and such that $f_1 = tf_2$ and $g_1 = tg_2$ $[\lambda]$. It is clear that the relation " $\equiv [\lambda]$" is indeed an equivalence; the equivalence class containing the admissible pair (f, g) will be called the *ratio* of f and g and will be denoted by $f \mid g$. (A ratio may accordingly be described as a measurable function from X to the real projective line.) For a ratio $f \mid g$ we shall write $f \mid g$ (ϵ) $T^{-1}(T)$ $[\lambda]$ if the equivalence class $f \mid g$ contains a pair (f_0, g_0) which is admissible $[\lambda]$ and for which f_0 (ϵ) $T^{-1}(T)$ and g_0 (ϵ) $T^{-1}(T)$.

LEMMA 8. *If μ, ν, λ_1, and λ_2 are measures on S such that $\mu + \nu \ll \lambda_1$ and $\mu + \nu \ll \lambda_2$, then the pairs $(d\mu/d\lambda_1, d\nu/d\lambda_1)$ and $(d\mu/d\lambda_2, d\nu/d\lambda_2)$ are admissible $[\mu + \nu]$ and equivalent $[\mu + \nu]$.*

PROOF. The admissibility of, for instance, $(d\mu/d\lambda_1, d\nu/d\lambda_1)$ follows from the fact that $d\mu/d\lambda_1 \neq 0$ $[\mu]$ and $d\nu/d\lambda_1 \neq 0$ $[\nu]$, whence

$$(\mu + \nu)\left(\left\{ x: \frac{d\mu}{d\lambda_1}(x) = \frac{d\nu}{d\lambda_1}(x) = 0 \right\} \right) = 0.$$

To prove equivalence, we write $\lambda_1 + \lambda_2 = \lambda$. Since

$$\frac{d\mu}{d\lambda_1}\frac{d\lambda_1}{d\lambda} = \frac{d\mu}{d\lambda} = \frac{d\mu}{d\lambda_2}\frac{d\lambda_2}{d\lambda}, \qquad \frac{d\nu}{d\lambda_1}\frac{d\lambda_1}{d\lambda} = \frac{d\nu}{d\lambda} = \frac{d\nu}{d\lambda_2}\frac{d\lambda_2}{d\lambda},$$

since also $d\lambda_1/d\lambda \neq 0$ $[\lambda_1]$ and therefore $d\lambda_1/d\lambda \neq 0$ $[\mu + \nu]$, and since, similarly, $d\lambda_2/d\lambda \neq 0$ $[\mu + \nu]$, the conditions of the definition of equivalence are satisfied by $t = (d\lambda_2/d\lambda)/(d\lambda_1/d\lambda)$.

If μ and ν are any two measures on S and if λ is any measure on S such that $\mu + \nu \ll \lambda$ (for instance if $\lambda = \mu + \nu$), then the ratio $d\mu/d\lambda \mid d\nu/d\lambda$, which according to Lemma 8 exists $[\mu + \nu]$ and is independent of λ, will be called the *likelihood ratio* of μ and ν and will be denoted by $d\mu \mid d\nu$. The result of Corollary 3 may be expressed in terms of likelihood ratios as follows.

THEOREM 2. *A necessary and sufficient condition that T be pairwise sufficient for a set $\mathfrak{M}$ of measures on S is that, for any two measures μ and ν in $\mathfrak{M}$, $d\mu \mid d\nu$ (ϵ) $T^{-1}(\mathbf{T})$.*

PROOF. If T is sufficient for μ and ν, then, by Corollary 3, $d\mu/d(\mu + \nu)$ (ϵ) $T^{-1}(\mathbf{T})$, $d\nu/d(\mu + \nu)$ (ϵ) $T^{-1}(\mathbf{T})$, and, by Lemma 8, $(d\mu/d(\mu + \nu), d\nu/d(\mu + \nu))$ is an admissible pair belonging to the equivalence class $d\mu \mid d\nu$. Suppose conversely that $f = d\mu/d(\mu + \nu)$, $g = d\nu/d(\mu + \nu)$, and let the real valued measurable functions t, f_0, and g_0 be such that $t \neq 0$ $[\mu + \nu]$, f_0 (ϵ) $T^{-1}(\mathbf{T})$, g_0 (ϵ) $T^{-1}(\mathbf{T})$, (f_0, g_0) is admissible $[\mu + \nu]$, and

$$f = t \cdot f_0, \qquad g = t \cdot g_0 \ [\mu + \nu].$$

Since f and g are non negative, it follows that $f = \mid t \mid \cdot \mid f_0 \mid$ and $g = \mid t \mid \cdot \mid g_0 \mid$ $[\mu + \nu]$, i.e. that there is no loss of generality in assuming that t, f_0, and g_0 are non negative. The relation $f + g = 1$ $[\mu + \nu]$ implies that $t \cdot (f_0 + g_0) = 1$ $[\mu + \nu]$; the fact that (f_0, g_0) is admissible $[\mu + \nu]$ then yields $t \epsilon T^{-1}(\mathbf{T})$. The proof is completed by comparing this result with the expressions for f and g in terms of f_0 and g_0 and applying Corollary 3.

8. Pairwise sufficiency versus sufficiency. In order to show that our results on pairwise sufficiency (in the preceding section and in the sequel) are not vacuous, we proceed now to exhibit a statistic which is, for a suitable set of measures, pairwise sufficient but not sufficient.

Let $X = \{(x, i): 0 \leq x \leq 1, i = 0, 1\}$ be the union of two unit intervals and let $Y = \{y: 0 \leq y \leq 1\}$ be a unit interval. In accordance with our basic convention, measurability in both X and Y is to be taken in the sense of Borel. The statistic T is defined by $T(x, i) = x$.

Write $X_0 = \{(x, 0): 0 \leq x \leq 1\}$ and $X_1 = \{(x, 1): 0 \leq x \leq 1\}$. Let μ be (linear) Lebesgue measure on the class S of Borel subsets of X, and define, whenever $E \epsilon S$ and $0 \leq \alpha \leq 1$,

$$\mu_\alpha(E) = \tfrac{1}{2}[\mu(E \cap X_0) + \chi_{E \cap X_1}(\alpha, 1)].$$

Let ν be (linear) Lebesgue measure on the class T of Borel subsets of Y, and define, whenever $F \epsilon T$ and $0 \leq \alpha \leq 1$,

$$\nu_\alpha(F) = \tfrac{1}{2}[\nu(F) + \chi_F(\alpha)].$$

Clearly $\nu_\alpha = \mu_\alpha T^{-1}$; we write $\mathfrak{M} = \{\mu_\alpha : 0 \leq \alpha \leq 1\}$.

If $\delta(y, \alpha)$ is defined to be 1 or 0 according as $y = \alpha$ or $y \neq \alpha$, if $\delta'(y, \alpha) = 1 - \delta(y, \alpha)$, and if

$$p_\alpha(E \mid y) = \delta'(y, \alpha)\chi_E(y, 0) + \delta(y, \alpha)\chi_E(y, 1),$$

then a straightforward computation shows that

$$\mu_\alpha(E \cap T^{-1}(F)) = \int_F p_\alpha(E \mid y) \, d\nu_\alpha(y),$$

so that $p_\alpha(E \mid y) = p_{\mu_\alpha}(E \mid y) \, [\nu_\alpha]$.

It is now easy to verify that T is pairwise sufficient for $\mathfrak{M}$. Indeed if α and β are any two different numbers in the closed unit interval, we may write

$$p(E \mid y) = \delta'(y, \alpha)\delta'(y, \beta)\chi_E(y, 0) + [\delta(y, \alpha) + \delta(y, \beta)]\chi_E(y, 1).$$

Since $\{y: p(E \mid y) \neq p_\alpha(E \mid y)\} = \{\beta\}$ and $\{y: p(E \mid y) \neq p_\beta(E \mid y)\} = \{\alpha\}$, it follows that $p(E \mid y) = p_\alpha(E \mid y) \, [\nu_\alpha]$ and $p(E \mid y) = p_\beta(E \mid y) \, [\nu_\beta]$.

To prove that T is not sufficient for $\mathfrak{M}$ we observe that $p_\alpha(X_1 \mid y) = \delta(y, \alpha)\chi_{X_1}(y, 1) = \delta(y, \alpha)$ and therefore

$$p_{\mu_\alpha}(X_1 \mid y) = \delta(y, \alpha) \, [\nu_\alpha].$$

Suppose that there is a conditional probability function p such that $p(E \mid y) = p_{\mu_\alpha}(E \mid y) \, [\nu_\alpha]$. Then, in particular,

$$p(X_1 \mid y) = \delta(y, \alpha) \, [\nu_\alpha].$$

Since $\nu_\alpha(\{\alpha\}) = \frac{1}{2} > 0$, it follows that

$$p(X_1 \mid \alpha) = \delta(\alpha, \alpha) = 1,$$

or, changing to a more suggestive notation, that $p(X_1 \mid y) = 1$ for all y. We have, however,

$$\nu_\alpha(\{y: p_\alpha(X_1 \mid y) = 0\}) = \nu_\alpha(\{y: \delta(y, \alpha) = 0\})$$

$$= \nu_\alpha(\{y: y \neq \alpha\}) = \tfrac{1}{2},$$

so that $\nu_\alpha(\{y: p_{\mu_\alpha}(X_1 \mid y) = 0\}) = \frac{1}{2}$. This contradiction shows the impossibility of the existence of a conditional probability function common to every μ in $\mathfrak{M}$.

This example shows also that, in a sense, sufficiency is more fundamental than pairwise sufficiency. If, for instance, we imagine that it is important to a statistician that he either estimate α sharply or refrain from estimating it altogether, then he is by no means as well off with the observation of y as with that of x.

9. Pairwise sufficiency for dominated sets. We now proceed to show that for dominated sets of measures no such example as the one in the preceding section exists, or, in other words, that for dominated sets the concepts of pairwise sufficiency and sufficiency do coincide.

LEMMA 9. *If T is pairwise sufficient for a set $\{\mu_0, \mu_1, \mu_2\}$ of three measures on S, then*[17]

$$\frac{d\mu_0}{d(\mu_0 + \mu_1 + \mu_2)} \; (\epsilon) T^{-1}(T).$$

PROOF. According to Corollary 3,

$$f_1 = \frac{d\mu_0}{d(\mu_0 + \mu_1)} \; (\epsilon) T^{-1}(T) \quad \text{and} \quad f_2 = \frac{d\mu_0}{d(\mu_0 + \mu_2)} \; (\epsilon) T^{-1}(T).$$

Since $d\mu_0 = f_1 d(\mu_0 + \mu_1) = f_2 d(\mu_0 + \mu_2)$, we have $f_1 d\mu_0 = f_1 f_2 d(\mu_0 + \mu_2)$ and $f_2 d\mu_0 = f_1 f_2 d(\mu_0 + \mu_1)$, so that

$$(f_1 + f_2 - f_1 f_2) d\mu_0 = f_1 f_2 d(\mu_0 + \mu_1 + \mu_2).$$

If we write $d\mu_0 = f d(\mu_0 + \mu_1 + \mu_2)$, then it follows that

$$(f_1 + f_2 - f_1 f_2) f = f_1 f_2 [\mu_0 + \mu_1 + \mu_2].$$

Since $0 \leq f_1 \leq 1$ and $0 \leq f_2 \leq 1$, the equation $f_1 + f_2 - f_1 f_2 = 0$ is equivalent to $f_1 = f_2 = 0$. Since $\mu_0(\{x: f_1(x) = f_2(x) = 0\}) = 0$, it follows that f may be redefined, if necessary, to be 0 on the set $\{x: f_1(x) = f_2(x) = 0\}$ without affecting the relation $d\mu_0 = f d(\mu_0 + \mu_1 + \mu_2)$; since outside this set $f = f_1 f_2 / (f_1 + f_2 - f_1 f_2)$, the proof of the lemma is complete.

LEMMA 10. *If T is pairwise sufficient for a finite set $\{\mu_0, \mu_1, \cdots, \mu_k\}$ of measures on S, then $d\mu_0/d(\sum_{i=0}^{k} \mu_i) \; (\epsilon) \; T^{-1}(T)$.*

PROOF. For $k = 1$ the conclusion is a restatement of the hypothesis; we proceed by induction. Given $\mu_0, \mu_1, \cdots, \mu_{k+1}$, we write $\mu = \sum_{i=1}^{k} \mu_i$. Then $d\mu_0/d(\mu_0 + \mu) \; (\epsilon) \; T^{-1}(T)$ by the induction hypothesis and $d\mu_0/d(\mu_0 + \mu_{k+1}) \; (\epsilon) \; T^{-1}(T)$ by Corollary 3. Lemma 9 may then be applied to $\{\mu_0, \mu, \mu_{k+1}\}$ and yields the desired conclusion.

LEMMA 11. *If $\{\mu_0, \mu_1, \mu_2 \cdots\}$ is a sequence of measures on S such that $\sum_{i=0}^{\infty} \mu_i(X) < \infty$; if, for every E in S, $\mu(E) = \sum_{i=0}^{\infty} \mu_i(E)$; and if λ is a measure S such that $\mu_i \ll \lambda$ for $i = 0, 1, 2, \cdots$, then*

$$\lim_k d(\textstyle\sum_{i=0}^{k} \mu_i)/d\lambda = d\mu/d\lambda \; [\lambda].$$

PROOF. Since $0 \leq d(\sum_{i=0}^{k} \mu_i)/d\lambda = \sum_{i=0}^{k} (d\mu_i/d\lambda) \leq d\mu/d\lambda \; [\lambda]$, the series $\sum_{i=0}^{\infty} (d\mu_i/d\lambda)$ does indeed converge to a measurable function $f \; [\lambda]$. Since, for every E in S,

$$\int_E f \, d\lambda = \sum_{i=0}^{\infty} \int_E \frac{d\mu_i}{d\lambda} \, d\lambda = \sum_{i=0}^{\infty} \mu_i(E) = \mu(E),$$

we have $f = d\mu/d\lambda \; [\lambda]$, as stated.

---

[17] In view of Theorem 1, Lemma 9 asserts that if T is pairwise sufficient for a set $\mathfrak{M}$ of three elements, then T is sufficient for $\mathfrak{M}$. Lemmas 10 and 12 extend this result to finite and countably infinite sets $\mathfrak{M}$ respectively. Since every countable set of measures is dominated, the final result, Theorem 3, contains all these preliminaries as special cases.

LEMMA 12. *If $\{\mu_0, \mu_1, \mu_2, \cdots\}$ is a sequence of measures on S such that $\sum_{i=0}^{\infty} \mu_i(X) < \infty$, and if, for every E in S, $\mu(E) = \sum_{i=0}^{\infty} \mu_i(E)$, then*

$$\lim_k d\mu_0/d(\textstyle\sum_{i=0}^{k} \mu_i) = d\mu_0/d\mu \; [\mu].$$

If, in addition, T is pairwise sufficient for the sequence $\{\mu_0, \mu_1, \mu_2 \cdots\}$, then $d\mu_0/d\mu \; (\epsilon) \; T^{-1}(T)$.

PROOF. We have, for $k = 0, 1, 2, \cdots$,

$$\frac{d\mu_0}{d(\sum_{i=0}^{k} \mu_i)} \cdot \frac{d(\sum_{i=0}^{k} \mu_i)}{d\mu} = \frac{d\mu_0}{d\mu}.$$

If we write $\lambda = \mu$, then the hypotheses of Lemma 11 are satisfied and, consequently, the second factor on the left side converges to 1 $[\mu]$; it follows that the first factor converges to $d\mu_0/d\mu \; [\mu]$. The second assertion of the lemma follows from Lemma 10.

THEOREM 3. *A necessary and sufficient condition that T be sufficient for a dominated set $\mathfrak{M}$ of measures on S is that T be pairwise sufficient for $\mathfrak{M}$.*

PROOF. The necessity of the condition is obvious. To prove its sufficiency, let $\mathfrak{N} = \{\mu_1, \mu_2, \cdots\}$ be a countable subset of $\mathfrak{M}$ which is equivalent to $\mathfrak{M}$ (Lemma 7), and let μ_0 be an arbitrary measure in $\mathfrak{M}$. Since the sufficiency or pairwise sufficiency of T remains unaltered if some or all of the measures in $\mathfrak{M}$ are replaced by positive constant multiples of themselves, we may assume that $\sum_{i=0}^{\infty} \mu_i(X) < \infty$. If we write, for every E in S, $\lambda(E) = \sum_{i=1}^{\infty} \mu_i(E)$, then the pairwise sufficiency of T and Lemma 12 imply that $d\mu_0/d(\mu_0 + \lambda) \cdot (\epsilon) \; T^{-1}(T)$. The relation

$$\frac{d\mu_0}{d\lambda} = \frac{d\mu_0}{d(\mu_0 + \lambda)} \cdot \frac{d(\mu_0 + \lambda)}{d\lambda} = \frac{d\mu_0}{d(\mu_0 + \lambda)} \cdot \left(\frac{d\lambda}{d(\mu_0 + \lambda)}\right)^{-1}$$

$$= \frac{d\mu_0}{d(\mu_0 + \lambda)} \left(1 - \frac{d\mu_0}{d(\mu_0 + \lambda)}\right)^{-1}$$

implies that $d\mu_0/d\lambda \; (\epsilon) \; T^{-1}(T)$; an application of Theorem 1 concludes the proof.

A comparison of Theorems 1 and 2 and Corollary 3 yields immediately the following consequence of Theorem 3.

COROLLARY 4. *A necessary and sufficient condition that the statistic T be sufficient for a dominated set $\mathfrak{M}$ of measures on S is that, for any two measures μ and ν in $\mathfrak{M}$, $d\mu/d(\mu + \nu) \; (\epsilon) \; T^{-1}(T)$, or, equivalently, $d\mu \mid d\nu \; (\epsilon) \; T^{-1}(T)$.*

10. The value of sufficient statistics in statistical methodology.

We gather from conversations with some able and prominent mathematical statisticians that there is doubt and disagreement about just what a sufficient statistic is sufficient to do, and in particular about in what sense if any it contains "all the information in a sample." We therefore conclude this paper with a brief explanation of a point of view which, while not original with us, has not received due publicity.

Suppose a statistician $\mathcal{S}$ is to be shown an observation x drawn at random from some sample space $(X, \mathbf{S})$ on which an unknown measure, μ, of a set $\mathfrak{M}$ of possible measures obtains, while for the same observation x another statistician $\mathcal{T}$ is only to be shown the value $T(x)$ of some statistic T sufficient for $\mathfrak{M}$. It is clear that $\mathcal{S}$ is as well off as $\mathcal{T}$; we shall argue that $\mathcal{T}$ is also as well off as $\mathcal{S}$.

Suppose $\mathcal{S}$ has decided how to use his datum, that, in other words, he has decided just what he will do (or, in particular, say) in the event of each possible x. His program can then be described schematically by saying that he has selected some function f (of the points x) which, without serious loss of generality, may be supposed to take real values. Now $\mathcal{S}$'s only real concern is for the probability distribution of f given μ, i.e. for the function φ of a real variable c, defined by

$$\varphi(c) = \mu(\{x : f(x) < c\}) = \mu(E(c)).$$

But $\mathcal{T}$ can if he wishes achieve exactly the same results as $\mathcal{S}$, in the following way. Let him, on learning the value of $T(x)$, select a real number f, with the aid of a "random machine" which produces numerical values according to the known distribution function ψ, defined by

$$\psi(c) = p(E(c) \mid T(x)).$$

Then, for any μ in $\mathfrak{M}$, the probability that $\mathcal{T}$ will select a value less than c is

$$\int p(E(c) \mid y) d\mu T^{-1}(y) = \mu(E(c)) = \varphi(c).$$

Thus $\mathcal{T}$ is at no disadvantage, save for the mechanical one of having to manipulate a random machine, and he may fairly be said to have as much information as $\mathcal{S}$.

As a matter of fact we know of no practical situation in which $\mathcal{T}$ would actually go to the trouble of using a random machine. There are some situations in which he should in principle do so, but in which practical statisticians have not, so far as we know, thought it worth while. If, for example, an outcome consists of a sequence of n heads and tails resulting from n spins of a coin the heads ratio of which is known to be either one half or one quarter, then a sufficient statistic is the number of heads which occur in the sequence. In basing a decision on the outcome of this program both $\mathcal{S}$ and, to a still greater extent, $\mathcal{T}$ have (according to Wald's theory of minimum risk) something to gain by recourse to a random machine. There are, on the other hand, many technical desiderata which sufficient statistics meet exactly without recourse to random machines. Thus, as Blackwell has shown,[18] if $\mathcal{S}$ has an unbiased estimate, R, of some parameter, $\mathcal{T}$ can find a function R^*, defined by $R^*(y) = e(R \mid y)$, which is an unbiased estimate of that parameter, with variance not greater than that of R. More generally, if R is *any* estimate with finite mean square deviation from a parameter, then it is easy to show with Blackwell's methods that R^*

---

[18] D. Blackwell, "Conditional expectation and unbiased sequential estimation," *Annals of Math. Stat.*, Vol. 18 (1947), pp. 105–110.

has no larger a mean square deviation than R.   Finally it is a well known fact that, under suitable hypotheses, if there exists a maximum likelihood estimate R of some parameter, then R depends only on y.

We think that confusion has from time to time been thrown on the subject by (a) the unfortunate use of the term "sufficient estimate," (b) the undue emphasis on the factorability of sufficient statistics, and (c) the assumption that a sufficient statistic contains all the information in only the technical sense of "information" as measured by variance.

Reprinted from the
SUMMA BRASILIENSIS MATHEMATICAE
Vol. II, Ano VI, pp. 125-134, Dec. 1950

NORMAL DILATIONS AND EXTENSIONS OF OPERATORS (*)

By Paul R. Halmos

1. INTRODUCTION. Suppose that a Hilbert space H is a subspace of a Hilbert space K (i.e. that H is a closed linear manifold in K), and suppose that A and B are operators (i.e. bounded linear transformations) on H and K respectively. The operator B is called an *extension* of the operator A, or A is called a *restriction* of B, if H is invariant under B and if $Ax = Bx$ whenever $x \in H$. It is frequently convenient to express the relation between A and B by means of the projection operator P with domain K and range H. The geometric condition of invariance, $BH \subseteq H$, is equivalent to the equation $BP = PBP$; the requirement that A be the restriction of B to H is equivalent to the equation $AP = BP$. Another way of describing what it means for one operator to be an extension of another makes use of the fact that K may be viewed as the direct sum of H and the orthogonal complement $H^\perp$ of H in K. If each element z of K is identified with the pair (x,y), where $x = Pz \in H$ and $y = (1-P)z \in H^\perp$, then every operator B on K may be identified with a matrix $\begin{pmatrix} Q & S \\ T & R \end{pmatrix}$. (More explicitly: if $z = (x,y)$ and $Bz = (u,v)$, then both u and v depend continuously and linearly on both x and y. This implies that $u = Qx + Sy$ and $v = Tx + Ry$, where Q and S are continuous linear transformations from H and $H^\perp$, respectively, into H, and T and R are continuous linear transformations from H and $H^\perp$, respectively, into $H^\perp$.) In this representation of K, the subspace H becomes identified with the set of all vectors of the form $(x, 0)$, with $x \in H$. It follows that the requirement that H be invariant under B is equivalent to the requirement that T be 0, and the requirement that B be an extension to K of the operator A is equivalent to the requirement that Q be A. In other

(*) Manuscrito recebido a 10 de Dezembro de 1950.

1 (Sum. Bras. Math. II — Fasc. 8, 125 — 1950)

words: extending an operator A on H to an operator B on K is the same as constructing a two-rowed matrix whose top left entry is A and whose bottom left entry is 0.

What does it mean to place an operator A on H into the top left position of a two-rowed matrix, with no conditions on the bottom left entry? Since in the direct sum representation of K the projection operator P from K onto H becomes the matrix $\begin{pmatrix} 1 & 0 \\ 0 & 0 \end{pmatrix}$, it follows easily that a necessary and sufficient condition for an operator B on K to have the form $\begin{pmatrix} A & S \\ T & R \end{pmatrix}$ is that $AP = PBP$. In geometric terms: if $x \in H$, then Ax is obtained by letting B act on x and projecting the result back into H. I shall refer to this relation between A and B by saying that B is a *dilation* of A to K or, equivalently, that A is a *compression* of B in H.

My purpose in this paper is to find more or less decent dilations and extensions of more or less arbitrary operators. In particular I answer completely the question: under what conditions does an operator have *normal* dilations and extensions? (Observe that study of the better known class of Hermitian operators is not very fruitful in this connection, since, obviously, every restriction and even every compression of a Hermitian operator is itself Hermitian).

2. DILATIONS. In this section I shall prove among other things that every operator A on every Hilbert space H has a normal dilation. The domain K of the dilation B can be taken to be the direct sum of H with itself, i.e. the set of all pairs (x,y), with $x \in H$ and $y \in H$, so that B may be viewed as a two-rowed matrix each of whose entries is an operator on H. (The prescribed Hilbert space H is identified, as before, with the set of all pairs of the form $(x,0)$, with $x \in H$.)

THEOREM 1. *If* $|A| \leqq 1$ *and if* S *and* T *are the positive square roots of* $1 - AA^*$ *and* $1 - A^*A$ *respectively, then the matrix* $B = \begin{pmatrix} A & S \\ T & -A^* \end{pmatrix}$ *is unitary.*

PROOF. Direct computation shows that

$$BB^* = \begin{pmatrix} 1 & AT - SA \\ TA^* - A^*S & 1 \end{pmatrix}$$

and

$$B^*B = \begin{pmatrix} 1 & A^*S - TA^* \\ SA - AT & 1 \end{pmatrix},$$

and that, consequently, it is sufficient to prove that $SA = AT$. Since the definitions of S and T imply that $S^2A = AT^2$, it follows that $p(S^2)A = Ap(T^2)$ for any polynomial p and that, therefore (by the Weierstrass approximation theorem), $f(S^2)A = Af(T^2)$ for every continuous function f. The desired result can be obtained by applying this comment to the function f defined by $f(t) = \sqrt{t}$ whenever $0 \le t \le 1$.

The geometric interpretation of this theorem is somewhat surprising and amusing; it asserts that if an operator does not properly increase the norm of any vector, then it can be obtained from a rotation by compression to a subspace. Since every scalar multiple of a unitary operator is normal, and since every operator is a scalar multiple of an operator of norm not greater than 1, it follows from Theorem 1 that every operator has a normal dilation. This conclusion has an interesting consequence. If A_1 and A_2 are arbitrary Hermitian operators on H, then the operator $A = A_1 + iA_2$ has a normal dilation B. If B_1 and B_2 are the real and imaginary parts of B, then B_1 and B_2 commute, and B_1 and B_2 are dilations of A_1 and A_2 respectively. In other words, any two Hermitian operators have commutative dilations.

COROLLARY 1. *If* A *is an operator of norm not greater than 1 on a Hilbert space* K, *if* H *is a subspace of* K *such that the dimension of* H⊥ *is at least as great as the dimension of* H, *and if* P *is the projection operator from* K *onto* H, *then there exists a unitary operator* U *on* K *such that* PAP = PUP.

PROOF. The dimensionality assumption implies that the direct sum of H with itself can be embedded into K in such a way that the restriction of P to the direct sum is the matrix $\begin{pmatrix} 1 & 0 \\ 0 & 0 \end{pmatrix}$. Since a unitary operator on a subspace of a Hilbert space can trivially be extended to a unitary operator on the entire space, an application of Theorem 1 (to the compression of A to H) yields the conclusion of Corollary 1.

COROLLARY 2. *The weak closure of the set of all unitary operators on an infinite dimensional Hilbert space is the set of all operators of norm not greater than* 1.

PROOF. If A is an operator on K such that $|A| \leqq 1$, and if $x_1, \ldots, x_n, y_1, \ldots, y_n$ are vectors in K, let H be the (finite dimensional) subspace spanned by these vectors, and apply Corollary 1 to find a unitary operator U on K such that $PAP = PUP$. Since $(Ax_j, y_j) = (APx_j, Py_j) = (PAPx_j, y_j) = (PUPx_j, y_j) = (UPx_j, Py_j) = (Ux_j, y_j)$, $j = 1, \ldots, n$, the conclusion of Corollary 2 follows from the definition of the weak topology.

COROLLARY 3. *The mapping* $A \to A^{-1}$ *(defined on the set of all invertible operators on an infinite dimensional Hilbert space) is not weakly continuous.*

PROOF. Let A be an arbitrary operator of norm not greater than 1 such that A is invertible but not unitary; for example A could be a scalar multiple of the identity. By Corollary 2, there exists a directed system (U_j) of unitary operators converging weakly to A. The weak continuity of the mapping $A \to A^*$ implies that $(U_j^{-1}) = (U_j^*)$ converges weakly to $A^* \neq A^{-1}$.

The methods of Theorem 1 and Corollaries 1, 2, and 3 extend immediately to prove the following results (which were first called to my attention by E. A. Michael).

THEOREM 2. *If* A *is a positive operator and* $|A| \leqq 1$, *and if* S *is the positive square root of* $A(1-A)$, *then the matrix* $B = \begin{pmatrix} A & S \\ S & 1-A \end{pmatrix}$ *is a projection.*

COROLLARY 4. *If* A *is a positive operator of norm not greater than 1 on a Hilbert space* K, *if* H *is a subspace of* K *such that the dimension of* $H^{\perp}$ *is at least as great as the dimension of* H, *and if* P *is the projection operator from* K *onto* H, *then there exists a projection* E *on* K *such that* $PAP = PEP$.

COROLLARY 5. *The weak closure of the set of all projections on an infinite dimensional Hilbert space is the set of all positive operators of norm not greater than 1.*

COROLLARY 6. *The mapping* $A \to A^2$ *(defined on the set of all positive operators on an infinite dimensional Hilbert space) is not weakly continuous.*

3. EXTENSIONS. The problem of extension is not as easy as that of dilation. To get a clue to the facts I shall begin by studying some of those properties of normal operators which are inherited by all their restrictions. If B is a normal operator on a Hilbert space K, if P is the projection from K onto a subspace H invariant under B, and if A is the restriction of B to H, then $AP = BP$ and $A^*P = PB^*$. It follows that if $x \in H$, then $|Ax| = |Bx| = |B^*x| \geq |PB^*x| = |A^*x|$). In other words, the operator A, instead of satisfying the identity $|Ax| = |A^*x|$ characteristic of normality, satisfies the inequality $|Ax| \geq |A^*x|$ for every vector x, or, equivalently, satisfies the relation $A^*A \geq AA$. I shall call an operator A *subnormal* if $A^*A \geq AA$; subnormality is thus seen to be a necessary condition for the possession of a normal extension.

It is pertinent to observe that if the Hilbert space H is finite dimensional, so that the concept of trace makes sense, then the trace of $A^*A - AA^*$ vanishes for every A. Since a positive operator with vanishing trace is 0, it follows that on a finite dimensional Hilbert space every subnormal operator is normal. From this in turn it follows that if an operator on a finite dimensional space has a normal extension, then it is normal.

If the results of the preceding paragraph were true without the hypothesis of finite dimensionality, the theory of normal extensions would be trivial. There are, however, many examples of non normal restrictions of normal operators (which are therefore subnormal without being normal). As one such example I mention the well known operator, $f(z) \to zf(z)$, defined on the Hilbert space of all square integrable functions analytic in the interior of the unit circle.

I have constructed an example of a subnormal operator which does not have a normal extension (see Section 4); it is necessary

5 (SUM. BRAS. MATH. II — Fasc. 8, 129 — 1950)

therefore to look for conditions stronger than subnormality. Observe that in studying a normal extension B of an operator A on H, there is no loss of generali y in assuming that the extension B is minimal (in the sense that the least subspace containing H and reducing B is K), since the restriction of the given B to that least subspace has all the properties (including normality) that B is required to have. (Recall that a subspace reduces an operator B if it is invariant under both B and B^*, or, equivalently, if both it and its orthogonal complement are invariant under B).

Suppose that B is a minimal normal extension of A. The set $\tilde{H}$ of all finite sums of the form $\Sigma_j B^{*j} x_j$, with $x_j \in H$, is obviously a linear manifold; since $B(\Sigma_j B^{*j} x_j) = \Sigma_j B^{*j}(B x_j)$ and $B^*(\Sigma_j B^{*j} x_j) = = \Sigma_j B^{*j+1} x_j$, it follows that the closure of $\tilde{H}$ reduces B. Since $H \subset \tilde{H}$, it follows from the assumed minimality that the set $\tilde{H}$ of all sums of the indicated form is dense in K. The inner product of two vectors $\Sigma_j B^{*j} x_j$ and $\Sigma_k B^{*k} y_k$ in $\tilde{H}$ can be calculated in terms of the operator A on the subspace H, since $(\Sigma_j B^{*j} x_j, \Sigma_k B^{*k} y_k) = \Sigma_j \Sigma_k (B^k B^{*j} x_j, y_k) = \Sigma_j \Sigma_k (B^{*j} B^k x_j, y_k) = \Sigma_j \Sigma_k (B^k x_j, B^j y_k) = \Sigma_j \Sigma_k (A^k x_j, A^j y_k)$. If both x_j and y_j are replaced by $c_j x_j$, where the c_j's are arbitrary complex numbers, it follows that $|\Sigma_j B^{*j}(c_j x_j)|^2 = = \Sigma_j \Sigma_k (A^k x_j, A^j x_k) c_j \overline{c_k}$. If the $n+1$ rowed matrix whose general term is $(A^k x_j, A^j x_k)$ is denoted by $A(x_0, \ldots, x)$, then two important consequences of the preceding computation may be expressed as follows: (1) since the norm of every vector is positive, every one of the matrices $A(x_0, \ldots, x)$ is positive and (2) since the operator B is bounded, there exists a positive constant c such that $A(Ax_0, \ldots, Ax_n) \leqq c A(x_0, \ldots, x)$ simultaneously for every finite sequence (x_j) of vectors in H.

Matrices such as $A(x_0, \ldots, x)$ are generalizations of the familiar Gramians, and I propose to call them *A-Gramians*. I shall also say that an operator A is *n-subnormal* ($n = 0, 1, 2, \ldots$) if $A(x_0, \ldots, x_n) \geqq 0$ for every sequence $x_0, \ldots, x_n$ of $n+1$ vectors, and that A is *n-bounded* if there exists a positive constant c such that $A(Ax_0, \ldots, Ax) \leqq c A(x_0, \ldots, x_n)$ for all such sequences. An operator is *completely subnormal* if it is *n*-subnormal for every n, and *completely bounded* if it is *n*-bounded for every n. The results of the preceding paragraphs may be summed up by saying that complete subnormality and complete boundedness are necessary conditions for the existence of a normal extension. (It is an open question whether or not complete subnormality implies complete boundedness.)

Since if $n=0$, then $A(x_0,\ldots,x_n)=(x_0,x_0)$, and $A(Ax_0,\ldots,Ax)=(Ax_0,Ax_0)$, it is clear that every operator is 0-subnormal and 0-bounded. The motivation for the terminology of n-subnormality is the fact that 1-subnormality coincides with what was earlier called subnormality. Indeed, to require A to be 1-subnormal means to require that the matrix $\begin{pmatrix} (x,x) & (Ax,y) \\ (y,Ax) & (Ay,Ay) \end{pmatrix}$ be positive for every pair of vectors x and y. Since the diagonal terms are positive in any case, A will be 1-subnormal if and only if $|(Ax,y)| \leq |x|.|Ay|$. Since $(Ax,y)=(x,A^*y)$, and since $\sup\{|(x,A^*y)|: |x|=1\}=|A^*y|$, the latter inequality is satisfied for all x and y if and only if $|A^*y| \leq .Ay|$ for all y, i.e. if and only if A is subnormal.

THEOREM 3. *A necessary and sufficient condition that an operator have a normal extension is that it be completely subnormal and completely bounded; if this condition is satisfied, then the minimal normal extension is unique (to within unitary equivalence).*

PROOF. Since the necessity of the condition has been proved already, it is sufficient to prove sufficiency. Suppose that A is completely subnormal and completely bounded on a Hilbert space H. Let $\tilde{H}$ be the complex vector space of all functions $\tilde{x}$ from the integers (positive, negative, or zero) into H, such that $\tilde{x}(n)=0$ whenever $n<0$ and such that $\tilde{x}(n)\neq 0$ for finitely many values of n only. (The consideration of negative n's is a technical notational convenience.) If $\tilde{x}$ and $\tilde{y}$ are in $\tilde{H}$, I write $(\tilde{x},\tilde{y})=\Sigma_j\Sigma_k(A^k\tilde{x}(j),A^j\tilde{x}(k))$. Direct verification shows that $(\tilde{x},\tilde{y})$ is linear in $\tilde{x}$ and conjugate linear in $\tilde{y}$, and that, in fact, $(\tilde{y},\tilde{x})=\overline{(\tilde{x},\tilde{y})}$. The assumed complete subnormality of A shows that the inner product so defined is positive; in general, however, it is not strictly positive. I introduce, as is usual in such situations, the vector space $\tilde{K}$ obtained from $\tilde{H}$ by identifying an element $\tilde{x}$ with 0 if and only if $(\tilde{x},\tilde{x})=0$; the relation of equality in $\tilde{K}$ will be denoted by $\equiv$.

7 (SUM. BRAS. MATH. II — Fasc. 8, 131 — 1950)

119

If $x \in \tilde{H}$, I write $B\tilde{x}(n) = A\tilde{x}(n)$. (Properly speaking I should write $(B\tilde{x})(n)$ instead of $B\tilde{x}(n)$, but since the more convenient notation leads to no confusion, I shall continue to use it.) If c is a positive constant, fulfilling the requirement described in the definition of complete boundedness, then $(B\tilde{x}, B\tilde{x}) = \Sigma_j \Sigma_k (A^{k+1}\tilde{x}(j),$ $A^{j+1}\tilde{x}(k)) \leq c \Sigma_j \Sigma_k (A^k \tilde{x}(j), A^j \tilde{x}(k)) = c(\tilde{x}, \tilde{x})$, so that the linear transformation B on $\tilde{H}$ is bounded. It follows that if $\tilde{x} \equiv 0$, then $B\tilde{x} = 0$, so that B may be viewed as a bounded linear transformation of the vector space $\tilde{K}$ into itself.

I proceed next to construct the transformation C on $\tilde{H}$ (and on $\tilde{K}$) which will presently turn out to play the role of the adjoint of the desired normal extension of A; I write $C\tilde{x}(n) = \tilde{x}(n-1)$. It is clear that C is a linear transformation of $\tilde{H}$ into itself; since $BC\tilde{x}(n) = B\tilde{x}(n-1) = A\tilde{x}(n-1)$ and $CB\tilde{x}(n) = CA\tilde{x}(n) = A\tilde{x}(n-1)$, it follows that B and C commute. The operator C wants to act like the adjoint of B; indeed $(\tilde{x}, C\tilde{y}) = \Sigma_j \Sigma_k (A^k \tilde{x}(j), A^j \tilde{y}(k-1)) = \Sigma_j \Sigma_k (A^{k+1}\tilde{x}(j), A^j \tilde{y}(k)) = (B\tilde{x}, \tilde{y})$. From this relation, together with the boundedness of B, it follows that C is bounded and hence that C may be viewed as a bounded linear transformation of the vector space $\tilde{K}$ into itself; considered on $\tilde{K}$, the transformations B and C are still commutative.

Let K be the completion of the scalar product space $\tilde{K}$; the boundedness of B and C guarantees that these transformations can be extended (uniquely) to operators on K. It follows from the preceding paragraph that, considered on K, $C = B^*$, so that B is a normal operator.

If $x \in H$, write $i(x)$ for the element $\tilde{x}$ of $\tilde{H}$ defined by $\tilde{x}(n) = x$ or 0 according as $n = 0$ or $n \neq 0$. It is trivial to verify that i is a norm preserving linear transformation from H into $\tilde{H}$. Since, in particular, $(i(x), i(x)) = 0$ if and only if $x = 0$, it follows that i may be viewed as a norm preserving linear transformation from H into $\tilde{K}$ and therefore into K. Since $Bi(x) \equiv i(Ax)$, it follows that the subspace $i(H)$ of K is invariant under B and that the restriction of B to $i(H)$ coincides with iAi^{-1}. If, in other words, i is interpreted as the identity mapping, or, equivalently, each x in H is

8 (Sum. Bras. Math. II — Fasc. 8, 132 — 1950)

120

identified with its image $i(x)$ in K, then B becomes a normal extension of A.

The uniqueness of a minimal B is a trivial consequence of the above construction and of the preliminary study which motivated the definitions of complete subnormality and complete boundedness; that study showed that (except for notation) every minimal normal extension is related to A in exactly the same way as the B constructed above.

4. EXAMPLES. Let Z be a Hilbert space, and let H be the set of all functions x from the integers (positive, negative, or zero) into Z, such that $\Sigma|x(n)|^2 < \infty$. If the inner product of two such functions x and y is defined by $(x,y) = \Sigma(x(n),y(n))$, then H is a Hilbert space. Let $(s(n))$ be a sequence of positive operators on Z such that the sequence $(|s(n)|)$ of their norms is bounded; and define, for every x in H, $Ux(n) = x(n+1)$ and $Sx(n) = s(n)x(n)$. It is easy to verify that U and S are operators on H; since U is unitary, it follows that $U^*x(n) = U^{-1}x(n) = x(n-1)$ and, just as obviously, $S^*x(n) = s^*(n)x(n) = s(n)x(n) = Sx(n)$, so that S is Hermitian (and, in fact, positive). If $A = US$, then $Ax(n) = s(n+1)x(n+1)$ and, since $A^* = SU^*$, $A^*x(n) = s(n)x(n-1)$. Since $A^*Ax(n) = s^2(n)x(n)$ and $AA^*x(n) = s^2(n+1)x(n)$, it follows that A is subnormal if and only if the sequence $(s^2(n))$ is decreasing. Virtually all the good and bad behavior of subnormal operators can be illustrated by examples of the type here described; special choices of Z and $(s(n))$ lead to many diverse properties.

Since $A^2x(n) = s(n+1)s(n+2)x(n+2)$ and $A^{*2}x(n) = s(n)s(n-1)x(n-2)$, it follows that $A^{*2}A^2x(n) = s(n)s^2(n-1)s(n)x(n)$ and $A^2A^{*2}x(n) = s(n+1)s^2(n+2)s(n+1)x(n)$. Consequently A^2 is subnormal if and only if the sequence $(s(n)s^2(n-1)s(n))$ is decreasing. Does the subnormality of A^2 follow from that of A? In other words: does $s^2(n) \geqq s(n+1)$ for all n imply that $s(n)s^2(n-1)s(n) \geqq s(n+1)s(n+2)s(n+1)$ for all n? The answer is no. The construction of a suitable example is based on the fact that there exist positive operators a and b such that $a \geqq b$ is true and $a^2 \geqq b^2$ is false. If, for instance, Z is two dimensional, so that operators on Z may be identified with two-rowed matrices, and if $a = \begin{pmatrix} 2 & 1 \\ 1 & 1 \end{pmatrix}$ and $b = \begin{pmatrix} 1 & 0 \\ 0 & 0 \end{pmatrix}$, then $a - b = \begin{pmatrix} 1 & 1 \\ 1 & 1 \end{pmatrix} \geqq 0$, but

$a^2 - b^2 = \begin{pmatrix} 5 & 3 \\ 3 & 2 \end{pmatrix} - \begin{pmatrix} 1 & 0 \\ 0 & 0 \end{pmatrix} = \begin{pmatrix} 4 & 3 \\ 3 & 2 \end{pmatrix}$, and therefore the determinant of

$a^2 - b^2$ is -1. If $s(n)$ is defined to be the positive square root of a wherever $n \leqq 0$ and the positive square root of b whenever $n \quad 0$, then $s^2(n) \geqq s^2(n+1)$ for all n, so that A is subnormal, but, if $n = 0$, then $s(n)s^2(n-1)s(n) = a^2$ and $s(n+1)s^2(n+2)s(n+1) = b^2$, so that A^2 is not subnormal. If A had a normal extension B, then B^2 would be a normal extension of A^2 and therefore A^2 would also be (completely) subnormal; since this is not the case, A cannot have a normal extension.

THE UNIVERSITY OF CHICAGO
CHICAGO, ILLINOIS, U. S. A.

10 (Sum. Bras. Math. II — Fasc. 8, 134 — 1950)

122

Reprinted from the
ACTA SCIENTIARUM MATHEMATICARUM
Tomus XII, pp. 153-156, 1950

Commutativity and spectral properties of normal operators.

By Paul R. Halmos in Chicago.

1. The results of this note grew out of a current investigation of spectral properties of operators on Hilbert space. While the characterization of the spectral manifolds of a normal operator (Theorem 1) appears to be new and may be considered to be of independent interest, I present it here main'y because it supplies an extremely easy proof of a theorem (Theorem 2) which was unknown until a few weeks ago. J. von Neumann has asked whether or not it is true that if an operator B commutes with a normal operator A, then B commutes with A^* also. Well known and quite elementary considerations show that in order to answer the question affirmatively it is sufficient to prove that, under the stated hypotheses, B is reduced by all the spectral manifolds of A. This has recently been proved by B. Fuglede — he communicated his proof to me at the Boulder meeting of the American Mathematical Society at the end of August, 1949. The proof I present below is somewhat different from his in spirit and in method. I should say also that Fuglede's proof is valid for not necessarily bounded transformations A and that, similarly, only minor modifications are needed to adapt my proof to this more general case.

For the orientation of the reader I present here the trivial proof of the theorem under discussion for the case in which A has pure point spectrum; the proof of the general theorem below uses essentially the same idea and method. If λ is a proper value of A and if $\mathfrak{F}$ is the subspace of all corresponding proper vectors, then the relations $A(Bx) = B(Ax) = B(\lambda x) = \lambda(Bx)$ show that $\mathfrak{F}$ is invariant under B. Since to say that A has pure point spectrum means that the entire Hilbert space is spanned by orthogonal subspaces such as $\mathfrak{F}$, it follows that the orthogonal complement of $\mathfrak{F}$ is also invariant under B, and this is exactly what was to be proved.

2. Throughout this note I shall deal with a fixed complex Hilbert space $\mathfrak{H}$. An *operator* is a bounded linear transformation of $\mathfrak{H}$ into itself; an operator A is *normal* if it commutes with its adjoint A^*. If A is normal, then

$\|Ax\|^2 = (Ax, Ax) = (A^*Ax, x) = (AA^*x, x) = (A^*x, A^*x) = \|A^*x\|^2$ for every vector x; it is easy to see that the identity $\|Ax\| = \|A^*x\|$ is not only necessary but also sufficient for the normality of A. A *subspace* is a closed linear manifold in $\mathfrak{H}$; a subspace $\mathfrak{M}$ *reduces* an operator A if both $\mathfrak{M}$ and $\mathfrak{M}^\perp$ (= the orthogonal complement of $\mathfrak{M}$) are invariant under A, i. e. if $A\mathfrak{M} \subset \mathfrak{M}$ and $A\mathfrak{M}^\perp \subset \mathfrak{M}^\perp$. There are two useful and elementary necessary and sufficient conditions that a subspace $\mathfrak{M}$ reduce an operator A; the first is that $\mathfrak{M}$ be invariant under both A and A^*, and the second is that A commute with the projection on $\mathfrak{M}$.

Lemma 1.[1]) *If A is a normal operator and if $\mathfrak{F}(A) = \{x : \|A^n x\| \le \|x\|,$ $n = 1, 2, \ldots\}$, then $\mathfrak{F}(A)$ is a subspace and $\mathfrak{F}(A)$ is invariant under every operator B which commutes with A.*

Proof. Write $\mathfrak{G}$ for the set of all those vectors x for which the sequence $\{\|A^n x\| : n = 1, 2, \ldots\}$ is bounded. Since $\|A^n(\alpha x)\| = |\alpha| \cdot \|A^n x\|$ and $\|A^n(x+y)\| \le \|A^n x\| + \|A^n y\|$, it follows that $\mathfrak{G}$ is a linear manifold; if an operator B commutes with A, then the relation $\|A^n(Bx)\| = \|B(A^n x)\| \le \|B\| \cdot \|A^n x\|$ implies that $\mathfrak{G}$ is invariant under B. Clearly $\mathfrak{F}(A)$ is a closed set and $\mathfrak{F}(A) \subset \mathfrak{G}$; the proof of the lemma will be completed by showing that $\mathfrak{F}(A) = \mathfrak{G}$. For this purpose it is sufficient to show that if x is a vector such that, for some positive integer p, $\|A^p x\| > \alpha \|x\|$, $\alpha > 1$, then the sequence $\{\|A^n x\|\}$ cannot be bounded. Since $\alpha^2 \|x\|^2 < \|A^p x\|^2 = (A^p x, A^p x) = (A^{*p} A^p x, x) \le \|A^{*p} A^p x\| \cdot \|x\| = \|A^{2p} x\| \cdot \|x\|$, it follows that $\|A^{2p} x\| > \alpha^2 \|x\|$. Since an inductive repetition of this argument shows that $\|A^{2^k} x\| > \alpha^{2^k} \|x\|$ for every positive integer k, the proof is complete.

3. A *spectral measure* is a function E from the class of all Borel subsets of the set $\varLambda$ of all complex numbers to projections on $\mathfrak{H}$, such that (i) $E(\varLambda) = 1$, (ii) $E(M \cap N) = E(M) E(N)$ whenever M and N are Borel sets, and (iii) $E(M) = \sum_{j=1}^{\infty} E(M_j)$ whenever $\{M_j\}$ is a disjoint sequence of Borel sets whose union is M (the series being understood to converge in the strong topology of operators).

Lemma 2. *If E is a spectral measure and if $\mathfrak{E}(M) = \{x : E(M)x = x\}$ for every Borel·set M, then $\mathfrak{E}(M)$ is the subspace spanned by the class of all subspaces of the form $\mathfrak{E}(N)$, where N is an arbitrary compact subset of M.*

Proof. The assertion of the theorem is that, in a sense well known in the theory of numerical measures, every spectral measure is regular. The proof may be given along lines entirely similar to the numerical case, or it

---

[1]) This lemma is proved for Hermitian operators by B. A. LENGYEL and M. H. STONE, Elementary proof of the spectral theorem, *Annals of Math.*, **37** (1936), pp. 853–864; cf. in particular p. 858. The following proof is a slight simplification of their proof.

may be reduced to that case as follows. All that it is necessary to prove is that if x is a vector in $\mathfrak{C}(M)$ such that x is orthogonal to $\mathfrak{C}(N)$ for every compact subset N of M, then $x = 0$. Since, however, by the regularity of numerical measures, $\|x\|^2 = \|E(M)x\|^2 = \sup_N \|E(N)x\|^2$, it follows that there exists a countable class $\{N_j\}$ of compact subsets of M such that $\|x\|^2 = \sup_j \|E(N_j)x\|^2$, and hence that indeed $x = 0$.

I shall make use below of the spectral theorem for normal operators in the following form. If A is a normal operator, then there exists a unique spectral measure E, called the spectral measure of A, such that $(Ax, y) = \int \lambda\, d(E(\lambda)x, y)$ for every pair of vectors x and y.

4. In this final section I shall assume that A is a fixed normal operator with spectral measure E. For every complex number λ and every positive real number ε, I shall write $\mathfrak{F}(\lambda, \varepsilon)$ for $\mathfrak{F}\left(\dfrac{A-\lambda}{\varepsilon}\right)$; for every set M of complex numbers and every positive real number ε, I shall write $\mathfrak{F}(M, \varepsilon)$ for the subspace spanned by all those $\mathfrak{F}(\lambda, \varepsilon)$ for which $\lambda \in M$; and, for every set M of complex numbers, I shall write $\mathfrak{F}(M) = \bigcap_{\varepsilon > 0} \mathfrak{F}(M, \varepsilon)$. Let $F(\lambda, \varepsilon)$, $F(M, \varepsilon)$, and $F(M)$ be the projections on the subspace $\mathfrak{F}(\lambda, \varepsilon)$, $\mathfrak{F}(M, \varepsilon)$, and $\mathfrak{F}(M)$, respectively.

Theorem 1. *For every compact set M, $\mathfrak{F}(M) = \mathfrak{C}(M)$.*

Proof. For any positive number ε, let $\{M_j\}$ be a disjoint sequence of non empty Borel sets of diameter not greater than ε and such that $\bigcup_j M_j = M$. If $x \in \mathfrak{C}(M)$, $x_j = E(M_j)x$, and $\lambda_j \in M_j$, then $\|(A - \lambda_j)^n x_j\|^2 = \int_{M_j} |(\lambda - \lambda_j)^n|^2\, d(E(\lambda)x_j, x_j) \leq \varepsilon^{2n} \|x_j\|^2$, so that, for each j, $x_j \in \mathfrak{F}(\lambda_j, \varepsilon) \subset \mathfrak{F}(M, \varepsilon)$. Since $x = E(M)x = \Sigma_j E(M_j)x = \Sigma_j x_j$, it follows that $x \in \mathfrak{F}(M, \varepsilon)$. The arbitrariness of ε implies that $x \in \mathfrak{F}(M)$, and the arbitrariness of x implies, consequently, that $\mathfrak{C}(M) \subset \mathfrak{F}(M)$. Note that this argument did not make use of compactness of M.

Suppose now that N is a compact subset of $\Lambda - M$, and let δ be the distance between M and N. If $\lambda_0 \in M$, if $0 < \varepsilon < \delta$, and if $x \in \mathfrak{F}(\lambda_0, \varepsilon)$, then $\|(A - \lambda_0)^n x\| \leq \varepsilon^n \|x\|$; if, on the other hand, $x \in \mathfrak{C}(N)$, then $\|(A - \lambda_0)^n x\|^2 = \int_N |(\lambda - \lambda_0)^n|^2\, d(E(\lambda)x, x) \geq \delta^{2n} \|x\|^2$. It follows that $\mathfrak{F}(\lambda_0, \varepsilon) \cap \mathfrak{C}(N) = \{0\}$. Since $E(N)$ commutes with A, it follows from Lemma 1 that $\mathfrak{F}(\lambda_0, \varepsilon)$ is invariant under $E(N)$ and hence, since $E(N)$ is Hermitian, that $E(N)$ commutes with $F(\lambda_0, \varepsilon)$. This in turn implies that $F(\lambda_0, \varepsilon)E(N)$ is the projection on $\mathfrak{F}(\lambda_0, \varepsilon) \cap \mathfrak{C}(N)$, i. e. that $F(\lambda_0, \varepsilon)E(N) = 0$, and it follows that $\mathfrak{F}(\lambda_0, \varepsilon)$ is orthogonal to $\mathfrak{C}(N)$. The validity of this assertion for every λ_0 in M shows that $\mathfrak{F}(M, \varepsilon)$ is orthogonal to $\mathfrak{C}(N)$ and therefore, *a fortiori*, that $\mathfrak{F}(M)$ is orthogonal to $\mathfrak{C}(N)$.

The result of the preceding paragraph implies, in view of Lemma 2, that $\mathfrak{F}(M)$ is orthogonal to $\mathfrak{C}(\Lambda - M)$. This means that $\mathfrak{F}(M) \subset (\mathfrak{C}(\Lambda - M))^{\perp} = \mathfrak{C}(M)$, and the proof of the theorem is complete. I remark that it is easy to construct examples to show that if M is not compact, then $\mathfrak{C}(M)$ may be a proper subset of $\mathfrak{F}(M)$.

Theorem 2. *If an operator B commutes with A, then* $\mathfrak{C}(M)$ *reduces B for every Borel set M.*

Proof. It follows from Lemma 1 that, for every complex number λ and every positive number ε, $\mathfrak{F}(\lambda, \varepsilon)$ is invariant under B, and hence that $\mathfrak{F}(M, \varepsilon)$ and $\mathfrak{F}(M)$ are invariant under B for every set M. Theorem 1 implies that $\mathfrak{C}(M)$ is invariant under B whenever M is compact and hence, by Lemma 2, that $\mathfrak{C}(M)$ is invariant under B for every Borel set M. Since $(\mathfrak{C}(M))^{\perp} = \mathfrak{C}(\Lambda - M)$, it follows automatically that $(\mathfrak{C}(M))^{\perp}$ is also invariant under B and hence that $\mathfrak{C}(M)$ reduces B.

University of Chicago.

(Received October 13, 1949.)

Reprinted from the
AMERICAN JOURNAL OF MATHEMATICS
Vol. 72, pp. 214-215, 1950

THE MARRIAGE PROBLEM.*

By Paul R. Halmos and Herbert E. Vaughan.

In a recent issue of this journal Weyl[1] proved a combinatorial lemma which was apparently considered first by P. Hall.[2] Subsequently Everett and Whaples[3] published another proof and a generalization of the same lemma. Their proof of the generalization appears to duplicate the usual proof of Tychonoff's theorem.[4] The purpose of this note is to simplify the presentation by employing the statement rather than the proof of that result. At the same time we present a somewhat simpler proof of the original Hall lemma.

Suppose that each of a (possibly infinite) set of boys is acquainted with a finite set of girls. Under what conditions is it possible for each boy to marry one of his acquaintances? It is clearly necessary that every finite set of k boys be, collectively, acquainted with at least k girls; the Everett-Whaples result is that this condition is also sufficient.

We treat first the case (considered by Hall) in which the number of boys is finite, say n, and proceed by induction. For $n = 1$ the result is trivial. If $n > 1$ and if it happens that every set of k boys, $1 \leq k < n$, has at least $k + 1$ acquaintances, then an arbitrary one of the boys may marry any one of his acquaintances and refer the others to the induction hypothesis. If, on the other hand, some group of k boys, $1 \leq k < n$, has exactly k acquaintances, then this set of k may be married off by induction and, we assert, the remaining $n - k$ boys satisfy the necessary condition with respect to the as yet unmarried girls. Indeed if $1 \leq h \leq n - k$, and if some set of h bachelors were to know fewer than h spinsters, then this set of h bachelors together with the k married men would have known fewer than $k + h$ girls. An

* Received June 6, 1949.

[1] H. Weyl, "Almost periodic invariant vector sets in a metric vector space," *American Journal of Mathematics*, vol. 71 (1949), pp. 178-205.

[2] P. Hall, "On representation of subsets," *Journal of the London Mathematical Society*, vol. 10 (1935), pp. 26-30.

[3] C. J. Everett and G. Whaples, "Representations of sequences of sets," *American Journal of Mathematics*, vol. 71 (1949), pp. 287-293. Cf. also M. Hall, "Distinct representatives of subsets," *Bulletin of the American Mathematical Society*, vol. 54 (1948), pp. 922-926.

[4] C. Chevalley and O. Frink, Jr., "Bicompactness of Cartesian products," *Bulletin of the American Mathematical Society*, vol. 47 (1941), pp. 612-614.

214

application of the induction hypothesis to the $n-k$ bachelors concludes the proof in the finite case.

If the set B of boys is infinite, consider for each b in B the set $G(b)$ of his acquaintances, topologized by the discrete topology, so that $G(b)$ is a compact Hausdorff space. Write G for the topological Cartesian product of all $G(b)$; by Tychonoff's theorem G is compact. If $\{b_1, \cdots, b_n\}$ is any finite set of boys, consider the set H of all those elements $g = g(b)$ of G for which $g(b_i) \neq g(b_j)$ whenever $b_i \neq b_j$, $i, j = 1, \cdots, n$. The set H is a closed subset of G and, by the result for the finite case, H is not empty. Since a finite union of finite sets is finite, it follows that the class of all sets such as H has the finite intersection property and, consequently, has a non empty intersection. Since an element $g = g(b)$ in this intersection is such that $g(b') \neq g(b'')$ whenever $b' \neq b''$, the proof is complete.

It is perhaps worth remarking that this theorem furnishes the solution of the celebrated problem of the monks.[5] Without entering into the history of this well-known problem, we state it and its solution in the language of the preceding discussion. A necessary and sufficient condition that each boy b may establish a harem consisting of $n(b)$ of his acqaintances, $n(b) = 1$, $2, 3, \cdots$, is that, for every finite subset B_0 of B, the total number of acquaintances of the members of B_0 be at least equal to $\Sigma n(b)$, where the summation runs over every b in B_0. The proof of this seemingly more general assertion may be based on the device of replacing each b in B by $n(b)$ replicas seeking conventional marriages, with the understanding that each replica of b is acquainted with exactly the same girls as b. Since the stated restriction on the function n implies that the replicas satisfy the Hall condition, an application of the Everett-Whaples theorem yields the desired result.

UNIVERSITY OF CHICAGO
AND
UNIVERSITY OF ILLINOIS.

---

[5] H. Balzac, *Les Cent Contes Drôlatiques*, IV, 9: *Des moines et novices*, Paris (1849).

SPECTRA AND SPECTRAL MANIFOLDS

By PAUL R. HALMOS (Chicago)

§ 1. Statement of the problem. Suppose that A is a (bounded) normal operator on a (complex) Hilbert space $\mathfrak{H}$ and that $\mathfrak{I}$ is a (closed) subspace of $\mathfrak{H}$ invariant under A. Let B be the restriction of A to $\mathfrak{I}$. In a preceding paper [1] I began a systematic study of operators such as B (operators that should perhaps be called *subnormal*) and presented an intrinsic (though somewhat complicated) characterization of subnormal operators. The main purpose of this note is to make a further contribution to the study of subnormal operators by investigating the relation between the spectrum of A and the spectrum of B.

Recall that, in general, the spectrum $\Lambda(A)$ of an operator A is the set of all those complex numbers λ for which the operator $A - \lambda$ is not invertible. A related and useful concept, which is not quite as well known, is the approximate point spectrum of A; it may be defined as the set $\Pi(A)$ of all those complex numbers λ for which

$$\inf \{\|Ax - \lambda x\| : \|x\| = 1\} = 0.$$

The two pertinent facts concerning these concepts are that $\Pi(A) \subset \Lambda(A)$ for every operator A and that, if A is normal, then $\Pi(A) = \Lambda(A)$ [2]. Since throughout this paper I shall be concerned with a fixed normal operator A and its restrictions (such as B) to invariant subspaces (such as $\mathfrak{I}$), it will be con-

---

[1] *Normal dilations and extensions of operators*, Summa Brasil. Math., fasc. 9, vol. 2 (1950).

[2] Cf., for instance, § 31 of my recent book, *Introduction to Hilbert space and the theory of spectral multiplicity*, New York, 1951.

venient to write $\Lambda(\Im)$ and $\Pi(\Im)$ instead of $\Lambda(B)$ and $\Pi(B)$ respectively. (It is important to note that the invertibility of an operator such as B is defined in terms of its action on its domain $\Im$, and has nothing to do with the part of $\mathfrak{H}$ outside $\Im$. In other words, the question of invertibility for, say, $B-\lambda$ is the question of existence of a two-sided inverse of $B-\lambda$ with domain $\Im$).

An important and interesting class of invariant subspaces (where "invariant" means, of course, "invariant under A") is the class of all those invariant subspaces whose orthogonal complement is also invariant. Subspaces of this class are said to *reduce A*. Since the intersection of every class of reducing subspaces is a reducing subspace, it follows that to every invariant subspace $\Im$ there corresponds a unique minimal reducing subspace $\mathfrak{K}$ containing it.

In the notation and terminology introduced above, the problem to be treated here becomes the study of the function Λ — a function which makes correspond a set $\Lambda(\Im)$ of complex numbers to every invariant subspace $\Im$ of $\mathfrak{H}$. More specifically: if $\Im$ and $\mathfrak{K}$ are invariant subspaces such that $\Im \subset \mathfrak{K}$, then what is the relation between $\Lambda(\Im)$ and $\Lambda(\mathfrak{K})$? It is an easy consequence of the positive results of this note that, in general, the sets $\Lambda(\Im)$ and $\Lambda(\mathfrak{K})$ are not comparable. The positive results concern (a) the case in which $\Im$ and $\mathfrak{K}$ are reducing subspaces (the facts here are essentially known and almost trivial), and (b) the case in which $\Im$ is an arbitrary invariant subspace and $\mathfrak{K}$ is the minimal reducing subspace containing it (here the facts lie considerably deeper and the full force of the spectral theorem is apparently necessary to get at them).

The spectral theorem will be used essentially in its classical form. According to that theorem, to the operator A there corresponds a *spectral measure E*, i. e. a function that assigns a projection $E(M)$ with domain $\mathfrak{H}$ to every Borel subset M of the complex plane. (The range of $E(M)$ will be denoted by $\mathfrak{E}(M)$). The relation between A and E may be indicated, as usual, by writing

$$A = \int \lambda dE(\lambda),$$

where the integration is extended over the entire complex plane. The complement with respect to the complex plane of a set M of complex numbers will be denoted by M'.

§ 2. Solution of the problem.

Theorem 1. *If $\mathfrak{J}$ and $\mathfrak{R}$ are reducing subspaces such that $\mathfrak{J} \subset \mathfrak{R}$, then $\Lambda(\mathfrak{J}) \subset \Lambda(\mathfrak{R})$.*

Proof. It is no loss of generality (but merely a change of notation) to assume that $\mathfrak{R} = \mathfrak{H}$. To say that $\mathfrak{J}$ reduces A is equivalent to saying that the projection with range $\mathfrak{J}$ commutes with A. If this is the case, and if $A - \lambda$ is invertible, then the projection with range $\mathfrak{J}$ commutes with $(A - \lambda)^{-1}$ also, or, equivalently, $\mathfrak{J}$ reduces $(A - \lambda)^{-1}$. Since the restriction of $(A - \lambda)^{-1}$ to $\mathfrak{J}$ is obviously the inverse of the restriction of $A - \lambda$ to $\mathfrak{J}$, it follows that $\lambda \in \Lambda(\mathfrak{J})'$ whenever $\lambda \in \Lambda(\mathfrak{H})'$, i. e. that $\Lambda(\mathfrak{J}) \subset \Lambda(\mathfrak{H})$ [3]).

Lemma 1. *If $M = \{\lambda : |\lambda| \leqq 1\}$ and $\mathfrak{D} = \{x : \|A^n x\| \leqq \|x\|,$ $n = 1, 2, \ldots\}$, then $\mathfrak{D} = \mathfrak{E}(M)$.*

Proof. If $x \in \mathfrak{E}(M)$, then

$$\|A^n x\|^2 = \int_M |\lambda^n|^2 d(E(\lambda) x, x) \leqq \|x\|^2$$

for every positive integer n, and therefore $\mathfrak{E}(M) \subset \mathfrak{D}$. The remainder of the proof will be devoted to establishing the reverse inclusion. If $x \in \mathfrak{E}(M')$, then

$$\|Ax\|^2 = \int_{M'} |\lambda|^2 d(E(\lambda) x, x) > \|x\|^2$$

unless $x = 0$, and therefore $\mathfrak{E}(M') \cap \mathfrak{D}$ contains no non-zero vector. According to a theorem of Stone and Lengyel, the set $\mathfrak{D}$ is a subspace of $\mathfrak{H}$, and the subspace $\mathfrak{D}$ is invariant under every operator which commutes with A [4]). Since $E(M')$ com-

---

[3]) An even easier proof of Theorem 1 can be given in terms of the approximate point spectrum. The proof above is preferable because it applies to not necessarily normal operators.

[4]) B. A. Lengyel and M. H. Stone, *Elementary proof of the spectral theorem*, Ann. Math. 37 (1936), pp. 853-864. Stone and Lengyel state the theorem for Hermitian operators only, but with a very slight modification

mutes with A, it follows that $\mathfrak{D}$ is invariant under $E(M')$ and therefore, since $E(M')$ is Hermitian, $\mathfrak{D}$ reduces $E(M')$. It follows that if D is the projection with range $\mathfrak{D}$, then $E(M')$ commutes with D and hence that $E(M')D$ is the projection with range $\mathfrak{E}(M')\cap\mathfrak{D}$. This implies that $E(M')D = 0$ and hence that $E(M)D = D$; in other words $\mathfrak{D}\subset\mathfrak{E}(M)$.

Theorem 2. *If $\mathfrak{J}$ is an invariant subspace and if $\mathfrak{R}$ is the minimal reducing subspace containing $\mathfrak{J}$, then $\Lambda(\mathfrak{R})\subset\Lambda(\mathfrak{J})$* [5]).

Proof. Since the restriction of a normal operator to a reducing subspace is normal, there is no loss of generality in assuming that $\mathfrak{R} = \mathfrak{H}$. Denoting the restriction of A to $\mathfrak{J}$ by B, I proceed to prove first that, if B is invertible, then so is A. Write $\beta = \|B^{-1}\|$ and let ε be a positive number such that $\varepsilon\beta < 1$. If

$$\mathfrak{D}_\varepsilon = \{x : \|A^n x\| \leqq \varepsilon^n \|x\|, \ n = 1, 2, \ldots\},$$

and if $x \in \mathfrak{D}_\varepsilon$ and $y \in \mathfrak{J}$, then

$$(x, y) = (x, B^n B^{-n} y) = (x, A^n B^{-n} y) = (A^{*n} x, B^{-n} y).$$

It follows that

$$|(x, y)| \leqq \|A^n x\| \cdot \|B^{-n} y\| \leqq (\varepsilon\beta)^n \|x\| \cdot \|y\|,$$

and hence, since this is true for every positive integer n, that $(x, y) = 0$. In other words $\mathfrak{D}_\varepsilon$ is contained in the orthogonal complement of $\mathfrak{J}$. If $M_\varepsilon = \{\lambda : |\lambda| \leqq \varepsilon\}$, then it follows from an application of Lemma 1 to A/ε in place of A that $\mathfrak{D}_\varepsilon = \mathfrak{E}(M_\varepsilon)$ and consequently that $\mathfrak{J}\subset\mathfrak{E}(M'_\varepsilon)$. Since the subspace $\mathfrak{E}(M'_\varepsilon)$ reduces A, it follows that $\mathfrak{E}(M'_\varepsilon) = \mathfrak{H}$ and hence that $E(M_\varepsilon) = 0$. This result, together with the spectral theorem, implies that A is invertible. The conclusion of the theorem can now be deduced as follows. If $\lambda \in \Lambda(\mathfrak{J})'$, so that $B - \lambda$ is invertible, then, since $\mathfrak{J}$ is invariant under the normal operator $A - \lambda$ and since the minimal subspace containing $\mathfrak{J}$ and reducing $A - \lambda$ is $\mathfrak{H}$,

---

their proof establishes the result for normal operators as well. The modification, together with a slight simplification of their proof, appears in my paper, *Commutativity and spectral properties of normal operators*, Acta Szeged 12 (1950), p. 153-156.

[5]) In terms of the language of subnormal operators, Theorem 2 asserts that in passing from a subnormal operator to its minimal normal extension the spectrum can only decrease.

it follows that $A-\lambda$ is invertible. In other words, if $\lambda \in \Lambda(\mathfrak{J})'$, then $\lambda \in \Lambda(\mathfrak{H})'$ and therefore $\Lambda(\mathfrak{H}) \subset \Lambda(\mathfrak{J})$.

It is worth noting that the inclusion relation asserted in Theorem 2 is not in general an equality. An example is obtained by considering a separable Hilbert space $\mathfrak{H}$ with a complete orthonormal sequence $\{x_n : n = 0, \pm 1, \pm 2, \ldots\}$ and defining A by $Ax_n = x_{n+1}$. If $\mathfrak{J}$ is the subspace spanned by the x_n's with $n > 0$ and if $\mathfrak{R} = \mathfrak{H}$, then it is a not particularly difficult exercise to show that $\Lambda(\mathfrak{J})$ is the closed unit disc, whereas $\Lambda(\mathfrak{R})$ is only the circumference.

§ 3. Application to spectral manifolds.

The main result of § 2 can be applied to obtain a new, rather simple, and completely geometric characterization of the spectral manifolds (i. e. the subspaces $\mathfrak{E}(M)$) of a normal operator. I present this characterization because I hope that it, or some reasonable analog of it, may turn out to be useful in the study of not necessarily normal operators, and because it sheds some light on the geometric behavior of normal operators also. (Note that in the sequel, although the statements of the results and the methods of the proofs refer to normal operators only, the definitions make sense in the most general case).

If M is any set of complex numbers, I write $\mathfrak{F}(M)$ for the subspace of $\mathfrak{H}$ spanned by the class of all those invariant subspaces $\mathfrak{J}$ for which $\Lambda(\mathfrak{J}) \subset M$. The consideration of subspaces such as $\mathfrak{F}(M)$ is rather natural from the geometric point of view; it is an attempt to sort out those parts of the space $\mathfrak{H}$ on which the "proper values" of A lie in a prescribed part of the complex plane. Clearly each subspace $\mathfrak{F}(M)$ is necessarily invariant. (Warning: the definition does not imply, and it is in fact not true, that the spectrum of the restriction of A to $\mathfrak{F}(M)$ is contained in M.)

The definition in the preceding paragraph can be given for reducing subspaces in place of invariant subspaces. I introduce the subspace $\mathfrak{G}(M)$ spanned by the class of all those reducing subspaces $\mathfrak{J}$ for which $\Lambda(\mathfrak{J}) \subset M$. The characterization theorem mentioned above is that $\mathfrak{E}(M) = \mathfrak{F}(M)$ whenever the left side of the equation is defined. The consideration of the subspaces $\mathfrak{G}(M)$ is a technical device. To prove that $\mathfrak{E} = \mathfrak{F}$, it is convenient to prove first that $\mathfrak{E} = \mathfrak{G}$ (this turns out to depend on a spectral-theoretic argument independent of the results of § 2), and second that $\mathfrak{F} = \mathfrak{G}$ (and this depends in an essential way on Theorem 2). The equation $\mathfrak{E} = \mathfrak{G}$ is itself a geometric characterization of the spectral manifolds; since, however, the requirement that a subspace be reducing is more stringent than that it be invariant, the equation $\mathfrak{E} = \mathfrak{F}$ is likely to be easier to apply. The preceding comment implicitly contains half the proof of the equation $\mathfrak{F} = \mathfrak{G}$. More precisely: since $\mathfrak{G}(M)$ is the span of a smaller class of subspaces than the one used to define $\mathfrak{F}(M)$, it follows that $\mathfrak{G}(M) \subset \mathfrak{F}(M)$ for every set M.

Lemma 2. *If M is a Borel set, then* $\mathfrak{E}(M) = \mathfrak{G}(M)$.

Proof. If N is a compact subset of the complex plane, if $\lambda_0 \in N'$, and if ε is the distance from λ_0 to N, then, for every vector x in $\mathfrak{E}(N)$, it follows that

$$\|Ax - \lambda_0 x\|^2 = \int_N |\lambda - \lambda_0|^2 \, d(E(\lambda)x, x) \geqq \varepsilon^2 \|x\|^2.$$

Since this implies that $\lambda_0 \in \Pi(\mathfrak{E}(N))'$, it follows that $\lambda_0 \in \Lambda(\mathfrak{E}(N))'$ and consequently, since λ_0 is arbitrary in N', that $\Lambda(\mathfrak{E}(N)) \subset N$. The definition of $\mathfrak{G}$ implies now that $\mathfrak{E}(N) \subset \mathfrak{G}(N)$. In other words, I have proved so far that $\mathfrak{E}(M) \subset \mathfrak{G}(M)$ whenever M is compact. If M is not compact, consider compact subsets N of M; since

$$\mathfrak{E}(N) \subset \mathfrak{G}(N) \subset \mathfrak{G}(M)$$

for every such N, it follows from the regularity of spectral measures [6] that $\mathfrak{E}(M) \subset \mathfrak{G}(M)$ for every Borel set M.

To prove the reverse inclusion, it is sufficient to prove that if $\mathfrak{I}$ is a reducing subspace such that $\Lambda(\mathfrak{I}) \subset M$, then $\mathfrak{I} \subset \mathfrak{E}(\mathfrak{M})$, or, equivalently, that under these conditions $\mathfrak{I}$ is orthogonal to $\mathfrak{E}(M')$. Let P be the projection on $\mathfrak{I}$. Since P is a Hermitian operator that commutes with A, it follows that P commutes with $E(N)$ for every Borel set N. Since, therefore, $E(N)P$ is the projection with range $\mathfrak{E}(N) \cap \mathfrak{I}$, the assertion that $\mathfrak{I}$ is orthogonal to $\mathfrak{E}(N)$ is equivalent to either of the assertions $E(N)P = 0$ or $\mathfrak{E}(N) \cap \mathfrak{I} = \mathfrak{O}$ (where $\mathfrak{O}$ denotes

---

[6] See my paper in Acta Szeged, referred to in footnote [4].

the trivial subspace containing 0 only). If now $\lambda_0 \, \epsilon \, M'$, then there exists a positive number ε such that

$$\|Ax - \lambda_0 x\| \geqq \varepsilon \|x\|$$

whenever $x \, \epsilon \, \mathfrak{I}$. If $N = \{\lambda : |\lambda - \lambda_0| < \varepsilon/2\}$, and if $x \, \epsilon \, \mathfrak{E}(N)$, then

$$\|Ax - \lambda_0 x\|^2 = \int_N |\lambda - \lambda_0|^2 \, d(E(\lambda)x, x) \leqq \frac{\varepsilon^2}{4} \|x\|^2$$

so that, unless $x = 0$, x does not belong to $\mathfrak{I}$. Since this means that $\mathfrak{E}(N) \cap \mathfrak{I} = \mathfrak{O}$, it follows that $E(N)P = 0$. In other words, each point λ_0 in M' has a neighborhood N such that $E(N)P = 0$. The separability of the complex plane and the countable additivity of E imply then that $E(M')P = 0$; this completes the proof.

Theorem 3. *If M is a Borel set, then $\mathfrak{E}(M) = \mathfrak{F}(M)$* [7].

Proof. It is sufficient to prove that $\mathfrak{F}(M) \subset \mathfrak{G}(M)$, and this is true, in fact, for an arbitrary set M. Suppose, indeed, that $\mathfrak{I}$ is an invariant subspace such that $\varLambda(\mathfrak{I}) \subset M$, and let $\mathfrak{R}$ be the minimal reducing subspace containing $\mathfrak{I}$. Since, by Theorem 2, $\varLambda(\mathfrak{R}) \subset M$, and since $\mathfrak{I} \subset \mathfrak{R} \subset \mathfrak{G}(M)$, it follows that whenever $\varLambda(\mathfrak{I}) \subset M$, then $\mathfrak{I} \subset \mathfrak{G}(M)$. The definition of $\mathfrak{F}(M)$ implies now that $\mathfrak{F}(M) \subset \mathfrak{G}(M)$.

---

[7] This characterization of $\mathfrak{E}(M)$ is somewhat similar to (but quite a bit simpler than) a characterization that is given in my Acta Szeged paper, referred to in footnote [4]. It is worth while to note that the commutativity theorem for normal operators can be deduced from the present characterization just as easily as from that earlier one.

University of Chicago.

Krakowska Drukarnia Naukowa

Reprinted from the
AMERICAN JOURNAL OF MATHEMATICS
Vol. LXXIV, No. 1, pp. 237-240, 1952

COMMUTATORS OF OPERATORS.*

By Paul R. Halmos.

If H is a (complex) Hilbert space and if P and Q are operators on H (i. e. bounded linear transformations of H into itself), the *commutator* $[P, Q]$ of P and Q is defined by

$$[P, Q] = PQ - QP.$$

The *self-commutator* $[P]$ of a single operator P is defined by

$$[P] = [P^*, P] = P^*P - PP^*.$$

My purpose in this note is to make a slight contribution to our as yet very meager knowledge of what the commutator of two operators on a Hilbert space can look like. Wintner [3] proved that if P and Q are Hermitian, then $[P, Q]$ cannot be a non-zero multiple of the identity; as Putnam [1] has pointed out, Wintner's method yields the same conclusion even without the assumption that P and Q are Hermitian. Wielandt [2] obtained (by entirely different methods) a somewhat more general result, applicable to normed algebras. Wintner then asked whether or not the negative assertion that $[P, Q]$ can never be equal to the identity can be strengthened by proving that

$$\inf\{\,|\,([P, Q]x, x)\,| : \|x\| = 1\} = 0.$$

Putnam showed that this is always true on a finite dimensional Hilbert space and that it remains true in the infinite dimensional case if at least one of the two operators P and Q is Hermitian, or even normal, or even semi-normal. (An operator P is *semi-normal* if P^*P and PP^* are comparable with respect to the usual partial ordering of Hermitian operators.) I propose to show that, in general, the answer to Wintner's question is no. This assertion follows easily from the fact (Theorem 2) that the real (i. e., Hermitian) part of a commutator on an infinite dimensional Hilbert space may be prescribed arbitrarily. Theorem 2, in turn, is a consequence of the assertion (Theorem 1) that every Hermitian operator on an infinite dimensional Hilbert space is the sum of two self-commutators.

---

* Received May 25, 1951.

For a fixed Hilbert space H, let K be the set of all sequences $x = \{x_n\}$ such that $x_n \, \varepsilon \, H$, $n = 1, 2, \cdots$, and such that $\sum_n \| x_n \|^2 < \infty$. If, for any two elements x and y of K, the inner product of x and y is defined by

$$(x, y) = \sum_n (x_n, y_n),$$

then K is a Hilbert space; K is, in fact, the direct sum of countably many copies of H. Suppose that A is a Hermitian operator on H and define an operator B on K by $(Bx)_n = Ax_n$. Define another operator U on K by writing $(Ux)_1 = 0$ and $(Ux)_n = x_{n-1}$ for $n > 1$.

LEMMA 1. *If* $P = BU$, *then* $([P]x)_1 = A^2x$ *and* $([P]x)_n = 0$ *for* $n > 1$.

Proof. It is easy to verify that the operator B is Hermitian and that the adjoint of U is defined by $(U^*x)_n = x_{n+1}$. It follows that

$$(P^*Px)_n = (U^*B^2Ux)_n = (B^2Ux)_{n+1} = A^2(Ux)_{n+1} = A^2x_n$$

and, if $n > 1$, that

$$(PP^*x)_n = (BUU^*Bx)_n = A(UU^*Bx)_n = A(U^*Bx)_{n-1} = A(Bx)_n = A^2x_n.$$

Since $(PP^*x)_1 = A(UU^*Bx)_1 = A(0) = 0$, the proof of the lemma is complete.

It is convenient to say that a subspace H of a Hilbert space K is *large* if H contains infinitely many orthogonal copies of its orthogonal complement, or, in other words, if $\dim(H) \geqq \aleph_0 \dim(K - H)$. Thus, for example, a subspace of a separable Hilbert space is large if and only if it is infinite dimensional.

LEMMA 2. *A Hermitian operator with a large null space is a self-commutator.*

Proof. Suppose first that the given Hermitian operator is positive, i. e. that it can be written in the form A^2 with a Hermitian A. Let H be the closure of the range of A. Since H is the orthogonal complement of the null space of A, there is no loss of generality in assuming that the originally given Hilbert space $\tilde{H}$ contains the direct sum K of countably many copies of H, and that, moreover, H is embedded in K so that it coincides with the set of all those sequences x in K for which $x_n = 0$ whenever $n > 1$. If an operator P is defined on K, as in Lemma 1, and extended to $\tilde{H}$ by defining it to be 0 (or, for that matter, any normal operator) on the orthogonal complement $\tilde{H} - K$, then Lemma 1 implies the desired result. If the given operator is negative, the representation can be achievd with P^* in place of P. The case of a general Hermitian operator can be treated by putting together the results of the positive and the negative cases. It suffices to note that every Hermitian operator is the *direct* sum of a positive and a negative operator, and, in case the original operator has a large null space, then the direct summands can be selected so that they too have that property.

LEMMA 3. *Every Hermitian operator on an infinite dimensional Hilbert space leaves invariant at least one large subspace with a large orthogonal complement.*

Proof. The underlying Hilbert space, if it is not already separable, can be expressed as a direct sum of separable, infinite dimensional subspaces invariant under the given operator. There is, therefore, no loss of generality in restricting attention to separable Hilbert spaces. If A is Hermitian and E is the spectral measure of A, and if, for every Borel subset M of the real line, $E(M) = 0$ or 1, then A is a scalar multiple of 1. It follows easily that if, for every M, the dimension of the range of $E(M)$ is finite or co-finite, then A differs from a scalar multiple of 1 by a finite dimensional operator. In the contrary case both $E(M)$ and $1 - E(M)$ have infinite dimensional ranges for some M. In either case the conclusion of the lemma is obvious.

THEOREM 1. *Every Hermitian operator on an infinite dimensional Hilbert space is the sum of two self-commutators.*

Proof. By Lemma 3, the given operator is the sum of two Hermitian operators with large null spaces, and the theorem follows from Lemma 2.

To apply these results to a general operator P, it is necessary to break up P into its real and imaginary parts, i. e. the uniquely determined Hermitian operators A and B for which $P = A + iB$. If $P = A + iB$, $Q = C + iD$ (with A, B, C, and D Hermitian), it is convenient to write $P' = P'(P, Q) = A + iD$, $Q' = Q'(P, Q) = B + iC$. It follows that $P'' = P'(P', Q') = A + iC$, $Q'' = Q'(P', Q') = D + iB$, and finally that $P''' = P'(P'', Q'') = P$, $Q''' = Q'(P'', Q'') = Q$. The reason for introducing P' and Q' is notational convenience; in terms of them it is easy to write down the commutator of P and Q. It is, in fact, a matter of automatic computation to verify that

$$[P, Q] = 2^{-1}([P'] + [Q']) + (2i)^{-1}([P''] + [Q'']).$$

Since a self-commutator is always Hermitian, and since an operator uniquely determines its real and imaginary parts, it follows that, for instance, the real part of $[P, Q]$ is $2^{-1}([P'] + [Q'])$. Since the transformation carrying P

and Q into P' and Q' is cyclic of order 3, it follows that any one of the three pairs $\{P, Q\}$, $\{P', Q'\}$, and $\{P'', Q''\}$ uniquely determines both others. These facts, combined with Theorem 1, yield the following result.

THEOREM 2. *Every Hermitian operator on an infinite dimensional Hilbert space is the real part of a commutator.*

COROLLARY. *There exist operators P and Q such that*

$$\inf\{|([P, Q]x, x)| : \|x\| = 1\} \geqq 1.$$

Proof. Theorem 2 yields the existence of two operators P and Q such that the real part of $[P, Q]$ is the identity. It follows that $([P, Q]x, x)$ where $\|x\| = 1$, is a complex number whose real part is 1, and that, consequently, $|([P, Q]x, x)| \geqq 1$.

It might be worth while, in closing, to call attention to another consequence of Theorem 1. Since a scalar multiple of a commutator is again a commutator, Theorem 1 and the decomposition of an operator into its real and imaginary parts imply that every operator on an infinite dimensional Hilbert space is the sum of four commutators. It follows that every additive functional of such operators, that vanishes on all commutators, vanishes identically, or, in other words, that the concept of *trace* cannot be extended to operators on infinite dimensional Hilbert spaces. (This comment was called to my attention by Irving Kaplansky.) Results of this type were known before, but only under additional assumptions of continuity or positiveness.

UNIVERSITY OF CHICAGO.

REFERENCES.

[1] C. R. Putnam, "On commutators of bounded matrices," this JOURNAL, vol. 73 (1951), pp. 127-131.

[2] H. Wielandt, "Ueber die Unbeschränktheit der Operatoren der Quantenmechanik," *Mathematische Annalen*, vol. 121 (1949), p. 21.

[3] A. Wintner, "The unboundedness of quantum-mechanical matrices," *Physical Review*, vol. 71 (1947), pp. 738-739.

Reprinted from the
PROCEEDINGS OF THE AMERICAN MATHEMATICAL SOCIETY
Vol. 4, No. 1, pp. 142-149, Feb. 1953

SQUARE ROOTS OF OPERATORS

PAUL R. HALMOS, GÜNTER LUMER, AND JUAN J. SCHÄFFER[1]

Introduction. If H is a complex Hilbert space and if A is an operator on H (i.e., a bounded linear transformation of H into itself), under what conditions does there exist an operator B on H such that $B^2 = A$? In other words, when does an operator have a square root? The spectral theorem implies that the normality of A is a sufficient condition for the existence of B; the special case of positive definite operators can be treated by more elementary means and is, in fact, often used as a step in the proof of the spectral theorem. As far as we are aware, no useful necessary and sufficient conditions for the existence of a square root are known, even in the classical case of finite-dimensional Hilbert spaces. The problem of finding some easily applicable conditions is of interest, in part because the use of square roots is frequently a helpful technique in the study of algebraic properties of operators, and in part because of the information that such conditions might yield about the hitherto rather mysterious behavior of non-normal operators.

If a non-zero, 2-rowed square matrix is nilpotent, then its index of nilpotence is equal to 2; this comment shows that no such matrix can have a square root. On the other hand, an elementary computation, based on the Jordan canonical form, shows that every invertible matrix does have a square root. Since the number 0 is known to have a special significance in the formation of square roots, it is not unreasonable to conjecture that its absence from the spectrum of an operator A is sufficient to ensure the existence of a square root of A, or, in other words, that even on not necessarily finite-dimensional Hilbert spaces, every invertible operator has a square root. (This conjecture was first called to our attention by Irving Kaplansky.) The main purpose of this paper is to prove that this conjecture is false. More precisely, we shall describe a small but interesting class of operators, derive a necessary and sufficient condition that an operator in this class have a square root, and achieve our announced purpose by exhibiting a relatively large subclass of invertible operators that do not satisfy the condition. We note in passing that our methods solve the analogous problem for nth roots, $n \geqq 2$, and that,

Presented to the Society, September 5, 1952; received by the editors April 4, 1952.
[1] The research on this paper was done while all the authors were in residence at the Instituto de Matemática y Estadística, Universidad de la República, Montevideo, Uruguay.

curiously, the condition for the existence of an nth root, within the class of operators we treat, is the same as that for the special case $n = 2$.

Analytic position operators. Let D be a bounded domain (i.e., a bounded, open, and connected subset of the complex plane), let μ be planar Lebesgue measure in D, and let H be the set of all complex-valued functions that are analytic throughout D and square-integrable with respect to μ. In other words, an element of H is an analytic function x of the complex variable t in D and is such that

$$\|x\|^2 = \int_D |x(t)|^2 d\mu(t) < \infty.$$

With respect to the pointwise linear operations, and the inner product defined by

$$(x, y) = \int_D x(t)\bar{y}(t)d\mu(t),$$

the set H is a complex inner product space. If $\lambda \varepsilon D$ and $x \varepsilon H$, we write

$$v_\lambda(x) = x(\lambda);$$

it is obvious that, for each fixed λ, the functional v_λ is linear.

LEMMA 1. *For each λ in D, the functional v_λ on H is bounded; if, in fact, $a^2(\lambda)$ is the area of the largest open circle C with center at λ that is contained in D, then*

$$|v_\lambda(x)| \leq \frac{1}{a(\lambda)}\|x\|.$$

PROOF. Since

$$\|x\|^2 = \int_D |x(t)|^2 d\mu(t) \geq \int_C |x(t)|^2 d\mu(t)$$

$$= \int_C \left| \sum_{n=0}^\infty a_n(t - \lambda)^n \right|^2 d\mu(t),$$

and since the powers of $(t-\lambda)$ constitute an orthogonal set in $L_2(C)$, it follows that

$$\|x\|^2 \geq \int_C |a_0|^2 d\mu(t) = a^2(\lambda) \cdot |x(\lambda)|^2.$$

LEMMA 2. *The inner product space H is a Hilbert space.*

PROOF. If $\{x_n\}$ is a Cauchy sequence in H, then, by Lemma 1,

$$| x_n(\lambda) - x_m(\lambda) | \leq \frac{1}{a(\lambda)} \| x_n - x_m \|$$

for every λ in D. It follows that if D_0 is a compact subset of D, so that $a(\lambda)$ is bounded away from 0 when $\lambda \varepsilon D_0$, then the sequence $\{x_n\}$ of functions is uniformly convergent on D_0. This implies that there exists an analytic function x on D such that $x_n(\lambda) \to x(\lambda)$ for all λ in D. At the same time, the completeness of $L_2(D)$ implies the existence of a complex-valued, square-integrable, but not necessarily analytic, function y on D such that $x_n \to y$ in the mean of order 2. It follows that a subsequence of $\{x_n\}$ converges to y almost everywhere and hence that $x = y$ almost everywhere. This implies that x is square-integrable, i.e., that $x \varepsilon H$, and hence that H is complete.[2]

The equation $(Ax)(t) = tx(t)$ defines a linear transformation A of H into itself; the boundedness of D implies that A is an operator. We shall call A the *analytic position operator* associated with the domain D. The class of analytic position operators is the class we mentioned in the introduction.[3]

Spectra. For our purposes we shall have need of an almost complete analysis of the spectrum of an analytic position operator; we devote this brief section to recalling the pertinent facts about spectra. If A is an operator on a Hilbert space H, the *spectrum* of A, in symbols $\Lambda(A)$, is the set of all those complex numbers λ for which $A - \lambda$ is not invertible. A well known and easy geometric argument shows that there are two (exhaustive but not exclusive) ways in which

---

[2] The results of Lemmas 1 and 2 are not new; they can be found, for instance, in Stefan Bergman, *Sur les fonctions orthogonales de plusieurs variables complexes avec les applications à la théorie des fonctions analytiques*, Paris, 1947, p. 24. Since, however, we have occasion below to make use of the notation and of the cornerstone (Lemma 1) of the reasoning above, we thought it appropriate to give the proof. It might also be remarked that since our proof makes explicit use of the Riesz-Fischer theorem, instead of proving it in the particular case at hand, it is somewhat simpler, from the point of view of the standard theory of Hilbert spaces, than the more analytic argument given by Bergman.

[3] If μ is a compact measure in the complex plane (i.e., a measure with compact support, in the terminology of Bourbaki), and if $(Ax)(t) = tx(t)$ for every complex-valued function x that is square-integrable with respect to μ, then we call A the position operator associated with μ. The adjective *analytic* serves as a reminder that the spectrum (support) of μ is the closure of a bounded open set and that the domain of A consists only of functions analytic in the given open set. The terminology is motivated by an analogy with certain operators considered in quantum mechanics.

$A - \lambda$ can fail to be invertible. One way is that the range of $A - \lambda$, in symbols $R(A - \lambda)$, fails to be dense in H; if this is the case, we say that λ belongs to the *compression spectrum* of A and we write $\lambda \varepsilon \Sigma(A)$. Another way is that, for suitable vectors x, the expression $\|(A - \lambda)x\|$ becomes arbitrarily small in comparison with $\|x\|$; if this is the case, we say that λ belongs to the *approximate point spectrum* of A and we write $\lambda \varepsilon \Pi(A)$. A more special way for $A - \lambda$ to fail to be invertible is for λ to belong to the *point spectrum* $\Pi_0(A)$; this means, of course, that $(A - \lambda)x = 0$ for a suitable non-zero vector x. We shall need to make use of the facts that (1) $\Lambda(A)$ is a closed (and in fact compact) subset of the complex plane, (2) $\Lambda(A) = \Sigma(A) \cup \Pi(A)$, and (3) $\Pi_0(A^*) = \Sigma^*(A)$. (In (3), A^* denotes, of course, the adjoint of the operator A, and $\Sigma^*(A)$ denotes the set of all complex numbers of the form $\bar{\lambda}$ with $\lambda \varepsilon \Sigma(A)$.)[4]

We shall also need a slightly more deep-lying fact about spectra, a special case of the so-called spectral mapping theorem.[5] The part of the theorem that is relevant to our work asserts that if p is any polynomial (and, in particular, if $p(\lambda) = \lambda^2$), then $\Sigma(p(A)) = p(\Sigma(A))$, $\Pi(p(A)) = p(\Pi(A))$, and $\Pi_0(p(A)) = p(\Pi_0(A))$. (A symbol such as $p(\Sigma(A))$ denotes the set of all complex numbers of the form $p(\lambda)$ with $\lambda \varepsilon \Sigma(A)$.) It is pertinent to remark that the first two of these three equations, together with (2) above, imply that $\Lambda(p(A)) = p(\Lambda(A))$.

The spectrum of an analytic position operator. Suppose now that A is the analytic position operator associated with a domain D; we propose to establish the connection between D and the spectrum of A. We begin with an auxiliary result.

LEMMA 3. *If $\lambda \varepsilon D$ and if y is a function in H such that $y(\lambda) = 0$, then there exists a unique function x in H such that $y(t) = (t - \lambda)x(t)$ for all t in D. If, moreover, δ is a positive number such that the open circle C with center λ and radius δ is contained in D, then*

$$\|y\|^2 \geqq \frac{\delta^2}{2} \|x\|^2.$$

PROOF. It is obvious that there exists a unique analytic function x on D such that $y(t) = (t - \lambda)x(t)$ for all t in D. Since the square-integrability of x, i.e., the relation $x \varepsilon H$, is a weaker condition than the asserted inequality between $\|x\|$ and $\|y\|$, it remains only to prove that inequality.

---

[4] For a treatment of the elementary properties of spectra see, for instance, Paul R. Halmos, *Introduction to Hilbert space and the theory of spectral multiplicity*, New York, 1951, p. 50.

[5] See Einar Hille, *Functional analysis and semi-groups*, New York, 1948, p. 123.

We observe first of all (cf. the proof of Lemma 1) that $\int_C (t-\lambda)^n(\bar{t}-\bar{\lambda})^m d\mu(t) = 0$ unless $m = n$ ($n, m = 0, 1, 2, \cdots$), and that

$$\int_C |(t-\lambda)^n|^2 d\mu(t) = \frac{\pi\delta^{2n+2}}{n+1}:$$

these facts are the results of an easily verified computation. If x is expanded in powers of $(t-\lambda)$, i.e.,

$$x(t) = \sum_{n=0}^{\infty} a_n(t-\lambda)^n, \; t \varepsilon C,$$

then it follows that

$$\int_C |x(t)|^2 d\mu(t) = \sum_{n=0}^{\infty} |a_n|^2 \cdot \frac{\pi\delta^{2n+2}}{n+1}$$

and

$$\int_C |(t-\lambda)x(t)|^2 d\mu(t) = \sum_{n=0}^{\infty} |a_n|^2 \cdot \frac{\pi\delta^{2n+4}}{n+2}.$$

Consequently

$$\int_C |(t-\lambda)x(t)|^2 d\mu(t) = \sum_{n=0}^{\infty} \frac{\delta^2(n+1)}{n+2} \cdot |a_n|^2 \cdot \frac{\pi\delta^{2n+2}}{n+1}$$

$$\geq \frac{\delta^2}{2} \int_C |x(t)|^2 d\mu(t).$$

The desired result now follows from the relations

$$\|y\|^2 = \int_D |(t-\lambda)x(t)|^2 d\mu(t)$$

$$\geq \frac{\delta^2}{2} \int_C |x(t)|^2 d\mu(t) + \delta^2 \int_{D-C} |x(t)|^2 d\mu(t)$$

$$\geq \frac{\delta^2}{2} \|x\|^2.$$

THEOREM 1. *If A is the analytic position operator associated with a domain D, and if $\lambda \varepsilon D$, then λ is not an approximate proper value of A, but $\bar{\lambda}$ is a simple proper value of A^*, and therefore, in particular, $D \subset \Sigma(A) - \Pi(A)$.*

PROOF. Lemma 1 implies that

$$\|(A - \lambda)x\|^2 \geq \frac{\delta^2}{2}\|x\|^2$$

for all x in H and hence that λ is not an approximate proper value of A. Lemma 1 implies also that a necessary and sufficient condition that a function y belong to $R(A - \lambda)$ is that $v_\lambda(y) = 0$. Since the range of an operator is dense in the orthogonal complement of the null space of its adjoint, it follows that $A^*z = \bar{\lambda}z$ if and only if z is orthogonal to the null space of v_λ. Since, by the standard Riesz representation of bounded linear functionals in terms of inner products, the dimension of the orthogonal complement of the null space of v_λ is equal to 1, the proof of the theorem is complete.

On the basis of this theorem it is easy to determine the various parts of the spectrum completely. For our purposes, however, it is sufficient to know Theorem 1 and the additional, global fact that $\Lambda(A) = \overline{D}$. Indeed, since $D \subset \Lambda(A)$ by Theorem 1, and since $\Lambda(A)$ is closed, it follows that $\overline{D} \subset \Lambda(A)$. If, on the other hand, λ does not belong to $\overline{D}$, and if

$$f_\lambda(t) = \frac{1}{t - \lambda},$$

then f_λ is a bounded analytic function on D and consequently the equation $(A_\lambda x)(t) = f_\lambda(t)x(t)$ defines an operator A_λ on H that is easily seen to be a two-sided inverse of $A - \lambda$. In other words $\Lambda(A)$ does not contain any point in the complement of $\overline{D}$, so that $\Lambda(A) = \overline{D}$.

Square roots of analytic position operators. In order to state the main theorem of this paper, it is convenient to introduce a new notation; if D is a domain, we shall write $\sqrt{D}$ for the (open) set of all complex numbers λ such that $\lambda^2 \varepsilon D$.

THEOREM 2. *If A is the analytic position operator associated with a domain D, then a necessary and sufficient condition that A have a square root is that $\sqrt{D}$ be disconnected.*

PROOF. Let U be a component of $\sqrt{D}$ and write $V = U \cup (-U)$, where $-U$ denotes the set of all complex numbers of the form $-\lambda$ with $\lambda \varepsilon U$. Since D is open, so also are V and $W = D - V$. Since V is the complete square root of V^2, and since V and W are disjoint, it follows that V^2 and W^2 are disjoint, where V^2, for instance, denotes the set of all complex numbers of the form λ^2 with $\lambda \varepsilon V$. Since an analytic function (and in particular the function f defined by $f(\lambda) = \lambda^2$) is an open mapping, V^2 and W^2 are open, and, clearly, $V^2 \cup W^2 = D$. The connectedness of D implies that W^2 is empty and hence that W

is empty. This proves that if $\sqrt{D}$ is not connected, then it consists of exactly two open components: U and its reflection $-U$. The correspondence f (where $f(\lambda) = \lambda^2$) is one-to-one between U and D, and therefore the inverse correspondence g between D and U is a (single-valued) bounded analytic function on D. The equation $(Bx)(t) = g(t)x(t)$ defines, therefore, an operator B on H; evidently $B^2 = A$.

Suppose now that $\sqrt{D}$ is connected; the same is then true of the set $\sqrt{D} - \{0\}$ obtained by deleting the origin from $\sqrt{D}$, in case it happened to belong to $\sqrt{D}$. We shall derive a contradiction from the assumption that there exists an operator B on H such that $B^2 = A$. Since $D \subset \Sigma(A) - \Pi(A)$, it follows from the spectral mapping theorem that no point of $\sqrt{D}$ can belong to $\Pi(B)$. Since $\Lambda^2(B) = \Lambda(A)$, it follows that if $\lambda \varepsilon D$, then at least one of the two square roots of λ belongs to $\Lambda(B)$ and hence, in view of the preceding sentence, to $\Sigma(B)$. If both the square roots of a non-zero number λ in D belonged to $\Sigma(B)$, then their conjugates would belong to $\Pi_0(B^*)$ and this would contradict the fact that $\bar{\lambda}$ is a simple proper value of A^*. What we have proved so far may be formulated as follows: if $\lambda \varepsilon \sqrt{D} - \{0\}$, then exactly one of the two numbers λ and $-\lambda$ belongs to $\Lambda(B)$. The set $(\sqrt{D} - \{0\}) \cap \Lambda(B)$ is a closed subset of $\sqrt{D} - \{0\}$ in the relative topology; the homeomorphism $\lambda \rightarrow -\lambda$ carries it onto its relative complement in $\sqrt{D} - \{0\}$, which is therefore also closed. Since $\sqrt{D} - \{0\}$ is connected, we have reached the desired contradiction, and the proof of the theorem is complete.[6]

Conclusion. (a) The condition of Theorem 2 is of a mixed algebraic and topological nature. It is not difficult, however, to replace it by a purely topological condition. Elementary reasoning, based on the information that the proof of Theorem 2 yields about the structure of $\sqrt{D}$, implies that $\sqrt{D}$ is connected if and only if D contains a Jordan curve surrounding the origin. Less elementary, topological, considerations can be used to show also that the latter condition is satisfied if and only if the origin does not belong to the infinite component of the complement of D.[7] Both these conditions may be expressed, in intuitive terms, by saying that D surrounds the origin.

---

[6] A sufficient condition for the existence of square roots of more general operators is given by Hille (op. cit., p. 276); in the special case of analytic position operators, however, the condition of Theorem 2 is considerably less restrictive than Hille's.

[7] The sufficiency of this condition follows also from Aurel Wintner, *On the logarithms of bounded matrices*, Amer. J. Math. vol. 74 (1952) pp. 360–364. Wintners' work appeared after this paper had been submitted for publication. Wintner, incidentally, also raises the question (answered in (c) below) of the existence of an invertible operator without a square root.

(b) For reasons of notational convenience we treated the case of square roots only; exactly the same method, however, solves the problem of nth roots for any $n = 2, 3, 4, \cdots$. The principal theorem in this case is obtained from Theorem 2 by replacing $\sqrt{D}$ by $\sqrt[n]{D}$. Since the purely topological reformulation of this condition, along the lines of (a) above, is independent of n, it follows that a necessary and sufficient condition that an analytic position operator have an nth root ($n = 2, 3, 4, \cdots$) is that it have a square root.

(c) What we now know about the spectra and the square roots of analytic position operators makes it very easy to construct an example of the type mentioned in the introduction, i.e., an example of an invertible operator without a square root. Indeed, the analytic position operator associated with a domain D is invertible if and only if 0 does not belong to $\overline{D}$ and it has a square root if and only if $\sqrt{D}$ is disconnected. The problem becomes then the construction of a domain D far from and surrounding the origin. This problem has many obvious solutions; one, for example, is an annulus with center at the origin.

University of Chicago and
    University of Montevideo

Reprinted from the
AMERICAN JOURNAL OF MATHEMATICS
Vol. LXXVI, No. 1, pp. 191-198, Jan. 1954

COMMUTATORS OF OPERATORS, II.*

By Paul R. Halmos.

---

1. Introduction. Suppose that A and B are linear transformations on a complex Hilbert space H and let C be their commutator, $C = AB - BA$. Problems concerning C can be studied on three levels of generality. On the lowest level, it is assumed that H is finite-dimensional; on the intermediate level, infinite-dimensional spaces are allowed, but only bounded linear transformations (here called *operators*) are considered; and, on the highest level, both the space and the transformations remain completely unrestricted. Since, roughly speaking, on the highest level anything can happen, whereas on the lowest level nothing can happen, the mathematically interesting and difficult problems are all likely to be on the intermediate level. Thus, for example, the identity operator is a commutator of unbounded transformations, but not of operators, and the identity operator is the sum of two commutators of operators, whereas the identity matrix (of finite size) is not the sum of any finite number of commutators of matrices.

The purpose of this note is to obtain some positive and some negative results concerning commutators on the middle level, and to call attention to some unsolved problems. The results are of three types. Section 2 studies the role which 0 plays in the spectrum of a commutator. Section 3 concentrates on the problem of approximation by commutators, and, in particular, on how near a commutator can be to the identity. In Section 4 a factorization theorem, valid for every operator, is proved and applied to the problem of studying the spectrum of a commutator.

The three levels of operatorial generality have their analogs in pure algebra; the lowest level corresponds to finite-dimensional algebras, the middle level to normed algebras, and the highest level to completly unrestricted algebras. At times the relation becomes more than an analogy; in particular, it will be seen that several of the proofs below can be transferred from (bounded) operators to (normed) algebras without any change. Nevertheless, in order to avoid unnecessary complications, the more special terminology and notation of operator theory will be preferred throughout.

It is occasionally convenient to use the differential notation for commu-

---

* Received October 1, 1953.

tators. Thus, to study $AB - BA$, the operator B will be fixed once and for all, and the commutator, regarded as a function of A, will be denoted by A'. The mapping $A \to A'$ is a derivation; i. e., it is a linear transformation from operators to operators that has many of the properties of ordinary differentiation. Thus

$$(A_1 A_2)' = A_1 A_2 B - B A_1 A_2$$
$$= A_1 (A_2 B - B A_2) + (A_1 B - B A_1) A_2 = A_1 A'_2 + A'_1 A_2,$$

and therefore the non-commutative versions of all the usual Leibniz formulas are valid. The product formula applied to A^k ($k = 1, 2, \cdot \cdot \cdot$) yields $(A^k)' = A^{k-1} A' + (A^{k-1})' A$; by an obvious induction argument it follows that

$$(A^k)' = \sum_{i=0}^{k-1} A^{k-1-i} A' A^i.$$

2. Divisors of zero. In the algebraic study of commutators an important role is played by Jacobson's lemma ([4], Lemma 2). That lemma asserts that, in an algebraic algebra over a field of characteristic zero, if A' commutes with A, then A' is nilpotent. Kaplansky conjectured that an analog of this result is valid for normed algebras, i. e., that if A' commutes with A, then A' is quasi-nilpotent. (Recall that, by definition, A' is quasi-nilpotent if $\| (A')^n \|^{1/n} \to 0$ as $n \to \infty$, or, equivalently, if the spectrum of A' consists of 0 alone. Still another formulation says that $r(A') = 0$, where r, here and below, denotes the spectral radius. For a discussion of such concepts and their role in normed algebras see [3], Chapter V.) Kaplansky's conjecture is not settled yet; the following results, however, are corroborative evidence.

THEOREM 1. *If A' commutes with A, then A' is a (two-sided) generalized divisor of zero.*

Proof. The equation for the "derivative" of A^n implies that under the present commutativity hypothesis $(A^n)' = n A^{n-1} A'$. It follows that if A is nilpotent (in the usual algebraic sense), then A' is a divisor of zero (also in the algebraic sense). If A is not nilpotent, then the relation

$$n \| A^{n-1} A' \| = \| A^n B - B A^n \| \leq 2 \| A \| \cdot \| B \| \cdot \| A^{n-1} \|$$

implies that $\| A^{n-1} A' \| / \| A^{n-1} \| \to 0$ as $n \to \infty$, i. e., that A' annihilates the sequence $\{A^{n-1}/\| A^{n-1} \|\}$ of operators of norm 1. Since this is the definition of a generalized divisor of zero, the proof is complete.[1]

The above proof is valid in any normed algebra. The method was first used by Wielandt in [6] to prove that A' cannot be equal to 1; for an earlier and different proof, see Wintner [7]. That result, of course, is an immediate consequence of Theorem 1. Equally immediate is the following corollary, which is being put on record here for convenience of reference.

COROLLARY. *If A' commutes with A, then A' is not invertible.*

The conclusion of Theorem 1 is expressed in normed-algebraic terms; it is sometimes convenient to be able to use operatorial language. This desideratum is achieved by the following slight weakening of Theorem 1.

THEOREM 2. *If A' commutes with A, then 0 belongs to the approximate point spectrum of A'.*

(For related results, see [5], especially Theorem II, p. 359.)

Proof. It is sufficient to prove that an operator C is a generalized left divisor of zero if and only if 0 belongs to its approximate point spectrum. Suppose therefore that $CD_n \to 0$, with $\| D_n \| = 1$, and let x_n be a vector such that $\| D_n x_n \| \geqq \frac{1}{2} \| x_n \|$. If $y_n = D_n x_n / \| D_n x_n \|$, then $\| y_n \| = 1$, and

$$\| C y_n \| = \| C D_n x_n \| / \| D_n x_n \| \leqq \| C D_n \| \cdot \| x_n \| / (\tfrac{1}{2}) \| x_n \| = 2 \| C D_n \| \to 0.$$

If, conversely, $Cx_n \to 0$ as $n \to \infty$, with $\| x_n \| = 1$, write $D_n x = (x, x_n) x_n$. It follows that D_n is the projection on the one-dimensional space spanned by x_n, so that $\| D_n \| = 1$. The relation

$$\| C D_n x \| = \| C (x, x_n) x_n \| \leqq \| C x_n \| \cdot \| x \|,$$

valid for all n and x, implies that $\| C D_n \| \leqq \| C x_n \| \to 0$ as $n \to \infty$.

3. Density. The results of the preceding section are essentially negative; since they say that commutators must satisfy certain conditions, their chief use is in proving that certain operators are not commutators. In this section and the next the emphasis will be positive; it will be shown that on an infinite-dimensional Hilbert space there are many commutators.

LEMMA 1. *If A is an operator on a Hilbert space H, and if M is a*

---

[1] I should like to express my appreciation to Günter Lumer and Juan J. Schäffer for many interesting conversations on the subject of commutators. Theorem 1, in particular, was obtained in collaboration with Schäffer.

13

subspace of H such that $\dim M \leqq \dim M^\perp$, *then there exists a commutator* C *on* H *such that* $Cx = Ax$ *whenever* $x \, \varepsilon \, M$.

Proof. Let Q be a partial isometry with initial space M and final space $N \subset M^\perp$. In other words, Q sends M isometrically onto N and annihilates $M^\perp$, whereas Q^* sends N isometrically onto M and annihilates $N^\perp$ (and therefore annihilates M). If $P = AQ^*$, then the desired commutator is $C = PQ - QP$; indeed, if $x \, \varepsilon \, M$, then

$$Cx = PQx - QPx = AQ^*Qx - QAQ^*x = Ax.$$

THEOREM 3. *If* H *is infinite-dimensional, then the set of commutators is dense, with respect to the strong topology, in the set of all operators on* H.

Proof. For any operator A and vectors $x_1, \cdots, x_n$, let M be the space spanned by the x's, and apply Lemma 1 to obtain a commutator C such that $Cx_j = Ax_j$, $j = 1, \cdots, n$. The existence of such a C implies that every strong neighborhood of A contains a commutator.

The preceding result makes it natural to ask whether or not the set of commutators is dense with respect to the uniform (norm) topology as well. The answer to that question is not known. In view of this ignorance, and in view of the quantum-mechanical history of the study of commutators, it becomes pertinent to ask: how well can the identity operator be approximated by commutators? In finite-dimensional spaces the answer is easy.

THEOREM 4. *If* H *is finite-dimensional, then* $r(1 - A') \geqq 1$ *for all* A.

Proof. For any operator A on H, let $t(A)$ denote the normalized trace of A, i. e., the average of the proper values of A. Since $t(A') = 0$ for all A, it follows that not all the proper values of A' can lie in the circle $\{\lambda : |1 - \lambda| < 1\}$.

COROLLARY. *If* H *is finite-dimensional, then* $\| 1 - A' \| \geqq 1$ *for all* A.

Proof. This follows immediately from Theorem 4, together with the fact that the normalized trace is dominated by the norm.

Note that, that since 0 is a commutator, the result of the corollary is best possible.

The main purpose of the next result is purely negative; it shows that the hypothesis of finite-dimensionality is indispensable in the corollary to Theorem 4.

THEOREM 5. *If* H *is infinite-dimensional, then there exists a commutator* C *on* H *such that* $\| 1 - C \|^2 \leqq .97$.

Proof. Let U and V be isometries on H with respective final spaces M and N, where $M = N^{\perp}$. (The existence of U and V is ensured by the fact that H is infinite-dimensional.) In other words, $U^*U - UU^* = 1 - UU^*$ is the projection on $M^{\perp}$, and $V^*V - VV^* = 1 - VV^*$ is the projection on $N^{\perp} = M$; it follows that $U^*U - UU^* + V^*V - VV^* = 1$. If P and Q are defined by

$$2^{\frac{1}{2}}P = (U + U^*) + i(V + V^*), \qquad 2^{\frac{1}{2}}Q = (U - U^*) - i(V - V^*),$$

then a straightforward computation shows that $C = PQ - QP = 1 + iR$, where R is the Hermitian operator $(U^*V^* - V^*U^*) - (UV - VU)$. It follows that $\| R \| \leqq 4$. (In my earlier paper [1] I showed that every operator, and in particular the identity, can be the Hermitian part of a commutator. Up to this point the present proof succeeded only in re-establishing that result, in a somewhat simpler and more explicit form than before, and in furnishing an estimate for the size of $\| R \|$.)

Every element of the spectrum of C is of the form $1 + i\alpha$, where α is a real number such that $| \alpha | \leqq \| R \| \leqq 4$. If δ is a positive number, and if λ belongs to the spectrum of δC, then $\lambda = \delta + i\delta\alpha$, so that

$$| 1 - \lambda |^2 = (1 - \delta)^2 + \delta^2\alpha^2.$$

It follows that if $\delta = .1$, then $| 1 - \lambda |^2 \leqq .97$. Three comments complete the proof. First: since the Hermitian part of C is 1, and therefore commutative with its skew-Hermitian part, the operator C is normal (and so also is δC). Second: since, for a normal operator, the spectral radius and the norm coincide, $\| 1 - \delta C \|^2 \leqq .97$. Third: since C is a commutator, so is δC.

In general the infimum of $\| 1 - C \|$, as C varies over all commutators, is not known. There are, however, two special situations in which the inequality of the corollary to Theorem 4 remains valid.

THEOREM 6. *If A' commutes with A, then $\| 1 - A' \| \geqq 1$.*

Proof. If $\| 1 - A' \| < 1$, then A' is invertible (see [3], Chapter V); the desired result follows from the corollary to Theorem 1.

THEOREM 7. *If the operator A is such that*

$$\limsup n^{-1} \sum_{i=0}^{n-1} \| A^{n-1-i} \| \cdot \| A^i \| / \| A^{n-1} \| \leqq 1,$$

then $\| 1 - A' \| \geqq 1$.

Remark. Since

$$n = \sum_{i=0}^{n-1} \| A^{n-1-i} \cdot A^i \| / \| A^{n-1} \| \leqq \sum_{i=0}^{n-1} \| A^{n-1-i} \| \cdot \| A^i \| / \| A^{n-1} \|,$$

the hypothesis trivially implies that the limit exists and is equal to 1.

Proof. The Wielandt trick can be applied again. Since

$$(A^n)' = \sum_{i=0}^{n-1} A^{n-1-i} A' A^i = n A^{n-1} \sum_{i=0}^{n-1} A^{n-1-i} (1 - A') A^i,$$

it follows that

$$n \, \| A^{n-1} \| \leqq \| A^n B - B A^n \| + \| \sum_{i=0}^{n-1} A^{n-1-i} (1 - A') A^i \|$$

$$\leqq 2 \, \| A \| \cdot \| B \| \cdot \| A^{n-1} \| + \| 1 - A' \| \sum_{i=0}^{n-1} \| A^{n-1-i} \| \cdot \| A^i \|.$$

Division by $\| A^{n-1} \|$ yields

$$n \leqq 2 \, \| A \| \cdot \| B \| + \| 1 - A' \| \sum_{i=0}^{n-1} \| A^{n-1-i} \| \cdot \| A^i \| / \| A^{n-1} \|.$$

It follows that the hypothesis of the theorem is not consistent with the relation $\| 1 - A' \| < 1$, and the proof is complete.

The hypothesis of Theorem 7 looks weird; its principal application is to subnormal (and a fortiori to normal) operators. An operator is subnormal if it has a normal extension (to a larger Hilbert space). The point is that if A is subnormal, then $\| A^k \| = \| A \|^k$ for every positive integer k. (For normal operators this is well known; for subnormal operators, it follows from the results of [2].) In view of these comments, and of the fact that an isometry is easily seen to be subnormal, the following assertion is an immediate consequence of Theorem 7.

COROLLARY. *If A is subnormal (and, in particular, if A is an isometry, or if A is normal), then $\| 1 - A' \| \geqq 1$.*

The considerations of this section have been guided mainly by the question: can a commutator be near to the identity in the sense of norm? A related question is: can a commutator C be near to the identity in the purely algebraic sense that $1 - C$ is nilpotent? In the finite-dimensional case the answer is obviously no; the trace of a nilpotent operator is 0, whereas the (normalized) trace of $1 - C$ is 1. In the general case this problem takes its place with the many other problems concerning commutators of operators that still remain unsolved.

4. Factorization. The main result (Theorem 8) of this final section is somewhat paradoxical; it asserts that (if H is infinite-dimensional) every operator behaves in some respects like a commutator.

If K is the direct sum of two copies of H, then every operator on K can be considered as a two-rowed matrix whose entries are operators on H. The following auxiliary result considers such matrices and proves that certain special ones among them are commutators. Since, in the infinite-dimensional case, H and K are isomorphic, the result could also be formulated in terms of H alone, but only at a considerable cost in clarity.

LEMMA 2. *If H is infinite-dimensional, then to every operator A on H there correspond two matrices*

$$X = \begin{pmatrix} 0 & P \\ 0 & R \end{pmatrix}, \, Y = \begin{pmatrix} 0 & 0 \\ Q & S \end{pmatrix} \, with \, \begin{pmatrix} A & 0 \\ 0 & 0 \end{pmatrix} = XY - YX,$$

where P, Q, R, and S are operators on H.

Proof. If K is the direct sum of a countably infinite number of copies of H, then every operator on K can be written as an infinite matrix whose entries are operators on H. Just as in the case of numerical (infinite) matrices, not every such matrix corresponds to an operator. If, however, X is the matrix whose entries in the diagonal just above the main diagonal are all equal to A, and if Y is the matrix whose entries in the diagonal just below the main diagonal are all equal to 1, all other entries in both cases being 0, then X and Y do correspond to operators on K. It is easy to verify that the commutator $XY - YX$ has A in the $(1,1)$ position and 0 everywhere else. Since the first "axis" of K is isomorphic to the direct sum of all the others, i. e., since K may also be considered as the direct sum of two copies of H, the result just obtained can also be expressed in terms of two-rowed matrices. The effect of expressing it in such terms is exactly the desired conclusion.

THEOREM 8. *Every operator on an infinite-dimensional space can be written as the product of two operators whose product in the reverse order is a commutator.*

Proof. An application of Lemma 2 to the prescribed operator A shows the existence of operators P, Q, R, and S such that

$$\begin{pmatrix} A & 0 \\ 0 & 0 \end{pmatrix} = \begin{pmatrix} PQ & PS \\ RQ & (RS - SR) - QP \end{pmatrix}.$$

It follows that $A = PQ$ and $QP = RS - SR$.

COROLLARY. *Every operator on an infinite-dimensional space is the sum of two commutators.*

Proof. If $A = PQ$ with $QP = RS - SR$, then $A = PQ - QP + RS - SR$.

This corollary is an improvement of one of the results of [1]; the methods of [1] yielded the same statement with *four* in place of *two*.

What can the spectrum of a commutator look like? Can it be an arbitrary compact subset of the complex plane? Here again the answer is not known. Easy finite-dimensional examples show that 0 need not belong to the spectrum of a commutator. If, however, 0 does belong to the spectrum, then the rest of the structure of the spectrum can be prescribed arbitrarily. The precise formulation of this somewhat vague statement is the following application of Lemma 2.

THEOREM 9. *If Λ is a compact subset of the complex plane, with $0 \, \varepsilon \, \Lambda$, then there exists a commutator C whose spectrum is Λ.*

Proof. Let A be an operator whose spectrum is Λ; the existence of an operator, and even of a normal operator, with prescribed spectrum is a standard exercise in elementary Hilbert space theory. If $C = \begin{pmatrix} A & 0 \\ 0 & 0 \end{pmatrix}$, then, since $0 \, \varepsilon \, \Lambda$, the spectrum of C is also Λ; the desired result follows from Lemma 2.

UNIVERSITY OF CHICAGO.

REFERENCES.

[1] P. R. Halmos, " Commutators of operators," *American Journal of Mathematics*, vol. 74 (1952), pp. 237-240.

[2] ———, " Spectra and spectral manifolds," *Annales de la Société Polonaise de Mathématique*, vol. 25 (1952), pp. 43-49.

[3] E. Hille, *Functional analysis and semi-groups*, New York, 1948.

[4] N. Jacobson, " Rational methods in the theory of Lie algebras," *Annals of Mathematics*, vol. 36 (1935), pp. 875-881.

[5] C. R. Putnam, " On normal operators in Hilbert space," *American Journal of Mathematics*, vol. 73 (1951), pp. 357-362.

[6] H. Wielandt, " Ueber der Unbeschränktheit der Operatoren der Quantenmechanik," *Mathematische Annalen*, vol. 121 (1949), p. 21.

[7] A. Wintner, " The unboundedness of quantum-mechanical matrices," *Physical Review*, vol. 71 (1947), pp. 738-739.

Reprinted from the
BULLETIN OF THE AMERICAN MATHEMATICAL SOCIETY
Vol. 64, No. 3, Part I, pp. 77-78
May, 1958

PRODUCTS OF SYMMETRIES

BY PAUL R. HALMOS AND SHIZUO KAKUTANI

Communicated January 25, 1958

A (bounded) operator Q on a (complex) Hilbert space H is a *symmetry* if it is a unitary involution, i.e., if $Q^*Q = QQ^* = 1$ (= the identity operator on H) and $Q^2 = 1$. In connection with his studies of the infinite-dimensional analogues of the classical groups, R. V. Kadison has asked us which operators can be represented as (finite) products of symmetries. The purpose of this note is to give a precise answer to Kadison's question.

THEOREM 1. *If H is infinite-dimensional, then every unitary operator on H is the product of four symmetries.*

PROOF. We need the auxiliary result that if U is a unitary operator on an infinite-dimensional Hilbert space H, then there exists a (closed) subspace H_0 of H such that H_0 reduces U and such that dim H_0 = dim $H_0^\perp$. This result holds, in fact, for an arbitrary normal operator on H. Since the proof is a straightforward application of the spectral theorem, and since the proof for a typical special case (namely, for Hermitian operators) has already appeared in the literature,[1] we do not present it here.

We apply the auxiliary result to the (unitary) operator on $H_0^\perp$ obtained by restricting U to $H_0^\perp$ and obtain thus a subspace H_1 (of $H_0^\perp$) such that H_1 reduces U and such that dim H_1 = dim $(H_0^\perp \cap H_1^\perp)$. Proceeding inductively, we obtain an infinite sequence $\{H_n\}$ of orthogonal subspaces (of H) such that each H_n reduces U and such that every H_n has the same dimension. If the intersection of the orthogonal complements of all the H_n is not trivial, it can be amalgamated to H_0; it follows that H is the direct sum of countably many equi-dimensional subspaces each of which reduces H. By suitably re-numbering the terms of this sequence, we may assume that the index n runs through all (not necessarily non-negative) integers.

Relative to the fixed direct sum decomposition $H = \sum_n H_n$, we

---

[1] Paul R. Halmos, *Commutators of operators*, Amer. J. Math. vol. 74 (1952) pp. 237–240; see Lemma 3 on p. 239.

define a *right shift* as a unitary operator S such that $SH_n = H_{n+1}$ for all n, and we define a *left shift* as a unitary operator T such that $TH_n = H_{n-1}$ for all n. The equi-dimensionality of all the H_n guarantees the existence of shifts. If S is an arbitrary right shift, we write $T = S^*U$. Since $TH_n = S^*UH_n = S^*H_n = H_{n-1}$ for all n, it follows that T is a left shift. Since $U = ST$, we have proved that every unitary operator on H is a product of two shifts; we shall complete the proof of the theorem by showing that every shift is the product of two symmetries.

Since the inverse (equivalently, the adjoint) of a left shift is a right shift, it is sufficient to consider right shifts. Suppose then that S is a right shift; let P be the operator that is equal to S^{1-2n} on H_n and let Q be the operator that is equal to S^{-2n} on H_n for all n. If $x \in H_n$, then $Qx = S^{-2n}x \in S^{-2n}H_n = H_{-n}$, so that $PQx = PS^{-2n}x = S^{1-2(-n)}S^{-2n}x = Sx$. The proof of Theorem 1 is complete.

To what extent is Theorem 1 the best possible result along these lines? The hypothesis of infinite-dimensionality clearly cannot be omitted. Indeed, if H is finite-dimensional, then the concept of determinant makes sense. Since the determinant of a symmetry is ± 1, it follows that no (unitary) operator with a nonreal determinant can be the product of symmetries. Equally clearly, the conclusion of the theorem cannot be strengthened so as to apply to nonunitary operators, because a product of unitary operators (and, in particular, of symmetries) must be unitary. The only conceivable improvement, therefore, is quantitative: possibly every unitary operator is a product of three symmetries. We conclude by showing that this is not so.

THEOREM 2. *On every Hilbert space H there exists a unitary operator U that is not the product of three symmetries.*

PROOF. Let c be a complex cube root of unity and let U be $c \cdot 1$. The operator U belongs to the center of the group of all unitary operators on H; the order of U in that group is exactly three. The remainder of our proof has nothing to do with operator theory; we shall show that, in every group G, a central element of order three is not the product of three elements of order two. More precisely, we show that if G is a group, if u is a central element in G, and if $u = xyz$ with $x^2 = y^2 = z^2 = 1$, then $u^4 = 1$. The proof consists of a simple computation:

$$u^4 = uxuyuz = u(xu) \cdot y(uz) = u(yz) \cdot y(xy)$$
$$= y(uz) \cdot y(xy) = yxy \cdot yxy = 1.$$

UNIVERSITY OF CHICAGO AND
    YALE UNIVERSITY

Reprinted from the
Journal für die reine und angewandte Mathematik
Sonderabdruck aus Band 208, Heft 1/2. 1961. Seite 102 bis 112

Shifts on Hilbert spaces[1].

By *Paul R. Halmos*, University of Chicago.

Introduction.

Does every operator on an infinite-dimensional Hilbert space have a non-trivial invariant subspace? The question is still unanswered. A possible approach is to classify all invariant subspaces of all known operators in the hope of getting an insight that will lead to a proof or a counterexample; this paper is a step in that direction. The operators selected for study are certain special isometries, called shifts. There are two reasons for this choice. First, shifts constitute perhaps the simplest and most natural class of operators for which the classification problem is solvable but not trivial. Second, shifts are known to be typically infinite-dimensional operators, and it is not unreasonable to hope that their study, even if it does not lead to a solution of the existence problem, might make further work on the classification problem easier.

Shifts can be discrete or continuous, they can go in one direction or in two, and they can have finite multiplicity or infinite. The study of invariant subspaces of shifts was begun by Beurling [1]; he considered discrete unilateral shifts of multiplicity one, and he used analytic function theory. The study was continued by Lax [4]; he considered continuous unilateral shifts of finite multiplicity, and he used the theory of Fourier transforms in the complex domain. This paper considers discrete shifts, both unilateral and bilateral, of countable multiplicity; the methods belong to the geometry of Hilbert space. The difference between discrete and continuous is not very important; it is more technical than conceptual. The difference between unilateral and bilateral is somewhat greater, but, as the geometric method shows, it is still not profound. The principal advantages claimed for this paper are the simplicity of the methods and the validity of the results for infinitely multiple shifts. Except for the difference between discrete and continuous, the latter settles a problem raised by Lax in connection with certain partial differential equations.

Discrete shifts can be defined in terms of sequences (functions of integers) or alternatively (dually) in terms of functions on the circle. In the sequential representation certain invariant subspaces of a rather simple structure become clearly visible. The pertinent concept is this: a subspace of a Hilbert space is a *wandering* subspace for an operator A if it is orthogonal to all its images under the (positive) powers of A. For any operator A, the span of the images of any subspace under the (non-negative) powers of A is invariant under A; if the operator is an isometry and if the subspace is a wandering one, the structure of the invariant subspace so spanned and the behavior of the operator

---

[1] Research supported in part by a grant from the National Science Foundation.

on that subspace are especially clear. Theorem 2 below says (for unilateral shifts) that every invariant subspace is obtained from a wandering subspace in the way just described. (For bilateral shifts the statement needs a slight modification.) Beurling and Lax prefer to represent shifts on function spaces (as opposed to sequence spaces) and to characterize invariant subspaces as the ranges of certain multiplication operators; the extension of that result to shifts of possibly infinite multiplicity appears below in Theorem 3 (existence) and Theorem 4 (uniqueness).

Wandering subspaces.

Fix an operator A and consider the correspondence that assigns to each wandering subspace $\mathfrak{N}$ the invariant subspace $\mathfrak{M} = \bigvee\limits_{n=0}^{\infty} A^n \mathfrak{N}$. It is important to know that this correspondence is one-to-one, in the sense that $\mathfrak{N}$ is uniquely determined by $\mathfrak{M}$. Indeed, since $A\mathfrak{M} = \bigvee\limits_{n=1}^{\infty} A^n \mathfrak{N}$, so that $A\mathfrak{M} \perp \mathfrak{N}$, it follows that $\mathfrak{N}$ is the orthogonal complement of $A\mathfrak{M}$ within $\mathfrak{M}$. Equivalently, $\mathfrak{N}$ is the orthogonal complement of the range of the operator obtained by restricting A to $\mathfrak{M}$.

The preceding argument suggests a way to construct examples of wandering subspaces: given the operator A, let $\mathfrak{K}$ be the orthogonal complement of the range of A. To say that $\mathfrak{K}$ is a wandering subspace for A means that if f and g are in $\mathfrak{K}$, and if $n > 0$, then $(A^n f, g) = 0$, and that is obviously true, even if f is not in $\mathfrak{K}$. To be sure, if the range of A is the entire space, then the example obtained in this way is a trivial one, and the same is true, though for a different reason, if $A = 0$. For most operators A, however, the construction yields a non-trivial example of a wandering space. Observe that the space $\mathfrak{K}$ could also have been described as the kernel (null space) of the adjoint operator A^*, but that description would not have been particularly helpful here.

For isometries, wandering subspaces behave better than usual: if U is an isometry, and if $\mathfrak{N}$ is a wandering subspace for U, then $U^m \mathfrak{N} \perp U^n \mathfrak{N}$ whenever m and n are distinct non-negative integers. In other words, if f and g are in $\mathfrak{N}$, then $U^m f \perp U^n g$. For the proof, assume that $m > n$ (this can always be achieved, by a change of notation if necessary); then

$$(U^m f, \ U^n g) = (U^{*n} U^m f, \ g) = (U^{m-n} f, \ g) = 0,$$

since $\mathfrak{N}$ is a wandering subspace. (Recall that since U is an isometry, $U^* U = 1$.) If U is unitary, even more is true: in that case $U^m \mathfrak{N} \perp U^n \mathfrak{N}$ (for a wandering subspace $\mathfrak{N}$) whenever m and n are any two distinct integers (possibly negative). Proof: given m and n, with $m \neq n$, find k so that $m + k$ and $n + k$ are positive; if f and g are in $\mathfrak{N}$, then $(U^m f, \ U^n g) = (U^{m+k} f, \ U^{n+k} g) = 0$, by the result for isometries.

Lemma 1. *If U is an isometry on a Hilbert space $\mathfrak{H}$ and if $\mathfrak{K} = (U\mathfrak{H})^{\perp}$ is the orthogonal complement of the range of U, then $\left(\bigvee\limits_{n=0}^{\infty} U^n \mathfrak{K} \right)^{\perp} = \bigcap\limits_{n=0}^{\infty} U^n \mathfrak{H}$.*

Remark. The lemma will be used in the form stated, but it will be proved in a slightly different form. The left side of the asserted equation is equal to $\bigcap\limits_{n=0}^{\infty} (U^n \mathfrak{K})^{\perp}$, and the proof will show that this is equal to the right side.

Proof. Assume that $f \in \bigcap\limits_{n=0}^{\infty} (U^n \mathfrak{K})^{\perp}$ and prove by induction that $f \in U^n \mathfrak{H}$ for all n. For $n = 0$ this is trivial. Assume therefore that $f \in U^n \mathfrak{H}$, so that $f = U^n g$ for some g.

165

Since $f \in (U^n \mathfrak{K})^\perp$, i. e., $U^n g \perp U^n \mathfrak{K}$, it follows that $g \perp \mathfrak{K}$. This implies that $g \in U\mathfrak{H}$, and hence that $f = U^n g \in U^{n+1} \mathfrak{H}$, as desired. Conclusion: $\bigcap_{n=0}^{\infty} (U^n \mathfrak{K})^\perp < \bigcap_{n=0}^{\infty} U^n \mathfrak{H}$.

To prove the reverse inclusion, use the geometric generality that $U\mathfrak{M}^\perp < (U\mathfrak{M})^\perp$ for all subspaces $\mathfrak{M}$. (Indeed, if $f \in \mathfrak{M}^\perp$, so that Uf is a typical element of $U\mathfrak{M}^\perp$, and if $g \in \mathfrak{M}$, so that Ug is a typical element of $U\mathfrak{M}$, then $Uf \perp Ug$ follows, since U is an isometry, from $f \perp g$.) Now compute:

$$U^{n+1} \mathfrak{H} = U^n (U\mathfrak{H}) = U^n \mathfrak{K}^\perp < (U^n \mathfrak{K})^\perp,$$

and the result follows.

Lemma 1 implies that if U is an isometry on $\mathfrak{H}$ and if $\mathfrak{K} = (U\mathfrak{H})^\perp$, then $\bigvee_{n=0}^{\infty} U^n \mathfrak{K}$ reduces U. Indeed, the space $\bigvee_{n=0}^{\infty} U^n \mathfrak{K}$ is obviously invariant under U; the lemma implies that the same is true of its orthogonal complement.

Unilateral shifts.

There are two extreme kinds of isometries on a Hilbert space $\mathfrak{H}$, the unitary ones and the pure ones; every isometry is a mixture of those two kinds. (A brisk summary of the main facts about isometries was given by Brown [2].) To motivate the definition of purity, recall that an isometry U is unitary if and only if $U\mathfrak{H} = \mathfrak{H}$, or, equivalently, $(U\mathfrak{H})^\perp = \{0\}$. A pure isometry behaves as differently as possible; the definition requires that $\bigvee_{n=0}^{\infty} U^n \mathfrak{K} = \mathfrak{H}$, where, as before, $\mathfrak{K} = (U\mathfrak{H})^\perp$. It is irrelevant for the purposes of this paper, but it might be of interest to observe, that, whenever U is an isometry, $f \in \bigvee_{n=0}^{\infty} U^n \mathfrak{K}$ if and only if $U^{*n} f \to 0$, and hence that U is pure if and only if $U^{*n} \to 0$ tends strongly to 0 as $n \to \infty$.

Here is the usual way to construct a pure isometry. Let $\mathfrak{B}$ be a Hilbert space, and let $\mathfrak{H}_+$ be the direct sum of a countable collection of copies of $\mathfrak{B}$ indexed by the non-negative integers. (The reason for the plus sign will become clear later; for the time being it may be regarded as a harmless decoration.) In other words, an element f of $\mathfrak{H}_+$ is a norm-square-summable sequence $\{f_0, f_1, f_2, \ldots\}$ whose terms are in $\mathfrak{B}$. (Norm and inner product in $\mathfrak{H}_+$ are given by $\|f\|^2 = \sum_{n=0}^{\infty} \|f_n\|^2$ and $(f, g) = \sum_{n=0}^{\infty} (f_n, g_n)$; the linear operations are defined coordinatewise, of course.) If an operator U_+ is defined on $\mathfrak{H}_+$ by

$$U_+ \{f_0, f_1, f_2, \ldots\} = \{0, f_0, f_1, f_2, \ldots\},$$

then it is clear that U_+ is an isometry. The range of U_+ is the set of f's with $f_0 = 0$; if $\mathfrak{K}$ is the orthogonal complement of that range, then $\mathfrak{K}$ is the set of f's with $f_n = 0$ for all $n \neq 0$. (Note that the projection $f \to f_0$, restricted to $\mathfrak{K}$, is an isomorphism between $\mathfrak{K}$ and $\mathfrak{B}$.) It follows that $U_m^+ \mathfrak{K}$ is the set of f's with $f_n = 0$ for all $n \neq m$, and hence that U_+ is a pure isometry. The operator U_+ is called a *shift*, or, more precisely, the *unilateral shift* based on $\mathfrak{B}$. Since, to within isomorphism, the dimension of $\mathfrak{B}$, say N, is the only thing that distinguishes U_+ from other unilateral shifts, U_+ may also be called the unilateral shift of *multiplicity N*.

Unilateral shifts are more than special examples of pure isometries; the definition of purity shows that each pure isometry is isomorphic to some unilateral shift. (If, in the notation consistently used above, $\bigvee\limits_{n=0}^{\infty} U^n \Re = \mathfrak{H}$, then each f in $\mathfrak{H}$ has a unique representation as an orthogonal series $\sum\limits_{n=0}^{\infty} U^n f_n$, with f_n in $\Re$ for all n.) Most of the remainder of this paper is devoted to the study of unilateral shifts, and the notation for them, introduced above, will remain fixed from now on. The essential fixed data are $\mathfrak{B}$, $\mathfrak{H}_+$, $\Re$ and U_+. To be sure, on two occasions below, $\mathfrak{H}_+$ and U_+ will be identified with certain isomorphic versions of themselves, but the identification will be natural and it will not call for a change of notation.

Commutativity.

The simplest invariant subspaces for an operator are the ones that reduce the operator. A subspace $\mathfrak{M}$ reduces an operator A if and only if the projection on $\mathfrak{M}$ commutes with A. Since a projection is Hermitian, it commutes with an operator A if and only if it commutes with both A and A^*. It follows that a natural way to generalize the determination of all subspaces that reduce A is to determine all operators that commute with both A and A^*. For unilateral shifts this turns out to be quite easy.

Each operator C on $\mathfrak{B}$ induces in a natural way an inflated operator $\widehat{C}$ on $\mathfrak{H}_+$; by definition

$$\widehat{C} \{f_0, f_1, f_2, \ldots\} = \{Cf_0, Cf_1, Cf_2, \ldots\}.$$

It is an elementary exercise to prove that the inflation correspondence $C \to \widehat{C}$ preserves the linear operations, products, adjoints, and norms; it is an embedding of the operator algebra on $\mathfrak{B}$ into the one on $\mathfrak{H}_+$.

Theorem 1. *An operator A commutes with both U_+ and U_+^* if and only if $A = \widehat{C}$ for some operator C on $\mathfrak{B}$.*

Proof. To prove something about U_+^*, it helps to know what it is. The answer (easily verified) is that

$$U_+^* \{f_0, f_1, f_2, \ldots\} = \{f_1, f_2, f_3, \ldots\}.$$

Now to verify that each $\widehat{C}$ commutes with both U_+ and U_+^* is just a straightforward computation. The product of U_+ and $\widehat{C}$, in either order, sends

$$\{f_0, f_1, f_2, \ldots\}$$

onto

$$\{0, Cf_0, Cf_1, Cf_2, \ldots\},$$

and the product of U_+^* and $\widehat{C}$, in either order, sends it onto

$$\{Cf_1, Cf_2, Cf_3, \ldots\}.$$

For the converse, assume that A commutes with both U_+ and U_+^*. Since $\Re$ is the kernel of U_+^*, it follows that if $f \in \Re$, then $U_+^* A f = A U_+^* f = 0$. This implies that $Af \in \Re$ whenever $f \in \Re$, i. e., that $\Re$ is invariant under A. Let C be the operator on $\mathfrak{B}$ defined as follows: if $f_0 \in \mathfrak{B}$ and if $f = \{f_0, 0, 0, 0, \ldots\}$ then $f \in \Re$, so that $Af \in \Re$, and, therefore, $Af = \{g_0, 0, 0, 0, \ldots\}$; put $Cf_0 = g_0$. (The operator C is isomorphic, under the natural isomorphism between $\mathfrak{B}$ and $\Re$, to the restriction of A to $\Re$.) It follows that $Af = \widehat{C}f$ at least when $f \in \Re$, and this implies that if $f \in \Re$, then

$$A U_+^n f = U_+^n A f = U_+^n \widehat{C} f = \widehat{C} U_+^n f$$

Journal für Mathematik. Bd. 208. Heft 1/2   14

167

(since both A and $\hat{C}$ commute with U_+). Since, finally, $\bigvee\limits_{n=0}^{\infty} U_+^n \mathfrak{K} = \mathfrak{H}$, it follows that $Af = \hat{C}f$ for all f, and the proof is complete.

Two special cases of the theorem are worthy of note. First, if U_+ is simple (that means that it has multiplicity one), then the only operators on $\mathfrak{B}$ are scalars; since the inflation of a scalar is a scalar, it follows that the only operators that commute with both the simple unilateral shift and its adjoint are the scalars. Second, since $\hat{C}$ is a projection if and only if C is a projection, the theorem gives a usable description of all reducing subspaces of U_+ (for all multiplicities). The result is that there is a one-to-one correspondence between all subspaces of $\mathfrak{B}$ and those subspaces of $\mathfrak{H}_+$ that reduce U_+; the reducing subspace of $\mathfrak{H}_+$ corresponding to a subspace of $\mathfrak{B}$ consists of the sequence whose terms belong to the given subspace. (This consequence of Theorem 1 appears in [2].) The two special cases combined yield the result that the simple unilateral shift is irreducible (i. e., it has no non-trivial reducing subspaces).

Bilateral shifts.

A subspace invariant under an operator A may reduce A, or it may be *irreducible*, in the sense that none of its non-zero subspaces reduces A. (Equivalently: a subspace is irreducible for A if the restriction of A to the subspace is an irreducible operator.) For every operator, each invariant subspace is the direct sum of a reducing subspace and an irreducible invariant subspace, and this decomposition is unique. (Just take the span of all the reducing subspaces included in the given subspace.) Since all the reducing subspaces for the unilateral shifts can be considered known, it remains only to determine all their irreducible invariant subspaces. This problem can be solved by methods that work just as well for a slightly generalized version; what follows is the formulation and solution of the generalized problem.

Consider, as before, a Hilbert space $\mathfrak{B}$, and let $\mathfrak{H}$ be the direct sum of a countable collection of copies of $\mathfrak{B}$ indexed by the set of all integers. In other words, an element f of $\mathfrak{H}$ is a norm-square-summable bilateral sequence

$$\{\ldots, f_{-2}, f_{-1}, (f_0), f_1, f_2, \ldots\}$$

whose terms are in $\mathfrak{B}$; linear operations, inner product, and norm are defined just as for unilateral sequences. (The parentheses in the sequence symbol are merely a notational device; they indicate the term that occupies position 0.) Let U be the operator on $\mathfrak{H}$ defined by

$$U\{\ldots, f_{-2}, f_{-1}, (f_0), f_1, f_2, \ldots\} = \{\ldots, f_{-3}, f_{-2}, (f_{-1}), f_0, f_1, \ldots\}.$$

It is clear that U is unitary; the adjoint (which is the same as the inverse) of U is defined by

$$U^*\{\ldots, f_{-2}, f_{-1}, (f_0), f_1, f_2, \ldots\} = \{\ldots, f_{-1}, f_0, (f_1), f_2, f_3, \ldots\}.$$

The operator U is called the *bilateral shift* based on $\mathfrak{B}$, or, in case $\dim \mathfrak{B} = N$, the bilateral shift of *multiplicity N*.

The space $\mathfrak{H}_+$ studied before is naturally isomorphic to a subspace of $\mathfrak{H}$; the isomorphism is the one that maps $\{f_0, f_1, f_2, \ldots\}$ onto $\{\ldots, 0, 0, (f_0), f_1, f_2, \ldots\}$. It is convenient to identify $\mathfrak{H}_+$ with isomorphic copy, and that is hereby done. With that understood, it makes sense to say and it is trivial to see that $\mathfrak{H}_+$ is invariant under the bilateral shift U, and that, in fact, the restriction of U to $\mathfrak{H}_+$ is just the unilateral shift U_+. The wandering subspace $\mathfrak{K}$ of $\mathfrak{H}_+$ that played a role before is still of some importance;

in the present notation $\Re$ consists of the set of bilateral sequences f with $f_n = 0$ for all $n \neq 0$. The fixed data from now on are the spaces $\mathfrak{V}$, $\mathfrak{H}$, $\mathfrak{H}_+$, and $\Re$, and the operators U and U_+.

The general problem to be solved is the determination of all irreducible invariant subspaces for bilateral shifts. The lemma that follows implies that this problem for bilateral shifts is indeed more general than the same problem for unilateral shifts. A "geometric" solution of the problem is given by Theorem 2; the more familiar "analytic" solution is in Theorem 3.

Lemma 2. *If $\Re$ is a wandering subspace for the bilateral shift U and if $\mathfrak{M} = \bigvee\limits_{n=0}^{\infty} U^n \Re$, then $\mathfrak{M}$ is an irreducible invariant subspace for U.*

Proof. Only the irreducibility of $\mathfrak{M}$ needs proof. If $f \in \mathfrak{M}$, $f \neq 0$, then $f = \sum\limits_{n=0}^{\infty} U^n f_n$, with $f_n \in \Re$ for all $n \geq 0$, and $f_m \neq 0$ for some $m \geq 0$. The non-zero component $U^m f_m$ may be characterized as the projection of f into the subspace $U^m \Re$; by the same token, f_m is equal to the projection of $U^{-(m+1)} f$ into the subspace $U^{-1} \Re$. Since $U^{-1} = U^*$, and since $U^{-1} \Re$ is orthogonal to $\mathfrak{M}$, it follows that every non-zero element of $\mathfrak{M}$ is mapped out of $\mathfrak{M}$ by a suitable power of U^*, and hence that $\mathfrak{M}$ cannot include a non-zero subspace invariant under U^*.

It is a special case of this lemma (put $\Re = \Re$) that $\mathfrak{H}_+$ is an irreducible invariant subspace for U. Since a subspace of $\mathfrak{H}_+$ is invariant under U_+ if and only if it is invariant under U, it follows that any determination of all irreducible invariant subspaces for U automatically serves to determine all invariant subspaces (irreducible or not) for U_+. Note in particular that the reducing subspaces for U_+ described after Theorem 1 above are included among the invariant subspaces described in Theorem 2 below. Thus the reducing subspaces for U_+ receive double coverage in this treatment, and the reducing subspaces for U none at all. The latter are covered by the well known commutativity theory of normal operators (the so-called functional calculus); nothing more needs to be said about them here.

Theorem 2. *If $\mathfrak{M}$ is an irreducible invariant subspace for U [an invariant subspace for U_+], then there exists a wandering subspace $\Re$ for U [for U_+] such that $\mathfrak{M} = \bigvee\limits_{n=0}^{\infty} U^n \Re$.*

Proof. The preceding discussion shows that it is sufficient to treat U alone; the theory for U_+ follows. Write $\Re = \mathfrak{M} \cap (U\mathfrak{M})^\perp$; in other words, $\Re$ is the orthogonal complement within $\mathfrak{M}$ of the range of the operator obtained by restricting U to $\mathfrak{M}$. Assertion: $\mathfrak{M} \cap \left(\bigvee\limits_{n=0}^{\infty} U^n \Re \right)^\perp$ reduces U. Indeed, Lemma 1, applied to the restriction of U to $\mathfrak{M}$, implies that that subspace is equal to $\bigcap\limits_{n=0}^{\infty} U^n \mathfrak{M}$; this makes it obvious that the subspace is invariant under U. (Since $\mathfrak{M} > U\mathfrak{M} > U^2\mathfrak{M} > \cdots$, the intersection of all the terms of this sequence is equal to the intersection of all but the first.) Invariance under $U^*(= U^{-1})$ also follows:

$$U^{-1} \bigcap\limits_{n=0}^{\infty} U^n \mathfrak{M} = U^{-1}\mathfrak{M} \cap \bigcap\limits_{n=1}^{\infty} U^{n-1}\mathfrak{M} < \bigcap\limits_{n=0}^{\infty} U^n \mathfrak{M}.$$

Since $\mathfrak{M}$ is irreducible, the subspace just proved to be reducing must be the zero subspace, and that implies the conclusion of the theorem.

It may be of interest to remark that if $\mathfrak{M} = \bigvee\limits_{n=0}^{\infty} U^n \Re$ is a reducing subspace for U_+, then $\Re < \Re$, and the corresponding subspace of $\mathfrak{V}$ (as described after Theorem 1) is the

14*

image of $\mathfrak{N}$ under the natural correspondence between $\mathfrak{K}$ and $\mathfrak{V}$; if, on the other hand, $\mathfrak{M}$ is given as $\hat{C}\mathfrak{H}_+$, for some projection C on $\mathfrak{V}$, then $\mathfrak{N} = \hat{C}\mathfrak{K}$.

Functional representation.

Theorem 2 provides some insight into the geometric structure of invariant subspaces for shifts, but it by no means solves all the problems of the theory. It does not, for instance, help in the construction of invariant subspaces, and it does not serve to answer even such a concrete question as this: are there any irreducible invariant subspaces for U other than the easy ones obtained by applying a fixed power of U to $\mathfrak{H}_+$? The next topic to be studied, the functional representation for shifts, furnishes the answer.

Let $Z = \{z: |z| = 1\}$ be the unit circle in the complex plane. Interpret measurability in Z in the sense of Borel and measure in the sense of Lebesgue, normalized so that the measure of Z itself is 1. The differential symbol "dz" will be used to indicate integration with respect to this measure.

Let $\mathfrak{V}$ be a Hilbert space; to avoid the pathology of the uncountable, assume from now on that $\mathfrak{V}$ is separable. Let $\mathfrak{H}$ be the set of all measurable functions f from Z to $\mathfrak{V}$ such that $\int \|f(z)\|^2 dz < \infty$; measurability here can be interpreted either weakly or strongly — in view of the separability of $\mathfrak{V}$, the two are equivalent. (See Hille and Phillips [3; Section 3.5.]) The functions in $\mathfrak{H}$ (modulo sets of measure zero) constitute a Hilbert space with the pointwise definitions of the linear operations and with the inner product given by $(f, g) = \int (f(z), g(z)) dz$. (Here $(f(z), g(z))$ is the inner product of the two elements $f(z)$ and $g(z)$ in $\mathfrak{V}$. Note that the separability of $\mathfrak{V}$ implies that $\mathfrak{H}$ is separable.) Consider on the Hilbert space $\mathfrak{H}$ the operator U defined by $(Uf)(z) = zf(z)$. It is easy to verify that U is unitary; $U^{-1}(= U^*)$ is defined by $(U^*f)(z) = \bar{z}f(z)$.

Since the elements of $\mathfrak{H}$ are functions, $\mathfrak{H}$ has some structure that is not definable in every Hilbert space. The most important special feature of $\mathfrak{H}$ is the subspace $\mathfrak{K}$ consisting of all constant functions. In other words, a function f belongs to $\mathfrak{K}$ if and only if there is a vector v in $\mathfrak{V}$ such that $f(z) = v$ for almost all z. Assertion: $\mathfrak{K}$ is a wandering subspace for U, and $\overset{+\infty}{\underset{n=-\infty}{\vee}} U^n \mathfrak{K} = \mathfrak{H}$. The first part of the assertion is easy to prove. Suppose, indeed, that $f(z) = u$ and $g(z) = v$ almost everywhere; if $n \neq 0$, then

$$(U^n f, g) = \int (z^n f(z), g(z)) dz = \int (z^n u, v) dz = (u, v) \int z^n dz = 0.$$

The second part of the assertion is somewhat deeper. Suppose that $f \in \mathfrak{H}$ and that $f \perp U^n \mathfrak{K}$ for all n; it is to be proved that $f = 0$. The assumption means that $\int (f(z), z^n v) dz = 0$ for all v in $\mathfrak{V}$ and all n. If $g(z) = (f(z), v)$, then g is an integrable complex-valued function and all the Fourier coefficients of g vanish. It follows that, for each v, the inner product $(f(z), v)$ vanishes for almost all z. The separability of $\mathfrak{V}$ implies that $f(z) = 0$ for almost all z, and that is the point at issue.

The preceding result implies that $\mathfrak{H}$ is naturally isomorphic to the space of all bilateral sequences of terms in $\mathfrak{V}$, with the usual linear operations and inner product; the natural isomorphism carries U onto the bilateral shift. It is convenient to identify the new $\mathfrak{H}$ (the function space) with the isomorphic old $\mathfrak{H}$ (the bilateral sequence space), and that is hereby done. Under the identification, the space $\mathfrak{H}_+$ is recognized as $\overset{\infty}{\underset{n=0}{\vee}} U^n \mathfrak{K}$, and the operator U_+ is, of course, the restriction of U to the invariant subspace $\mathfrak{H}_+$. All the previous considerations are thereby recaptured in the new setting, and the use of the same notation is justified.

The functional representation will presently provide a characterization of wandering subspaces for U that will greatly facilitate the application of Theorem 2. Here are the first two steps in that direction.

Lemma 3. *If f and g are in $\mathfrak{H}$ and if $(U^n f, g) = 0$ for all $n \neq 0$, then $(f(z), g(z))$ is a constant almost everywhere.*

Proof. The assumption says that $\int z^n (f(z), g(z)) \, dz = 0$ for all $n \neq 0$. This means that the inner product part of the integrand, which is a complex-valued integrable function, is such that all its Fourier coefficients, except possibly the constant term, are equal to zero.

Lemma 4. *If $\mathfrak{N}$ is a wandering subspace for U, then $\dim \mathfrak{N} \leq \dim \mathfrak{B}$.*

Proof. Let $\{f_i\}$ be an orthonormal set in $\mathfrak{N}$. Since $\mathfrak{H}$ is separable, the set is countable. Since $\mathfrak{N}$ is a wandering subspace, $(U^n f_i, f_j) = 0$ for all $n \neq 0$. It follows from Lemma 3 that $(f_i(z), f_j(z))$ is a constant almost everywhere. Since

$$\int (f_i(z), f_j(z)) \, dz = (f_i, f_j) = \delta_{ij},$$

it follows that $(f_i(z), f_j(z)) = \delta_{ij}$ for almost all z. Since there are only countably many exceptional sets of measure zero to discard, there exists at least one z such that $(f_i(z), f_j(z)) = \delta_{ij}$. This implies that the space $\mathfrak{B}$ includes an orthonormal set of the same cardinality as $\{f_i\}$; the proof is complete.

Operator functions.

The functional structure of $\mathfrak{H}$ suggests the consideration of certain special operators. Suppose that F is a function on Z whose values are operators on $\mathfrak{B}$. If $f \in \mathfrak{H}$, then $F(z) f(z)$ makes sense for each z in Z; write $(Ff)(z) = F(z) f(z)$. Does Ff belong to $\mathfrak{H}$? The answer is yes if F is measurable (weakly or strongly — as before, it is all the same) and essentially bounded. The latter means that the essential (i. e., almost everywhere) supremum of $\| F(z) \|$, for z in Z, is finite; the value of that essential supremum is then the norm $\| \hat{F} \|$ of the operator $\hat{F}$. The mapping $F \to \hat{F}$ preserves the linear operations, products, adjoints, and norms; it is an embedding of the algebra of bounded measurable $\mathfrak{B}$-operator functions (modulo sets of measure zero, of course) into the operator algebra on $\mathfrak{H}$. If F is a constant almost everywhere, then the operator $\hat{F}$ here defined coincides (at least on $\mathfrak{H}_+$) with the inflation of that constant as defined before; the passage from F to $\hat{F}$, for any bounded measurable operator function F, may be regarded as a generalized inflation. If F is a scalar multiple of the identity almost everywhere, then $\hat{F}$ commutes with $\hat{G}$ for every G. This applies, in particular, to the function given by $F(z) = z \cdot 1$; in that case $\hat{F} = U$, of course.

Certain operator functions can be used to give examples of wandering spaces. Suppose, to begin with, that $F(z)$ is an isometry on $\mathfrak{B}$ for almost all z. (Measurability is always assumed.) Assertion: the subspace $\hat{F}\mathfrak{K}$ is a wandering subspace for U. (Recall that $\mathfrak{K}$ consists of those vector functions in $\mathfrak{H}$ that are constant almost everywhere.) Proof: if $f(z) = F(z) u$ and $g(z) = F(z) v$, with u and v in $\mathfrak{B}$, then

$$(U^n f, g) = \int (z^n F(z) u, F(z) v) \, dz = \int z^n (F(z) u, F(z) v) \, dz = \int z^n (u, v) \, dz$$
$$\big(\text{since } F(z) \text{ is an isometry}\big) \quad = (u, v) \int z^n \, dz = 0$$

for all $n \neq 0$. The same proof works for an unavoidable technical generalization of the result. Suppose, in fact, that $\mathfrak{U}$ is a subspace of $\mathfrak{B}$ and that, for almost all z, the operator

171

$F(z)$ is a partial isometry on $\mathfrak{B}$ with initial space $\mathfrak{U}$. (This means that $\| F(z) u \| = \| u \|$ when $u \in \mathfrak{U}$, and $F(z) u = 0$ when $u \perp \mathfrak{U}$.) It follows that $(F(z) u, F(z) v)$ is a constant (independent of z), and that is all that was needed in the proof just given. For a concise summary of the result it is convenient to make the following definition: an operator function F is *rigid* if there exists a subspace $\mathfrak{U}$ of $\mathfrak{B}$ such that F is almost everywhere an isometry on $\mathfrak{U}$ and zero on $\mathfrak{U}^\perp$. Conclusion: if F is rigid, then $\widehat{F}\mathfrak{K}$ is a wandering subspace for U. This conclusion implies that if F is rigid, then $\widehat{F}\mathfrak{H}_+$ is an irreducible invariant subspace for U. Indeed, $\widehat{F}\mathfrak{H}_+ = \widehat{F} \bigvee_{n=0}^{\infty} U^n \mathfrak{K} = \bigvee_{n=0}^{\infty} U^n(\widehat{F}\mathfrak{K})$, and the result follows from Lemma 2. Note that invariance is easy to prove directly (from the commutativity of $\widehat{F}$ and U); the result for wandering subspaces, and Lemma 2, are needed to prove irreducibility.

The preceding discussion of U on $\mathfrak{H}$ leaves unanswered one question about U_+ on $\mathfrak{H}_+$. If $\widehat{F}\mathfrak{K}$ is a wandering subspace for U_+, then $\widehat{F}\mathfrak{K}$ is included in $\mathfrak{H}_+$ (or, equivalently, $\widehat{F}\mathfrak{H}_+$ is included in $\mathfrak{H}_+$); under what conditions does that happen? Here is one answer: for a bounded measurable operator function F, the operator $\widehat{F}$ maps $\mathfrak{K}$ into $\mathfrak{H}_+$ if and only if $\int (F(z) u, v)\overline{z}^n dz = 0$ for all u and v in $\mathfrak{B}$ and for all $n < 0$. Indeed: $\widehat{F}\mathfrak{K}$ is orthogonal to $\mathfrak{H}_+^\perp$ if and only if $(\widehat{F}f, U^n g) = 0$ for all f and g in $\mathfrak{K}$ and for all $n < 0$, and this is equivalent to the vanishing of $\int (F(z) u, z^n v) dz$ for all u and v in $\mathfrak{B}$ and for all $n < 0$. This characterization of the F's with $\widehat{F}\mathfrak{K} < \mathfrak{H}_+$ says, roughly speaking, that all the Fourier coefficients of F with negative index are equal to zero. For this reason, an operator function F such that $\widehat{F}\mathfrak{K} < \mathfrak{H}_+$ will be called a *Taylor* function below; the idea is that it behaves the way Taylor series behave (as opposed to Laurent series). Conclusion: if F is a rigid Taylor function, then $\widehat{F}\mathfrak{K}$ is a wandering subspace and $\widehat{F}\mathfrak{H}_+$ is an invariant subspace for U_+.

It is easy to construct some rigid operator functions, and even rigid Taylor functions, and thus to construct non-trivial examples of invariant subspaces for U_+ and irreducible invariant subspaces for U. Suppose, for instance, that $\mathfrak{B}$ is the two-dimensional complex coordinate space; in that case an operator on $\mathfrak{B}$ is naturally identifiable with a two-rowed matrix. If, for a concrete example, $F(z) = \dfrac{1}{2}\begin{pmatrix} z^2 + 1 & 0 \\ z^2 - 1 & 0 \end{pmatrix}$, then it is easy to verify that $F(z)$ is, for each z, a partial isometry whose initial space is the set of all vectors with vanishing second coordinate, and hence that F is a rigid operator function. (It is even a Taylor function.)

If dim $\mathfrak{B} = 1$, then operator functions may be regarded as ordinary complex numerical functions. An arbitrary measurable function of constant absolute value 1 is a rigid function. If, in particular, $e_n(z) = z^n$, then each e_n is a rigid function; for $n \geqq 0$, the e_n's are even Taylor functions. More interesting examples of one-dimensional rigid Taylor functions exist, but they are not quite so easy to see. There is a bit of a theory associated with the subject; an elegant recent discussion of the basic facts has been given by Rudin [5].

Invariant subspaces.

The purpose of what follows is to prove that the preceding construction of wandering and invariant subspaces succeeds in constructing them all.

Lemma 5. *If $\mathfrak{N}$ is a wandering subspace for U [for U_+], then there exists a rigid function F [a rigid Taylor function F] such that $\mathfrak{N} = \widehat{F}\mathfrak{K}$.*

Proof. The preceding discussion shows that it is sufficient to treat U alone; the theory for U_+ follows. Since, by Lemma 4, dim $\mathfrak{N} \leqq$ dim $\mathfrak{B}$, there exists a subspace $\mathfrak{U}$ of $\mathfrak{B}$ with dim $\mathfrak{U} =$ dim $\mathfrak{N}$. Let A be an operator from $\mathfrak{B}$ to $\mathfrak{N}$ such that A maps $\mathfrak{U}$ isometrically onto $\mathfrak{N}$ and such that A maps $\mathfrak{U}^\perp$ onto 0. If F is defined by $F(z)\, v = (Av)\,(z)$ for each v in $\mathfrak{B}$, then F is a measurable operator function. Since $Av \in \mathfrak{N}$, so that $(U^n Av,\, Av) = 0$ for all $n \neq 0$, Lemma 3 implies that $\|\,F(z)\,v\,\|$ is a constant almost everywhere, and hence, in particular, that F is bounded. If $v \in \mathfrak{U}$, then

$$\|\,v\,\|^2 = \|\,Av\,\|^2 = \int \|\,(Av)\,(z)\,\|^2 dz = \int \|\,F(z)\,v\,\|^2 dz,$$

so that the constant $\|\,F(z)\,v\,\|$ is equal to $\|\,v\,\|$ almost everywhere; this means that almost every $F(z)$ is an isometry on $\mathfrak{U}$. (To see that the "almost every" is independent of v, argue separately for each v in a basis.) If $v \in \mathfrak{U}^\perp$, then $Av = 0$, and, again, it follows that almost every $F(z)$ annihilates $\mathfrak{U}$. This proves that almost every $F(z)$ is a partial isometry with initial space $\mathfrak{U}$, and hence, in particular, that F is rigid. If $f \in \mathfrak{R}$, so that $f(z) = v$ almost everywhere, for some v in $\mathfrak{B}$, then $(\hat{F}f)\,(z) = F(z)\,v = (Av)\,(z)$, so that $\hat{F}\mathfrak{R} < \mathfrak{N}$. If, conversely, $g \in \mathfrak{N}$, so that $g = Av$ for some v in $\mathfrak{B}$ (in fact for some v in $\mathfrak{U}$), and if $f(z) = v$ for all z, then $f \in \mathfrak{R}$ and $g(z) = (Av)\,(z) = (\hat{F}f)\,(z)$, so that $\mathfrak{N} < \hat{F}\mathfrak{R}$. The proof of the lemma is complete.

Theorem 3. *If $\mathfrak{M}$ is an irreducible invariant subspace for U [an invariant subspace for U_+], then there exists a rigid function F [a rigid Taylor function F] such that $\mathfrak{M} = \hat{F}\mathfrak{H}_+$.*

Proof. By Theorem 2, there exists a wandering subspace $\mathfrak{N}$ such that $\mathfrak{M} = \bigvee\limits_{n=0}^{\infty} U^n \mathfrak{N}$; by Lemma 5, $\mathfrak{N} = \hat{F}\mathfrak{R}$ for some rigid function F (which is a rigid Taylor function in case $\mathfrak{M} < \mathfrak{H}_+$). The conclusion follows by inverting an argument used earlier:

$$\mathfrak{M} = \bigvee_{n=0}^{\infty} U^n (\hat{F}\mathfrak{R}) = \hat{F} \bigvee_{n=0}^{\infty} U^n \mathfrak{R} = \hat{F}\mathfrak{H}_+.$$

The proof of Lemma 5 involves two arbitrary choices ($\mathfrak{U}$ and A); this makes it difficult to see to what extent, if any, the rigid function F is uniquely determined by the wandering subspace $\mathfrak{N}$ in Lemma 5 or by the invariant subspace $\mathfrak{M}$ in Theorem 3. Since the correspondence between wandering subspaces and the appropriate kind of invariant subspaces is one-to-one, there is only one uniqueness problem here, not two; the purpose of what follows is to solve it.

Uniqueness.

Suppose that F_1 and F_2 are rigid operator functions with (constant) initial spaces $\mathfrak{U}_1$ and $\mathfrak{U}_2$; under what conditions does it happen that $\hat{F}_1\mathfrak{H}_+ < \hat{F}_2\mathfrak{H}_+$? A trivial necessary condition is that $\hat{F}_1\mathfrak{R} < \hat{F}_2\mathfrak{H}_+$. This means that for each v in $\mathfrak{B}$ there exists an f in $\mathfrak{H}_+$ such that $F_1(z)\,v = F_2(z)\,f(z)$ almost everywhere. The function f is not uniquely determined, but an additional requirement makes it so. For each z, replace $f(z)$ by its projection into $\mathfrak{U}_2$. The result is a $\mathfrak{U}_2$-valued function whose value at each z is $F_1(z)\,v$; there is no loss of generality in assuming that f has these properties in the first place. Since $F_2(z)$ is an isometry on $\mathfrak{U}_2$, the value $f(z)$ is now uniquely determined, and, therefore, it makes sense to define an operator function G by $G(z)\,v = f(z)$. If $v \in \mathfrak{U}_1$, then

$$\|\,v\,\| = \|\,F_1(z)\,v\,\| = \|\,F_2(z)\,f(z)\,\| = \|\,f(z)\,\| = \|\,G(z)\,v\,\|,$$

so that $G(z)$ is an isometry on $\mathfrak{U}_1$; if $v \perp \mathfrak{U}_1$, then (almost everywhere) $F_1(z) \, v = 0$, so that $f(z) = 0$, and therefore $G(z) \, v = 0$. This implies that G is rigid, with initial space $\mathfrak{U}_1$; the range of $G(z)$, for almost every z, is included in $\mathfrak{U}_2$. Since, moreover, G maps $\mathfrak{K}$ into $\mathfrak{H}_+$, it follows (by definition) that G is a Taylor function. Conclusion: if $\widehat{F}_1 \mathfrak{H}_+ < \widehat{F}_2 \mathfrak{H}_+$, then $F_1 = F_2 G$, where G is a rigid Taylor function with initial space $\mathfrak{U}_1$ and range in $\mathfrak{U}_2$. It is convenient to say that F_1 is *divisible* by F_2 exactly when a G of this sort exists.

    Lemma 6. *If F_1 and F_2 are rigid operator functions, then a necessary and sufficient condition that $\widehat{F}_1 \mathfrak{H}_+ < \widehat{F}_2 \mathfrak{H}_+$ is that F_1 be divisible by F_2.*

    Proof. Necessity was just proved. Sufficiency is trivial: if $F_1 = F_2 G$, with G Taylor, then $\widehat{F}_1 \mathfrak{H}_+ = \widehat{F}_2 \widehat{G} \mathfrak{H}_+ < \widehat{F}_2 \mathfrak{H}_+$.

    How near does the divisibility relation come to being a partial order? Reflexivity and transitivity are easy to prove. The only delicate question is this: what happens when each of F_1 and F_2 is divisible by the other? Suppose that $F_1 = F_2 G_2$ and $F_2 = F_1 G_1$; here the values of G_2 are partial isometries with initial space $\mathfrak{U}_1$ and range in $\mathfrak{U}_2$, the values of G_1 are partial isometries with initial space $\mathfrak{U}_2$ and range in $\mathfrak{U}_1$, and both G_1 and G_2 are rigid Taylor functions. The first consequence of these assumptions is that almost every value of $G_1 G_2$ is equal to the projection on $\mathfrak{U}_1$. Indeed, since $F_1(z) \, G_1(z) \, G_2(z) \, v = F_1(z) \, v$ for all v, it follows that if $v \in \mathfrak{U}_1$, then $G_1(z) \, G_2(z) \, v = v$ (because $F_1(z)$ is one-to-one on $\mathfrak{U}_1$); if $v \perp \mathfrak{U}_1$, then $G_1(z) \, G_2(z) \, v = 0$ (because $G_2(z) \, v = 0$).

    Observe now that almost every value of $G_1 G_1^*$ is also equal to the projection on $\mathfrak{U}_1$ (just because almost every value of G_1 is a partial isometry with initial space $\mathfrak{U}_1$); it follows that $G_1 G_2 = G_1 G_1^*$. Since the ranges of $G_2(z)$ and $G_1^*(z)$ are in $\mathfrak{U}_1$ and since $G_1(z)$ is one-to-one on $\mathfrak{U}_1$, it follows that $G_2(z) = G_1^*(z)$ on $\mathfrak{U}_1$. Since on $\mathfrak{U}_1^\perp$ both $G_2(z)$ and $G_1^*(z)$ vanish, it follows that $G_2(z) = G_1^*(z)$ on all $\mathfrak{B}$ for almost every z.

    Suppose now that G is an operator function (such as G_1 above) with the property that both G and G^* are Taylor functions. It follows that if u and v are in $\mathfrak{B}$, then $\int (G(z) \, u, v) \bar{z}^n \, dz = 0$ and $\int (G^*(z) \, v, u) \bar{z}^n \, dz = 0$ for all $n < 0$. (The interchange of u and v in the second integral is deliberate.) The second equation is equivalent to $\int (G(z) \, u, v) z^n \, dz = 0$ for all $n < 0$ and hence to $\int (G(z) \, u, v) \bar{z}^n \, dz = 0$ for all $n > 0$. It follows (compare Lemma 3) that $(G(z) \, u, v)$ is a constant almost everywhere, and hence (u and v are arbitrary) that $G(z)$ is a constant almost everywhere. The conclusion is that, in the notation of the preceding paragraph, both G_1 and G_2 are constants. This conclusion can be expressed as follows.

    Theorem 4. *If F is a rigid operator function, then $\widehat{F} \mathfrak{H}_+$ uniquely determines F to within a constant partially isometric factor on the right.*

References.

[1] A. *Beurling*, On two problems concerning linear transformations in Hilbert space, Acta Math. **81** (1949), 239—255.
[2] A. *Brown*, On a class of operators, Proc. A. M. S. **4** (1953), 723—728.
[3] E. *Hille* and R. S. *Phillips*, Functional analysis and semi-groups, A. M. S., Providence (1957).
[4] P. D. *Lax*, Translation invariant spaces, Acta Math. **101** (1959), 163—178.
[5] W. *Rudin*, The closed ideals in an algebra of analytic functions, Can. J. Math. **9** (1957), 426—434.

Eingegangen 18. November 1960.

Reprinted from the
PACIFIC JOURNAL OF MATHEMATICS
Vol. 13, No. 2, pp. 585-596, 1963

PARTIAL ISOMETRIES

P. R. Halmos and J. E. McLaughlin

0. Introduction. For normal operators on a Hilbert space the problem of unitary equivalence is solved, in principle; the theory of spectral multiplicity offers a complete set of unitary invariants. The purpose of this paper is to study a special class of not necessarily normal operators (partial isometries) from the point of view of unitary equivalence.

Partial isometries form an attractive and important class of operators. The definition is simple: a partial isometry is an operator whose restriction to the orthogonal complement of its null-space is an isometry. Partial isometries play a vital role in operator theory; they enter, for instance, in the theory of the polar decomposition of arbitrary operators, and they form the cornerstone of the dimension theory of von Neumann algebras. There are many familiar examples of partial isometries: every isometry is one, every unitary operator is one, and every projection is one. Our first result serves perhaps to emphasize their importance even more; the assertion is that the problem of unitary equivalence for completely arbitrary operators is equivalent to the problem for partial isometries. Next we study the spectrum of a partial isometry and show that it can be almost anything; in the finite-dimensional case even the multiplicities can be prescribed arbitrarily. In a special (finite) case, we solve the unitary equivalence problem for partial isometries. After that we ask how far a partial isometry can be from the set of normal operators and obtain a very curious answer. Generalizing and simplifying a result of Nagy, we show also that if two partial isometries are sufficiently near, then some natural cardinal numbers (dimensions) associated with them are the same. This result yields a partitioning of the metric space of all partial isometries into open-closed sets, and we conclude by proving that these sets are exactly the components.

For any operator A with null-space $\mathfrak{N}$ we write $\nu(A) = \dim \mathfrak{N}$ and we call $\nu(A)$ the *nullity* of A. If A is a partial isometry with range $\mathfrak{R}$, we write $\rho(A) = \dim \mathfrak{R}$ and $\rho'(A) = \dim \mathfrak{R}^\perp$; the cardinal numbers $\rho(A)$ and $\rho'(A)$ are the *rank* and the *co-rank* of A. The subspace $\mathfrak{N}^\perp$ is the *initial space* of A; the range $\mathfrak{R}$ (which is the same as the image $A\mathfrak{N}^\perp$) is the *final space* of A. If A is a partial isometry then so is A^*; the initial space of A^* is the final space of A, and vice versa. It follows that $\nu(A^*) = \rho'(A)$ and

---
Received February 21, 1963. Research supported in part by grants from the National Science Foundation.

585

$\rho'(A^*) = \nu(A)$.

It is natural to define a partial order for partial isometries as follows: $A \leq B$ in case B agrees with A on the initial space of A. (This implies that the initial space of A is included in the initial space of B.) A partial isometry is maximal with respect to this order if and only if either its initial space or its final space is the entire underlying Hilbert space. It follows that every partial isometry can be enlarged to either an isometry or a co-isometry (the adjoint of an isometry). A necessary and sufficient condition that a partial isometry possess a unitary enlargement (i.e., that there exist a unitary operator that dominates it) is that its nullity be equal to its co-rank. If the underlying Hilbert space is finite-dimensional, this condition is always satisfied; in the infinite-dimensional case it may not be.

1. **Reduction.** If A is a construction (i.e., if $\| A \| \leq 1$) on a Hilbert space $\mathfrak{H}$, then $1 - AA^*$ is positive, and, consequently, $1 - AA^*$ has a unique positive square root A'. Assertion: if $M = M(A)$ is the operator matrix $\begin{pmatrix} A & A' \\ 0 & 0 \end{pmatrix}$, interpreted as an operator on $\mathfrak{H} \oplus \mathfrak{H}$, then M is a partial isometry. One quick proof is to compute MM^* and observe that it is a projection; this can happen if and only if M is a partial isometry. Consequence: every contraction on a Hilbert space can be extended to a larger Hilbert space so as to become a partial isometry.

THEOREM 1. *If A and B are unitarily equivalent contractions, then $M(A)$ and $M(B)$ are unitarily equivalent; if, conversely, A and B are invertible contractions such that $M(A)$ and $M(B)$ are unitarily equivalent, then A and B are unitarily equivalent.*

Proof. If U is a unitary operator that transforms A onto B, then U transforms A^* onto B^*, and therefore U transforms A' onto B'; it follows that $\begin{pmatrix} U & 0 \\ 0 & U \end{pmatrix}$ transforms $M(A)$ onto $M(B)$.

Suppose next that A and B are invertible and that $M(A)$ and $M(B)$ are unitarily equivalent. The range of $M(A)$ consists of all column vectors of the form $\begin{pmatrix} Af + A'g \\ 0 \end{pmatrix}$. This set is included in the set of all column vectors with vanishing second coordinate; the invertibility of A implies that the range of $M(A)$ consists exactly of all column vectors with vanishing second coordinate. Since the same is true for $M(B)$, it follows that every unitary operator matrix that transforms $M(A)$ onto $M(B)$ maps the subspace of all vectors of the form $\begin{pmatrix} f \\ 0 \end{pmatrix}$ onto itself. This implies that that subspace reduces every such unitary operator matrix, and hence that every such unitary operator matrix is diagonal. Since the diagonal entries of a diagonal unitary matrix are unitary operators, it follows that A and B are unitarily equivalent, as asserted.

The theorem implies that the problem of unitary equivalence for partial isometries is equivalent to the problem for invertible contractions. The latter problem, in turn, is equivalent to the problem for arbitrary operators. The reason is that by a translation $(A \rightarrow A + \alpha)$ and a change of scale $(A \rightarrow \beta A)$ every operator becomes an invertible contraction, and translations and changes of scale do not affect unitary equivalence.

Here is a comment on the technique used in the proof. There are many ways that a possibly "bad" operator A can be used to manufacture a "good" one $\left(\text{e.g., } A + A^* \text{ and } \begin{pmatrix} 0 & A \\ A^* & 0 \end{pmatrix}\right)$. None of these ways has ever yielded sufficiently many usable unitary invariants for A. It is usually easy to prove that if A and B are unitarily equivalent, then so are the various constructs in which they appear. It is, however, usually false that if the constructs are unitarily equivalent, then so are A and B. In the case treated by Theorem 1 this converse is true, and its proof is the less trivial part of the argument.

2. Spectrum. What can the spectrum of a partial isometry be? Since a partial isometry is a contraction, its spectrum is included in the closed unit disc. If the partial isometry is invertible (i.e., if 0 is not in the spectrum), then it is unitary, and therefore the spectrum is a non-empty compact subset of the unit circle; well known constructions prove that every such set is the spectrum of some unitary operator. If the partial isometry is not invertible, then its spectrum contains 0; what else can be said about it? The answer is, nothing else. This answer was pointed out to us by Arlen Brown; its precise formulation is as follows.

THEOREM 2. *If a compact subset of the closed unit disc contains the origin, then it is the spectrum of some partial isometry.*

Proof. It is sufficient to prove that if A is a contraction, then the spectrum of $M(A)$ is the union of the spectrum of A and the singleton $\{0\}$. (This is sufficient because every non-empty compact subset of the closed unit disc is the spectrum of some contraction.) It is easy enough to see that 0 always belongs to the spectrum of $M(A)$; indeed every vector of the form $\begin{pmatrix} 0 \\ f \end{pmatrix}$ is in the null-space of $M(A)^*$. It remains to prove that if $\lambda \neq 0$, then a necessary and sufficient condition that $\begin{pmatrix} A - \lambda & A' \\ 0 & -\lambda \end{pmatrix}$ be invertible is that $A - \lambda$ be invertible. This assertion belongs to the theory of formal determinants of operator matrices. Here is a sample theorem from that theory: if C and D commute and if D is invertible, then a necessary and sufficient condition that $\begin{pmatrix} A & B \\ C & D \end{pmatrix}$ be invertible is that $AD - BC$ be invertible. For our present purpose it is sufficient to consider the special case $C = 0$, in which case the commutativity hypothesis is automatically satisfied; we proceed to give the proof for that case. If A is invertible, then $\begin{pmatrix} A & B \\ 0 & D \end{pmatrix}$ can be proved to be invertible by exhibiting its inverse: it is $\begin{pmatrix} A^{-1} & A^{-1}BD^{-1} \\ 0 & D^{-1} \end{pmatrix}$. (Recall that the invertibility hypothesis on D is in force throughout.) If, conversely, $\begin{pmatrix} A & B \\ 0 & D \end{pmatrix}$ is invertible, with inverse $\begin{pmatrix} P & Q \\ R & S \end{pmatrix}$ say, then

$$\begin{pmatrix} AP + BR & AQ + BS \\ DR & DS \end{pmatrix} = \begin{pmatrix} PA & PB + QD \\ RA & RB + SD \end{pmatrix} = \begin{pmatrix} 1 & 0 \\ 0 & 1 \end{pmatrix}.$$

It follows that $DR = 0$; since D is invertible, this implies that $R = 0$, and hence that $AP = PA = 1$. The proof is complete.

3. **Multiplicity.** For finite sets what the preceding argument proves is this: if $\lambda_1, \cdots, \lambda_n$ are distinct complex numbers with $|\lambda_i| \leq 1$ for all i, and if $\lambda_i = 0$ for at least one i, then there exists a partial isometry whose spectrum is the set $\{\lambda_1, \cdots, \lambda_n\}$. The partial isometry that the proof yields acts on a space of dimension $2n$ and has a large irrelevant null-space. There is an alternative proof that yields much more for finite sets.

THEOREM 3. *If $\lambda_1, \cdots, \lambda_n$ are complex numbers (not necessarily distinct) with $|\lambda_i| \leq 1$ for all i, and if $\lambda_i = 0$ for at least one i, then there exists a partial isometry on a space of dimension n, whose characteristic roots are exactly the λ's, each with the algebraic multiplicity equal to the number of times it occurs in the list.*

Proof. The proof can be given by induction on n. For $n = 1$, the operator 0 on a space of dimension 1 satisfies all the conditions. The induction step is implied by the following assertion: if an $n \times n$ matrix U with 0 in its spectrum is a partial isometry, and if $|\lambda| \leq 1$, then there exists a column vector f with n coordinates such that $\begin{pmatrix} U & f \\ 0 & \lambda \end{pmatrix}$ is a partial isometry. To prove this, observe that, since 0 is in the spectrum of U, the column-rank of U is less than n. This makes it possible to find a non-zero vector f orthogonal to all the columns of U; to finish the construction, normalize f so that $\|f\|^2 = 1 - |\lambda|^2$.

4. Equivalence. In at least one case, a very special case, the unitary equivalence problem for partial isometries has a simple and satisfying solution.

THEOREM 4. *If two partial isometries on a finite-dimensional space are such that 0 is a simple root of each of their characteristic equations, then a necessary and sufficient condition that they be unitarily equivalent is that they have the same characteristic equation (i.e., that they have the same characteristic roots with the same algebraic multiplicities).*

REMARK. The principal hypothesis is that 0 is a root of multiplicity 1 of the characteristic equation. If this were replaced by the hypothesis that 0 is not a root of the characteristic equation at all (i.e., is a root of multiplicity 0), then the statement would become the classical solution of the unitary equivalence problem for normal operators on a finite-dimensional space.

Proof. The necessity of the condition is trivial. Sufficiency can be proved by induction on the dimension. If the dimension is 1, the assertion is trivial. For the induction step, if the dimension is $n + 1$, represent the given partial isometries by triangular matrices with 0 in the northwest corner, and write the results in the form

$$U = \begin{pmatrix} U_0 & f \\ 0 & \lambda \end{pmatrix}, \qquad V = \begin{pmatrix} V_0 & g \\ 0 & \lambda \end{pmatrix},$$

where U_0 and V_0 are $n \times n$ matrices, and f and g are n-rowed column vectors. Since both U and V are partial isometries with first column 0 and rank n, it follows that, in both cases, the remaining n columns constitute an orthonormal set, and hence, in particular, that f is orthogonal to the columns of U_0 and g is orthogonal to the columns of V_0. The thing to prove is that if U_0 and V_0 are unitarily equivalent, then so also are U and V. Suppose therefore that W_0 is unitary and $W_0 U_0 W_0^* = V_0$. Assertion: there exists a complex number θ of modulus 1 such that $W = \begin{pmatrix} W_0 & 0 \\ 0 & \theta \end{pmatrix}$ transforms U onto V. Indeed, if $|\theta| = 1$, then

$$WUW^* = \begin{pmatrix} V_0 & \bar{\theta} W_0 f \\ 0 & \lambda \end{pmatrix},$$

Since this matrix is a partial isometry with first column 0 and rank n, it follows that $\bar{\theta} W_0 f$ is orthogonal to the span (of dimension $n - 1$) of the columns of V_0. Since g also is orthogonal to the columns of V_0, it follows that θ can indeed be chosen so that $\bar{\theta} W_0 f = g$. The only case that gives a moment's pause is the one in which $W_0 f = 0$. In that case $f = 0$, and therefore $|\lambda| = 1$; this implies that $g = 0$, and all is well.

5. Distance. Since the unitary equivalence problem is solved for normal operators, it is reasonable to approach its solution in the general case by asking how far any particular operator is from normality. The figurative "how far" can be interpreted literally, and its literal interpretation yields a curious unitary invariant. Let N be the set of all normal operators, and for each (not necessarily normal) operator A consider the distance $d(A, N)$ from A to N. The distance here is meant in the usual sense appropriate to subsets of metric spaces: $d(A, N) = \inf \{ \| A - N \| : N \in N \}$. The definition makes sense for all operators, and, in particular, for partial isometries. We proceed to study one of the simplest questions that the definition suggests: as U varies over the set P of partial isometries, what possible values can $d(U, N)$ attain? The answer we obtain is rather peculiar.

THEOREM 5. *The set of all possible values of $d(U, N)$, for U in P, is the closed interval $[0, 1/2]$ together with the single number 1.*

Proof. We begin with the assertion that if a partial isometry U has a unitary enlargement, then $d(U, N) \leq 1/2$. The proof consists in verifying that if W is a unitary enlargement of U, then

$$\left\| U - \frac{1}{2} W \right\| = \frac{1}{2} .$$

Indeed, if $\mathfrak{N}$ is the null-space of U, then U is equal to 0 on $\mathfrak{N}$ and to W on $\mathfrak{N}^{\perp}$; it follows that $U - \frac{1}{2} W$ is equal to $-\frac{1}{2} W$ on $\mathfrak{N}$ and to $\frac{1}{2} W$ on $\mathfrak{N}^{\perp}$. This implies that $U - \frac{1}{2} W$ is $1/2$ times a unitary operator and hence that its norm is $1/2$.

It is easy to exhibit a partial isometry U such that $d(U, N) = 1/2$; in fact this can be done on a two-dimensional Hilbert space. A simple example is the operator U_0 given by the matrix $\begin{pmatrix} 0 & 1 \\ 0 & 0 \end{pmatrix}$. That U_0 is a partial isometry can be verified at a glance. (Its matrix is obtained from a unitary matrix by "erasing" a column.) The preceding paragraph implies that $d(U_0, N) \leq 1/2$; it remains to prove that if N is normal, then $\| U_0 - N \| \geq 1/2$. For this purpose, let f be an arbitrary unit vector and note that

$$\big| \| U_o f \| - \| U_o^* f \| \big| \leq \big| \| U_o f \| - \| Nf \| \big| + \big| \| N^* f \| - \| U_o^* f \| \big|$$
$$\leq 2 \| U_o - N \| .$$

(Recall that, by normality, $\| Nf \| = \| N^* f \|$.) If f is the column vector $\binom{0}{1}$, then $\| U_o f \| = 1$ and $\| U_o^* f \| = 0$; the proof is complete.

For each number t in the interval $[0, 1]$ write $t' = \sqrt{1 - t^2}$. The mapping $t \to U_t = \begin{pmatrix} t & t' \\ 0 & 0 \end{pmatrix}$ is a continuous path in the metric space P, which joins the partial isometry U_o to the projection $\begin{pmatrix} 1 & 0 \\ 0 & 0 \end{pmatrix}$ (a normal partial isometry). Conclusion: as U varies over all partial isometries, $d(U, N)$ can take (at least) all values between 0 and $1/2$ inclusive. $\big($The technique of the preceding paragraph can be used to show that $d(U_t, N) = \frac{1}{2} t'.\big)$

For the next step we need the following lemma: if P is a projection, and if A is an operator such that $P + A$ is one-to-one, then $\nu(A) \leq \rho(P)$. To prove this, observe that the null-spaces of P and A have only 0 in common, so that the restriction of P to the null-space $\mathfrak{N}$ of A is one-to-one. It follows that the dimension of $\mathfrak{N}$ is less than or equal to the dimension of the entire range of P, which is the desired conclusion. (We use here the assertion that one-to-one bounded linear transformations do not lower dimension; cf. [2, Lemma 3].)

Suppose now that U is a partial isometry such that $\nu(U) < \rho'(U)$. Assertion: no operator at a distance less than 1 from U can be invertible. Suppose, indeed, that $\| U - A \| < 1$, so that $\| U^* U - U^* A \| < 1$. Write $P = 1 - U^* U$; since U is a partial isometry, P is the projection onto the null-space of U. Since $U^* U - U^* A = 1 - (P + U^* A)$, it follows that $P + U^* A$ is invertible, and hence, from the lemma of the preceding paragraph, that $\nu(U^* A) \leq \rho(P) = \nu(U)$. If A were invertible, then $U^* A$ and U^* would have the same nullity, and it would follow that $\nu(U^*) \leq \nu(U)$. This contradicts the assumption on U, and it follows that A cannot be invertible.

Since the closure of the set of all invertible operators includes N, it follows from the preceding paragraph that if U is a partial isometry with $\nu(U) < \rho'(U)$, then $d(U, N) \geq 1$. This result quickly implies some minor improvements of itself. To begin with, the hypothesis $\nu(U) < \rho'(U)$ can be replaced by $\nu(U) \neq \rho'(U)$. (If $\nu(U) > \rho'(U)$, then $\rho'(U^*) > \nu(U^*)$, and the original formulation is applicable to U^*.) Next, the conclusion $d(U, N) \geq 1$ can be replaced by $d(U, N) = 1$. (Since 0 is normal, no partial isometry is at a distance greater than 1 from N.) Finally, the result implies the principal assertion: if U is a partial isometry such that $d(U, N) > 1/2$, then $d(U, N) = 1$. Indeed, if $d(U, N) > 1/2$, then $\nu(U) \neq \rho'(U)$, for otherwise U would have a unitary enlargement, and therefore, by the first paragraph of this proof, U would be at a distance not more than $1/2$ from N. The proof of Theorem 5 is complete.

6. **Continuity.** Associated with each partial isometry U there are three cardinal numbers: the rank $\rho(U)$, the nullity $\nu(U)$, and the co-rank $\rho'(U)$. Our next purpose is to prove that the three functions ρ, ν, and ρ' are continuous. For the space P of partial isometries we use the topology induced by the norm; for cardinal numbers we use the discrete topology. With this explanation the meaning of the continuity assertion becomes unambiguous: if U is sufficiently near to V, then U and V have the same rank, the same nullity, and the same co-rank. The following assertion is a precise quantitative formulation of the same result.

THEOREM 6. *If U and V are partial isometries such that $\| U - V \| < 1$, then $\rho(U) = \rho(V)$, $\nu(U) = \nu(V)$, and $\rho'(U) = \rho'(V)$.*

Proof. The null-space of U and the initial space of V can have only 0 in common. Indeed, if f is a nonzero vector such that $Uf = 0$ and $\| Vf \| = \| f \|$, then $\| Uf - Vf \| = \| f \|$, and this contradicts the hypothesis $\| U - V \| < 1$. It follows that the restriction of U to the initial space of V is one-to-one, and hence (see [2] again) that the dimension of the initial space of V is less than or equal to the dimension of the entire range of U. In other words, the result is that $\rho(V) \leqq \rho(U)$; the assertion about ranks follows by symmetry. This part of the theorem generalizes (from projections to arbitrary partial isometries) a theorem of Nagy (see [4, § 105]), and, at the time, considerably shortens its proof. The original proof is, in a sense, more constructive; it not only proves that two subspaces have the same dimension, but it exhibits a partial isometry for which the first is the initial space and the second the final space.

The assertion about ν can be phrased this way: if $\nu(U) \neq \nu(V)$, then $\| U - V \| \geqq 1$. Indeed, if $\nu(U) \neq \nu(V)$, say, for definiteness, $\nu(U) < \nu(V)$, then there exists at least one unit vector f in the null-space of V that is orthogonal to the null-space of U. To say that f is orthogonal to the null-space of U is the same as to say that f belongs to the initial space of U. It follows that $1 = \| f \| = \| Uf \| = \| Uf - Vf \| \leqq \| U - V \|$, and the proof of the assertion about nullities is complete.

The assertion about co-ranks is an easy corollary: if $\| U - V \| < 1$, then $\| U^* - V^* \| < 1$, and therefore $\rho'(N) = \nu(U^*) = \nu(V^*) = \rho'(V)$.

If the dimension of the underlying Hilbert space is δ, then the rank, nullity, and co-rank of each partial isometry are cardinal numbers

ρ, ν, and ρ' such that $\rho + \nu = \rho + \rho' = \delta$. If, conversely, ρ, ν, and ρ' are any three cardinal numbers satisfying these equations, then there exist partial isometries with rank ρ, nullity ν, and co-rank ρ'. Let $P(\rho, \nu, \rho')$ be the set of all such partial isometries. Clearly the sets of the form $P(\rho, \nu, \rho')$ constitute a partition of the space P of all partial isometries; it is a consequence of Theorem 6 that each set $P(\rho, \nu, \rho')$ is both open and closed.

7. **Connectivity.** We proved in § 5 that there is a continuous path in the space P joining a normal partial isometry (in fact a projection) to one whose distance from N is $1/2$. On the other hand, § 6 shows that P is not connected, and this suggests the question of just how disconnected P is. The following assertion is the answer.

THEOREM 7. *For each ρ, ν, and ρ', the set $P(\rho, \nu, \rho')$ of all partial isometries of rank ρ, nullity ν, and co-rank ρ' is arcwise connected.*

Proof. The principal tool is the theorem that the set $P(\rho, 0, 0)$ of all unitary operators is arcwise connected. This is a consequence of the functional calculus. Indeed, if U is unitary, then there exists a Hermitian operator A such that $U = e^{iA}$, If $U_t = e^{itA}$, $0 \leqq t \leqq 1$, then $t \to U_t$ is a continuous path of unitary operators joining $1 \,(= U_0)$ to $U(= U_1)$. Since each unitary operator can be joined to 1, it follows that any two can be joined to each other. This settles the case $P(\rho, 0, 0)$. A useful consequence is that if two partial isometries are unitarily equivalent, then they can be joined by a continuous path. Indeed if U_0 and U_1 are partial isometries, and if V is a unitary operator such that $V^* U_0 V = U_1$, then let $t \to V_t$ be a continuous path joining 1 to V, and observe that $t \to V_t^* U_0 V_t$ is a continuous path joining U_0 to U_1.

For the next step we need to recall the basic facts about shifts (see [1] or [3]). A simple shift (more precisely, a simple unilateral shift) is an isometry V for which there exists a unit vector f such that the vectors f, Vf, V^2f, $\cdots$ form an orthonormal basis for the space. A shift (not necessarily simple) is, by definition, the direct sum of simple ones. It is easy to see that every shift is an isometry whose co-rank is the number of simple direct summands. Two shifts are unitarily equivalent if and only if they have the same co-rank. The fundamental theorem about shifts is that every element of $P(\rho, 0, \rho')$ (i.e., every isometry of co-rank ρ') is either unitary (in which case $\rho' = 0$), or a shift of co-rank ρ', or the direct sum of a unitary operator and a shift of co-rank ρ'.

Suppose now that U_0 and U_1 are in $P(\rho, 0, \rho')$, with $\rho' \neq 0$. If both U_0 and U_1 are shifts, then (since they have the same co-rank) they are unitarily equivalent, and, therefore, they can be joined by a continuous path.

Suppose next that U_0 is a shift (of co-rank ρ') and that $U_1 = V_1 \oplus W_1$, where V_1 is a shift (of co-rank ρ') and W_1 is unitary. Since the dimension of the domain of U_0 is $\rho' \cdot \aleph_0$, and since U_0 and U_1 have the some domain, it follows that the dimension of the domain of W_1 is not more than $\rho' \cdot \aleph_0$. If $\rho' > \aleph_0$, then break up W_1 into ρ' direct summands, each on a space of dimension $\aleph_0$, and match these summands with the ρ' simple direct summands of U_0 and U_1. The result of this procedure is to reduce the problem to the problem of joining a simple shift U to the direct sum of a simple shift V and a unitary operator W on a space of dimension $\aleph_0$ or smaller.

If the dimension of the domain of W is n ($< \aleph_0$), the problem is easy to describe and to solve in terms of matrices. The shift U is unitarily equivalent to (and therefore it can be joined to) an operator with matrix

$$
\begin{pmatrix}
0 & 0 & 0 & 0 & \\
1 & 0 & 0 & 0 & \\
0 & 1 & 0 & 0 & \\
0 & 0 & 1 & 0 & \\
& & & & \ddots
\end{pmatrix},
$$

and, similarly, the direct sum $V \oplus W$ can be joined to an operator with matrix

$$
\begin{pmatrix}
1 & 0 & 0 & & 0 & 0 & & & & & \\
0 & 1 & 0 & \cdots & 0 & 0 & & & & & \\
0 & 0 & 1 & & 0 & 0 & & 0 & & & \\
& \vdots & & & \vdots & & & & & & \\
0 & 0 & 0 & \cdots & 1 & 0 & & & & & \\
0 & 0 & 0 & & 0 & 1 & & & & & \\
& & & & & & 0 & 0 & 0 & 0 & \\
& & & & & & 1 & 0 & 0 & 0 & \\
& & 0 & & & & 0 & 1 & 0 & 0 & \\
& & & & & & 0 & 0 & 1 & 0 & \\
& & & & & & & & & & \ddots
\end{pmatrix}
$$

It remains to prove that the first of these two matrices can be joined to the second. For this purpose, note that the (unitary) permutation matrix (with $n + 1$ rows and columns)

$$\begin{pmatrix} 0 & 0 & 0 & & 0 & 1 \\ 1 & 0 & 0 & \cdots & 0 & 0 \\ 0 & 1 & 0 & & 0 & 0 \\ & \vdots & & & & \vdots \\ 0 & 0 & 0 & & 0 & 0 \\ 0 & 0 & 0 & \cdots & 1 & 0 \end{pmatrix}$$

can be joined to the identity matrix (with $n + 1$ rows and columns). Let $t \to M_t$ be a continuous path of unitary matrices that joins them, and let P be the projection matrix (with $n + 1$ rows and columns)

$$\begin{pmatrix} 1 & 0 & 0 & & 0 & 0 \\ 0 & 1 & 0 & \cdots & 0 & 0 \\ 0 & 0 & 1 & & 0 & 0 \\ & \vdots & & & & \vdots \\ 0 & 0 & 0 & & 1 & 0 \\ 0 & 0 & 0 & \cdots & 0 & 0 \end{pmatrix}$$

The "product" path $t \to M_t P$ joins

$$\begin{pmatrix} 0 & 0 & 0 & & 0 & 0 \\ 1 & 0 & 0 & \cdots & 0 & 0 \\ 0 & 1 & 0 & & 0 & 0 \\ & \vdots & & & & \vdots \\ 0 & 0 & 0 & & 0 & 0 \\ 0 & 0 & 0 & \cdots & 1 & 0 \end{pmatrix}$$

to P. Use this path in the northwest corner (of size $n + 1$) of the infinite matrices to obtain a path joining the matrix of U to the matrix of $V \oplus W$.

If the dimension of the domain of W is $\aleph_0$, the solution is easier. It is easy to verify that the operator matrix $\begin{pmatrix} 0 & U \\ 1 & 0 \end{pmatrix}$ (considered as an operator on the direct sum of the underlying space with itself) is unitarily equivalent to U, and the operator matrix $\begin{pmatrix} W & 0 \\ 0 & U \end{pmatrix}$ is unitarily equivalent to $V \oplus W$. Since W can be joined to the identity by a continuous path, it remains to prove that $\begin{pmatrix} 0 & U \\ 1 & 0 \end{pmatrix}$ can be joined to $\begin{pmatrix} 1 & 0 \\ 0 & U \end{pmatrix}$. If $t \to \begin{pmatrix} \alpha_t & \beta_t \\ \gamma_t & \delta_t \end{pmatrix}$ is a continuous path of numerical unitary matrices that joins $\begin{pmatrix} 0 & 1 \\ 1 & 0 \end{pmatrix}$ to $\begin{pmatrix} 1 & 0 \\ 0 & 1 \end{pmatrix}$, then $t \to \begin{pmatrix} \alpha_t & \beta_t U \\ \gamma_t & \delta_t U \end{pmatrix}$ is a continuous path of partial isometries that joins $\begin{pmatrix} 0 & U \\ 1 & 0 \end{pmatrix}$ to $\begin{pmatrix} 1 & 0 \\ 0 & U \end{pmatrix}$.

What we have proved so far (after successive reductions) implies that any two isometries can be joined by a continuous path, i.e., that the set $P(\rho, 0, \rho')$ is arcwise connected.

To prove that $P(\rho, \nu, \rho')$ is always arcwise connected, it is sufficient to consider the case $\nu \leq \rho'$. (Argue by adjoints.) If U_0 and U_1 are in $P(\rho, \nu, \rho')$, then they can be enlarged to isometries V_0 and V_1. Such enlargements are far from unique; what is important for our purposes is that V_0 and V_1 can be found so that they have the same co-rank. If P_0 and P_1 are the projections onto the initial spaces of U_0 and U_1 (i.e., $P_0 = U_0^* U_0$ and $P_1 = U_1^* U_1$), then P_0 and P_1 have the same rank and co-rank. It follows that there exist paths $t \to V_t$ and $t \to P_t$ joining V_0 to V_1 and P_0 to P_1. Since $U_0 = V_0 P_0$ and $U_1 = V_1 P_1$, this implies that $t \to V_t P_t$ is a continuous path joining U_0 to U_1. The proof of Theorem 7 is complete.

The following consequence of Theorems 6 and 7 is trivial, but worth making explicit: the components of P are exactly the sets $P(\rho, \nu, \rho')$.

REFERENCES

1. Arlen Brown, *On a class of operators*, Proc. Amer. Math. Soc., **4** (1953), 723–728.
2. P. R. Halmos and Günter Lumer, *Square roots of operators*, II, Proc. Amer. Math. Soc., **5** (1954), 589–595.
3. P. R. Halmos, *Shifts on Hilbert spaces*, J. reine u. angew. Math., **208** (1961), 102–112.
4. F. Riesz and B. Sz.-Nagy, *Leçons d'analyse fonctionelle*, Budapest (1952).

UNIVERSITY OF MICHIGAN

Journal für die reine und angewandte Mathematik

Herausgegeben von **Helmut Hasse** und **Hans Rohrbach**

Verlag Walter de Gruyter & Co., Berlin W 30

Sonderabdruck aus Band 213, Heft 1/2. 1963. Seite 89 bis 102

Algebraic Properties of Toeplitz operators

By *Arlen Brown*[1]) and *P. R. Halmos*[2]) at Ann Arbor (Michigan, USA)

Introduction

Let μ be normalized Lebesgue measure on the Borel sets of the unit circle in the complex plane. If $e_n(z) = z^n$ for $|z| = 1$ and $n = 0, \pm 1, \pm 2, \ldots$, then the bounded measurable functions e_n constitute an orthonormal basis for $\mathfrak{L}^2 = \mathfrak{L}^2(\mu)$. It will be convenient to say that a function f in $\mathfrak{L}^2$ (or possibly even in $\mathfrak{L}^1$) is *analytic* if all its Fourier coefficients with negative index vanish, i. e., if $\int \bar{f} e_n d\mu = 0$ for $n = -1, -2, -3, \ldots$; the complex conjugate of an analytic function will be called *co-analytic*. The analytic functions in $\mathfrak{L}^2$ constitute (by definition) the class $\mathfrak{H}^2$; the class $\bar{\mathfrak{H}}^2$ consists (by definition) of the corresponding complex conjugates.

A bounded measurable function φ on the circle induces in a natural way two operators, one on $\mathfrak{L}^2$ and one on $\mathfrak{H}^2$, as follows. The *Laurent operator* $L = L_\varphi$ is just multiplication by φ:

$$Lf = \varphi \cdot f$$

for every f in $\mathfrak{L}^2$. The *Toeplitz operator* $T = T_\varphi$ is defined, in terms of the orthogonal projection P from $\mathfrak{L}^2$ onto $\mathfrak{H}^2$, as the compression of L to $\mathfrak{H}^2$:

$$Tf = PLf = P(\varphi \cdot f)$$

for every f in $\mathfrak{H}^2$. The theory of Laurent operators is relatively easy and essentially known. The mapping $\varphi \to L_\varphi$ does the right thing by scalars, sums, complex conjugates, products, and norms, and this makes it possible to predict all algebraic properties of L_φ by the behavior of φ. The theory of Toeplitz operators is much more difficult and much less well understood. The purpose of this paper is to study some of the elementary algebraic properties of Toeplitz operators and to study the extent to which they parallel the classical properties of Laurent operators.

Laurent operators

What is a usable abstract characterization of Laurent operators? This question does not appear to have been explicitly raised before, but the answer to it is accessible by known techniques. The problem is to find necessary and sufficient conditions on an

---

[1]) Research supported in part by the U.S. Air Force Office of Scientific Research.
[2]) Research supported in part by a grant from the National Science Foundation.

Journal für Mathematik. Bd. 213. Heft 1/2

12

operator defined on an abstract Hilbert space in order that it be unitarily equivalent to some Laurent operator L defined on the concrete Hilbert space $\mathfrak{L}^2$. Two necessary conditions, one on the space and one on the operator, are close to the surface: the space must have dimension $\aleph_0$ (since $\mathfrak{L}^2$ has), and the operator must be normal (since L is). These two conditions together come very near to being sufficient. Indeed, every $\aleph_0$-dimensional Hilbert space is unitarily equivalent to every other, and every normal operator is unitarily equivalent to some multiplication. (The latter assertion is essentially the spectral theorem [3], which, however, is frequently not stated that way.) The only question that remains is what, if anything, distinguishes a multiplication on our particular $\mathfrak{L}^2$ $(= \mathfrak{L}^2(\mu))$ from an arbitrary multiplication on the space of square-integrable functions of an arbitrary measure space.

If a set X of positive measure is exactly the set where a bounded measurable function takes some particular value, say λ, then λ is a proper value of the induced multiplication, with multiplicity equal to the dimension of $\mathfrak{L}^2(X)$. A necessary and sufficient condition that that multiplicity be equal to 1 is that X be an atom. More generally, a necessary and sufficient condition that that multiplicity be equal to a positive integer n is that X be the union of exactly n atoms. If the multiplicity is infinite, then the Boolean structure of X is not uniquely determined by the Hilbert space structure of $\mathfrak{L}^2(X)$; in that case X can be non-atomic, or it can consist of infinitely many atoms, or it can be the union of a non-atomic piece and a set of (finitely or infinitely many) atoms. Since the measure algebra of μ has no atoms, it follows that a multiplication on $\mathfrak{L}^2(\mu)$ can have no proper values of finite multiplicity, and hence the same is true of each operator unitarily equivalent to some Laurent operator.

The necessary conditions just derived were discovered by Hartman and Wintner [4]; we proceed to prove that they are not only necessary but also sufficient.

Theorem 1. *A necessary and sufficient condition that a normal operator on an $\aleph_0$-dimensional Hilbert space be unitarily equivalent to a Laurent operator is that it have no proper values of finite multiplicity.*

Proof. Necessity was proved above; it remains to prove sufficiency. In that direction what is already known is that an operator satisfying the stated conditions is unitarily equivalent to a multiplication on the $\mathfrak{L}^2$-space of some non-atomic measure space. Because of the separability of that $\mathfrak{L}^2$-space, there is no loss of generality in assuming that the measure space has finite total measure, which then might as well be normalized. (The reduction argument goes as follows. If ν is a σ-finite measure, if ϱ is a strictly positive function with $\int \varrho d\nu = 1$, and if $d\nu' = \varrho d\nu$, then the mapping $f \to \frac{1}{\sqrt{\varrho}} \cdot f$ is an isometry from $\mathfrak{L}^2(\nu)$ onto $\mathfrak{L}^2(\nu')$ that transforms the multiplication operator induced on $\mathfrak{L}^2(\nu)$ by a bounded measurable function φ onto the multiplication operator induced by φ on $\mathfrak{L}^2(\nu')$.) The desired transfer to the circle is now an immediate corollary of the known characterization (separable, non-atomic, normalized) of the measure algebra of μ (see [2, Theorem C, p. 173]).

It would be interesting to have a characterization of Toeplitz operators similar to the one Theorem 1 gives for Laurent operators; nothing like that is known so far.

The Laurent operator L_{e_1} is of particular interest; it will be denoted by W throughout the sequel. The operator W is called the *bilateral shift* (or, more precisely, the *simple bilateral shift*, or the bilateral shift of *multiplicity* 1). The name "shift" comes from the relations

$$We_n = e_{n+1} \qquad (n = 0, \pm 1, \pm 2, \ldots);$$

"bilateral" refers to the set of indices, which contains all integers, on both sides of 0. It is easy to see that W is a unitary operator on $\mathfrak{L}^2$.

There is a "concrete" characterization of Laurent operators in terms of W. The adjective is intended to indicate a characterization of Laurent operators themselves on the space $\mathfrak{L}^2$ itself, and not a characterization of operators unitarily equivalent to Laurent operators on arbitrary $\aleph_0$-dimensional spaces. The result (Theorem 2) is known to many students of Hilbert space, but we could find no explicit reference. We offer here a proof that extends (with only minor notational changes) to any finite measure with compact support in the role of μ, and that applies to $\mathfrak{H}^2$ (and its generalizations defined with respect to other measures) as well as to $\mathfrak{L}^2$.

Theorem 2. *A necessary and sufficient condition that an operator on $\mathfrak{L}^2$ be a Laurent operator is that it commute with the bilateral shift.*

Proof. Since any two multiplication operators commute, the necessity of the condition is clear. It remains to prove that if an operator A on $\mathfrak{L}^2$ commutes with W, then A is a Laurent operator. If this were already known, if, say, A were known to be L_φ, then it would follow that

$$\varphi(z) = \varphi(z) \cdot e_0(z) = (Ae_0)(z).$$

Motivated by this equation, we put

$$\varphi = Ae_0,$$

so that, clearly, $\varphi \in \mathfrak{L}^2$; the problem is to prove that φ is bounded and $A = L_\varphi$.

Bounded or not, the function φ induces a multiplication operator B on $\mathfrak{L}^2$; the domain of B is the set of all those f's in $\mathfrak{L}^2$ for which $\varphi \cdot f \in \mathfrak{L}^2$. It is very easy to show that the domain of B is dense and that B is closed. If $n \geq 0$, then

$$Ae_n = AW^n e_0 = W^n Ae_0 = W^n \varphi = e_n \cdot \varphi = \varphi \cdot e_n;$$

if $n \leq 0$, then the same argument works with W^* in place of W. (This step uses the Fuglede commutativity theorem [1].) It follows that $Ae_n = Be_n$ for all n, and hence that $Ap = Bp$ whenever p is a finite linear combination of the e_n's. The closed transformation B agrees with the bounded operator A on a dense set; this implies that the domain of B is the entire space $\mathfrak{L}^2$ and that $B = A$ on $\mathfrak{L}^2$.

We conclude by proving that φ is bounded. Since $\varphi = Ae_0$, it follows that $\|\varphi\| \leq \|A\| \cdot \|e_0\|$. Since $\varphi^2 = \varphi \cdot \varphi = A\varphi$, it follows that $\|\varphi^2\| \leq \|A\| \cdot \|\varphi\| \leq \|A\|^2 \cdot \|e_0\|$. Inductive repetition of this argument shows that $\|\varphi^n\| \leq \|A\|^n \cdot \|e_0\|$ for every positive integer n. This norm inequality can be expressed (except in the trivial case $A = 0$) in the form $\int \left| \dfrac{\varphi}{\|A\|} \right|^{2n} d\mu \leq \|e_0\|^2$. It follows that the modulus of the quotientient $\varphi / \|A\|$ can be greater than 1 on a set of measure zero only, i. e., that $|\varphi| \leq \|A\|$ almost everywhere. (This ingenious argument was shown to us by Morris Schreiber.)

Every operator A on $\mathfrak{L}^2$ has a matrix $\langle a_{ij} \rangle$ with respect to the basis $\{e_n: n = 0, \pm 1, \pm 2, \ldots\}$; the matrix entries are given by

$$a_{ij} = (Ae_j, e_i).$$

If $A = L_\varphi$ for some bounded measurable function φ, then the matrix of A has a simple expression in terms of the Fourier expansion

$$\varphi = \sum_i \alpha_i e_i,$$

where, of course,

$$\alpha_i = (\varphi, e_i).$$

Indeed,

$$a_{ij} = (Ae_j, e_i) = (\varphi \cdot e_j, e_i) = (e_j \cdot \varphi, e_i) = (W^j\varphi, e_i)$$
$$= (\varphi, W^{*j}e_i) = (\varphi, e_{i-j}) = \alpha_{i-j}.$$

Note the corollary that

$$(Ae_{j+k}, e_{i+k}) = (Ae_j, e_i).$$

This motivates the definition of a *Laurent matrix* as a two-way infinite matrix $\langle a_{ij} \rangle$ (i. e., $i, j = 0, \pm 1, \pm 2, \ldots$) such that

$$a_{i+1, j+1} = a_{ij}.$$

Theorem 3. *A necessary and sufficient condition that an operator on $\mathfrak{L}^2$ be a Laurent operator is that its matrix (with respect to the orthonormal basis $\{e_n : n = 0, \pm 1, \pm 2, \ldots\}$) be a Laurent matrix.*

Proof. The proof of necessity is included in the remarks preceding the statement of the theorem. To prove sufficiency, it is enough (by Theorem 2) to prove that if A is an operator on $\mathfrak{L}^2$ such that

$$(Ae_{j+1}, e_{i+1}) = (Ae_j, e_i) \qquad (i, j = 0, \pm 1, \pm 2, \ldots)$$

then A commutes with W. The proof is immediate:

$$(AWe_j, e_i) = (Ae_{j+1}, e_i) = (Ae_j, e_{i-1}) = (Ae_j, W^*e_i) = (WAe_j, e_i).$$

Note incidentally that if $A = L_\varphi$, it is easy to recapture φ from the matrix of A. Since, $\varphi = Ae_0$, the Fourier coefficients of φ are the numbers (Ae_0, e_i), i. e., the terms of the 0-column of the matrix.

Theorem 2 is a commutativity result; one of its applications is the matrix characterization Theorem 3. We are indebted to Philip Hartman for the observation that the techniques used to prove Theorem 2 serve to prove Theorem 3 directly, and that Theorem 2 can then be recaptured from it. (Compare Theorems 4 and 6 below). This procedure avoids the use of Fuglede's theorem; its disadvantage is that it is restricted to Lebesgue measure in the circle.

Theorem 3 is the standard basic fact about Laurent operators; for some authors, in fact, it is the definition.

Toeplitz operators

The main reason the preceding results about Laurent operators on $\mathfrak{L}^2$ were derived here was to establish the proper background for a view of Toeplitz operators on $\mathfrak{H}^2$; we proceed now to the study of Toeplitz operators.

Some of the natural questions about the correspondence $\varphi \to T_\varphi$ are easy to answer and some are exceedingly difficult. On the easiest level are such assertions as these: if $\varphi = 1$, then T_φ is the identity; if $\varphi = \alpha\chi + \beta\psi$, then $T_\varphi = \alpha T_\chi + \beta T_\psi$ (i. e., T_φ depends linearly on φ). Only slightly harder is the assertion about adjoints: $T_\varphi^* = T_{\bar\varphi}$. The proof of this latter assertion depends on the fundamental identity

$$PT_\varphi P = PL_\varphi P,$$

which can also be expressed by saying that $(T_\varphi f, g) = (L_\varphi f, g)$ whenever f and g are in

$\mathfrak{H}^2$. Using this, we compute as follows:

$$(T_\varphi^* f, g) = (f, T_\varphi g) = (f, L_\varphi g) = (L_\varphi^* f, g) = (L_{\bar\varphi} f, g) = (T_{\bar\varphi} f, g).$$

From the simple facts obtained so far we can infer that the set of all Toeplitz operators is a self-adjoint vector space containing the identity.

Is the correspondence $\varphi \to T_\varphi$ one-to-one? In other words, if $T_\varphi = 0$, does it follow that $\varphi = 0$? It is easy to see that the answer is yes. Indeed, if $T_\varphi = 0$, then $P(\varphi \cdot f) = 0$ whenever $f \in \mathfrak{H}^2$, and therefore, in particular, $P(\varphi \cdot e_n) = 0$ whenever $n \geq 0$. If $\varphi = \sum_i \alpha_i e_i$, then $\varphi \cdot e_n = \sum_i \alpha_i e_{i+n}$; to say that $P(\varphi \cdot e_n) = 0$ means that $\alpha_i = 0$ whenever $i + n \geq 0$. Since $i + n$ can always be made positive by proper choice of a positive n, it follows that $\alpha_i = 0$ for all i, and hence that $\varphi = 0$.

What we have proved so far implies that a Toeplitz operator T_φ is Hermitian if and only if φ is real; indeed $T_\varphi = T_\varphi^*$ if and only if $\varphi = \bar\varphi$. It is also true that T_φ is positive if and only if φ is positive. Indeed, since $(T_\varphi f, f) = (L_\varphi f, f)$ whenever $f \in \mathfrak{H}^2$, it follows that T_φ is positive if and only if $(L_\varphi f, f) \geq 0$ for all f in $\mathfrak{H}^2$. The latter condition is equivalent to this one: $(W^n L_\varphi f, W^n f) \geq 0$ whenever $f \in \mathfrak{H}^2$ (and n is an integer). Since W commutes with L_φ, the condition can also be expressed in this form: $(L_\varphi W^n f, W^n f) \geq 0$ whenever $f \in \mathfrak{H}^2$. Since the set of all $W^n f'$s, with f in $\mathfrak{H}^2$, is dense in $\mathfrak{L}^2$, the condition is equivalent to $L_\varphi \geq 0$ and hence to $\varphi \geq 0$.

To get deeper results we need sharper tools; one useful tool is the Toeplitz analogue of Theorem 3. The matrix of a Toeplitz operator with respect to the basis $\{e_n: n = 0, 1, 2, \ldots\}$ has a special form closely related to and easily derivable from the matrix of the associated Laurent operator. Indeed if φ is a bounded measurable function, then $T_\varphi f = PL_\varphi f$ for all f in $\mathfrak{H}^2$, and consequently, for $i, j \geq 0$,

$$(T_\varphi e_j, e_i) = (PL_\varphi e_j, e_i) = (L_\varphi e_j, e_i) = (L_\varphi e_{j+1}, e_{i+1})$$
$$= (PL_\varphi e_{j+1}, e_{i+1}) = (T_\varphi e_{j+1}, e_{i+1}).$$

This motivates the definition of a *Toeplitz matrix* as a one-way infinite matrix $\langle a_{ij} \rangle$ (i. e., $i, j = 0, 1, 2, \ldots$) such that

$$a_{i+1, j+1} = a_{ij}.$$

Theorem 4. *A necessary and sufficient condition that an operator on $\mathfrak{H}^2$ be a Toeplitz operator is that its matrix (with respect to the orthonormal basis $\{e_n: n = 0, 1, 2, \ldots\}$) be a Toeplitz matrix.*

Proof. The proof of necessity is included in the remarks preceding the statement of the theorem. To prove sufficiency, we assume that A is an operator on $\mathfrak{H}^2$ such that

$$(A e_{j+1}, e_{i+1}) = (A e_j, e_i) \qquad (i, j = 0, 1, 2, \ldots);$$

we are to prove that A is a Toeplitz operator.

Consider for each non-negative integer n the operator on $\mathfrak{L}^2$ given by

$$A_n = W^{*n} A P W^n.$$

If $i, j \geq 0$, then

$$(A_n e_j, e_i) = (A_0 e_{j+n}, e_{i+n}) = (A e_j, e_i).$$

Something like this is true even for negative indices. Indeed, for n sufficiently large both $j + n$ and $i + n$ are positive, and from then on $(A_0 e_{j+n}, e_{i+n})$ is independent of n.

Conclusion: if p and q are finite linear combinations of e_i's, then the sequence $\{(A_n p, q)\}$ is convergent. Since

$$\| A_n \| \leq \| A_0 \| = \| A \|,$$

it follows on easy general grounds that the sequence $\{A_n\}$ of operators on $\mathfrak{L}^2$ is weakly convergent to a bounded operator A_∞ on $\mathfrak{L}^2$.

Since, for all i and j,

$$\begin{aligned}
(A_\infty e_j, e_i) &= \lim_n (W^{*n} A P W^n e_j, e_i) \\
&= \lim_n (W^{*n+1} A P W^{n+1} e_j, e_i) = \lim_n (W^{*n} A P W^n e_{j+1}, e_{i+1}) \\
&= (A_\infty e_{j+1}, e_{i+1}),
\end{aligned}$$

it follows that the operator A_∞ has a Laurent matrix and hence that it is a Laurent operator. If f and g are in $\mathfrak{H}^2$, then

$$(P A_\infty f, g) = (A_\infty f, g) = \lim_n (W^{*n} A P W^n f, g) = (A P f, g) = (A f, g),$$

so that $P A_\infty f = A f$ for each f in $\mathfrak{H}^2$. Conclusion: A is the compression to $\mathfrak{H}^2$ of Laurent operator, and hence, by definition, A is a Toeplitz operator. (This proof is modelled after one given by Hartman and Wintner [4].)

How can we recapture the function φ that induces A from the matrix of A? If $A = T_\varphi$, then $A_\infty = L_\varphi$, and therefore the Fourier coefficients of φ are the entries in the 0-column of the matrix of A_∞. This is an answer, but not a satisfying one; it is natural to wish for an answer expressed in terms of A instead of A_∞. That turns out to be easy. If $i, j \geq 0$, then

$$(A e_j, e_i) = (A_\infty e_j, e_i) = (\varphi, e_{i-j});$$

this implies that

$$(\varphi, e_i) = (A e_0, e_i) \text{ for } i \geq 0,$$

and

$$(\varphi, e_{-j}) = (A e_j, e_0) \text{ for } j \geq 0.$$

Conclusion: φ is the function whose forward Fourier coefficients (the ones with positive index) are the terms of the 0-column of the matrix of A and whose backward Fourier coefficients are the terms of the 0-row of that matrix.

Corollary. *The only completely continuous Toeplitz operator is* 0.

Proof. If φ is a bounded measurable function, and if both n and $n + k$ are non-negative integers, then

$$(\varphi, e_k) = (T_\varphi e_n, e_{n+k}).$$

If T_φ is completely continuous, then $\| T_\varphi e_n \| \to 0$ (since $e_n \to 0$ weakly); it follows that $(\varphi, e_k) = 0$ for all k (positive, negative, or zero), and hence that $\varphi = 0$.

It is worth while to observe that the weak convergence established in the course of the proof of Theorem 4 is, in fact, strong convergence.

Theorem 5. *If φ is any bounded measurable function, then*

$$W^{*n} T_\varphi P W^n \to L_\varphi \qquad \qquad (\text{as } n \to \infty)$$

in the strong operator topology.

Proof. If $\varphi = 1$, then both T_φ and L_φ are equal to the identity operator (on their respective domains), and the assertion becomes

$$\| W^{*n} P W^n f - f \| \to 0$$

for each f in $\mathfrak{L}^2$. Since W is unitary, an application of W^n does not change norms; the desired assertion is the same as

$$\| P W^n f - W^n f \| \to 0,$$

or, equivalently,

$$\| (1 - P) W^n f \| \to 0.$$

In the latter form the result is obvious; after an application of W^n, with n large, only a small part of the norm remains to be carried by the Fourier coefficients with negative index.

For an arbitrary φ we use the preceding result, the relation $P T_\varphi P = P L_\varphi P$, and the strong sequential continuity of operator multiplication, as follows:

$$W^{*n} T_\varphi P W^n = W^{*n} P L_\varphi P W^n = (W^{*n} P W^n)(W^{*n} L_\varphi W^n)(W^{*n} P W^n),$$

where the first and the last factor on the right tend strongly to 1 and the middle factor is constantly equal to L_φ.

Theorem 5 can be used to improve the assertion that the correspondence $\varphi \to T_\varphi$ is one-to-one; in fact, that correspondence is norm-preserving, in the sense that $\| T_\varphi \| = \| \varphi \|_\infty$ (the essential supremum of $| \varphi |$). For Laurent operators this is well known; we shall prove it for Toeplitz operators by showing that $\| T_\varphi \| = \| L_\varphi \|$. Since T_φ is the restriction of $P L_\varphi$ to $\mathfrak{H}^2$, it is clear that $\| T_\varphi \| \leq \| P L_\varphi \| \leq \| L_\varphi \|$. The reverse inequality comes from Theorem 5; since $W^{*n} T_\varphi P W^n$ tends strongly to L_φ, and since $\| W^{*n} T_\varphi P W^n \| \leq \| T_\varphi \|$, it follows that $\| L_\varphi \| \leq \| T_\varphi \|$.

We are now ready to turn to the Toeplitz analogue of Theorem 2. The Toeplitz operator T_{e_1} is of particular interest; it will be denoted by U throughout the sequel. The operator U is called the *unilateral shift* (or, more precisely, the *simple* unilateral shift, or the unilateral shift of *multiplicity* 1). It is easy to see that U is an isometry on $\mathfrak{H}^2$; it is the restriction to $\mathfrak{H}^2$ of the bilateral shift.

Theorem 6. *A necessary and sufficient condition that an operator A on $\mathfrak{H}^2$ be a Toeplitz operator is that $U^* A U = A$.*

Proof. The result is an immediate consequence of the matricial characterization of Toeplitz operators. Indeed, A is a Toeplitz operator if and only if

$$(A e_{j+1}, e_{i+1}) = (A e_j, e_i) \qquad (i, j = 0, 1, 2, \ldots).$$

Since the left term is equal to $(A U e_j, U e_i) = (U^* A U e_j, e_i)$, the asserted conclusion follows.

It is a corollary of Theorem 6 that topologically the set of all Toeplitz operators is well behaved. Since the product $U^* A U$ is weakly continuous in its middle factor, the set of all Toeplitz operators is weakly closed, and therefore, a fortiori, it is strongly and uniformly closed.

In Theorem 6 (compare Theorem 2) the distinction between $U^* A U = A$ and $A U = U A$ appears to play a role, and it is indeed an important one; the chief point is that U is not unitary. The condition $U^* A U = A$ characterizes Toeplitz operators; what about the condition $AU = UA$? The answer is surprisingly easy. To formulate it concisely, we define a Toeplitz operator T_φ to be *analytic* or *co-analytic* according as the function φ is analytic or co-analytic.

A preliminary remark about analytic Toeplitz operators is in order. If φ is analytic and $f \in \mathfrak{H}^2$, then it seems intuitively obvious (because of the way series are multiplied) that $\varphi \cdot f$ is not only in $\mathfrak{L}^2$ but in fact in $\mathfrak{H}^2$, and hence that $T_\varphi f$, which is equal to $P(\varphi \cdot f)$ by definition, turns out to be equal to $\varphi \cdot f$. This conclusion is valid, and all is well, but the reader should be warned that the conclusion needs proof. The proof is not difficult, and it (and wide generalizations of it) are an established part of Fourier theory [11, vol. 1, p. 157 ff.].

Theorem 7. *A necessary and sufficient condition that an operator on $\mathfrak{H}^2$ be an analytic [or co-analytic] Toeplitz operator is that it commute with the unilateral shift [or the adjoint of the unilateral shift].*

Proof. If $A = T_\varphi$ with $(\varphi, e_{-n}) = 0$ whenever $n > 0$, and if $f \in \mathfrak{H}^2$, then

$$AUf = \varphi \cdot e_1 \cdot f = e_1 \cdot \varphi \cdot f = UAf,$$

so that A commutes with U. If, conversely, A commutes with U, i. e., $AU = UA$, then $U^*AU = A$ (since U is an isometry), and it follows from Theorem 6 that $A = T_\varphi$ for some bounded measurable function φ. To prove that $(\varphi, e_{-n}) = 0$ whenever $n > 0$, .compute as follows:

$$(\varphi, e_{-n}) = (Ae_n, e_0) = (AU^n e_0, e_0) = (U^n Ae_0, e_0) = (Ae_0, U^{*n} e_0) = 0.$$

(The reason for the last step is that $U^* e_0 = 0$.) This proves the theorem about analytic Toeplitz operators; the result about co-analytic ones follows by the formation of adjoints.

Note that if φ is analytic and if $i, j \geq 0$, then $(T_\varphi e_j, e_i) = (\varphi, e_{i-j}) = 0$ whenever $i < j$. In other words, the matrix of an analytic Toeplitz operator is lower triangular: it has 0's above the main diagonal. Co-analytic Toeplitz operators correspond, similarly, to upper triangular Toeplitz matrices.

Multiplicative properties

The set of all Toeplitz operators is certainly not commutative and certainly not closed under multiplication. A counterexample for both assertions is given by the unilateral shift and its adjoint. Both U and U^* are Toeplitz operators, but the product U^*U (which is equal to the Toeplitz operator 1) is not the same as the product UU^* (which is not a Toeplitz operator). One way to prove that UU^* is not a Toeplitz operator is to use Theorem 6: since $U^*(UU^*) U = (U^*U)(U^*U) = 1 \ (\neq UU^*)$, everything is settled. Alternatively, this negative result could have been obtained via Theorem 4 by a direct look at the matrix of UU^*.

When is the product of two Toeplitz operators a Toeplitz operator? The answer is: rarely. The proof (and the precise formulation) depends on a simple but useful equation involving matrix entries. Suppose that T_φ and T_ψ are Toeplitz operators. Write $C = T_\varphi T_\psi$ and let $\langle c_{ij} \rangle$ be the (not necessarily Toeplitz) matrix of C. If the Fourier expansions of φ and ψ are

$$\varphi = \sum_i \alpha_i e_i \quad \text{and} \quad \psi = \sum_j \beta_j e_j,$$

so that the matrices of T_φ and T_ψ are $\langle \alpha_{i-j} \rangle$ and $\langle \beta_{i-j} \rangle$ respectively, then

$$c_{i+1,\,j+1} = c_{ij} + \alpha_{i+1}\beta_{-j-1}$$

whenever $i, j \geqq 0$. (We shall refer to this equation as the *product matrix formula*.) The proof is straightforward. Since

$$c_{ij} = \sum_{k=0}^{\infty} \alpha_{i-k}\beta_{k-j},$$

it follows that

$$c_{i+1,\,j+1} = \sum_{k=0}^{\infty} \alpha_{i+1-k}\beta_{k-j-1}$$

$$= \alpha_{i+1}\beta_{-j-1} + \sum_{k=1}^{\infty} \alpha_{i+1-k}\beta_{k-j-1}$$

$$= \alpha_{i+1}\beta_{-j-1} + \sum_{k=0}^{\infty} \alpha_{i-k}\beta_{k-j}$$

$$= \alpha_{i+1}\beta_{-j-1} + c_{ij}.$$

Theorem 8. *A necessary and sufficient condition that the product $T_\varphi T_\psi$ of two Toeplitz operators be a Toeplitz operator is that either φ be co-analytic or ψ be analytic; if the condition is satisfied, then $T_\varphi T_\psi = T_{\varphi\psi}$.*

Proof. If ψ is analytic, then

$$T_\varphi T_\psi f = T_\varphi(\psi \cdot f) = P(\varphi \cdot \psi \cdot f) = T_{\varphi\psi}f$$

for all f in $\mathfrak{H}^2$, so that $T_\varphi T_\psi = T_{\varphi\psi}$; if φ is co-analytic, then $\overline{\varphi}$ is analytic, and therefore

$$T_\varphi T_\psi = (T_{\overline{\psi}} T_{\overline{\varphi}})^* = T_{\overline{\psi}\,\overline{\varphi}}^* = T_{\varphi\psi}.$$

This proves the sufficiency of the condition and the last assertion of the theorem. If conversely, the product $T_\varphi T_\psi$ is a Toeplitz operator, then its matrix is a Toeplitz matrix (Theorem 4); the product matrix formula then implies that $\alpha_{i+1}\beta_{-j-1} = 0$ whenever $i, j \geqq 0$. From this, in turn, it follows that either $\alpha_{i+1} = 0$ for all $i \geqq 0$ or else $\beta_{-j-1} = 0$ for all $j \geqq 0$, which is equivalent to the desired conclusion.

Corollary 1. *A necessary and sufficient condition that the product of two Toeplitz operators be zero is that at least one factor be zero. (There are no zero divisors.)*

Proof. Sufficiency is trivial. If, conversely, $T_\varphi T_\psi = 0$, then, since 0 is a Toeplitz operator, it follows from Theorem 8 that either $\overline{\varphi}$ or ψ is analytic and that $\varphi\psi = 0$. A celebrated theorem of F. and M. Riesz [7] applies: since a non-zero analytic function, cannot vanish on a set of positive measure, it follows that if $\overline{\varphi}$ is analytic, then $\psi = 0$, and if ψ is analytic, then $\varphi = 0$.

Corollary 2. *If a Toeplitz operator T_φ is invertible, then a necessary and sufficient condition that T_φ^{-1} be a Toeplitz operator is that φ be either analytic or co-analytic. (The only Toeplitz matrices with Toeplitz inverses are the triangular ones.)*

Proof. Suppose that T_φ is invertible. If φ is analytic [or co-analytic], then (Theorem 7) T_φ commutes with U [or U^*], and it follows that T_φ^{-1} commutes with U [or U^*]. From this, in turn, it follows that T_φ^{-1} is an analytic [or co-analytic] Toeplitz operator; the proof of sufficiency is complete. Suppose now that T_φ^{-1} is known to be a Toeplitz operator, say $T_\varphi^{-1} = T_\psi$. Since $T_\varphi T_\psi = 1$, it follows from Theorem 8 that either $\overline{\varphi}$ or ψ is analytic. Since, similarly, $T_\psi T_\varphi = 1$, it follows also that either $\overline{\psi}$ or φ is analytic. If

$\overline{\varphi}$ is not analytic, then ψ is analytic and not constant; this implies that $\overline{\psi}$ is not analytic and hence that φ is.

The substance of Corollary 2 is not that it is hard for a Toeplitz operator to have an inverse, but that it is hard for the inverse of a Toeplitz operator to be a Toeplitz operator itself.

Corollary 3. *A necessary and sufficient condition that a Toeplitz operator T_φ be an isometry is that φ be analytic and that $|\varphi|$ be the constant function e_0.*

Proof. Sufficiency is obvious. To prove necessity, suppose that $T_\varphi^* T_\varphi = 1$. This means that $T_{\overline{\varphi}} T_\varphi = 1$, and it follows from Theorem 8 that either $\overline{\varphi}$ is co-analytic (i. e., φ is analytic) or φ is analytic, and that $\overline{\varphi} \cdot \varphi = 1$.

Corollary 4. *The only unitary Toeplitz operators are the scalars of modulus 1.*

Proof. If both T_φ and T_φ^* are isometries, then, by Corollary 3, both φ and $\overline{\varphi}$ are analytic.

Corollary 5. *The only idempotent Toeplitz operators are 0 and 1.*

Proof. If $T_\varphi = T_\varphi^2$, then $T_\varphi(1 - T_\varphi) = 0$ and therefore, by Corollary 1, either $T_\varphi = 0$ or $T_\varphi = 1$.

Note that since UU^* is a projection, Corollary 5 proves again that UU^* is not a Toeplitz operator. The same reasoning that proved Corollary 5 proves also that the only Toeplitz operators that satisfy a quadratic equation are the scalar roots of that equation. This is but a small fragment of the truth; we shall prove below that scalars are the only Toeplitz operators that satisfy any one of a large class of functional equations.

Theorem 8 and its corollaries are concerned with pairs of Toeplitz operators whose product is a Toeplitz operator. What can be said about commutative pairs of Toeplitz operators? We know that we cannot expect their product to be a Toeplitz operator; how likely are we to encounter the situation at all?

Theorem 9. *A necessary and sufficient condition that two Toeplitz operators commute is that either both be analytic, or both be co-analytic, or one be a linear function of the other.*

Proof. Sufficiency is clear. To prove necessity, suppose that $T_\varphi T_\psi = T_\psi T_\varphi$, where φ and ψ are bounded measurable functions with Fourier expansions

$$\varphi = \sum_i \alpha_i e_i \quad \text{and} \quad \psi = \sum_j \beta_j e_j.$$

The product matrix formula is applicable to the product of T_φ and T_ψ in both orders; the assumed commutativity implies that

$$\alpha_{i+1}\beta_{-j-1} = \beta_{i+1}\alpha_{-j-1}$$

whenever $i, j \geq 0$. Both φ and ψ are analytic if and only if $\alpha_{-j-1} = \beta_{-j-1} = 0$ for $j \geq 0$; both φ and ψ are co-analytic if and only if $\alpha_{i+1} = \beta_{i+1} = 0$ for $i \geq 0$. If $\varphi = 0$, the conclusion is trivial. In all other cases there exist non-negative integers i_0 and j_0 such that $\alpha_{i_0+1} \neq 0$ and $\alpha_{-j_0-1} \neq 0$. If the common value of $\beta_{-j_0-1} / \alpha_{-j_0-1}$ and $\beta_{i_0+1} / \alpha_{i_0+1}$ is denoted by λ, then $\beta_{i+1} = \lambda\alpha_{i+1}$ for $i \geq 0$ and $\beta_{-j-1} = \lambda\alpha_{-j-1}$ for $j \geq 0$. Conclusion: $\beta_k = \lambda\alpha_k$ whenever $k \neq 0$, and consequently

$$T_\psi - \beta_0 = \lambda(T_\varphi - \alpha_0);$$

the proof is complete.

Corollary. *The only normal Toeplitz operators are linear functions of Hermitian ones.*

Proof. If $\varphi = \sum_i \alpha_i e_i$, then $\overline{\varphi} = \sum_i \overline{\alpha}_i e_{-i} = \sum_i \overline{\alpha}_{-i} e_i$. It follows that if φ is real, then $\alpha_i = \overline{\alpha}_{-i}$ and hence that no real φ can be analytic or co-analytic unless it is a constant. This implies that if T_φ and T_ψ are commutative Hermitian Toeplitz operators, then (Theorem 9) either one of the two is a scalar, or one of them is a linear function of the other. It follows, in any case, that $T_{\varphi+i\psi}$ is a linear function either of T_φ or of T_ψ.

Spectral properties

The most difficult questions about Toeplitz operators concern their spectral behavior. In some special cases (notably when φ is analytic and when φ is real) the spectrum of T_φ is known [5, 9, 10]. In the general case very little is known: we do not even know whether the spectrum of a Toeplitz operator is always connected.

One general result is the spectral inclusion theorem of Hartman and Wintner [4]; it says that the approximate point spectrum of T_φ includes the essential range of φ (which is known to be equal to the spectrum of the Laurent operator L_φ). This result has several pleasant consequences.

It is, for instance, a consequence of the spectral inclusion theorem that $r(L_\varphi) \leqq r(T_\varphi)$ (where r stands for spectral radius). It follows that $\| L_\varphi \| \leqq r(T_\varphi) \leqq \| T_\varphi \|$, and this is the less trivial half of the assertion (which we proved earlier by non-spectral means) that $\| T_\varphi \| = \| L_\varphi \|$. (Note the corollary $r(T_\varphi) = r(L_\varphi)$.)

Another consequence is that there are no quasinilpotent Toeplitz operators other than 0. Indeed, if the spectrum of T_φ consists of 0 alone, then the same is true of L_φ, and it follows that $\varphi = 0$. Similar reasoning shows that a Toeplitz operator with a real spectrum must be Hermitian. Indeed, if the spectrum of T_φ is real, then the same is true of L_φ, and it follows that φ is real.

A slightly more sophisticated consequence of the spectral inclusion theorem is that $\overline{W(T_\varphi)} = \overline{W(L_\varphi)}$. (Here W stands for numerical range; $W(A)$, for any operator A, is the set of all complex numbers of the form (Af, f) with $\| f \| = 1$. The three principal facts about $W(A)$ are that (i) it is always convex, (ii) its closure includes the spectrum of A, and (iii) for normal A its closure is equal to the convex hull of the spectrum of A [8].) If $f \in \mathfrak{H}^2$, then $(T_\varphi f, f) = (L_\varphi f, f)$; this proves that $W(T_\varphi) < W(L_\varphi)$. In the reverse direction: $\overline{W(T_\varphi)}$ includes the spectrum of T_φ, and therefore it includes the spectrum of L_φ; since $\overline{W(T_\varphi)}$ is convex and L_φ is normal, it follows that $\overline{W(T_\varphi)} > \overline{W(L_\varphi)}$. The result implies that in this respect Toeplitz operators are like normal ones: the closure of the numerical range is the convex hull of the spectrum.

Another result of Hartman and Wintner [5] is that a Hermitian Toeplitz operator that is not a scalar has no proper values. Here is a simple proof. It is sufficient to show that if φ is a real-valued bounded measurable function, and if $T_\varphi f = 0$ for some f in $\mathfrak{H}^2$, then either $f = 0$ or $\varphi = 0$. Since $\varphi \cdot \overline{f} = \overline{\varphi} \cdot \overline{f} \in \mathfrak{H}^2$ (because $P(\varphi \cdot f) = 0$), and since $f \in \mathfrak{H}^2$, it follows that $\varphi \cdot \overline{f} \cdot f$ is an analytic element of $\mathfrak{L}^1$ (in standard notation, $\varphi \cdot \overline{f} \cdot f \in \mathfrak{H}^1$); since, however, $\varphi \cdot \overline{f} \cdot f$ is real, it follows that $\varphi \cdot \overline{f} \cdot f$ is a constant. Since $\int \varphi \cdot \overline{f} \cdot f \, d\mu = (\varphi \cdot f, f) = (T_\varphi f, f) = 0$ (because $T_\varphi f = 0$), the constant must be 0. The theorem of F. and M. Riesz implies that either $f = 0$ or $\varphi \cdot \overline{f} = 0$. If $f \neq 0$, then f can vanish on a set of measure zero only, and then $\varphi = 0$.

What can be said about the spectrum of a Toeplitz operator on the basis of the facts presented so far? We know that the spectrum of a non-scalar Toeplitz operator

13*

cannot consist of just one point (for otherwise, by translation, we would have a quasi-nilpotent Toeplitz operator different from 0). We know also that if the spectrum of a Toeplitz operator is included in a line, then the operator is a linear function of a Hermitian operator. This implies that the spectrum of a Toeplitz operator cannot consist of exactly two points (for otherwise we would have a Hermitian Toeplitz operator with a two-point spectrum, and hence with proper values). Can the spectrum of a Toeplitz operator consist of exactly three points? The answer is almost certainly no, but no one can prove it. Our next two theorems can be viewed as steps toward the answer.

Theorem 10. *If the numerical range of a Toeplitz operator lies in the upper half plane and contains the origin, then the operator is Hermitian.*

Proof.[3]) We are to prove that if $\operatorname{Im} T_\varphi \geqq 0$ (equivalently: $\operatorname{Im} \varphi \geqq 0$), and if there exists a unit vector f in $\mathfrak{H}^2$ with $(T_\varphi f, f) = 0$, then $\operatorname{Im} \varphi = 0$. Since $0 = (T_\varphi f, f) = \int \varphi \cdot |f|^2 d\mu$, it follows that $\int \operatorname{Im} \varphi \cdot |f|^2 d\mu = 0$, and hence that $\operatorname{Im} \varphi \cdot |f|^2 = 0$ almost everywhere. (Here is where we use the assumption $\operatorname{Im} \varphi \geqq 0$.) Since $f \in \mathfrak{H}^2$, the theorem of F. and M. Riesz and the assumption $\| f \| = 1$ enable us to conclude that f vanishes on a set of measure zero only, and hence that $\operatorname{Im} \varphi = 0$.

Corollary 1. *If a line of support of the numerical range of a Toeplitz operator contains a point of the numerical range, then it includes the entire spectrum (and hence the entire numerical range).*

Proof. Apply an appropriate linear function so as to make the line of support the real axis, with the numerical range in the upper half plane and the crucial point of the numerical range at the origin, and then invoke Theorem 10.

Corollary 2. *If the spectrum of a Toeplitz operator consists of a finite number of proper values, then the operator is a scalar.*

Proof. Find a line of support of the numerical range that contains exactly one proper value and invoke Corollary 1.

It is perhaps of interest to observe that these techniques yield another proof of the corollary to Theorem 4 (about completely continuous operators).

Some incidental light is shed on the numerical range of a Toeplitz operator T_φ. We know that $\overline{W(T_\varphi)} = \overline{W(L_\varphi)}$; can we make the same assertion for the numerical ranges (instead of their closures)? The answer, in general, is no. If, for instance, φ is a (non-constant) characteristic function, then $W(L_\varphi)$ is the closed unit interval, whereas, by Corollary 1, the closed unit interval cannot be the numerical range of T_φ. (Consider the imaginary axis in the role of the line of support.)

For any operator A on any Hilbert space (and, in fact, for any element of any Banach algebra), and for any function F holomorphic in an open set that includes the spectrum of A, a natural generalization of the Cauchy integral formula gives an unambiguously defined operator $F(A)$; the properties of the correspondence $F \to F(A)$ (the functional calculus) are a standard part of modern analysis. (See, for instance, [6, Section 5. 2].) Our next result (a sweeping generalization of the assertion that the only Toeplitz operators that can satisfy a quadratic equation are scalars) refers to this functional calculus.

Corollary 3. *If T is a Toeplitz operator, and if F is a non-zero function holomorphic in an open set that includes the spectrum of T, such that $F(T) = 0$, then T is a scalar.*

---

[3]) This proof is much shorter than our first one; the shortening is due to Philip Hartman.

Proof. The spectral mapping theorem [6, Section 5.3] implies that $F(\lambda) = 0$ whenever λ is in the spectrum of T. Since F is not identically zero, the spectrum of T must be finite. It follows [6, Section 5.6] that $\mathfrak{H}^2$ is the algebraic (but not necessarily orthogonal) direct sum of a finite number of subspaces invariant under T such that the restriction of T to each one has a singleton spectrum. If $\mathfrak{M}_0$ is any one of these subspaces, and if the spectrum of $T \mid \mathfrak{M}_0$ consists of λ_0, then λ_0 is a zero of F, so that $F(\lambda) = (\lambda - \lambda_0)^k G(\lambda)$, where $G(\lambda_0) \neq 0$ and G is holomorphic on the domain of F. The subspace $\mathfrak{M}_0$ is invariant under $G(T)$ and $G(T) \mid \mathfrak{M}_0$ is invertible (because its spectrum consists of just $G(\lambda_0)$). It follows that if f_0 is any non-zero element of $\mathfrak{M}_0$, then $G(T) f_0 \neq 0$; since, however, $(T - \lambda_0)^k G(T) f_0 = 0$, some non-zero vector of the form $(T - \lambda_0)^i G(T) f_0$ is annihilated by $T - \lambda_0$. Conclusion: the spectrum of T consists of a finite number of proper values and Corollary 2 is applicable.

Corollary 4. *If T is a Toeplitz operator, and if F is a function holomorphic in an open set that includes the spectrum of T, such that $F(T)$ is a Toeplitz operator, then either T is analytic, or T is co-analytic, or F is linear.*

Proof. Since both $F(T)$ and T are Toeplitz operators, and since they commute, Theorem 9 implies that if T is neither analytic nor co-analytic, then $F(T)$ is a linear function of T, say $F(T) = \alpha T + \beta$. If $G(\lambda) = F(\lambda) - (\alpha\lambda + \beta)$, then $G(T) = 0$; Corollary 3 implies therefore that $G = 0$, and hence that $F(\lambda) = \alpha\lambda + \beta$.

The next statement appears in [4] for real φ, but the proof does not use the assumption that φ is real. The proof is short, and, for the sake of completeness, we include it.

Theorem 11. *If φ is a non-constant bounded measurable function, the operators L_φ and T_φ have no proper values in common.*

Proof. It is sufficient to prove that if $\varphi \neq 0$, and if 0 is a proper value of L_φ (i. e., φ vanishes on a set of positive measure), then 0 is not a proper value of T_φ. Suppose, indeed, that $f \in \mathfrak{H}^2$ and $P(\varphi \cdot f) = 0$; we are to prove that $f = 0$. The assumption $P(\varphi \cdot f) = 0$ implies that $\varphi \cdot f \in \bar{\mathfrak{H}}^2$. Since $\varphi \cdot f$ vanishes where φ does, the theorem of F. and M. Riesz implies that $\varphi \cdot f = 0$. Since $\varphi \neq 0$, the function f must vanish where φ does not; the F. and M. Riesz theorem yields the desired conclusion.

Minimal normal dilations

We conclude this study of Toeplitz operators by answering a natural question about their relation to the corresponding Laurent operators.

Lemma. *If χ is a non-constant characteristic function, then $L_\chi(\mathfrak{H}^2)$ is dense in $L_\chi(\mathfrak{L}^2)$.*

Proof. Clearly $L_\chi(\mathfrak{H}^2) < L_\chi(\mathfrak{L}^2)$; it is sufficient to prove that if $f \in L_\chi(\mathfrak{L}^2)$ (i. e., $L_\chi f = f$), and if $f \perp L_\chi(\mathfrak{H}^2)$, then $f = 0$. For the proof, take an arbitrary g in $\mathfrak{H}^2$, and observe that

$$0 = (f, L_\chi g) = (L_{\bar{\chi}} f, g) = (f, g),$$

so that $f \perp \mathfrak{H}^2$. This implies that $f \in \bar{\mathfrak{H}}^2$. Since $f (= \chi \cdot f)$ vanishes at least when χ does, the F. and M. Riesz theorem implies that $f = 0$.

Theorem 12. *If φ is a non-constant bounded measurable function, then L_φ is a minimal normal dilation of T_φ.*

Proof. Let F be the characteristic function of a Borel set in the complex plane such that $\chi = F \circ \varphi$ is a non-constant characteristic function on the circle; such an F exists because φ is not constant. If a (closed) subspace of $\mathfrak{L}^2$ includes $\mathfrak{H}^2$ and reduces L_φ, then it is invariant under both L_χ and $L_{1-\chi}$. By the lemma, however, $L_\chi(\mathfrak{H}^2)$ is dense in $L_\chi(\mathfrak{L}^2)$, and $L_{1-\chi}(\mathfrak{H}^2)$ is dense in $L_{1-\chi}(\mathfrak{L}^2)$ $\left(= (L_\chi(\mathfrak{L}^2))^\perp\right)$. This implies that the only subspace of $\mathfrak{L}^2$ that includes $\mathfrak{H}^2$ and reduces L_φ is $\mathfrak{L}^2$ itself, and that is what it means to say that L_φ is a minimal normal dilation of T_φ.

References

[1] *B. Fuglede*, A commutativity theorem for normal operators, Proc. N. A. S. **36** (1950), 35—40.

[2] *P. R. Halmos*, Measure theory, New York 1950.

[3] *P. R. Halmos*, What does the spectral theorem say? Amer. Math. Monthly **70** (1963), 241—247.

[4] *P. Hartman* and *A. Wintner*, On the spectra of Toeplitz's matrices, Amer. J. Math. **72** (1950), 359—366.

[5] *P. Hartman* and *A. Wintner*, The spectra of Toeplitz's matrices, Amer. J. Math. **76** (1954), 867—882.

[6] *E. Hille* and *R. S. Phillips*, Functional analysis and semi-groups, Providence 1957, A. M. S.

[7] *F. Riesz* and *M. Riesz*, Über die Randwerte einer analytischen Funktion, Quatrième Congrés des Mathématiciens Scandinaves, Stockholm 1916, 27—44.

[8] *M. H. Stone*, Linear transformations in Hilbert space, New York 1932, A. M. S.

[9] *H. Widom*, Inversion of Toeplitz matrices, II. Ill. J. Math. **4** (1960) 88—99.

[10] *A. Wintner*, Zur Theorie der beschränkten Bilinearformen, Math. Z. **30** (1929) 228—282.

[11] *A. Zygmund*, Trigonometrical series, Cambridge 1959.

Eingegangen 18. Dezember 1962

Reprinted from the
ACTA SCIENTIARUM MATHEMATICARUM
Tomus XXV, pp. 1-5, 1964

Numerical ranges and normal dilations *)

By P. R. HALMOS in Ann Arbor (Michigan, U. S. A.)

Each operator A on a Hilbert space $\mathfrak{H}$ induces a quadratic form Q_A; by definition

$$Q_A(x) = (Ax, x)$$

for every x in $\mathfrak{H}$. (In what follows all Hilbert spaces are complex and all operators are bounded.) The *numerical range* of A, in symbols $W(A)$, is the range of Q_A on the unit sphere; explicitly

$$W(A) = \{(Ax, x) : \|x\| = 1\}.$$

The Toeplitz—Hausdorff theorem says that the numerical range of every operator is a convex subset of the complex plane ([1], [3], [4]). It is disappointing that all the known proofs of this elegant statement are ugly. The methods are elementary, but the arguments are computational. The purpose of this paper is to give a new insight into the geometric structure of numerical ranges, which seems to be interesting in its own right, and which may some day lead to a clean conceptual proof of the Toeplitz—Hausdorff theorem.

Suppose that $\mathfrak{H}$ is a subspace (closed linear manifold) of a Hilbert space $\mathfrak{K}$, and suppose that A and B are operators on $\mathfrak{H}$ and on $\mathfrak{K}$ respectively. If $Q_A(x) = = Q_B(x)$ for every vector x in $\mathfrak{H}$, then the operator A is called the *compression* of B to $\mathfrak{H}$, and B is called a *dilation* of A to $\mathfrak{K}$ (see [2]). Compression and dilation for operators are the same as restriction and extension for the corresponding quadratic forms. Usually the most convenient way to study a dilation of A to $\mathfrak{K}$ is to regard it as an operator matrix $\begin{pmatrix} A & X \\ Y & Z \end{pmatrix}$, where X maps $\mathfrak{H}^\perp$ into $\mathfrak{H}$, Y goes in the other direction, and Z operates on $\mathfrak{H}^\perp$. The easiest dilations are from $\mathfrak{H}$ to $2\mathfrak{H}$ (the direct sum of $\mathfrak{H}$ with itself); for such dilations all entries in the corresponding operator matrices may be regarded as operating on $\mathfrak{H}$.

There is a well known and easy argument that leads from the Toeplitz—Hausdorff theorem for two-by-two matrices to the most general version. In abbreviated form the argument is this: given any two unit vectors, restrict Q_A to their span, and apply the two-dimensional theorem to that restriction. The reason the argument works is that convexity is a condition on only two vectors at a time.

For normal matrices (two-by-two, or, for that matter, any size) the Toeplitz—Hausdorff theorem is an immediate consequence of diagonability (the spectral

---

*) Research supported in part by a grant from the National Science Foundation.

A 1

theorem). Indeed, since each normal matrix is unitarily equivalent to a diagonal one, it is sufficient to prove the theorem in case $A = \text{diag} \langle \lambda_1, \ldots, \lambda_n \rangle$. If $x = \langle \xi_1, \ldots, \xi_n \rangle$ is a unit vector, then $(Ax, x) = \sum_{i=1}^{n} \lambda_i |\xi_i|^2$; since the ξ's vary over all n-tuples satisfying $\sum_{i=1}^{n} |\xi_i|^2 = 1$, it follows that the numerical range $W(A)$ is exactly the set of all convex linear combinations of the λ's. This proves more than just the convexity of $W(A)$; it proves that if A is a normal matrix, then $W(A)$ is the convex hull of the spectrum of A. The proof extends, with only trivial symbolic changes, to normal operators on infinite-dimensional spaces. Since, however, an integral is not a finite sum but a limit of finite sums, the conclusion is that $\overline{W(A)}$ (the closure of $W(A)$) is the convex hull of the spectrum of A.

The simplicity of the Toeplitz—Hausdorff theorem for normal matrices makes it natural to try to reduce the general case to the normal one. In principle such a reduction is possible; this is the statement of Theorem 1 below. Existing proofs do not, however, become simpler thereby; the exasperating fact is that the proof of Theorem 1 uses the Toeplitz—Hausdorff theorem.

Theorem 1. *The numerical range of every operator on a finite-dimensional Hilbert space $\mathfrak{H}$ is the intersection of the numerical ranges of its normal dilations to $\mathfrak{H}$.*

Proof. If A is an operator on $\mathfrak{H}$ and if B is a dilation of A (normal or not), then Q_B is an extension of Q_A, and therefore $W(A) \subset W(B)$. It follows that

$$W(A) \subset \bigcap_{N \in \mathfrak{N}(A)} W(N),$$

where $\mathfrak{N}(A)$ is the set of all normal dilations of A to $2\mathfrak{H}$. It remains to prove the converse inclusion. Since $W(A)$ is convex, it is sufficient to prove that to each closed half plane that includes $W(A)$ there corresponds an N in $\mathfrak{N}(A)$ such that $W(N)$ is included in the same half plane. (Observe that $W(A)$ is closed: it is a continuous image of the unit sphere.) By a translation and a rotation (i. e., by a substitution $\to \alpha A + \beta$ with $|\alpha| = 1$) the desired assertion becomes this: if $W(A)$ is included in the closed right half plane, then so is $W(N)$ for some N in $\mathfrak{N}(A)$. To say of an operator that its numerical range is included in the closed right half plane is the same as to say that its real part (i. e., the arithmetic mean of it and its adjoint) is positive. In these terms the desired assertion is this: if $\text{Re } A \geqq 0$, then $\text{Re } N \geqq 0$ for some N in $\mathfrak{N}(A)$.

The proof of the last assertion is explicit: put $N = \begin{pmatrix} A & A^* \\ A^* & A \end{pmatrix}$. Since $N^* = \begin{pmatrix} A^* & A \\ A & A^* \end{pmatrix}$, it follows that

$$N^*N = \begin{pmatrix} A^*A + AA^* & A^{*2} + A^2 \\ A^2 + A^{*2} & AA^* + A^*A \end{pmatrix};$$

since this is symmetric in A and A^*, it follows that N is normal. It remains to prove that if $A + A^* \geqq 0$, then $N + N^* \geqq 0$. Since

$$N + N^* = \begin{pmatrix} A + A^* & A + A^* \\ A + A^* & A + A^* \end{pmatrix},$$

the problem reduces to proving that if T is positive, then so is $\begin{pmatrix} T & T \\ T & T \end{pmatrix}$, and this follows from the simple identity

$$\left(\begin{pmatrix} T & T \\ T & T \end{pmatrix} \begin{pmatrix} x \\ y \end{pmatrix}, \begin{pmatrix} x \\ y \end{pmatrix} \right) = \left(\begin{pmatrix} Tx + Ty \\ Tx + Ty \end{pmatrix}, \begin{pmatrix} x \\ y \end{pmatrix} \right) = (T(x+y), (x+y)).$$

The proof of the theorem is complete.

The normality of N and the positiveness of $N + N^*$ can be proved also by an amusing matrical computation. If $U = \dfrac{1}{\sqrt{2}} \begin{pmatrix} 1 & 1 \\ -1 & 1 \end{pmatrix}$, then

$$U^* \begin{pmatrix} A & A^* \\ A^* & A \end{pmatrix} U = \begin{pmatrix} A + A^* & O \\ O & A - A^* \end{pmatrix},$$

and, for every operator X on $\mathfrak{H}$,

$$U^* \begin{pmatrix} X & X \\ X & X \end{pmatrix} U = \begin{pmatrix} 2X & O \\ O & O \end{pmatrix};$$

in other words, N is unitarily equivalent to something obviously normal, and $N + N^*$ is unitarily equivalent to something obviously positive. These observations have the advantage that they clearly exhibit the spectra and the norms of N and $N + N^*$.

The proof of Theorem 1 used finite-dimensionality in one place only; that is what was needed to guarantee that the set under consideration (the numerical range of A) was closed. It is therefore a corollary of the proof that *the closure of the numerical range of every operator, on every Hilbert space, is the intersection of the closures of the numerical ranges of its normal dilations.* Whether or not the conclusion of Theorem 1, as is, is valid for infinite-dimensional spaces is an open question.

If A is a contraction (i. e., if $\|A\| \leq 1$), then A has not only normal but even unitary dilations; it is natural to ask whether Theorem 1 remains true if "normal" is replaced by "unitary". If $A = 0$ (on, say, a one-dimansional space), the answer is yes. Indeed, if $|\lambda| = 1$, then

$$U_\lambda = \begin{pmatrix} 0 & \lambda \\ \lambda & 0 \end{pmatrix}$$

is a unitary dilation of A, and the intersection of all the $W(U_\lambda)$'s (in fact, the intersection of any two of them) is $\{0\}$, which is just what $W(A)$ is. Here is a more interesting example: write $A = \begin{pmatrix} 0 & 0 \\ 1 & 0 \end{pmatrix}$ and

$$U_\lambda = \begin{pmatrix} 0 & 0 & \lambda \\ 1 & 0 & 0 \\ 0 & 1 & 0 \end{pmatrix}.$$

The spectrum of U_λ consists of the three cube roots of λ, and, consequently, $W(U_\lambda)$ is an equilateral triangle (interior and boundary). The intersection of all the $W(U_\lambda)$'s is the disc with center 0 and radius $\frac{1}{2}$, which is just what $W(A)$ is. (The determination of this $W(A)$ is an amusing exercise; it was explicitly carried out by Toeplitz himself [5].) The experimental evidence is favorable, but the general result it indicates is not known; the following result about normal operators is a step in that direction.

Theorem 2. *The closure of the numerical range of every normal contraction on a finite-dimensional Hilbert space $\mathfrak{H}$ is the intersection of the closures of the numerical ranges of its unitary dilations to $2\mathfrak{H}$.*

Proof. Given A on $\mathfrak{H}$, with $\|A\| \leq 1$, let $\mathfrak{U}(A)$ be the set of all unitary dilations of A to $2\mathfrak{H}$. As before, it is trivial that

$$\overline{W(A)} \subset \bigcap_{U \in \mathfrak{U}(A)} \overline{W(U)};$$

it remains to prove the reverse inclusion. Translations may push the norm of A beyond 1, but rotations are still permissible; it is therefore sufficient to prove that if $W(A)$ is included in a vertical half plane (i. e., one whose boundary is parallel to the imaginary axis), then there exists a U in $\mathfrak{U}(A)$ such that $W(U)$ is included in the same half plane. Equivalently, the desired assertion is this: if $\operatorname{Re} A \geq \alpha$, then $\operatorname{Re} U \geq \alpha$ for some U in $\mathfrak{U}(A)$.

The first step of the construction makes sense for any contraction, normal or not. Write S for the unique positive square root of $1 - AA^*$ and T for the unique positive square root of $1 - A^*A$. It is known (and easy to recompute) that if

$$U = \begin{pmatrix} A & -S \\ T & A^* \end{pmatrix},$$

then U is unitary. Since

$$U^* = \begin{pmatrix} A^* & T \\ -S & A \end{pmatrix},$$

the real part of U is given by

$$\operatorname{Re} U = \frac{1}{2} \begin{pmatrix} A+A^* & T-S \\ T-S & A+A^* \end{pmatrix}.$$

If A is normal, then $T = S$; it follows that

$$\left((\operatorname{Re} U) \begin{pmatrix} x \\ y \end{pmatrix}, \begin{pmatrix} x \\ y \end{pmatrix} \right) = ((\operatorname{Re} A)x, x) + ((\operatorname{Re} A)y, y).$$

If $\operatorname{Re} A \geq \alpha$, and if $\left\| \begin{pmatrix} x \\ y \end{pmatrix} \right\| = 1$, then

$$\left((\operatorname{Re} U) \begin{pmatrix} x \\ y \end{pmatrix}, \begin{pmatrix} x \\ y \end{pmatrix} \right) \geq \alpha(\|x\|^2 + \|y\|^2) = \alpha.$$

The proof of the theorem is complete.

It is perhaps worth while to remark that more is true about U than was needed in the proof. It can be shown that the spectrum of U (for normal A) consists exactly of those complex numbers of modulus 1 whose real parts are in the spectrum of A. If A is not normal, then both this assertion and the weaker one about $\operatorname{Re} U$ may be false. Nevertheless, the conclusion of Theorem 2 is valid for many non-normal contractions and may be valid for all. $\left(\text{Recall the example } \begin{pmatrix} 0 & 0 \\ 1 & 0 \end{pmatrix}.\right)$ One more remark along these lines is called for. A dilation of a dilation is a dilation; this may be thought to indicate that Theorems 1 and 2 could be combined to derive the conclusion of Theorem 2 for all contractions. The argument has a serious flaw, however: the normal dilations that Theorem 1 uses may not have norms less than or equal to 1, even if the operator to which Theorem 1 is being applied does, and this means that when Theorem 2 becomes needed it is not applicable. This does not mean that the proposed argument is worthless, but only that its use must be restricted to operators of small norm. An examination of the proof of Theorem 1 shows that the norms of the normal dilations of A that are introduced there need never exceed $3\|A\|$. Conclusion: *the numerical range of every operator A, with $\|A\| \leqq 1/3$, on a finite-dimensional Hilbert space $\mathfrak{H}$, is the intersection of the numerical ranges of its unitary dilations to $4\,\mathfrak{H}$.* If "numerical range" is changed to "closure of numerical range", the conclusion is valid for infinite-dimensional spaces also.

In conclusion it seems appropriate to mention a possible generalization of the preceding considerations that is interesting and non-trivial. Suppose that k is a positive integer and that A is an operator on a Hilbert space of dimension at least k. If P is a projection with $\mathrm{r}(P) = k$ ("r" stands for rank), then $\mathrm{r}(PAP) \leqq k$, and, consequently, it is possible to form $\mathrm{tr}(PAP)$ ("tr" stands for trace). Write

$$W_k(A) = \left\{ \frac{1}{k}\, \mathrm{tr}\,(PAP) \colon \mathrm{r}\,(P) = k \right\}.$$

(The normalizing factor $1/k$ is not essential, but it serves to make some of the formulas more elegant and more familiar.) It is easy to verify that $W_1(A)$ is the same as the numerical range of A. Question 1: is $W_k(A)$ always convex? Question 2: is $W_k(A)$ the intersection of all $W_k(N)$'s for N in $\mathfrak{N}(A)$? Question 3: if $\|A\| \leqq 1$, is $W_k(A)$ the intersection of all $W_k(U)$'s for U in $\mathfrak{U}(A)$? None of the answers is known. Conjecturally they are all affirmative, but the proofs may be difficult.*)

References

[1] W. F. Donoghue, On the numerical range of a bounded operator, *Michigan Math. J.*, **4** (1957), 261—263.
[2] P. R. Halmos, Normal dilations and extensions of operators, *Summa Brasil. Math.*, **2** (1950), 125—134.
[3] F. Hausdorff, Der Wertvorrat einer Bilinearform, *Math. Zeitschr.*, **3** (1919), 314—316.
[4] M. H. Stone, Hausdorff's theorem concerning Hermitian forms, *Bull. Amer. Math. Soc.*, **36** (1930), 259—261.
[5] O. Toeplitz, Das algebraische Analogon zu einem Satze von Fejér, *Math. Zeitschr.*, **2** (1918), 187—197.

UNIVERSITY OF MICHIGAN

(Received November 7, 1962)

*) *Note added March 30, 1964.* All three questions have recently been answered by C. A. Berger (Ph. D. thesis, Cornell University, 1963). The answers are yes, yes, and (for $k \geqq 2$) no. For $k=1$, on an infinite-dimensional space, E. Durszt has shown that the answer to Question 3 is no; the finite-dimensional case is of interest, and remains open.

Reprinted from the
ACTA SCIENTIARUM MATHEMATICARUM
Tomus XXVI, pp. 125-137, 1965

Cesàro operators*)

By ARLEN BROWN, P. R. HALMOS, A. L. SHIELDS in Ann Arbor (Michigan, U. S. A.)

Introduction

If f is a sequence of complex numbers, $f = \langle f(0), f(1), f(2), \ldots \rangle$, the sequence $C_0 f$ of averages plays a role in the theory of Cesàro limits; by definition

$$(C_0 f)(n) = \frac{1}{n+1} \sum_{i=0}^{n} f(i)$$

for $n = 0, 1, 2, \ldots$. Our study of Cesàro operators began with the following questions. Is it true that if $f \in l^2$, then $C_0 f \in l^2$? If it is true, is the linear transformation C_0 bounded? If C_0 is bounded, what is its spectrum? Along with these discrete questions, it is natural to ask the corresponding continuous ones; they concern the operator C_1 defined on $L^2(0, 1)$ by

$$(C_1 f) = \frac{1}{x} \int_0^x f(y)\, dy$$

for $0 < x < 1$, and the operator C_∞ defined on $L^2(0, \infty)$ by

$$(C_\infty f)(x) = \frac{1}{x} \int_0^x f(y)\, dy$$

for $0 < x < \infty$.

It turns out that all three Cesàro operators (that is, C_0, C_1, and C_∞) are everywhere defined bounded linear transformations on their respective Hilbert spaces (that is, on l^2, $L^2(0, 1)$, and $L^2(0, \infty)$). For C_0 and C_∞ this fact is proved by HARDY, LITTLEWOOD, and PÓLYA [5, Chapter IX]; the proof below (Theorem 1) is somewhat more conceptual and less computational than theirs.

For C_0 we completely determine the norm, the spectrum, and the various parts of the spectrum (Theorem 2). There is, however, much about C_0 that remains unknown. Thus, for instance, very little is known about the structure of the lattice of invariant subspaces of C_0 — a problem that belongs to a subject of great current

*) Research supported in part by a grant from the National Science Foundation.

interest. Another instance: while we prove that C_0 is hyponormal (Theorem 3), the problem of whether or not it is subnormal remains open.

In view of our incomplete information about C_0, it may be surprising to learn that the structures of C_1 and C_∞ are completely known. We prove that $1 - C_1^*$ is a unilateral shift of multiplicity 1 (Theorem 4), and $1 - C_\infty^*$ is a bilateral shift of multiplicity 1 (Theorem 5). (The operator C_1 has been studied by DE BRANGES also [3]; our methods are completely different from his.) From these facts, via the Beurling theory [1], it is easy to determine the spectra of C_1 and C_∞, and to derive a satisfactory description of their invariant subspace lattices.

Boundedness

The proof that the Cesaro operators are bounded can be made to depend on a criterion due essentially to I. SCHUR [7]. (In the notation of the statement below, SCHUR discusses the case $p(x) \equiv 1$ only; his proof is different from ours. Cf. also [6, Chapter X].) Since this criterion does not seem to be explicit in the literature, we proceed to state and to prove it with sufficent generality to make it appropriate for most applications.

Schur test. *If X is a measure space, if $k(\geqq 0)$ is a measurable function on $X \times X$, if $p(>0)$ is a measurable function on X, and if α and β are constants such that*

$$\int k(x, y) p(y)\, dy \leqq \alpha p(x)$$

and
$$\int k(x, y) p(x)\, dx \leqq \beta p(y),$$

then the equation

$$(Af)(x) = \int k(x, y) f(y)\, dy$$

defines an operator (a bounded linear transformation) on L^2, and $\|A\|^2 \leqq \alpha\beta$.

Proof. If f is a bounded measurable function that vanishes outside some measurable set of finite measure, then

$$\int \left| \int k(x, y) f(y)\, dy \right|^2 dx = \int \left| \left(\int \sqrt{k(x, y)} \sqrt{p(y)} \right) \cdot \left(\frac{\sqrt{k(x, y)}}{\sqrt{p(y)}} f(y) \right) dy \right|^2 dx \leqq$$

$$\leqq \int \left(\int k(x, y) p(y)\, dy \right) \cdot \left(\int \frac{k(x, y)}{p(y)} |f(y)|^2 \, dy \right) dx \leqq$$

$$\leqq \int \alpha p(x) \left(\int \frac{k(x, y)}{p(y)} |f(y)|^2 \, dy \right) dx =$$

$$= \alpha \int \frac{|f(y)|^2}{p(y)} \left(\int k(x, y) p(x)\, dx \right) dy \leqq \alpha \int \frac{|f(y)|^2}{p(y)} \beta p(y)\, dy = \alpha\beta \|f\|^2.$$

Since the functions such as f are dense in L^2, the proof is complete.

Theorem 1. *Each of the Cesàro operators C_0, C_1, and C_∞ is bounded.*

Proof. For C_0 consider the measure space $\{0, 1, 2, ...\}$ with the counting measure, and let the kernel k_0 be defined by

$$k_0(i,j) = \begin{cases} 0 & \text{if } 0 \leq i < j, \\ \dfrac{1}{i+1} & \text{if } 0 \leq j \leq i. \end{cases}$$

If $p_0(n) = \dfrac{1}{\sqrt{n+1}}$, then

$$\sum_j k_0(i,j)p_0(j) = \sum_{j=0}^{i} \frac{1}{i+1} \frac{1}{\sqrt{j+1}} <$$

$$< \frac{1}{i+1} \int_0^i \frac{dx}{\sqrt{x}} = \frac{1}{i+1} 2\sqrt{i} < \frac{1}{i+1} 2\sqrt{i+1} = 2p_0(i).$$

If $j \neq 0$, then

$$\sum_i k_0(i,j)p_0(i) = \sum_{i=j}^{\infty} \frac{1}{i+1} \frac{1}{\sqrt{i+1}} <$$

$$< \int_{j-1}^{\infty} \frac{dx}{(x+1)^{3/2}} = \frac{2}{\sqrt{j}} = \frac{2}{\sqrt{j+1}} \frac{\sqrt{j+1}}{\sqrt{j}} \leq 2\sqrt{2}p_0(j).$$

Since also

$$\sum_i k_0(i,0)p_0(i) = 1 + \sum_{i=1}^{\infty} k_0(i,0)p_0(i) < 1 + 2 = 3p_0(0),$$

it follows that

$$\sum_i k_0(i,j)p_0(i) < 3p_0(j)$$

for all j, and the Schur test implies the boundedness of C_0.

For C_1 the measure space is $(0, 1)$ with Lebesgue measure and the kernel is defined by

$$k_1(x, y) = \begin{cases} 0 & \text{if } 0 < x \leq y, \\ \dfrac{1}{x} & \text{if } 0 < y < x. \end{cases}$$

If $p_1(x) = \dfrac{1}{\sqrt{x}}$, then

$$\int_0^1 k_1(x, y)p_1(y)\, dy = \frac{1}{x} \int_0^x \frac{dy}{\sqrt{y}} = \frac{1}{x} 2\sqrt{x} = 2p_1(x),$$

and

$$\int_0^1 k_1(x, y)p_1(x)\, dx = \int_y^1 \frac{dx}{x^{3/2}} = \frac{2}{\sqrt{y}} - 2 < 2p_1(y),$$

and the Schur test applies again.

For C_∞ the measure space is $(0, \infty)$ with Lebesgue measure, and the kernel k_∞ is defined formally the same way as k_1; the difference is that x and y now vary in $(0, \infty)$ instead of $(0, 1)$. If, as before, $p_\infty(x) = \frac{1}{\sqrt{x}}$, then

$$\int_0^\infty k_\infty(x, y) p_\infty(y)\, dy = \frac{1}{x} \int_0^x \frac{dy}{\sqrt{y}} = \frac{2}{\sqrt{x}} = 2p_\infty(x),$$

and

$$\int_0^\infty k_\infty(x, y) p_\infty(x)\, dx = \int_y^\infty \frac{dx}{x^{3/2}} = \frac{2}{\sqrt{y}} = 2p_\infty(y),$$

and, once more, the Schur test yields the desired result.

An examination of the proof of Theorem 1 yields (via the last assertion of the Schur test) estimates for the norms of C_0, C_1, and C_∞. For C_0 this estimate turns out to be quite crude, and even for C_1 and C_∞, where it is sharp, the method is not sharp enough to tell what the norms of the operators actually are. To settle this question, and others, we turn now to detailed separate examinations of the three Cesàro operators.

The discrete Cesàro operator

Since C_0 is defined on a sequence space, it is naturally associated with a matrix, which is in fact just the kernel k_0. Since

$$k_0 = \begin{pmatrix} 1 & 0 & 0 \\ \frac{1}{2} & \frac{1}{2} & 0 \\ \frac{1}{3} & \frac{1}{3} & \frac{1}{3} \\ & & & \ddots \end{pmatrix}, \quad k_0^* = \begin{pmatrix} 1 & \frac{1}{2} & \frac{1}{3} \\ 0 & \frac{1}{2} & \frac{1}{3} \\ 0 & 0 & \frac{1}{3} \\ & & & \ddots \end{pmatrix},$$

it follows that

$$k_0 k_0^* = \begin{pmatrix} 1 & \frac{1}{2} & \frac{1}{3} \\ \frac{1}{2} & \frac{1}{2} & \frac{1}{3} \\ \frac{1}{3} & \frac{1}{3} & \frac{1}{3} \\ & & & \ddots \end{pmatrix}.$$

It turns out therefore that the product $C_0 C_0^*$ is almost the same as the sum $C_0 + C_0^*$; the difference $C_0 + C_0^* - C_0 C_0^*$ is the diagonal operator D_0 with matrix

$$\begin{pmatrix} 1 & 0 & 0 \\ 0 & \frac{1}{2} & 0 \\ 0 & 0 & \frac{1}{3} \\ & & & \ddots \end{pmatrix}.$$

Since $(1 - C_0)(1 - C_0^*) = 1 - D_0$, it follows that

$$\|1 - C_0\| = 1,$$

and hence that $\|C_0\| \leq 2$.

It is perhaps worth while to remark that there are other ways of proving the last inequality. One way is to compute $C_0 C_0^*$ immediately, and then apply the Schur test to it (with the same p_0 as in the proof of Theorem 1). Since $C_0 C_0^*$ is Hermitian, only half the computation is necessary, and, moreover, the inequalities do yield the sharp result $\|C_0 C_0^*\| \leq 4$. To infer, via this approach, that C_0 itself is bounded, one more step is necessary; we need to know that if k is an infinite matrix with rows in l^2 such that kk^* is bounded, then k itself is bounded (cf. [7] and [5, Chapter VIII]). The proof of this can be carried out by looking at the n-th section $k^{(n)}$ of k and showing that the n-th section of kk^* domaintes $k^{(n)} k^{(n)*}$. (Recall that an infinite matrix is bounded if and only if its sections are uniformly bounded.)

It is easy to prove that the inequality $\|C_0\| \leq 2$ cannot be improved:

$$\|C_0\| = 2.$$

Indeed if $f_\alpha(n) = \dfrac{1}{(n+1)^a}$ $\left(\alpha > \dfrac{1}{2}, \; n = 0, 1, 2, ... \right)$, then $f_\alpha \in l^2$ and $\|C_0^* f_\alpha\| \to 2 \|f_\alpha\|$ as $\alpha \to \frac{1}{2} +$. The proof of the latter assertion is a straightforward computation. Since $(C_0^* f_\alpha)(m) = \sum\limits_{n=m}^{\infty} \dfrac{1}{(n+1)^{\alpha+1}}$, $m = 0, 1, 2, ...,$ it follows that

$$\|C_0^* f_\alpha\|^2 = \sum_{m=0}^{\infty} \left(\sum_{n=m}^{\infty} \frac{1}{(n+1)^{\alpha+1}} \right)^2 > \sum_{m=0}^{\infty} \left(\int_{m+1}^{\infty} \frac{dx}{x^{\alpha+1}} \right)^2 = \sum_{m=0}^{\infty} \left(\frac{1}{\alpha} \frac{1}{(m+1)^\alpha} \right)^2 =$$

$$= \frac{1}{\alpha^2} \sum_{m=0}^{\infty} \frac{1}{(m+1)^{2\alpha}} = \frac{1}{\alpha^2} \|f_\alpha\|^2,$$

and this implies the limit assertion.

For our next purpose we need the following lemma: if A is an operator such that $\|A\| \leq 1$ and if $\|Af\| = \|f\|$ for some nonzero vector f, then $\|A^* g\| = \|g\|$ for some non-zero vector g. For the proof, write $g = Af$, so that $\|g\| = \|f\|$, and observe that

$$\|f\|^2 = (A^* A f, f) \leq \|A^* A f\| \cdot \|f\| \leq \|f\|^2.$$

It follows that $\|A^* A f\| = \|f\|$, so that $\|A^* g\| = \|g\|$.

We know that the supremum of $\|C_0 f\|$ (and hence of $\|C_0^* f\|$) for vectors f on the unit sphere is 2; we shall show that the supremum is not attained. Since $\|(1 - D_0) f\| < \|f\|$ unless $f = 0$, it follows that

$$\|(1 - C_0^*) f\|^2 = ((1 - C_0)(1 - C_0^*) f, f) \leq \|(1 - C_0)(1 - C_0^*) f\| \cdot \|f\| < \|f\|^2$$

unless $f = 0$. The preceding paragraph is applicable, and we may infer that both $\|(1 - C_0) f\|$ and $\|(1 - C_0^*) f\|$ are strictly less than $\|f\|$, except when $f = 0$. It follows of course that $\|C_0 f\|$ and $\|C_0^* f\|$ are strictly less than $2 \|f\|$, except when $f = 0$. (Proof: $\|C_0 f\| = \|f - (1 - C_0) f\| \leq \|f\| + \|(1 - C_0) f\|$.)

9 A

The following statement sums up what we have just proved about norms and what we shall go on to prove about spectra.

Theorem 2. (1) $\|1 - C_0\| = 1$ *and* $\|C_0\| = 2$. (2) *If* $\|f\| = 1$, *then* $\|(1 - C_0)f\| < 1$ *and* $\|(1 - C_0^*)f\| < 1$. (3) *The point spectrum of* C_0 *is empty.* (4) *If* $|1 - \lambda| < 1$, *then* λ *is a simple proper value of* C_0^*. (5) *The point spectrum of* C_0^* *is the open disc* $\{\lambda : |1 - \lambda| < 1\}$: (6) *The spectrum of* C_0 *is the closed disc* $\{\lambda : |1 - \lambda| \leq 1\}$.

Proof. (1) and (2) were proved above. To prove (3), observe first that if $C_0 f = g$, then $f(0) = g(0)$, and if $n \geq 1$, then $f(n) = (n + 1) g(n) - ng(n - 1)$. Consequently, if $C_0 f = \lambda f$, then $f(n) = \lambda((n + 1)f(n) - nf(n - 1))$ or $(\lambda(n + 1) - 1)f(n) = \lambda n f(n - 1)$ whenever $n \geq 1$. If m is the smallest integer for which $f(m) \neq 0$, then $\lambda = \dfrac{1}{m + 1}$, so that $0 < \lambda \leq 1$. It follows that if $n \geq 1$, then

$$|f(n)| = \left| \frac{\lambda n}{\lambda n - (1 - \lambda)} f(n - 1) \right| \geq |f(n - 1)|,$$

which, for a non-zero f in l^2, is impossible.

To prove (4), observe first that $(C_0^* f)(n) = \sum_{i=n}^{\infty} \dfrac{1}{i + 1} f(i)$ (cf. the matrix k_0^*). If $C_0^* f = g$, then $f(n) = (n + 1)(g(n) - g(n + 1))$ for $n = 0, 1, 2, \ldots$. Consequently if $C_0^* f = \lambda f$, then $f(n) = \lambda(n + 1)(f(n) - f(n + 1))$ or $\lambda(n + 1)f(n + 1) = (\lambda(n + 1) - 1)f(n)$. It follows that 0 is not a proper value of C_0^* (if $\lambda = 0$, then $f(n) = 0$ for all n), and it follows also that $f(n + 1) = \left(1 - \dfrac{1}{\lambda(n + 1)}\right) f(n)$. This implies that if $n \geq 1$, then

$$f(n) = \prod_{j=1}^{n} \left(1 - \frac{1}{j\lambda}\right) f(0),$$

and we can conclude, even before we know which values of λ can be proper values of C_0^*, that all the proper values are simple.

Suppose now that $|1 - \lambda| < 1$, or, equivalently, that $\text{Re} \dfrac{1}{\lambda} > \dfrac{1}{2}$. It is convenient to rewrite the condition once more; if $\mu = \dfrac{1}{\lambda}$, then the condition is that $2 \operatorname{Re} \mu = 1 + \varepsilon$ for some positive number ε. Our task is to prove that if this condition is satisfied, and if

$$f(n) = \prod_{j=1}^{n} \left(1 - \frac{\mu}{j}\right),$$

for $n \geq 1$, then $f \in l^2$. Since

$$\left|1 - \frac{\mu}{j}\right|^2 = 1 - \frac{2 \operatorname{Re} \mu}{j} + \frac{|\mu|^2}{j^2} = 1 - \frac{1 + \varepsilon}{j} + \frac{|\mu|^2}{j^2} \leq \exp\left(\frac{|\mu|^2}{j^2} - \frac{1 + \varepsilon}{j}\right),$$

it follows that

$$|f(n)|^2 \leq \frac{\exp\left(|\mu|^2 \sum_{j=1}^{n} \frac{1}{j^2}\right)}{\exp\left((1+\varepsilon) \sum_{j=1}^{n} \frac{1}{j}\right)} < \frac{c}{\exp\left((1+\varepsilon)\log n\right)} = \frac{c}{n^{1+\varepsilon}},$$

where $c = \exp\left(|\mu|^2 \sum_{j=1}^{\infty} \frac{1}{j^2}\right)$. This completes the proof of (4). (We note in passing that if f is a proper vector of C_0^* with proper value λ, then $\sum_{n=0}^{\infty} f(n)z^n = (1-z)^{\frac{1}{\lambda}-1}$ whenever $|z| < 1$.)

Since $\|1 - C_0\| = 1$, the spectrum of $1 - C_0$ is included in the closed disc $\{\lambda : |\lambda| \leq 1\}$, and, consequently, the spectrum of C_0 is included in the closed disc $\{\lambda : |1 - \lambda| \leq 1\}$. The preceding paragraph implies that the spectrum of $1 - C_0^*$ includes the open disc $\{\lambda : |\lambda| < 1\}$, and hence that the same is true of the spectrum of $1 - C_0$. This, in turn, implies that the spectrum of C_0 includes the open disc $\{\lambda : |1 - \lambda| < 1\}$, and the proof of (6) is complete.

In view of what was just proved, the proof of (5), and hence of the theorem, can be completed by showing that if $|1 - \lambda| = 1$, then λ is not a proper value of C^*, or, equivalently, $1 - \lambda$ is not a proper value of $1 - C_0^*$. This, however, is an immediate consequence of (2): if $\|f\| = 1$ and $(1 - C_0^*)f = (1 - \lambda)f$, then $\|(1 - C_0^*)f\| = |1 - \lambda|$, and therefore $|1 - \lambda|$ cannot be equal to 1.

We conclude our discussion of the discrete Cesàro operator by reporting a fact that may not be important but that is at least an interesting curiosity.

Theorem 3. *The operator C_0 is hyponormal, that is, $C_0^* C_0 - C_0 C_0^*$ is positive.*

Proof. The matrix $k_0^* k_0$ is "L-shaped", meaning that it is of the form

$$\begin{pmatrix} \alpha_0 & \alpha_1 & \alpha_2 & \\ \alpha_1 & \alpha_1 & \alpha_2 & \\ \alpha_2 & \alpha_2 & \alpha_2 & \\ & & & \ddots \end{pmatrix},$$

with $\alpha_n = \sum_{j=n}^{\infty} \frac{1}{(j+1)^2}$. Since $k_0 k_0^*$ is also L-shaped $\left(\text{with } \alpha_n = \frac{1}{n+1}\right)$, and since the difference of two L-sharped matrices is another one, the problem of proving the hyponormality of C_0 reduces to the problem of deciding when an L-shaped matrix is positive. An infinite matrix is positive if and only if all its finite sections have positive determinants; the problem has reduced to the evaluation of the determinant of

$$\begin{vmatrix} \alpha_0 & \alpha_1 & \alpha_2 & \cdots & \alpha_n \\ \alpha_1 & \alpha_1 & \alpha_2 & \cdots & \alpha_n \\ \alpha_2 & \alpha_2 & \alpha_2 & \cdots & \alpha_n \\ \vdots & \vdots & \vdots & & \vdots \\ \alpha_n & \alpha_n & \alpha_n & \cdots & \alpha_n \end{vmatrix}$$

This is easy. Subtract the second column from the first, then subtract the third column from the second, and continue this way through the columns. The resulting matrix has the same determinant as the original one and is triangular; its determinant therefore is the product of its diagonal elements. The diagonal elements are $\alpha_0 - \alpha_1, \alpha_1 - \alpha_2, \ldots, \alpha_{n-1} - \alpha_n$, and α_n. Conclusion: an finite L-shaped matrix is positive if and only if its determining sequence is positive and decreasing. The proof of the theorem is completed by verifying that the sequence $\left\{ \sum\limits_{j=n}^{\infty} \dfrac{1}{(j+1)^2} - \dfrac{1}{n+1} \right\}$ has these properties.

The finite continuous Cesàro operator

For C_1 the facts are simpler and the proofs are easier than for C_0; to get at those facts, it is convenient to recall a few simple results about unilateral shifts. An operator U on a Hilbert space H is a unilateral shift of multiplicity 1 if H has an orthonormal basis $\{e_0, e_1, e_2, \ldots\}$ such that $Ue_n = e_{n+1}$, $n = 0, 1, 2, \ldots$. A unilateral shift of multiplicity m (here m can be any cardinal number, finite or infinite) is the direct sum of m unilateral shifts of multiplicity 1. Each unilateral shift is an isometry, and so therefore is the direct sum of a unilateral shift and a unitary operator. Conversely, every isometry is a direct sum of a unilateral shift and a unitary operator, it being understood that either summand may be absent. If U is an isometry, then $U^*U - UU^*$ is the projection on the co-range of U (the orthogonal complement of the range of U), and consequently the rank of $U^*U - UU^*$ (the co-rank of U) is the multiplicity of the shift component of U.

If U is a unilateral shift, then the spectrum of U is the closed unit disc, the point spectrum of U is empty, and the point spectrum of U^* is the open unit disc. Each number in the open unit disc is a proper value of U^* of multiplicity equal to the multiplicity of U. The proper vectors of U^* form a total set (that is, they span the entire underlying Hilbert space). All these facts are known; see [1, 2, 4].

There are several ways of characterizing simple unilateral shifts (that is, unilateral shifts of multiplicity 1). For our purposes the most convenient one is this: an operator U is a simple unilateral shift if and only if (1) U is an isometry, (2) the co-rank of U is 1, and (3) U^* has a total set of proper vectors with proper values of modulus strictly less than 1. Indeed, a unilateral shift has these three properties. If, conversely, U is an operator satisfying (1), (2), and (3), then, by (1), it is he direct sum of a unilateral shift and a unitary operator, and, by (2), its shift component is simple. It remains only to use (3) to prove that its unitary component is absent. Suppose therefore that W is a unitary direct summand of U. If $U^*f = \lambda f$ with $|\lambda| < 1$, and if g is the component of f in the domain of W, then $W^*g = \lambda g$; since W^* is unitary, it follows that $g = 0$. Thus each proper vector of U^* corresponding to a proper value of modulus strictly less than 1 belongs to the domain of the shift component of U; if such vectors span the whole space, then the unitary component of U cannot be present.

Theorem 4. *The operator* $1 - C_1^*$ *is a simple unilateral shift.*

Proof. Since C_1 is given by the kernel k_1, where $k_1(x, y) = 1/x$ if $0 < y \leq x$ and $k_1(x, y) = 0$ otherwise, it follows that C_1^* is given by the kernel k_1^*, where

$$k_1^*(x, y) = \begin{cases} 0 & \text{if} \quad 0 < y \leq x, \\ \dfrac{1}{y} & \text{if} \quad 0 < x < y. \end{cases}$$

In other words if $f \in L^2(0, 1)$, then

$$(C_1^* f)(x) = \int_x^1 \frac{1}{y} f(y)\, dy.$$

The operator $C_1 C_1^*$ is given by the kernel

$$\int_0^1 k_1(x, u) k_1^*(u, y)\, du = \int_0^{\min(x, y)} \frac{1}{x} \frac{1}{y}\, du = \frac{\min(x, y)}{xy}.$$

Since

$$k_1(x, y) + k_1^*(x, y) = \begin{cases} \dfrac{1}{x} & \text{if} \quad 0 < y \leq x, \\ \dfrac{1}{y} & \text{if} \quad 0 < x < y, \end{cases}$$

it follows that $C_1 C_1^* = C_1 + C_1^*$, and hence that

$$(1 - C_1)(1 - C_1^*) = 1.$$

Conclusion: $1 - C_1^*$ is an isometry.

If we write $1 - C_1^* = U$, then $U^* U - U U^* = C_1 C_1^* - C_1^* C_1$. Since $C_1^* C_1$ is given by the kernel

$$\int_0^1 k_1^*(x, u) k_1(u, y)\, du = \int_{\max(x, y)}^1 \frac{du}{u^2} = \frac{1}{\max(x, y)} - 1,$$

it follows that the kernel of $C_1 C_1^* - C_1^* C_1$ is the constant function 1. Conclusion: the co-rank of $1 - C$ is equal to 1.

Before completing the proof of the theorem, we remark on the kernel techniques used in the proof so far. Since the kernels in question are neither in L^2 (that is, the operators are not in the Hilbert—Schmidt class), nor symmetric (the two textbook cases), it is not quite automatic that if an operator is given by a kernel, then its adjoint is given by the conjugate transpose kernel, and that the product of two operators given by kernels is given by the product kernel. Since, however, the kernels k in question (that is, k_1 and k_1^*) have positive values, and have the property that if f and g are in L^2, then the function given on the unit square by $k(x, y) f(x) g(y)$ is in L^1, no unboundedness or infinity pathology can occur; the necessary changes in 'the order of integration are immediate consequences of FUBINI's theorem.

To complete the proof of the theorem it is sufficient to show that $1 - C_1$ has a total set of proper vectors corresponding to proper values of modulus strictly less than 1. This is trivial modulo the Weierstrass approximation theorem. If $f_n(x) = x^n$, $n = 0, 1, 2, \ldots$, then the set $\{f_0, f_1, f_2, \ldots\}$ is total in $L^2(0, 1)$. Since $(C_1 f_n)(x) =$

$$= \frac{1}{x} \int_0^x y^n \, dy = \frac{x^n}{n+1} = \frac{1}{n+1} f_n(x), \quad \text{it follows that} \quad (1 - C_1) f_n = \left(1 - \frac{1}{n+1}\right) f_n,$$

and the proof is complete.

It may be worth while to remark that Theorem 4 implies that all the spectral assertions of Theorem 2 $((3), (4), (5),$ and $(6))$ remain true, word for word, if in their statement C_0 is replaced by C_1^*. The norm assertion (1) is also invariant under this change; the only part of the theorem that changes is (2). Since $1 - C_1^*$ is an isometry, $\|(1 - C_1^*)f\| = \|f\|$ always and $\|(1 - C_1)f\| = \|f\|$ often. What can be said, however, is that if $\|f\| = 1$, then $\|C_1 f\| < 2$ and $\|C_1^* f\| < 2$. This follows either by an examination of the cases of equality in the Schur test, or by a direct argument valid for isometries with no proper values.

Here is another useful comment about unilateral shifts, and hence about $1 - C_1^*$. The basis that a simple unilateral shift shifts is uniquely determined to within a multiplicative constant. The reason is that the co-range is one-dimensional and e_0 is in the co-range. Since the projection on the co-range of $1 - C_1^*$ is $C_1 C_1^* - C_1^* C_1$, and since, as we have seen, this projection is given by the kernel that is identically 1, it follows that the co-range of $1 - C_1^*$ is the set of all constant functions. The most natural choice for e_0 is the constant function 1. Once e_0 is chosen, the other terms of the shifted basis are determined; they are the successive images of e_0 under iterations of $1 - C_1^*$.

There is another approach to Theorem 4, more analytic than the one given above; we proceed to sketch it. If $U = 1 - C_1^*$ and $f_\alpha(x) = x^\alpha$ whenever $\operatorname{Re} \alpha > -\frac{1}{2}$,

then $U^* f_\alpha = \dfrac{\alpha}{\alpha + 1} f_\alpha$. A change of parameters is convenient: if $\beta = \bar{\alpha} + \frac{1}{2}$ and

$g_\beta = f_{\bar{\beta} - \frac{1}{2}}$ whenever $\operatorname{Re} \beta > 0$, then $U^* g_\beta = \overline{\varphi(\beta)} g_\beta$, where $\varphi(\beta) = \dfrac{\beta - \frac{1}{2}}{\beta + \frac{1}{2}}$.

By means of these proper vectors, the operator U can be represented as a multiplication on a Hilbert space of analytic functions on the right half plane, as follows. For f in $L^2(0, 1)$ define $\hat{f}$ by

$$\hat{f}(\beta) = (f, g_\beta) = \int_0^1 f(t) t^{\beta - \frac{1}{2}} \, dt;$$

the transform of U by the mapping $f \to \hat{f}$ is multiplication by φ. Indeed,

$$(Uf)\hat{\,}(\beta) = (Uf, g_\beta) = (f, U^* g_\beta) = \Phi(\beta) \hat{f}(\beta).$$

Making the change of variables $t = e^{-u}$ $(0 < u < \infty)$, we obtain

$$\hat{f}(\beta) = \int_0^\infty f(e^{-u}) e^{-u/2} e^{-u\beta} \, du = \int_0^\infty g(u) e^{-u\beta} \, du,$$

where g is the element of $L^2(0, \infty)$ defined by $g(u)=f(e^{-u})e^{-u/2}$. Thus the space of functions $\hat{f}$ is the space of Laplace transforms of functions in $L^2(0, \infty)$. By the Paley—Wiener theorem [6, Chapter VIII] this is precisely the space H^2 of the right half plane, and therefore the preceding paragraph exhibits U as multiplication by φ on that H^2 space. Switching to the unit disc via the conformal mapping $w = \varphi(z)$, we obtain a representation of U as multiplication by the independent variable on H^2 of the disc, and Theorem 4 follows.

We conclude our discussion of the finite continuous Cesàro operator by mentioning a curious by-product of Theorem 4. One of our earlier proofs of that theorem made use of the completeness of the set of Laguerre functions in $L^2(0, \infty)$. The proof actually offered above is independent of such considerations; since it turns out that our earlier argument is reversible, Theorem 4 can be used to prove that the Laguerre functions span $L^2(0, \infty)$. Here is how it goes. If $f \in L^2(0, 1)$, write

$$(Tf)(x) = f(e^{-x})e^{-x/2}$$

for $0 < x < \infty$, and verify that T is an isometry from $L^2(0, 1)$ onto $L^2(0, \infty)$. Transform the shift $1 - C_1^*$ by T; that is, consider on $L^2(0, \infty)$ the operator $V = T(1 - C_1^*)T^{-1}$. If $f \in L^2(0, \infty)$, then Vf can be calculated explicitly:

$$(Vf)(x) = f(x) - e^{-x/2}\int_0^x f(y)e^{y/2}\, dy.$$

If, as usual, the Laguerre polynomials are defined by

$$L_n(x) = \frac{1}{n!}e^x \frac{d^n}{dx^n}(x^n e^{-x}),$$

and the Laguerre functions by

$$f_n(x) = e^{-x/2}L_n(x), \qquad n = 0, 1, 2, \ldots,$$

then the f_n's form an orthonormal set in $L^2(0, \infty)$. A straightforward argument, based on the standard identity

$$L_n(x) = \frac{d}{dx}\left(L_n(x) - L_{n+1}(x)\right)$$

(see [8, Chapter VI]) implies that $Vf_n = f_{n+1}$. Since $Te_0 = f_0$, it follows that $Te_n = f_n$ for $n = 0, 1, 2, \ldots$, and the completeness of the f_n's follows from that of the e_n's.

The infinite continuous Cesàro operator

We shall get at the facts about C_∞ by reducing its study to that of C_1. It is convenient to begin by establishing a simple result about the relation between unilateral shifts and bilateral shifts. An operator W on a Hilbert space K is a simple bilateral shift if K has an orthonormal basis $\{\ldots, e_{-2}, e_{-1}, e_0, e_1, e_2, \ldots\}$ such that $We = e_{n+1}$ for all n. It follows from this definition that a simple bilateral shift is a unitary operator. If H is the span of $\{e_0, e_1, e_2, \ldots\}$, then H is invariant under

W and the restriction of W to H is a unilateral shift. If R is the operator on K such that $Re_n = e_{-n-1}$ for all n, then R is a symmetry (a unitary involution). The symmetry R is related to the shift W in the following three ways:

(1) $Re_0 = W^{-1}e_0$,   (2) $RH = H^{\perp}$,   (3) $RW = W^{-1}R$.

What makes these assertions important is that they serve to characterize simple bilateral shifts, in the following sense. Suppose that K is a Hilbert space, W is a unitary operator on K, R is a symmetry on K, H is a subspace of K invariant under W, and e_0 is a vector in H. If the vectors $W^n e_0$, $n = 0, 1, 2, \ldots$, form an orthonormal basis for H, and if the conditions (1), (2), and (3) are satisfied, then W is a simple bilateral shift.

The proof is straightforward. We begin by writing $e_n = W^n e_0$ for all n $(= 0, \pm 1, \pm 2, \ldots)$. If n and m are arbitrary integers, find a positive integer j such that both $n+j$ and $m+j$ are positive; it follows that

$$(e_n, e_m) = (W^n e_0, W^m e_0) = (W^{n+j} e_0, W^{m+j} e_0) = (e_{n+j}, e_{m+j}) = \delta_{n+j, m+j} = \delta_{nm},$$

and hence that the e_n's form an orthonormal set in K. By assumption $\{e_0, e_1, e_2, \ldots\}$ spans H; it follows that $\{Re_0, Re_1, Re_2, \ldots\}$ spans $H^{\perp}$. Since $Re_n = RW^n e_0 = W^{-n}Re_0 = W^{-n}W^{-1}e_0 = e_{-n-1}$, it follows that $\{e_{-1}, e_{-2}, e_{-3}, \ldots\}$ spans $H^{\perp}$, and hence that the e_n's form an orthonormal basis for K. Since the definition of the e_n's makes it obvious that W shifts them, the proof of the characterization of simple bilateral shifts is complete.

Theorem 5. *The operator* $1 - C_{\infty}^{*}$ *is a simple bilateral shift.*

Proof. We apply the preceding characterization of simple bilateral shifts with $K = L^2(0, \infty)$, $W = 1 - C_{\infty}^{*}$, and

$$(Rf)(x) = -\frac{1}{x} f\left(\frac{1}{x}\right)$$

whenever $f \in K$. The role of H is played by those elements of K that vanish on $(1, \infty)$, and the role of e_0 is played by the characteristic function of $(0, 1)$. We observe that H differs from $L^2(0, 1)$ in notation only.

If $f \in K$, then

$$(Wf)(x) = f(x) - \int_x^{\infty} \frac{1}{y} f(y)\, dy$$

for $0 < x < \infty$. With this explicit representation of W, the verifications needed to justify the application of the characterization theorem for bilateral shifts become a matter of routine integrations. They are not only routine, but they are almost identical with the integrations indicated in our study of C_1. (Note that if H is identified with $L^2(0, 1)$, then the restriction of W to H must be identified with $1 - C_1^{*}$.) With these remarks we consider the proof of Theorem 5 complete.

It follows from Theorem 5 (just as the corresponding facts for C_1 followed from Theorem 4) that $\|1 - C_{\infty}\| = 1$ and $\|C_{\infty}\| = 2$; if $\|f\| = 1$, then $\|C_{\infty}f\| < 2$ and

$\|C_{\infty}^{*}f\| < 2$. Using in addition well known (and easily recaptured) facts about the spectrum of a bilateral shift, we obtain the following description of the spectrum of C_{∞}: the point spectra of both C_{∞} and C_{∞}^{*} are empty, and the spectrum of C_{∞} is the circle $\{\lambda: |1 - \lambda| = 1\}$.

References

[1] A. BEURLING, On two problems concerning linear transformations in Hilbert space, *Acta Math.*, **81** (1949), 239—255.

[2] A. BROWN, On a class of operators, *Proc. Amer. Math. Soc.*, **4** (1953), 723—728.

[3] L. DE BRANGES, Some Hilbert spaces of entire functions. III, *Transactions Amer. Math. Soc.*, **100** (1961), 73—115.

[4] P. R. HALMOS, Shifts on Hilbert spaces, *J. reine angew. Math.*, **208** (1961), 102—112.

[5] G. H. HARDY, J. E. LITTLEWOOD, and G. PÓLYA, *Inequalities* (Cambridge, 1934),

[6] K. HOFFMAN, *Banach spaces of analytic functions* (Englewood Cliffs, 1962).

[7] I. SCHUR, Bemerkungen zur Theorie der beschränkten Bilinearformen mit unendlich vielen Veränderlichen, *J. reine angew. Math.*, **140** (1911), 1—28.

[8] F. G. TRICOMI, *Vorlesungen über Orthogonalreihen* (Berlin, 1955).

UNIVERSITY OF MICHIGAN

(Received September 14, 1964)

PACIFIC JOURNAL OF MATHEMATICS
Vol. 16, No. 3, 1966

INVARIANT SUBSPACES OF POLYNOMIALLY COMPACT OPERATORS

P. R. HALMOS

This paper is a comment on the solution of an invariant subspace problem by A. R. Bernstein and A. Robinson [2]. The theorem they prove can be stated as follows: if A is an operator on a Hilbert space H of dimension greater than 1, and if p is a nonzero polynomial such that $p(A)$ is compact, then there exists a nontrivial subspace of H invariant under A. ("Operator" means bounded linear transformation; "Hilbert space" means complete complex inner product space; "compact" means completely continuous; "subspace" means closed linear manifold; "nontrivial", for subspaces, means distinct from $\{0\}$ and from H.) The Bernstein-Robinson proof has two aspects: it is an ingenious adaptation of the proof by N. Aronszajn and K. T. Smith of the corresponding theorem for compact operators [1], and it makes strong use of metamathematical concepts such as nonstandard models of higher order predicate languages. The purpose of this paper is to show that by appropriate small modifications the Bernstein-Robinson proof can be converted (and shortened) into one that is expressible in the standard framework of classical analysis.

A quick glance at the problem is sufficient to show that there is no loss of generality in assuming the existence of a unit vector e such that the vectors $e, Ae, A^2e, \cdots$ are linearly independent and have H for their (closed linear) span. (This comment appears in both [1] and [2].) The Gram-Schmidt orthogonalization process applied to the sequence $\{e, Ae, A^2e, \cdots\}$ yields an orthonormal basis $\{e_1, e_2, e_3, \cdots\}$ with the property that the span of $\{e, \cdots, A^{n-1}e\}$ is the same as the span of $\{e_1, \cdots, e_n\}$ for each positive integer n. It follows that if $a_{mn} = (Ae_n, e_m)$, then $a_{mn} = 0$ unless $m \leq n + 1$; in other words, in the matrix of A all entries more than one step below the main diagonal must vanish. The matrix entries of the kth power of A are given by $a_{mn}^{(k)} = (A^k e_n, e_m)$. A straightforward induction argument, based on matrix multiplication, yields the result that $a_{mn}^{(k)} = 0$ unless $m \leq n + k$, and

$$a_{n+k,n}^{(k)} = \Pi_{1 \leq j \leq k} a_{n+j,\, n+j-1}.$$

(With the usual understanding about an empty product having the value 1, the result is true for $k = 0$ also.) This result for powers has an implication for polynomials. If the degree of p (the only polynomial

---

Received October 10, 1964. Research supported in part by a grant from the National Science Foundation.

needed) is k (≥ 1), and if the matrix entries of $p(A)$ are given by $a_{mn}^{(p)} = (p(A)e_n, e_m)$, then $a_{n+k,n}^{(p)}$ is a constant multiple (by the leading coefficient of p) of $a_{n+k,n}^{(k)}$. Since $\| p(A)e_n \| \to 0$ as $n \to \infty$ (because of the compactness of $p(A)$), there exists an increasing sequence $\{k(n)\}$ of positive integers (in fact a sequence with no gaps of length greater than the degree of p) such that the corresponding subdiagonal terms $a_{k(n)+1,k(n)}$ tend to 0 as n tends to ∞. (This very useful conclusion is one of the analytic tools used in [2], where it is described in terms of "infinite positive integers".)

If H_n is the span of $\{e_1, \cdots, e_{k(n)}\}$, then $\{H_n\}$ is an increasing sequence of finite-dimensional subspaces of H whose span is H. If P_n is the projection with range H_n, then $P_n \to 1$ (the identity operator) strongly. Since, for each n, the operator $P_n A P_n$ leaves H_n invariant, it follows that, for each n, there exists a chain of subspaces invariant under $P_n A P_n$,

$$\{0\} = H_n^{(0)} \subset H_n^{(1)} \subset \cdots \subset H_n^{(k(n))} = H_n ,$$

with dim $H_n^{(i)} = i$, $i = 0, 1, \cdots, k(n)$. (The consideration of such chains is essential in both [1] and [2].)

If $\{f_n\}$ and $\{g_n\}$ are sequences of vectors in H, it is convenient to write $f_n \sim g_n$ to mean that $\| f_n - g_n \| \to 0$ as $n \to \infty$. Assertion: if $\{f_n\}$ is a bounded sequence of vectors in H, then

$$(1) \qquad\qquad A P_n f_n \sim P_n A P_n f_n .$$

(Intuitively: H_n is approximately invariant under A.) The proof is a straightforward computation, based on the fact that $P_n f = \sum_{j=1}^{k(n)} (f, e_j)e_j$ whenever $f \in H$. Since $A P_n f_n - P_n A P_n f_n = \sum_{j=1}^{k(n)} (f_n, e_j) \sum_{i=k(n)+1}^{\infty} a_{ij}e_i$, since the largest j here is $k(n)$ and the smallest i is $k(n) + 1$, and since $a_{ij} = 0$ unless $i \leq j + 1$, it follows that $\| A P_n f_n - P_n A P_n f_n \| \leq \| f_n \| \cdot | a_{k(n)+1,k(n)} |$.

The conclusion (1) can be generalized to higher exponents:

$$(2) \qquad\qquad A^k P_n f_n \sim (P_n A P_n)^k f_n , \qquad\qquad k = 1, 2, 3, \cdots ;$$

the proof is by induction on k and is omitted. For $k = 0$, (2) says that $\| P_n f_n - f_n \| \to 0$, which is a stringent condition on the bounded sequence $\{f_n\}$; if that condition is satisfied, then (2) implies that

$$(3) \qquad\qquad p(A)P_n f_n \sim p(P_n A P_n)f_n .$$

Return now to the unit vector e. Since $P_n e = e$ for each n, it follows that $p(P_n A P_n)e \sim p(A)e$. Since $p(A)e \neq 0$ (because the vectors $e, Ae, A^2 e, \cdots$ are linearly independent), it follows that

$$\varepsilon = \lim_n \| p(P_n A P_n)e \| = \| p(A)e \| > 0 .$$

Consider, for each n, the numbers

$$\| p(P_n AP_n)e - p(P_n AP_n)P_n^{(0)}e \| ,$$
$$\| p(P_n AP_n)e - p(P_n AP_n)P_n^{(1)}e \| ,$$
$$\cdots$$
$$\| p(P_n AP_n)e - p(P_n AP_n)P_n^{(k(n))}e \| ,$$

where $P_n^{(i)}$ is the projection with range $H_n^{(i)}$. Since $P_n^{(0)}$ is the zero projection, the first of these numbers tends to ε. Since, on the other hand, $P_n^{(k(n))} = P_n$, the last of these numbers is always 0. In view of these facts it is possible to choose for each n (with possibly a finite number of exceptions) a positive integer $i(n)$, $1 \leq i(n) \leq k(n)$, such that

(4) $$\| p(P_n AP_n)e - p(P_n AP_n)P_n^{(i(n)-1)}e \| \geq \frac{\varepsilon}{2} ,$$

and

(5) $$\| p(P_n AP_n)e - p(P_n AP_n)P_n^{(i(n))}e \| < \frac{\varepsilon}{2} ;$$

the simplest way to do it is to let $i(n)$ be the smallest positive integer for which these inequalities are true. (The construction of this particular "infinite positive integer" i is the second major analytic insight in [2].)

Since both $\{P_n^{(i(n)-1)}\}$ and $\{P_n^{(i(n))}\}$ are bounded sequences of operators, there exists an increasing sequence $\{n_j\}$ of positive integers such that both $\{P_{n_j}^{(i(n_j)-1)}\}$ and $\{P_{n_j}^{(i(n_j))}\}$ are weakly convergent. Write, for typographical convenience, $Q_j^- = P_{n_j}^{(i(n_j)-1)}$ and $Q_j^+ = P_{n_j}^{(i(n_j))}$. Let M^- be the set of all those vectors f in H for which $Q_j^- f \to f$ (strongly), and, similarly, let M^+ be the set of those vectors f for which $Q_j^+ f \to f$ (strongly). The purpose of what follows is to prove that both M^- and M^+ are subspaces of H, that both are invariant under A, and that at least one of them is nontrivial.

Since linear combinations are continuous, it follows that M^- is a linear manifold. To prove that M^- is closed, suppose that g is in the closure of M^-; it is to be proved that $g \in M^-$, i.e., that $Q_j^- g \to g$. Given a positive number δ, find f in M^- so that $\| f - g \| < \delta/3$, and then find j_0 so that $\| Q_j f - f \| < \delta/3$ whenever $j \geq j_0$. It follows that if $j \geq j_0$, then $\| Q_j^- g - g \| \leq \| Q_j^- g - Q_j^- f \| + \| Q_j^- f - f \| + \| f - g \| < \delta$. This proves that M^- is closed; the proof for M^+ is the same.

To prove that M^- is invariant under A, suppose that $f \in M^-$, so that $Q_j^- f \to f$, and infer, first, that $AQ_j^- f \to Af$, just because A is bounded, and, second, that $Q_j^- AQ_j^- f \sim Q_j^- Af$, because Q_j^- is uniformly bounded. Then reason as follows: $Q_j^- Af \sim Q_j^- AQ_j^- f = Q_j^- P_{n_j} AP_{n_j} Q_j^- f$ (because $Q_j^- \leq P_{n_j}$) $= P_{n_j} AP_{n_j} Q_j^- f$ (because the range of Q_j^- is invariant under $P_{n_j}AP_{n_j}) \sim AP_{n_j}Q_j^- f$ (by (1)) $= AQ_j^- f \to Af$. This proves that M^- is invariant; the proof for M^+ is the same.

The next step is to prove that $M^- \neq H$; this is done by proving that e does not belong to M^-. For this purpose observe first that the operators $p(P_n A P_n)$ are uniformly bounded. (Observe that

$$\| (P_n A P_n)^k \| \leq \| P_n A P_n \|^k \leq \| A \|^k$$

and use the polynomial whose coefficients are the absolute values of the coefficients of p.) Now use (4):

$$\frac{\varepsilon}{2} \leq \| p(P_{n_j} A P_{n_j}) \| \cdot \| e - Q_j^- e \|.$$

Since $\| p(P_{n_j} A P_{n_j}) \|$ is bounded from above, its reciprocal is bounded away from zero, and, consequently, $\| e - Q_j^- e \|$ is bounded away from zero, which makes the convergence $Q_j^- e \to e$ impossible.

The corresponding step for M^+ says that $M^+ \neq \{0\}$; the proof is quite different. The choice of the sequence $\{n_j\}$ implies that the sequence $\{Q_j^+ e\}$ is weakly convergent; the compactness of $p(A)$ implies, therefore, that the sequence $\{p(A)Q_j^+ e\}$ is strongly convergent to, say, f. The proof that follows consists of two parts: (i) $f \neq 0$, (ii) $f \in M^+$. Part (i): $p(A)Q_j^+ e \sim p(P_{n_j}AP_{n_j})Q_j^+ e$ (by (3)), which is within $\varepsilon/2$ of $p(P_{n_j}AP_{n_j})e$ (by (5)), whose norm tends to ε; it follows that $\| p(A)Q_j^+ e \|$ cannot tend to 0, and hence that $f \neq 0$. Part (ii): $Q_j^+ f \sim Q_j^+ p(A)Q_j^+ e$ (since Q_j^+ is uniformly bounded) $\sim Q_j^+ p(P_{n_j}AP_{n_j})Q_j^+ e$ (by (3), and, again, uniform boundedness) $= p(P_{n_j}AP_{n_j})Q_j^+ e$ (because the range of Q_j^+ is invariant under $p(P_{n_j}AP_{n_j})) \sim p(A)Q_j^+ e$ (by (3)) $\to f$ (by definition).

If $M^+ \neq H$, all is well; it remains to be proved that if $M^+ = H$, then $M^- \neq \{0\}$. If $M^+ = H$, then $Q_j^+ f \to f$ for all f, and, a fortiori, $Q_j^+ f \to f$ weakly. At the same time the sequence $\{Q_j^-\}$ is known to be weakly convergent to, say, Q^-. The operators Q_j^- and Q_j^+ are projections such that $Q_j^- \leq Q_j^+$ and such that $Q_j^+ - Q_j^-$ has rank 1. It follows that, for each j, there exists a unit vector f_j such that $(Q_j^+ - Q_j^-)f = (f, f_j)f_j$ for all f. Observe now that $Q_j^- e$ cannot tend weakly to e, for, if it did, then it would tend strongly to e (an elementary property of projections), and that was proved to be not so. This implies that $Q^- e \neq e$, or, equivalently, that $(1 - Q^-)e \neq 0$. Can the numbers $|(e, f_j)|$ be arbitrarily small? Since $|((Q_j^- - Q_j^-)e, g)| \leq |(e, f_j)| \cdot \| g \|$ for all g, an affirmative answer would imply that $((1 - Q^-)e, g) = 0$ for all g, so that $(1 - Q^-)e = 0$—a contradiction. The fact so obtained (that the numbers $|(e, f_j)|$ are bounded away from zero) makes it possible to prove that $M^- \neq \{0\}$; it turns out that if $g \perp (1 - Q^-)e$, then $g \in M^-$. Indeed, since $(e, f_j)(f_j, g) \to ((1 - Q^-)e, g) = 0$, it follows that $(f_j, g) \to 0$, and hence that $(f, f_j)(f_j, g) \to 0$ for all

f. This implies that $((1 - Q^-)f, g) = 0$ for all f, and hence that $(1 - Q^-)g = 0$. In other words, $Q_j^- g \rightarrow g$ weakly, and therefore strongly (the same property of projections that was alluded to above); from this it follows, finally, that $g \in M^-$.

I am grateful to Professor Robinson for a prepublication copy of [2] and for a kind letter helping me over some metamathematical difficulties.

REFERENCES

1. N. Aronszajn and K. T. Smith, *Invariant subspaces of completely continuous operators*, Ann. Math. **60** (1954), 345–350.
2. A. R. Bernstein and A. Robinson, *Solution of an invariant subspace problem of K. T. Smith and P. R. Halmos.*

UNIVERSITY OF MICHIGAN

Reprinted from the
MICHIGAN MATHEMATICAL JOURNAL
Vol. 15, pp. 215-233, 1968

IRREDUCIBLE OPERATORS

P. R. Halmos

Dedicated to Marshall Harvey Stone on his 65th birthday.

THEOREM. *On a separable Hilbert space, the set of irreducible operators is a dense* G_δ.

Remarks. The theorem says that the set of irreducible operators is topologically large: most operators are irreducible. (The separability assumption is obviously necessary; on a non-separable space every operator is reducible.) The proof rests on several analytic and algebraic lemmas; they occur in Section 1 below. Section 2 contains a few related remarks on finite-dimensional spaces (everything is easier) and on the set of normal operators (it is topologically small). The topological considerations needed to show that the set of irreducible operators is a G_δ occur in Section 3. Although the principal theorem gives some information about the size of the set of reducible operators, it does not answer all questions about that set. For instance, it is still not known whether the set of reducible operators is dense; Section 4 contains some comments on that subject. Closely related to this whole circle of ideas is the possibility of a topological attack on the problem of invariant subspaces. It is, after all, not inconceivable that the existence of an operator with only trivial invariant subspaces could be proved by showing that the set of all such operators is topologically large. Section 5 contains a result that seems to kill that hope, and it suggests that in fact the set is topologically small. An appendix contains a theorem, a special case of which is used in Section 2; the result (rank is weakly lower semi-continuous) may be of interest in its own right.

Terminology. Hilbert spaces are complex, subspaces are closed linear manifolds, operators are bounded linear transformations, and, when it is not otherwise indicated, all topological concepts (for both vectors and operators) refer to the norm topology. The *commutant* of a set of operators is the set of all those operators that commute with each operator in the given set. An operator is *irreducible* if its commutant contains no projections other than 0 and 1.

Notation. The underlying Hilbert space is H. If E is a subset of H, then $\bigvee E$ is the span of E (the smallest subspace that includes E). If E is an orthonormal basis for H, then $\mathbb{D}(E)$ is the set of all operators that are diagonal with respect to E (that is, the operators for which each element of E is an eigenvector). If A is an operator on H, then

$$\Re A = \frac{1}{2}(A + A^*) \quad \text{and} \quad \Im A = \frac{1}{2i}(A - A^*).$$

The set of all those operators A for which both $\Re A$ and $\Im A$ are simple diagonal operators is $\mathbb{D}$. That is: $A \in \mathbb{D}$ if and only if there exist orthonormal bases E and F such that $\Re A \in \mathbb{D}(E)$, $\Im A \in \mathbb{D}(F)$, and all eigenvalues of both $\Re A$ and $\Im A$ have multiplicity 1. If $\mathbb{K}$ is a set of operators, its commutant is $\mathbb{K}'$. The set of irreducible operators is $\mathbb{I}$, the set of reducible operators is $\mathbb{R}$, and the set of scalar multiples of the identity is $\mathbb{O}$.

Received September 22, 1967.
Research supported in part by a grant from the National Science Foundation.

1. DENSITY

LEMMA 1. *A necessary and sufficient condition that an operator* A *be irreducible is that* $\{\Re A\}' \cap \{\Im A\}' = \mathbf{O}$.

Proof. If A is reducible, then there is a non-trivial projection P in $\{A\}'$. Since P is Hermitian, $P \in \{A^*\}'$, and therefore $P \in \{\Re A\}' \cap \{\Im A\}'$; it follows that if $\{\Re A\}' \cap \{\Im A\}' = \mathbf{O}$, then $A \in \mathbf{I}$. If, conversely, $\{\Re A\}' \cap \{\Im A\}' \neq \mathbf{O}$, then that intersection contains a non-trivial projection, and therefore A is reducible; it follows that if $A \in \mathbf{I}$, then $\{\Re A\}' \cap \{\Im A\}' = \mathbf{O}$.

LEMMA 2. *If* E *is an orthonormal basis and* A *is an operator in* $\mathbf{D}(E)$ *with simple spectrum, then* $\{A\}' = \mathbf{D}(E)$.

Proof. A standard and elementary computation.

LEMMA 3. *If* E *and* F *are orthonormal bases, then a necessary and sufficient condition that* $\mathbf{D}(E) \cap \mathbf{D}(F) \neq \mathbf{O}$ *is that there exist non-trivial subsets* E_0 *and* F_0 *of* E *and* F, *respectively, such that* $\bigvee E_0 = \bigvee F_0$.

Here non-trivial means not empty and not the whole set.

Proof. If $A \in \mathbf{D}(E) \cap \mathbf{D}(F)$ and $A \neq \mathbf{O}$, consider an eigenvalue α_0 of A. Let E_0 and F_0 be the sets of eigenvectors corresponding to α_0 in E and in F, respectively. Clearly, $\bigvee E_0 = \bigvee F_0 = \{f: Af = \alpha_0 f\}$. Conversely, if E_0 and F_0 are non-trivial subsets of E and F with $\bigvee E_0 = \bigvee F_0$, let A be the projection whose range is that common span. Then $A \neq \mathbf{O}$ and $A \in \mathbf{D}(E) \cap \mathbf{D}(F)$.

LEMMA 4. *If* A *is an operator such that both* $\Re A$ *and* $\Im A$ *are diagonal operators with simple spectrum (that is, if* $A \in \mathbf{D}$), *and if* $\Im A$ *has an eigenvector* f *such that* $(e, f) \neq 0$ *for every eigenvector* e *of* $\Re A$, *then* $A \in \mathbf{I}$.

Proof. Let E and F be orthonormal bases such that $\Re A \in \mathbf{D}(E)$ and $\Im A \in \mathbf{D}(F)$. If E_0 and F_0 are subsets of E and F such that $\bigvee E_0 = \bigvee F_0$, then

$$\bigvee(E - E_0) = \left(\bigvee E_0\right)^{\perp} = \left(\bigvee F_0\right)^{\perp} = \bigvee(F - F_0).$$

There is no loss of generality in assuming that $f \in F$ (the simplicity assumption implies that some scalar multiple of f belongs to F), and, in view of the preceding comment, there is no loss of generality in assuming that $f \in F_0$ (since f belongs either to F_0 or to $F - F_0$, and the difference between the two cases is merely notational). Since, by assumption, f has a non-zero projection on every e in E, the only way it can happen that $f \in \bigvee E_0$ is that $E_0 = E$. The desired conclusion follows from Lemmas 3, 2, and 1.

LEMMA 5. $\mathbf{D}$ *is dense.*

Proof. It is sufficient to prove that the set of Hermitian operators in $\mathbf{D}$ is dense in the set of all Hermitian operators. To do this, represent any given Hermitian operator as a multiplication on L^2 over a finite measure space. (This is where the separability assumption comes in.) The multiplier can be uniformly approximated by simple functions. Multiplication by a real-valued simple function is the direct sum of a finite set of real scalars, and consequently it is a diagonal Hermitian operator. A diagonal Hermitian operator can obviously be approximated by one with simple spectrum.

The next auxiliary result is easy, but it has a certain interest in its own right. It is the solution of an approximation problem: given two unit vectors, find a unitary operator that maps one onto the other and is as near as possible to the identity.

LEMMA 6. *If f and g are unit vectors, then the infimum of* $\|1 - U\|$ *over all unitary operators* U *such that* Uf = g *is equal to* $\|f - g\|$, *and it is always attained.*

The proof shows that U can be chosen so that Uh = h whenever $h \perp \{f, g\}$; this is sometimes useful.

Proof. If Uf = g, then $\|1 - U\| \geq \|f - Uf\| = \|f - g\|$, so that the infimum is not smaller than $\|f - g\|$. What follows is the proof that $\|f - g\|$ can always be attained.

If $g = \alpha f$, put $Uf = \alpha f$ and Uh = h whenever $h \perp f$. In all other cases the span of f and g has dimension 2. Define Uh = h whenever $h \perp \{f, g\}$; the problem is thereby reduced to an elementary matrix computation. Choose coordinates so that $f = \langle 1, 0 \rangle$. If $g = \langle \alpha, \beta \rangle$ (with $|\alpha|^2 + |\beta|^2 = 1$, of course), put

$$U = \begin{pmatrix} \alpha & -\beta^* \\ \beta & \alpha^* \end{pmatrix}.$$

Clearly, U is unitary. Since $\|1 - U\|^2 = \|(1 - U)(1 - U^*)\| = \|2 - U - U^*\|$, since $2 - U - U^*$ is $2 - \alpha - \alpha^*$ times the identity, and since $\|f - g\|^2 = 2 - \alpha - \alpha^*$, the proof is complete.

Proof of density. Since $\mathbb{D}$ is dense (Lemma 5), it is sufficient to prove that if $\varepsilon > 0$ and $A \in \mathbb{D}$, then there exists an irreducible A_0 in $\mathbb{D}$ with $\|A - A_0\| < \varepsilon$. Write $B = \Re A$ and $C = \Im A$, and let E and F be orthonormal bases such that $B \in \mathbb{D}(E)$ and $C \in \mathbb{D}(F)$.

Consider an arbitrary f in F. The Fourier expansion of f with respect to E may have some zero coefficients; let g be a unit vector obtained from f by varying those coefficients slightly so that $(e, g) \neq 0$ for all e in E. Let the variation be so slight that $\|f - g\| < \varepsilon/2\|C\|$. Lemma 6 yields a unitary operator U such that Uf = g and $\|1 - U\| < \varepsilon/2\|C\|$. If $C_0 = UCU^*$, then C_0 is a diagonal Hermitian operator with simple spectrum and

$$\|C - C_0\| = \|C - UCU^*\| \leq \|C - UC\| + \|UC - UCU^*\|$$

$$\leq \|C\| \cdot \|1 - U\| + \|UC\| \cdot \|1 - U^*\| < \varepsilon.$$

If $A_0 = B + iC_0$, then $A_0 \in \mathbb{D}$. Since C_0 has an eigenvector (namely g) that has a non-zero inner product with every eigenvector of B, it follows from Lemma 4 that A_0 is irreducible. Since, finally,

$$\|A - A_0\| = \|C - C_0\| < \varepsilon,$$

the proof is complete.

R. G. Douglas has observed that in the proof of Lemma 5 it is possible to invoke the von Neumann approximation (each Hermitian operator when suitably perturbed by an operator of arbitrarily small Hilbert-Schmidt norm becomes diagonal) in place of the more obvious norm approximation; the result is that the principal theorem is true in the sense of Hilbert-Schmidt approximation also. The same result was obtained, later but independently, by J. G. Stampfli.

2. FINITE-DIMENSIONALITY AND NORMALITY

If H is finite-dimensional, the theorem has a relatively simple geometric proof, and it can be significantly strengthened.

Here is a possible proof. (a) The operators all whose eigenvalues have algebraic multiplicity 1 are dense. (b) If all the eigenvalues of an operator have algebraic multiplicity 1, then its eigenvectors span H, and consequently there exists a linear basis of H consisting of eigenvectors. (c) By a small perturbation an operator whose eigenvectors span H can be transmuted into another one of the same kind such that some particular element of a basis consisting of eigenvectors is *not* orthogonal to any other. (Consider a basis of eigenvectors, form the span of all but one, note that that one is not in the span, and perturb it slightly, if necessary, so as to push it out of the orthogonal complement of the span.) (d) If a basis consisting of eigenvectors of an operator is such that some element of it is not orthogonal to any other, then that operator is irreducible. (For each reducing subspace, every eigenvector must belong either to it or to its orthogonal complement; whichever one the distinguished "non-orthogonal" vector belongs to must contain all others.)

The strengthening is that $\mathbb{I}$ is not only a G_δ (the proof of this in the general case is in Section 3) but open.

PROPOSITION 1. *On a finite-dimensional Hilbert space the set of reducible operators is closed and nowhere dense.*

Proof. Suppose that A_n is reducible and $A_n \to A$, and, for each n, let P_n be a non-trivial projection that commutes with A_n. Finite-dimensionality implies the compactness of the unit ball in the space of operators. There is, therefore, no loss of generality in assuming that $P_n \to P$, where P is, of course, a projection, and, clearly, AP = PA. If dim H = k, then $1 \le \operatorname{rank} P_n \le k - 1$; since rank is lower semicontinuous (see the Appendix), it follows that $P \ne 0, 1$. This proves that $\mathbb{R}$ is closed; that it is nowhere dense follows from the already proved density of its complement.

Similar easy techniques give information about the size of the set of normal operators on spaces of arbitrarily large dimension.

PROPOSITION 2. *On a Hilbert space of dimension greater than 1, the set of normal operators is closed and nowhere dense.*

Proof. Closure is obvious; since the mapping $A \to A^*A - AA^*$ is continuous, the set $\{A: A^*A - AA^* = 0\}$ is closed. A closed set is nowhere dense if and only if its complement is dense. Since an irreducible operator can be normal only if the space has dimension 0 or 1, the conclusion for separable spaces follows from the density of the set of irreducible operators.

There is a better, direct proof, independent of the previous density theorem, that works for non-separable spaces just as well as for separable ones. It is sufficient to prove that every normal operator is arbitrarily near non-normal ones. For scalars this is obvious. (Find a non-normal T, and consider $\lambda + \varepsilon T$, where λ is the given scalar and ε is small.) If A is normal but not a scalar, then either $\Re A$ or $\Im A$ is not a scalar; say $\Re A$ is not. Then there exists a Hermitian T that does not commute with $\Re A$ (this is where it is necessary that the dimension be greater than 1). If $\varepsilon > 0$, then $A + i\varepsilon T$ is not normal but can be arbitrarily near to A.

3. TOPOLOGY

This section contains the second half of the proof of the principal theorem, that is, the proof that $\mathbb{I}$ is a G_δ.

Let $\mathbb{P}$ be the set of all those Hermitian operators P on H for which $0 \le P \le 1$. Recall that $\mathbb{P}$ is exactly the weak closure of the set of projections. Let $\mathbb{P}_0$ be the subset of those elements of $\mathbb{P}$ that are *not* scalar multiples of the identity. Since $\mathbb{P}$ is a weakly closed subset of the unit ball, it is weakly compact, and hence the weak topology for $\mathbb{P}$ is metrizable. Since the set of scalars is weakly closed, it follows that $\mathbb{P}_0$ is weakly locally compact. Since the weak topology for $\mathbb{P}$ has a countable base, the same is true for $\mathbb{P}_0$, and therefore $\mathbb{P}_0$ is weakly σ-compact. Let $\mathbb{P}_1$, $\mathbb{P}_2$, $\cdots$ be weakly compact subsets of $\mathbb{P}_0$ such that $\bigcup_{n=1}^{\infty} \mathbb{P}_n = \mathbb{P}_0$.

It is to be proved that $\mathbb{R}$ is an F_σ (norm topology). Let $\hat{\mathbb{P}}_n$ be the set of all those operators A on H for which there exists a P in $\mathbb{P}_n$ such that AP = PA ($n = 1, 2, 3, \cdots$); the spectral theorem implies that $\bigcup_{n=1}^{\infty} \hat{\mathbb{P}}_n = \mathbb{R}$.

The proof can be completed by showing that each $\hat{\mathbb{P}}_n$ is (norm) closed. Suppose, therefore, that $A_k \in \hat{\mathbb{P}}_n$ and that $A_k \to A$ (norm). For each k, find a P_k in $\mathbb{P}_n$ such that $A_k P_k = P_k A_k$. Since $\mathbb{P}_n$ is weakly compact and metrizable, there is no loss of generality in assuming that the sequence $\{P_k\}$ is weakly convergent to P, say. (This is the point where it is advantageous to consider all the operators in $\mathbb{P}$, and not just projections; there is no guarantee that P is a projection even if the P_k's are. Note that $P \in \mathbb{P}_n$, so that, in particular, P is not a scalar.)

Assertion: AP = PA. This follows from an easy lemma: if $A_k \to A$ (norm) and $P_k \to P$ (weak), then $A_k P_k \to AP$ and $P_k A_k \to PA$ (weak). Indeed:

$$|(A_k P_k f, g) - (APf, g)| \le |(A_k P_k f, g) - (A P_k f, g)| + |(A P_k f, g) - (APf, g)|$$

$$\le \|A_k - A\| \cdot \|f\| \cdot \|g\| + |((P_k - P)f, A^* g)|.$$

(It is important that the sequence $\{P_k\}$ is bounded.) These inequalities imply that $A_k P_k \to AP$ (weak); the other order follows from the consideration of adjoints. Once this is done, everything is done: $A \in \hat{\mathbb{P}}_n$, hence $\hat{\mathbb{P}}_n$ is closed (norm), hence $\mathbb{R}$ is an F_σ (norm).

4. REDUCIBILITY

By the principal theorem, the set $\mathbb{R}$ of reducible operators is always an F_σ, and, by Proposition 1, in the finite-dimensional case $\mathbb{R}$ is closed. Could it be that $\mathbb{R}$ is always closed? The answer is no. Reason: on an infinite-dimensional space every operator of finite rank is reducible, so that every compact operator is in the closure of $\mathbb{R}$, but it is easy to construct compact operators (weighted shifts) that are irreducible.

There is another example, which shows that $\mathbb{R}$ is not closed in a more surprising way. For each positive integer n, let $\mathbb{R}_n$ be the set of all operators that have a reducing subspace of dimension n.

PROPOSITION 3. *Every isometry is in the closure of* $\mathbb{R}_1$.

Proof. Observe to begin with that for an operator to be nearly irreducible is the same as to be near to a reducible operator. This auxiliary assertion can be stated as follows, in precise terms: if $\|A - A_0\| < \varepsilon$ and P is a projection such that $A_0 P = PA_0$, then $\|AP - PA\| < 2\varepsilon$; if, conversely, $\|AP - PA\| < \varepsilon$ and $A_0 = PAP + (1 - P)A(1 - P)$, then $A_0 P = PA_0$ and $\|A - A_0\| < 2\varepsilon$. The proof of the first assertion is implied by the inequality

$$\|AP - PA\| \leq \|AP - A_0 P\| + \|A_0 P - PA_0\| + \|PA_0 - PA\|.$$

The proof of the second assertion is implied by

$$\|A - A_0\| = \|(1 - P)AP + PA(1 - P)\|$$

$$= \|(AP - PA)P - P(AP - PA)\| \leq 2\|AP - PA\|.$$

The motivation for the definition of A_0 is the consideration of the matrix of A corresponding to the direct sum decomposition of H into ran P and $(\text{ran } P)^\perp$: throw away the off corners.

In view of the preceding paragraph, it is sufficient to prove that if U is an isometry, then there exist projections of rank 1 that nearly commute with U. To prove it, let λ be a number of modulus 1 that is an approximate eigenvalue of U, that is, an element of the approximate point spectrum. It follows, by definition, that corresponding to each positive number ε there is a unit vector e such that $\|Ue - \lambda e\| < \varepsilon$, and hence such that

$$\|U^* e - \lambda^* e\| = \|-\lambda^* U^* (Ue - \lambda e)\| < \varepsilon.$$

If P is the projection onto e, that is, if $Pf = (f, e)e$ for all f, then

$$\|(UP - PU)f\| = \|(f, e)Ue - (Uf, e)e\| \leq \|(f, e)Ue - (f, e)\lambda e\| + \|(f, e)\lambda e - (f, U^* e)e\|$$

$$\leq |(f, e)| \cdot \|Ue - \lambda e\| + \|f\| \cdot \|U^* e - \lambda e\| \leq 2\varepsilon \|f\|,$$

so that $\|UP - PU\| \leq 2\varepsilon$. The proof is complete.

Since there exist irreducible isometries (for example, the unilateral shift), Proposition 3 implies again that $\mathbb{R}$ is not closed.

Proposition 3 raises the hope that $\mathbb{R}_1$ is dense, but it is not. If H is a Hilbert space, and if A is the operator on $H \oplus H$ with matrix $\begin{pmatrix} 0 & 0 \\ 1 & 0 \end{pmatrix}$, then A is not in the closure of $\mathbb{R}_1$. (It is easy to see that A is in $\mathbb{R}_2$; in fact, A is the direct sum of operators of rank 2, equal in number to the dimension of H.)

To prove that A is not in the closure of $\mathbb{R}_1$, it is sufficient to prove that the infimum of $\|A - B\|$, as B varies over $\mathbb{R}_1$, is positive. Given B in $\mathbb{R}_1$, let $\langle f, g \rangle$ (with f and g in H) be a reducing eigenvector of B of norm 1; that is,

$$B\langle f, g \rangle = \lambda \langle f, g \rangle, \quad B^*\langle f, g \rangle = \lambda^* \langle f, g \rangle, \quad \|f\|^2 + \|g\|^2 = 1.$$

Since $A - \lambda = (A - B) + (B - \lambda)$, it follows that

$$\|(A - \lambda)\langle f, g \rangle\| \leq \|A - B\| \quad \text{and} \quad \|(A^* - \lambda^*)\langle f, g \rangle\| \leq \|A - B\|;$$

since $(A - \lambda)\langle f, g \rangle = \langle -\lambda f, f - \lambda g \rangle$ and $(A^* - \lambda^*)\langle f, g \rangle = \langle g - \lambda^* f, -\lambda^* g \rangle$, it follows that

$$|\lambda|^2 \|f\|^2 + \|f - \lambda g\|^2 \leq \|A - B\|^2 \quad \text{and} \quad \|g - \lambda^* f\|^2 + |\lambda|^2 \|g\|^2 \leq \|A - B\|^2 .$$

The latter inequalities imply that

$$1 + 2|\lambda|^2 - 4 \Re \lambda^* (f, g) \leq 2 \|A - B\|^2 ,$$

and hence that $1 + 2|\lambda|^2 - 2|\lambda| \leq 2\|A - B\|^2$. Since $1 - 2|\lambda|^2 - 2|\lambda| \geq 1/2$ for all λ, the proof is complete.

There is at least one question along these lines that seems to be of interest and that is unanswered: is $\mathbb{R}$ dense?

5. INVARIANCE

An operator is *transitive* if $\{0\}$ and H are the only subspaces it leaves invariant. The problem of invariant subspaces is to decide whether there exist transitive operators on Hilbert spaces of dimension greater than 1. (The word "transitive" was suggested by W. B. Arveson. Its present meaning is not identical with its meanings in group theory and ergodic theory, but it is in close harmony with them.) Let $\mathbb{T}$ be the set of transitive operators. One possible approach to the problem is to try to prove that $\mathbb{T}$ is not empty by proving that it is topologically large, *i. e.*, that it is (or includes) a dense G_δ. As it stands, this is doomed to failure: $\mathbb{T}$ is not dense.

PROPOSITION 4. *If* U *is a non-invertible isometry, and if* $\|U^* - A\| < 1$, *then* ker $A \neq \{0\}$.

Proof. Since $\|U^* - A\| < 1$, it follows that $\|U^* U - AU\| < 1$, *i. e.*, that $\|1 - AU\| < 1$. From this, in turn, it follows that AU is invertible, and hence that ran $A = H$. If it were true that ker $A = \{0\}$, then it would follow from the closed graph theorem that A in invertible. Since AU is already known to be invertible, it would then follow that U is invertible, and this contradicts the assumption.

Since there exist non-invertible isometries (for example, the unilateral shift), Proposition 4 implies that $\mathbb{T}$ is not dense.

There is an open question along these lines that has at least some curiosity value: is the complement of $\mathbb{T}$ topologically large? The following result is pertinent.

PROPOSITION 5. *The set of all operators with an eigenvalue is dense.*

Proof. The result is an easy consequence of the existence of approximate eigenvalues (compare the proof of Proposition 3). Given an operator A, let λ be an approximate eigenvalue of A; it follows that corresponding to each positive number ε there is a unit vector e such that $\|Ae - \lambda e\| < \varepsilon$. If P is the projection onto e, that is, if $Pf = (f, e)e$ for all f, and if $A_0 = A - (1 - P)AP$, then a direct verification proves that e is an eigenvector of A_0 with eigenvalue (Ae, e). Since $|(Ae, e) - \lambda| < \varepsilon$, it follows that

$$\|(A - A_0)f\| = \|(1 - P)APf\| \leq |(f, e)| \cdot \|(1 - P)Ae\|$$

$$\leq \|f\| \cdot \|Ae - (Ae, e)e\| \leq 2\varepsilon \|f\|,$$

so that $\|A - A_0\| \leq 2\varepsilon$. The proof is complete.

(In matrix terms the proof could have been phrased this way: choose an approximate eigenvector e for A with approximate eigenvalue λ; use e as the first term of an orthonormal basis; replace the first column of the resulting matrix for A by $\langle \lambda, 0, 0, \cdots \rangle$; the new matrix A_0 has the eigenvalue λ and is near to A.)

APPENDIX

THEOREM. *Rank is weakly lower semicontinuous.*

The rank of an operator is the dimension of the closure of its range. The statement means that for each operator A_0 there exists a weak neighborhood N of A_0 such that rank $A \geq$ rank A_0 for all A in N. Equivalently, in terms of convergence: if $\{A_n\}$ is a net that converges weakly to A_0, then $\lim \inf_n$ rank $A_n \geq$ rank A_0.

The possible values of rank in this context are the non-negative integers, together with ∞; no distinction is made among different infinite cardinals. Were such a distinction to be made, the result would become false. Here is an example. Let H be a non-separable Hilbert space with an orthonormal basis $\{e_j\}$. Let D be the set of all countable subsets of the index set, ordered by inclusion: for each n in D, write A_n for the projection onto $\bigvee \{e_j: j \in n\}$. Since for each f_0 in H there exists an n_0 in D such that $f_0 \perp e_j$ whenever $j \notin n_0$, it follows that $A_n \to 1$ (not only weakly, but, in fact, strongly). Since rank $A_n = \aleph_0$ and rank $1 > \aleph_0$, the cardinal version of semicontinuity is false.

LEMMA 1. *If $\{e_1, \cdots, e_n\}$ is an orthonormal set and $\|f_i - e_i\| < 1/\sqrt{n}$ $(i = 1, \cdots, n)$, then the set $\{f_1, \cdots, f_n\}$ is linearly independent.*

Proof. If $\xi_i \neq 0$ for at least one i, then

$$\left\| \sum_{i=1}^{n} \xi_i (f_i - e_i) \right\| \leq \sum_{i=1}^{n} |\xi_i| \cdot \|f_i - e_i\| < \left(\sum_{i=1}^{n} |\xi_i| \right) \cdot (1/\sqrt{n}) \leq \sqrt{n} \sqrt{\sum_{i=1}^{n} |\xi_i|^2} \Big/ \sqrt{n},$$

and therefore

$$\left\| \sum_{i=1}^{n} \xi_i f_i \right\| \geq \left\| \sum_{i=1}^{n} \xi_i e_i \right\| - \left\| \sum_{i=1}^{n} \xi_i (f_i - e_i) \right\| > \sqrt{\sum_{i=1}^{n} |\xi_i|^2} - \sqrt{\sum_{i=1}^{n} |\xi_i|^2}.$$

LEMMA 2. *Rank is strongly lower semicontinuous.*

Proof. To prove: if rank $A_0 \geq n$ ($= 1, 2, 3, \cdots$), then there exists a strong neighborhood N of A_0 such that rank $A \geq n$ for all A in N. Let $\{e_1, \cdots, e_n\}$ be an orthonormal set in ran A_0; find $f_1, \cdots, f_n$ such that $A_0 f_i = e_i$ $(i = 1, \cdots, n)$. Write

$$N = \{A: \|Af_i - A_0 f_i\| < 1/\sqrt{n}, i = 1, \cdots, n\};$$

it follows from Lemma 1 that, for each A in N, the set $\{Af_1, \cdots, Af_n\}$ is linearly independent.

Proof of the theorem. To prove: if rank $A_0 \geq n$ ($= 1, 2, 3, \cdots$), then there exists a weak neighborhood N of A_0 such that rank $A \geq n$ for all A in N. Let $\{e_1, \cdots, e_n\}$ be an orthonormal set in ran A_0; find $f_1, \cdots, f_n$ such that $A_0 f_i = e_i$ $(i = 1, \cdots, n)$. Write

$$N = \{A: |((A - A_0)f_j, e_i)| < \varepsilon; \; i, j = 1, \cdots, n\},$$

where ε is an as yet unspecified positive number. Given A in N, write $\alpha_{ij} = (Af_j, e_i)$. Note that $(A_0 f_j, e_i) = (e_j, e_i) = \alpha_{ij}$. Since (by Lemma 2) rank is strongly lower semicontinuous, and since for finite-dimensional spaces all the usual operator topologies coincide, it follows that if the matrix α is sufficiently near the identity matrix (that is, if ε is sufficiently small), then rank $\alpha \geq n$, and therefore rank $\alpha = n$. In other words, if ε is sufficiently small, then α is invertible. This implies that if $\sum_{j=1}^{n} \xi_j Af_j = 0$, so that

$$\sum_{j=1}^{n} \alpha_{ij}\xi_j = \sum_{j=1}^{n} (Af_j, e_i)\xi_j = 0,$$

then $\xi_1 = \cdots = \xi_n = 0$, so that the set $\{Af_1, \cdots, Af_n\}$ is linearly independent.

The University of Michigan
Ann Arbor, Mich. 48104

Reprinted from the
ACTA SCIENTIARUM MATHEMATICARUM
Tomus XXIX, pp. 259-293, 1968

Quasitriangular operators

By P. R. HALMOS in Ann Arbor (Michigan, U.S.A.)*)

Every square matrix with complex entries is unitarily equivalent to a triangular one. In other words, if A is an operator on a finite-dimensional Hilbert space H, then there exists an increasing sequence $\{M_n\}$ of subspaces such that $\dim M_n = n$ ($n = 0, \ldots, \dim H$), and such that each M_n is invariant under A. On a Hilbert space of dimension $\aleph_0$ the appropriate definition is this: A is *triangular* if there exists an increasing sequence $\{M_n\}$ of finite-dimensional subspaces whose union spans H such that each M_n is invariant under A. It is easy, but not obviously desirable, to fill in the dimension gaps, and hence to justify the added assumption that $\dim M_n = n$ ($n = 0, 1, 2, \ldots$).

In many considerations of invariant subspaces ($AM \subset M$) it is convenient to treat their projections instead ($AE = EAE$). In terms of projections a necessary and sufficient condition that an operator A on a separable Hilbert space H be triangular is that

(Δ) *there exists an increasing sequence $\{E_n\}$ of projections of finite rank such that $E_n \to 1$ (strong topology) and such that $AE_n - E_n A E_n = 0$ for all n.*

This formulation suggests an asymptotic generalization of itself. An operator A is *quasitriangular* if

(Δ_1) *there exists an increasing sequence $\{E_n\}$ of projections of finite rank such that $E_n \to 1$ (strong topology) and such that $\|AE_n - E_n A E_n\| \to 0$.*

(Informally: E_n is approximately invariant under A.) The concept (but not the name) has been seen before; it plays a central role in the proofs of the Aronszajn—Smith theorem [1] on the existence of invariant subspaces for compact operators, and in the proofs of its various known generalizations [2], [3], [5].

---

*) Research supported in part by a grant from the National Science Foundation.

237

It is interesting and useful to examine a variant of the condition (Δ_1); the variant requires that

(Δ_2) *there exists a sequence $\{E_n\}$ of projections of finite rank such that $E_n \to 1$ (strong topology) and such that $\|AE_n - E_nAE_n\| \to 0$.*

The only difference between Δ_1 and Δ_2 is that the latter does not require the sequence $\{E_n\}$ to be increasing.

There is still another pertinent condition. The set of all projections of finite rank, ordered by range inclusion, is a directed set. Since $E \to \|AE - EAE\|$ is a net on that directed set, it makes sense to say that

(Δ_0) $$\liminf_{E \to 1} \|AE - EAE\| = 0.$$

What it means is that for every positive number ε and for every projection E_0 of finite rank there exists a projection E of finite rank such that $E_0 \leqq E$ and $\|AE - EAE\| < \varepsilon$.

The purpose of this paper is to initiate a study of quasitriangular operators. The study begins with the observation that approximately invariant projections that are large (in the sense of having large ranks) always exist (Section 1). The main result is the characterization of quasitriangular operators; it asserts (for separable spaces) that the conditions Δ_0, Δ_1, and Δ_2 are mutually equivalent (Section 2). This characterization is applied to show that there exist operators that are *not* quasitriangular. On the other hand the set of quasitriangular operators is quite rich (Section 3); it is closed under the formation of polynomials, it is closed in the norm topology of operators, it is closed under the formation of countable direct sums, and it contains, for example, all operators of the form $N + K$ where N is normal and K is compact. The paper concludes with a few questions (Section 4). Sample: is it true for every operator A that either A or A^* is quasitriangular?

Section 1

Sequences of approximately invariant projections that are not required to be "large" always exist. A precise statement is this: for each operator A there exists a sequence $\{E_n\}$ of non-zero projections of finite rank such that $\|AE_n - E_nAE_n\| \to 0$; in fact the E_n's can be chosen to have rank 1. The proof is immediate from the existence of approximate eigenvalues and eigenvectors. Let λ be a scalar and $\{e_n\}$ a sequence of unit vectors such that $\|Ae_n - \lambda e_n\| \to 0$. If the projections E_n are defined by $E_n f = (f, e_n)e_n$, then

$$(AE_n - E_nAE_n)f = (f, e_n)(Ae_n - (Ae_n, e_n))e_n.$$

Since $(Ae_n, e_n) \to \lambda$, it follows that $\|AE_n - E_nAE_n\| \to 0$.

Since every operator has approximately invariant projections of rank 1, it is tempting to conclude, via the formation of finite spans, that every operator on an infinite-dimensional Hilbert space has approximately invariant projections of arbitrarily large finite ranks. The theory of approximate invariance turns out, however, to be surprisingly delicate. It is, for instance, not true that the span of two approximate eigenvectors is approximately invariant. More precisely, there exists a 3×3 matrix A and there exist two projections F and G of rank 1 such that F is invariant under A, G is nearly invariant under A, but if $E = F \vee G$, then $\|AE - EAE\| = 1$. In detail: put

$$A = \begin{pmatrix} 0 & 0 & 0 \\ 0 & 0 & 0 \\ 1 & 0 & 0 \end{pmatrix},$$

let F be the projection onto $\langle 0, 1, 0 \rangle$, and let G be the projection onto $\langle a, b, 0 \rangle$, where $|a|^2 + |b|^2 = 1$ and a is "small" (but not 0). It is easy to verify that

$$F = \begin{pmatrix} 0 & 0 & 0 \\ 0 & 1 & 0 \\ 0 & 0 & 0 \end{pmatrix}, \quad G = \begin{pmatrix} |a|^2 & ab^* & 0 \\ a^*b & |b|^2 & 0 \\ 0 & 0 & 0 \end{pmatrix}, \quad E = \begin{pmatrix} 1 & 0 & 0 \\ 0 & 1 & 0 \\ 0 & 0 & 0 \end{pmatrix},$$

$$\|AF - FAF\| = 0, \quad \|AG - GAG\| = |a|, \quad \text{and} \quad \|AE - EAE\| = 1.$$

This example, informal in its interpretation of "nearly invariant", can be used to construct an example of two sequences of approximately invariant projections, in the precise technical sense, such that the sequence of their spans is not approximately invariant, as follows. Let H be the direct sum $H_1 \oplus H_2 \oplus \ldots$ of 3-dimensional spaces such as played a role in the preceding paragraph, and let the operator A on H be the direct sum $A_1 \oplus A_2 \oplus \ldots$ of the corresponding operators. Let F_n be the direct sum projection whose summand with index n is the previous F and whose other summands are 0; let G_n be the direct sum projection whose summand with index n is the previous G with $a = \dfrac{1}{n}$ and whose other summands are 0. It follows that $\|AF_n - F_nAF_n\| = 0$ for all n, $\|AG_n - G_nAG_n\| \to 0$, and, if $E_n = F_n \vee G_n$, then $\|AE_n - E_nAE_n\| = 1$.

It is slightly surprising that, despite the evidence of the preceding example, approximately invariant projections of arbitrarily large ranks always exist.

Theorem 1. *If A is an operator on an infinite-dimensional Hilbert space, ε is a positive number, and n is a positive integer, then there exists a projection E of rank n such that $\|AE - EAE\| < \varepsilon$.*

Proof. For $n = 1$, the result was derived from the existence of approximate eigenvectors. The idea of the inductive proof that follows is that although near invariance is not preserved by the formation of spans, it is preserved by the formation of orthogonal spans. Given ε and n, assume the result for n, and let F be a projection of rank n such that $\|AF - FAF\| < \varepsilon/2$. Since the compression of A to ran $(1 - F)$ (i.e., the restriction of $(1 - F)A(1 - F)$ to ran $(1 - F)$) has approximately invariant projections of rank 1, it follows that there exists a projection G of rank 1 such that $G \perp F$ and

$$\|(1 - F)A(1 - F)G - G(1 - F)A(1 - F)G\| < \varepsilon/2.$$

(Find G on ran $(1 - F)$ first and then extend it by definining it to be 0 on ran F.) Since $G(1 - F) = (1 - F)G = G$, the last inequality is equivalent to

$$\|(1 - F)(1 - G)AG\| < \varepsilon/2.$$

If $E = F + G$, then E is a projection of rank $n + 1$ and

$$\|AE - EAE\| = \|(1 - E)AE\| = \|(1 - F)(1 - G)A(F + G)\| =$$

$$= \|(1 - G)(1 - F)AF + (1 - F)(1 - G)AG\| \leqq \|(1 - F)AF\| + \|(1 - F)(1 - G)AG\| < \varepsilon.$$

Section 2

It is trivial that the definition of quasitriangularity (Δ_1) implies the weakened form (Δ_2) (obtained from (Δ_1) by omitting the word "increasing"). It is also quite easy to prove that if, on a separable Hilbert space, $\liminf_{E \to 1} \|AE - EAE\| = 0 \ (\Delta_0)$, then A is quasitriangular (Δ_1). Indeed, let $\{e_1, e_2, \ldots\}$ be an orthonormal basis for the space. By (Δ_0) there exists a projection E_1 of finite rank such that $e_1 \in$ ran E_1 and $\|AE_1 - E_1 AE_1\| < 1$. Again, by (Δ_0), there exists a projection E_2 of finite rank such that $E_1 \leqq E_2$, $e_2 \in$ ran E_2, and $\|AE_2 - E_2 AE_2\| < \frac{1}{2}$. In general, inductively, use (Δ_0) to get a projection E_{n+1} of finite rank such that $E_n \leqq E_{n+1}$, $e_{n+1} \in$ ran E_{n+1}, and $\|AE_{n+1} - E_{n+1}AE_{n+1}\| < \frac{1}{n+1}$. Conclusion: $\{E_n\}$ is an increasing sequence of projections of finite rank such that $E_n \to 1$ and such that $\|AE_n - E_n AE_n\| \to 0$; in other words A is quasitriangular, as promised.

The non-trivial implication along these lines is the one from (Δ_2) to (Δ_0). The proof depends on a lemma according to which if two projections have the same finite rank and are near, then there is a "small" unitary operator that transforms one onto the other. (For unitary operators "small" means "near to 1".) A possible quantitative formulation goes as follows.

Lemma 1. *If E and F are projections of the same finite rank such that $\|E - F\| = = \varepsilon < 1$, then the infimum of $\|1 - W\|$, extended over all unitary operators W such that $W^* EW = F$, is not more than $2\varepsilon^{\frac{1}{2}}$.*

The lemma can be improved, but the improvement takes considerably more work and for present purposes it is not needed. A trivial improvement is to drop the assumption that E and F have the same rank and recapture it from the known result [7, p. 58] that the inequality $\|E - F\| < 1$ implies rank $E =$ rank F. Another qualitative improvement is to drop the assumption that the ranks are finite and pay for it by introducing partial isometries instead of unitary operators. The best kind of improvement is quantitative; the estimate $2\varepsilon^{\frac{1}{2}}$ can be sharpened to $2^{\frac{1}{2}}[1 - (1 - \varepsilon^2)^{\frac{1}{2}}]^{\frac{1}{2}}$. For a discussion of such results and references to related earlier work see [4]. Conjecturally the sharpened estimate is best possible, but the proof of that does not seem to be in the literature.

Proof. The equality of rank E and rank F implies the existence of a unitary operator W_0 such that $W_0^* E W_0 = F$. Write E, F, and W_0 as operator matrices, according to the decomposition $1 = E + (1 - E)$, so that, for instance, $E = \begin{pmatrix} 1 & 0 \\ 0 & 0 \end{pmatrix}$.

If $W_0 = \begin{pmatrix} A & B \\ C & D \end{pmatrix}$, then $W_0^* = \begin{pmatrix} A^* & C^* \\ B^* & D^* \end{pmatrix}$, and therefore

$$W_0^* W_0 = \begin{pmatrix} A^*A + C^*C & A^*B + C^*D \\ B^*A + D^*C & B^*B + D^*D \end{pmatrix} = \begin{pmatrix} 1 & 0 \\ 0 & 1 \end{pmatrix}.$$

The matrix of $F (= W_0^* E W_0)$ can now be computed; it turns out to be $\begin{pmatrix} A^*A & A^*B \\ B^*A & B^*B \end{pmatrix}$. Since the norm of each entry of a matrix is dominated by the norm of the matrix, it follows that

$$\|C^*C\| = \|1 - A^*A\| \le \varepsilon \quad \text{and} \quad \|1 - D^*D\| = \|B^*B\| \le \varepsilon.$$

Observe next that if U and V are unitary operators on ran E and ran $(1 - E)$ respectively, and if $W_1 = \begin{pmatrix} U & 0 \\ 0 & V \end{pmatrix}$, then W_1 commutes with E, and, therefore, $W_1 W_0$ transforms E onto F (just as W_0 does). The purpose of the rest of the proof is to choose U and V so as to make $\|1 - W_1 W_0\|$ small. Since

$$1 - W_1 W_0 = \begin{pmatrix} 1 - UA & -UB \\ -VC & 1 - VD \end{pmatrix},$$

and since the norm of a matrix is dominated by the square root of the sum of the squares of the norms of its entries, it is sufficient to prove that by appropriate choices of U and V the entries of the last written matrix can be made to have small norms. The off-diagonal entries of $1 - W_1 W_0$ are easy to estimate:

$$\| - VC\|^2 = \|C\|^2 = \|C^*C\| \le \varepsilon \quad \text{and} \quad \| - UB\|^2 = \|B\|^2 = \|B^*B\| \le \varepsilon.$$

In these estimates U and V are arbitrary unitary operators; it is only in the next step that they have to be chosen so as to make something small.

Observe that since the ranks of E and F are finite, the lemma loses no generality if it is stated for finite-dimensional spaces only; the infinite-dimensional case is recaptured by applying the finite-dimensional lemma to E and F restricted to $\operatorname{ran} E \vee \operatorname{ran} F$ and extending the resulting unitary operator by defining it to be the identity on the orthogonal complement of $\operatorname{ran} E \vee \operatorname{ran} F$. In the finite-dimensional case A is the product of a unitary operator and $(A^*A)^{\frac{1}{2}}$ (polar decomposition); let U be the inverse of the unitary factor. With this choice $1 - UA$ becomes $1 - P$, where $P = (A^*A)^{\frac{1}{2}}$. Since $\|A\| \leq 1$, so that $0 \leq P^2 \leq P \leq 1$, it follows that $0 \leq 1 - P \leq 1 - P^2$, and hence that

$$\|1 - UA\| = \|1 - P\| \leq \|1 - P^2\| = \|1 - A^*A\|.$$

A similar argument for D produces a unitary V such that $\|1 - VD\| \leq \|1 - D^*D\|$. Conclusion:

$$\|1 - W_1 W_0\|^2 \leq 2(\varepsilon + \varepsilon^2) \leq 4\varepsilon,$$

and the proof of the lemma is complete.

The ground is now prepared for the proof of the principal result.

Theorem 2. *If $\{E_n\}$ is a sequence of projections of finite rank such that $E_n \to 1$ (strong topology) and such that $\|AE_n - E_n AE_n\| \to 0$, then $\liminf_{E \to 1} \|AE - EAE\| = 0$.*

Proof. It is to be proved that if $\varepsilon > 0$ and if E_0 is a projection with rank $E_0 = n_0 < \infty$, then there exists a projection E of finite rank such that $E_0 \leq E$ and $\|AE - EAE\| < \varepsilon$.

Let δ be a temporarily indeterminate positive number; it will be specified, in terms of ε, later. Suppose that $\{e_1, \ldots, e_{n_0}\}$ is an orthonormal basis for $\operatorname{ran} E_0$. The two limiting assumptions imply the existence of a positive integer n such that $\|e_j - E_n e_j\| < \delta/\sqrt{n_0}$ $(j = 1, \ldots, n_0)$ and $\|AE_n - E_n AE_n\| < \delta$. The first of these inequalities implies that if δ is sufficiently small, then the set $\{E_n e_1, \ldots, E_n e_{n_0}\}$ is linearly independent. (The proof is easy and is omitted here; it is explicitly carried out in [6].) Let F_0 be the projection (of rank n_0) onto their span; note that $F_0 \leq E_n$. (The n here used will remain fixed from now on.)

If $f \in \operatorname{ran} E_0$, so that $f = \sum_{j=1}^{n_0} \xi_j e_j$, then

$$\|f - F_0 f\|^2 = \|\sum_j \xi_j (e_j - E_n e_j)\|^2 \leq (\sum_j |\xi_j| \cdot \|e_j - E_n e_j\|)^2 \leq$$

$$\leq \sum_j |\xi_j|^2 \cdot \sum_j \|e_j - E_n e_j\|^2 \leq \|f\|^2 \cdot n_0 (\delta/\sqrt{n_0})^2,$$

and therefore

$$\|E_0 - F_0 E_0\| \leq \delta.$$

This shows that E_0 is approximately dominated by F_0; what is needed for the rest of the proof is the stronger assertion that E_0 is approximately equal to F_0.

By definition, ran F_0 is spanned by the vectors $F_0 e_j (= E_n e_j)$, $j = 1, \ldots, n_0$; it follows that ran $F_0 = $ ran $F_0 E_0$. In other words, the restriction of F_0 to ran E_0 maps ran E_0 onto ran F_0. Call that restriction T; then T is a linear transformation from a space of dimension n_0 onto a space of dimension n_0, and, consequently, T is invertible. Since the spaces involved are finite-dimensional, the transformation T^{-1} is bounded, but that is not enough information; what is needed is an effective estimate of $\|T^{-1}\|$. That turns out to be easy to get. If $f \in$ ran E_0, then

$$\|F_0 f\| \geqq \|f\| - \|f - F_0 f\| \geqq \|f\| - \delta\|f\| = (1 - \delta)\|f\|,$$

and therefore $\|T^{-1}\| \leqq \dfrac{1}{1 - \delta}$.

The inequality $\|E_0 - F_0 E_0\| \leqq \delta$ shows that F_0 is near to E_0 on ran E_0; the next step is to show that F_0 is near to E_0 on ran$^\perp E_0$. Suppose therefore that $f \perp$ ran E_0, i.e., that $E_0 f = 0$, and write $g = T^{-1} F_0 f$. Since $g \in$ ran E_0, it follows that $F_0 g = Tg = F_0 f$, or $F_0 E_0 g = F_0 f$; note that $\|g\| \leqq \dfrac{1}{1-\delta}\|f\|$. Since $\|F_0 f - E_0 g\| \leqq$

$$\leqq \|F_0 f - F_0 E_0 g\| + \|F_0 E_0 g - E_0 g\| \leqq \delta\|g\| \leqq \frac{\delta}{1-\delta}\|f\|, \text{ it follows that}$$

$$\|F_0 f\|^2 = (F_0 f, f) \leqq |F_0 f - E_0 g, f)| + |(E_0 g, f)| \leqq \frac{\delta}{1-\delta}\|f\|^2$$

$((E_0 g, f) = 0$ because $E_0 f = 0)$, and hence that

$$\|F_0(1 - E_0)\| \leqq \left(\frac{\delta}{1-\delta}\right)^{1/2}.$$

This inequality together with $\|E_0 - F_0 E_0\| \leqq \delta$ yields

$$\|E_0 - F_0\| \leqq \delta + \left(\frac{\delta}{1-\delta}\right)^{1/2} = \gamma.$$

Lemma 1 is now applicable. Choose δ small enough to make sure that $\gamma < 1$ and conclude that there exists a unitary operator W such that $W^* E_0 W = F_0$ and $\|1 - W\| \leqq 2\sqrt{\gamma}$. Write $E = W E_n W^*$. Since $F_0 \leqq E_n$, it follows that $E_0 \leqq E$; all that remains is to verify that E can be forced to be within ε of being invariant under A. That is easy; since

$$\|AE - EAE\| = \|A(W E_n W^*) - (W E_n W^*)A(W E_n W^*)\|,$$

and since the right hand term depends continuously on W, it follows that if W is chosen sufficiently near to 1 (i.e., if δ is chosen sufficiently small), then the right

19 A

hand term can be made arbitrarily near to $\|AE_n - E_nAE_n\|$, within $\varepsilon/2$ of it, say. Since $\|AE_n - E_nAE_n\| < \delta$, it might now be necessary to make δ a little smaller still, so as to guarantee $\delta < \varepsilon/2$; after this modification it will follow that, indeed, $\|AE - EAE\| < \varepsilon$.

The first definition of quasitriangularity (Δ_1) is quite hard ever to disprove; how does one show that there does not exist a sequence with the required properties? Theorem 2 makes the job easier. For an example suppose that $\{e_0, e_1, e_2, \ldots\}$ is an orthonormal basis and let U be the corresponding unilateral shift. The properties of U that will be needed are that it is an isometry ($U^*U = 1$) whose adjoint has a non-trivial kernel ($U^*e_0 = 0$).

Theorem 3. *The unilateral shift is not quasitriangular.*

Proof. Let E_0 be the projection (of rank 1) onto e_0. The proof will show that if E is a projection of finite rank such that $E_0 \leqq E$ (i.e., $e_0 \in \operatorname{ran} E$), then $\|UE - EUE\| = 1$.

Put $D = UE - EUE = (1 - E)UE$. Clearly $\|D\| \leqq 1$; the problem is to prove the reverse inequality. Observe that $D^*D = EU^*(1 - E) \cdot (1 - E)UE = EU^*UE - EU^*EUE = E - (EUE)^*(EUE)$. The finite-dimensional space $\operatorname{ran} E$ reduces both E and EUE, and on its orthogonal complement both those operators vanish. It follows that if T is the restriction of EUE to $\operatorname{ran} E$, then $\|D^*D\| = \|1 - T^*T\|$; the symbol "1" here refers, of course, to the identity operator on $\operatorname{ran} E$.

Now use the assumption that $e_0 \in \operatorname{ran} E$ and observe that $T^*e_0 = EU^*Ee_0 = EU^*e_0 = 0$. Since T^* is an operator on a finite-dimensional space and has a non-trivial kernel, the same is true of T^*T. (The falsity of this implication on infinite-dimensional spaces is shown by U itself.) If f is a unit vector in $\ker T^*T$, then $\|(1 - T^*T)f\| = 1$, and therefore $\|1 - T^*T\| \geqq 1$; the proof of the theorem is complete.

Section 3

It is not difficult to see that a polynomial in a quasitriangular operator is quasitriangular. Suppose indeed that $\{E_n\}$ is a sequence of projections such that $\|AE_n - E_nAE_n\| \to 0$, and let p be a polynomial. Since $AE_n - E_nAE_n$ is linear in A, it is sufficient to prove the assertion for monomials, $p(z) = z^k$, and that can be done by induction. The case $k = 1$ is covered by the hypothesis. (Note icidentally that constant terms can come and go with impunity: $(A + \lambda)E_n - E_n(A + \lambda)E_n = AE_n - E_nAE_n$.) The induction step from k to $k + 1$ is implied by the identity:

$$(1 - E_n)A^{k+1}E_n = \left((1 - E_n)A^{k+1}E_n - (1 - E_n)AE_nA^kE_n\right) + (1 - E_n)AE_nA^kE_n =$$

$$= (1 - E_n)A\left((1 - E_n)A^kE_n\right) + \left((1 - E_n)AE_n\right)A^kE_n.$$

W. B. ARVESON has proved that an operator similar to a quasitriangular one is also quasitriangular. The result of the preceding paragraph and ARVESON's result are closure properties of the set of all quasitriangular operators. The next two results are of the same kind.

Theorem 4. *A countable direct sum of quasitriangular operators is quasi-triangular.*

Proof. Suppose that for each $j(=1, 2, 3, \ldots)$ $A^{(j)}$ is an operator and $\{E_n^{(j)}\}$ is a sequence of projections of finite rank such that $\|A^{(j)} E_n^{(j)} - E_n^{(j)} A^{(j)} E_n^{(j)}\| \to 0$ as $n \to \infty$. Write $A = A^{(1)} \oplus A^{(2)} \oplus \ldots$. For each fixed k, find n_k so that $\|A^{(j)} E_n^{(j)} - E_n^{(j)} A^{(j)} E_n^{(j)}\| < \dfrac{1}{k}$ when $1 \leq j \leq k$ and $n \geq n_k$; write $E_k = E_{n_k}^{(1)} \oplus \ldots \oplus E_{n_k}^{(k)} \oplus 0 \oplus 0 \oplus \ldots$.

The E_k's are projections of finite rank. Since $E_{n_k}^{(j)} \to 1$ as $k \to \infty$ (strong topology) for each j, it follows that

$$E_k \langle f^{(1)}, f^{(2)}, f^{(3)}, \ldots \rangle \to \langle f^{(1)}, f^{(2)}, f^{(3)}, \ldots \rangle$$

whenever the vector $\langle f^{(1)}, f^{(2)}, f^{(3)}, \ldots \rangle$ is finitely non-zero. The boundedness of the sequence $\{E_k\}$ implies that $E_k \to 1$ (strong topology). Since $\|AE_k - E_k A E_k\| =$

$$= \max \{\|A^{(j)} E_{n_k}^{(j)} - E_{n_k}^{(j)} A^{(j)} E_{n_k}^{(j)}\|: \ j = 1, \ldots, k\} < \dfrac{1}{k}, \quad \text{the proof is complete.}$$

Theorem 5. *The set of quasitriangular operators is closed in the norm topology.*

Proof. Suppose that A_n is quasitriangular and $\|A_n - A\| \to 0$. Given a positive number ε and a projection E_0 of finite rank, find n_0 so that $\|A - A_{n_0}\| < \varepsilon/3$, and then find a projection E of finite rank such that $E_0 \leq E$ and $\|A_{n_0} E - E A_{n_0} E\| < \varepsilon/3$. It follows that $\|AE - EAE\| \leq \|AE - A_{n_0} E\| + \|A_{n_0} E - E A_{n_0} E\| + \|E A_{n_0} E - EAE\| < \varepsilon$.

Theorem 4 implies (and it is obvious anyway) that (on a separable Hilbert space, as always) every diagonal operator is quasitriangular. Since every normal operator is in the closure of the set of diagonal operators, Theorem 5 implies that every normal operator is quasitriangular.

A similar application of Theorem 5 shows that every compact operator is quasitriangular; what is needed is the easy observation that every operator of finite rank is quasitriangular. For compact operators, however, more is true; not only does there exist a well behaved sequence of projections, but in fact all "large" sequences are well behaved. That is: if A is compact and if $\{E_n\}$ is a sequence of projections such that $E_n \to 1$ (strong topology), then $\|AE_n - E_n A E_n\| \to 0$. The following formulation in terms of the directed set of projections of finite rank is more elegant; the assertion is that lim inf can be replaced by lim.

Lemma 2. *If A is compact, then $\lim\limits_{E \to 1} \|AE - EAE\| = 0$.*

Proof. Given a positive number ε, find an operator F of finite rank such that $\|A - F\| < \varepsilon/2$, and then find a projection E_0 of finite rank such that $FE_0 = E_0 F = F$. If E is a projection of finite rank such that $E_0 \leqq E$, then

$$\|AE - EAE\| \leqq \|AE - FE\| + \|FE - EFE\| + \|EFE - EAE\| < \varepsilon.$$

Lemma 2 implies that an operator of the form $A + K$, where A is quasitriangular and K is compact, is quasitriangular; in particular so is every operator of the form $N + K$, where N is normal and K is compact.

Still other quasitriangular operators of interest have arisen in the various generalizations of the Aronszajn—Smith theorem on invariant subspaces of compact operators. Thus, for instance, a crucial step in the treatment of polynomially compact operators [5] is the proof that every polynomially compact operator with a cyclic vector is quasitriangular. In their generalization of the invariant subspace theorem for polynomially compact operators, ARVESON and FELDMAN [2] need and prove the statement that every quasinilpotent operator with a cyclic vector is quasitriangular.

Section 4

Quasitriangular operators first arose in connection with the invariant subspace problem, but their status in that connection is still not settled.

Question 1. *Does every quasitriangular operator have a non-trivial invariant subspace?*

Experience with compact and polynomially compact operators suggests that the answer to Question 1 is yes. On the other hand, if the answer is yes, then it follows that every quasinilpotent operator has a non-trivial invariant subspace. Since it is a not unreasonable guess that the general invariant subspace question is equivalent to the one for quasinilpotent operators, and since the answer to the general invariant subspace question is more likely no than yes, the compact and polynomially compact experience comes under suspicion.

PETER ROSENTHAL suggested a more concrete way of connecting Question 1 with quasinilpotent operators. It is quite a reasonable conjecture that the spectrum of every unicellular operator is a singleton. (An operator is unicellular if its lattice of invariant subspaces is a chain.) Every transitive operator is obviously unicellular. (An operator is transitive if it has no non-trivial invariant subspaces.) The truth of the conjecture would imply therefore that, except for an additive scalar, every transitive operator is quasinilpotent, and hence, once again, an affirmative answer to Question 1 would imply an affirmative answer to the general invariant subspace question.

Question 2. *If the direct sum of two operators is quasitriangular, are both summands quasitriangular?*

This question is due to CARL PEARCY. He has proved that if $A \oplus 0$ is quasi-triangular, then A must be, but the general case is open. An interesting related question concerns the unilateral shift U: is $U \oplus U^*$ quasitriangular? If the answer to Question 2 is yes, then the answer to this question about U must be no. What is known, as a special case of PEARCY's result, is that $U \oplus 0$ is not quasitriangular.

Question 3. *Is it true for every operator that either it or its adjoint is quasi-triangular?*

The only example presented above of an operator that is not quasitriangular is the unilateral shift U; a glance at the matrix of U proves that U^* is quasitriangular. If the answer to Question 3 is yes, then Question 1 is equivalent to the general in-variant subspace question. Since $U \oplus U^*$ is unitarily equivalent to its own adjoint, it follows that an affirmative answer to Question 3 would imply that $U \oplus U^*$ is quasi-triangular, and, therefore, that the answer to Question 2 is no. There are other interesting and unknown special cases of Question 3. Thus, for instance, by an improvement of the argument that proved that U is not quasitriangular, PEARCY has obtained a large class of operators that are not quasitriangular; one of them is $3U + U^*$. It is not known whether the adjoint of that operator is quasitriangular.

References

[1] N. ARONSZAJN and K. T. SMITH, Invariant subspaces of completely continuous operators, *Ann. of Math.*, **60** (1954), 345—350.

[2] W. B. ARVESON and J. FELDMAN, A note on invariant subspaces, *Mich. Math. J.*, **15** (1968), 61—64.

[3] A. R. BERNSTEIN and A. ROBINSON, Solution of an invariant subspace problem of K. T. Smith and P. R. Halmos, *Pac. J. Math.*, **16** (1966), 421—341.

[4] H. O. CORDES, On a class of C*-algebras, *Math. Ann.*, **170** (1967), 283—313.

[5] P. R. HALMOS, Invariant subspaces of polynomially compact operators, *Pac. J. Math.*, **16** (1966), 433—437.

[6] P. R. HALMOS, Irreducible operators, *Mich. Math. J.*, **15** (1968), 215—223.

[7] B. SZ.-NAGY, *Spektraldarstellung linearer Transformationen des Hilbertschen Raumes* (Berlin, 1942).

UNIVERSITY OF MICHIGAN

(Received February 6, 1968)

INVARIANT SUBSPACES 1969[1]

P. R. HALMOS

CONTENTS

PREFACE

This is not a book. It is a set of lecture notes. I hope that with them the lectures will be easier to understand and to remember than without them. They are not, however, designed to be a complete exposition of anything. Readers who were not present at the lectures (and who are therefore not supposed to be reading this anyway) are hereby forbidden to complain about anything.

Many hard theorems are casually stated between commas (on the way to something else), and several easy lemmas are proved in perhaps exhausting detail (to illustrate a method). I tried not to make any mistakes, but I must have, and, sight unseen, I apologize for them.

The subject of these lectures is a part of operator theory on Hilbert space. The basic prerequisites are standard linear algebra (through Jordan form theory); measure theory (through the completeness of L^2); general topology (through convergence for nets); analytic function theory (the elements of the convergence theory for power series); and the most elementary definitions and facts of lattice theory. The special prerequisite is friendship with Hilbert space. The listener (or reader) will be lost if he doesn't know about orthonormal bases, Fourier coefficients, the formation of intersections and spans of subspaces, projections, Hermitian, unitary, and, more generally, normal operators,

---

[1]Lectures delivered at the 7th Colóquio Brasileiro de Matemática, Poços de Caldas, July, 1969.

isometries, spectra, and the general notion of unitary equivalence; these notions pervade everything. Knowledge of the spectral theorem would be a big help some of the time. More special topics (e.g., the weak, strong, and norm topologies for operators, and compact operators) are needed in one or two important places; isolated references to such refinements as, say, the Schatten–von Neumann classes may be skipped with no harm done.

There are few references to the literature: the bibliography of (7) is intended to take their place. The bibliography at the end of these notes is disjoint from that of (7).

P. R. H.

1. MOSTLY ABOUT NORMAL OPERATORS

The main reason for the success of finite-dimensional complex linear algebra is the existence of eigenvalues and eigenvectors. An eigenvector of a transformation spans a 1-dimensional subspace invariant under the transformation; it makes possible a study of the transformation on the whole space via a study of its behavior on smaller and therefore more manageable subspaces. The existence of eigenvalues is a deep fact, derived by techniques far from the spirit of linear algebra. What it comes down to is that an eigenvalue, a geometric concept, is the same as a zero of the characteristic polynomial, an algebraic concept, and the existence of such zeroes is guaranteed by the fundamental theorem of algebra, an analytic tool.

Neither the methods nor the results hinted at by the preceding paragraph extend to Hilbert space, which is the simplest, most natural, and most useful infinite-dimensional generalization of finite-dimensional vector spaces. There is no simple generalization of characteristic polynomials; what there is has no useful relation to eigenvalues; there are transformations that have no eigenvectors at all, and, for all anyone knows, non-trivial invariant subspaces may fail to exist.

That, in fact, is the invariant subspace problem. If A is an operator (bounded linear transformation) on a separable infinite-dimensional Hilbert space H, does there exist a subspace M of H (closed linear manifold) different from both 0 and H and such that M is invariant under A ($AM \subset M$)? If the topological restrictions (boundedness and closure) are removed, the answer (yes) becomes easy. If the space is too small (finite-dimensional), the answer is classical; if the space is too large (non-separable), the answer (yes again) is trivial.

It is convenient to have a single word to describe an operator whose only invariant subspaces are the trivial ones; such operators are called *transitive*. The invariant subspace problem asks about the existence of transitive operators. The most probable answer is that they do exist. What is wanted is not just an example, but a structure theory: how many transitive operators are there, what do they all look like, and how is an arbitrary operator made up of transitive pieces? The first step, however, is to find one, and, failing that, to gather as much experimental evidence about where such operators can and cannot be found as possible. Partial positive results and detailed studies of concrete special operators have constituted a large part of operator theory in the last few decades. Like Fermat's last conjecture, the invariant subspace problem has already proved itself to be of major importance by the new concepts and results that it has suggested.

For Hermitian operators the spectral theorem yields the existence of many non-trivial invariant subspaces, and multiplicity theory gives complete insight into their structure. For normal operators, the spectral theorem is still applicable, and it goes far enough to solve the invariant subspace problem, but many questions about the structure of the invariant subspaces of normal operators remain unanswered.

There are two natural ways to make small operators out of large ones, called compression and restriction; their application even to the best-behaved operators, the normal ones, yields operators for which the invariant subspace problem is unsolved.

The compression of an operator A on H to a subspace H_0 of H is the operator A_0 on H_0 defined as follows: for each f in H_0, define $A_0 f$ by forming Af and, if it is not in H_0, projecting it back there. If, in other words, P_0 is the projection with range H_0, then $A_0 = P_0 A | H_0$ (the vertical bar denotes restriction), or, equivalently, $A_0 = P_0 A P_0 | H_0$. It is not surprising that there is no satisfactory theory for the compressions of normal operators: the fact is that every operator can be obtained that way. It might be hoped, however, that if A, H, and H_0 are sufficiently special, endowed with additional structure not commonly encountered in operator theory, then A_0 may be tractable.

A classically important H is $L^2(C)$, where C is the unit circle $\{z: |z| = 1\}$ with normalized Lebesgue measure. In this Hilbert space, the functions defined by $e_n(z) = z^n$, $n = 0, \pm 1, \pm 2, \ldots$, constitute an orthonormal basis; a distinguished role is played by the subspace H_0 spanned by $\{e_0, e_1, e_2, \ldots\}$. The *Toeplitz operator* T_φ induced by a bounded measurable function φ on C is, by definition, the compression to H_0 of the multiplication operator induced by φ on H. In other words, $T_\varphi f = P_0(\varphi f)$ for each f in H. Is there a transitive Toeplitz operator? The answer is not known.

Restriction is the best known special case of compression, the special case in which H_0 is invariant under A. If that is so, then the projection P_0 is not necessary, and A_0 ($= A | H_0$) is called a *part* of A. A *subnormal operator* is, by definition, a part of a normal operator. Is there a transitive subnormal operator? The answer probably depends on some quite delicate considerations in classical analytic function theory; at present, it is not known. The invariant subspace problem for subnormal operators is especially important because some vague plausibility considerations suggest that it may be equivalent to the problem for arbitrary operators.

Normal, and, in particular, Hermitian, operators are not transitive; for Toeplitz and subnormal operators the problem is unsolved. Two further items deserve mention here, a known theorem and an unknown counterexample. The theorem is of a curious, isolated kind; it says that if $\|A\| \le 1$ (or, more generally, if A is power-bounded, in the sense that the set $\{\|A^n\|: n = 1, 2, 3, \ldots\}$ is bounded), and if neither A^n nor A^{*n} tends strongly to 0, then A is *not* transitive [(14), p. 74]. The example (due to Errett Bishop) is the operator A on $L^2(0, 1)$ defined by $Af(x) = xf(Tx)$, where T is an ergodic measure-preserving transformation on $(0, 1)$, e.g., $Tx = x + \alpha \bmod 1$ for some irrational α. What makes the example noteworthy is that it is a candidate for a transitive operator; so far, in any case, no one has found a non-trivial subspace that is invariant under it.

2. COMPACTNESS ASSUMPTIONS

Except for the special cases mentioned above, all intransitivity (= non-transitivity) theorems have something to do with compact (= completely continuous) operators. The first theorem of this kind was found, and then lost, by von Neumann; it asserts that every compact operator on a separable infinite-dimensional Hilbert space has a non-trivial invariant subspace. The proof seemed, for a long time, to be inelastic; every attempt to push it in a new direction made it spring a leak somewhere. Thus, for instance, the square root conjecture remained open for many years: according to that conjecture not only are all compact operators intransitive, but so are all their square roots. (In other words: if A^2 is compact, then A is intransitive.) The conjecture turned out to be true, and it is, in fact, a special case of the following interesting theorem: if p is a non-zero polynomial such that $p(A)$ is compact, then A is intransitive. The proof (Bernstein–Robinson) makes use of metamathematical concepts such as non-standard models of higher order predicate languages, but, for-tunately, it turned out to be possible to translate it into standard analysis.

The Bernstein–Robinson theorem can be generalized in two directions: one generalizes the polynomial and the other the Hilbert space. If, for instance, $p(A)$ is compact for some non-zero entire function p, then A is intransitive. This interesting statement has an almost trivial proof. Indeed, since $p \ne 0$, p can have only finitely many zeroes in the spectrum of A. Factor p into a polynomial q that has just those zeroes, with the appropriate multiplicities, and the quotient r that does not vanish on the spectrum of A. It follows that $r(A)$ is invertible, so that $q(A) = p(A)r(A)^{-1}$, and therefore $q(A)$ is compact; the assertion for the entire function p now follows from the known result for the polynomial q. A similar conclusion can be proved even if all that is assumed is the existence of a function p that is analytic and non-vanishing on some neighborhood of the spectrum of A and is such that $p(A)$ is compact; but that takes a little more argument (P. Meyer–Nieberg, unpublished).

As for generalizing the Hilbert space, the first step was taken by Bernstein himself; he proved the Bernstein–Robinson theorem for Banach spaces (4). Other proofs of the same result were given by Meyer–Nieberg (13) and Hsu (9). The two generalizations (analytic functions instead of polynomials and Banach spaces instead of Hilbert spaces) can be carried out at the same time; Meyer–Nieberg's unpublished note does that too.

A different direction of generalization was first indicated by J. Feldman and later proved, by an improved technique, by W. B. Arveson (3). The theorem is that if A is quasinilpotent ($\|A^n\|^{1/n} \to 0$) and if the norm-closed algebra generated by A and 1 contains a non-zero compact operator (i.e., there exists a sequence $\{p_n\}$ of polynomials such that $p_n(A)$ is norm-convergent to a non-zero compact operator), then A is not transitive. It is not immediately obvious that the Bernstein–Robinson theorem can be derived from this result, but there is a short and elementary argument that does just that. (To see the idea, assume that A is compact and derive Aronszajn–Smith from Arveson–Feldman as follows: if the spectrum of A contains a number $\lambda \neq 0$, then λ is an eigenvalue of A and a corresponding eigenvector yields an invariant subspace; if the spectrum of A consists of 0 alone, then Arveson–Feldman applies.) The Arveson–Feldman theorem can, in turn, be extended to Banach spaces (1), (6).

The basic ideas of all these generalizations are the same as the ones used in the Aronszajn–Smith proof for compact operators. What makes the proof work is that compact operators can be approximated by operators of finite rank. What follows is a proof of the Aronszajn–Smith theorem.

If $A = 0$, everything is trivial. If $A \neq 0$, let e be a unit vector such that $(Ae, e) \neq 0$. The span of the vectors $e, Ae, A^2e, \ldots$ is an invariant subspace; hence there is no loss of generality in assuming that their span is H. There is also no loss of generality in assuming that they are linearly independent; in the contrary case H is finite-dimensional. Let $\{e_1, e_2, e_3, \ldots\}$ be the orthonormalized version of $\{e, Ae, A^2e, \ldots\}$. The matrix of A (with respect to the orthonormal basis $\{e_1, e_2, e_3, \ldots\}$ of course) is quasitriangular in the sense that only zero entries occur below the diagonal that is just below the main one. That sub-main diagonal, moreover, tends to 0. Reason: $a_{n+1,n} = (Ae_n, e_{n+1})$, $e_n \to 0$ (weakly), and therefore, since A is compact, $Ae_n \to 0$ (strongly). This implies that if H_n is the span of $\{e_1, \ldots, e_n\}$, then the sequence $\{H_1, H_2, \ldots\}$ is approximately invariant under A in the following sense: if $E_n = \operatorname{pro} H_n$ (= the projection with range H_n), then $\|AE_n - E_nAE_n\| \to 0$. Proof: the matrix of $AE_n - E_nAE_n$ has only one non-zero entry, namely $a_{n+1,n}$.

The next step is a finite-dimensional lemma that will presently be applied to the operator E_nAE_n restricted to the space H_n. Note that $e \in H_n$ for all n, and $E_nAE_ne = Ae$ as soon as $n > 1$. For typographical simplicity (in this paragraph only), denote the operator by A and the space by H. The lemma asserts that if $n > 1$, then there exist projections P and Q such that (i) $AP = PAP$, $AQ = QAQ$, (ii) $P \leq Q$, $\operatorname{rank}(Q - P) = 1$, and (iii) $|(APe, e)| \leq \frac{1}{2}|(Ae, e)| \leq |(AQe, e)|$. Condition (i) says that P and Q are invariant under A; (ii) says that P and Q are near to each other; (ii) and (iii) together say that P and Q are not too near the trivial extremes—they are in some sense halfway between. To prove the lemma, put A in triangular form. In geometric terms, find a chain of subspaces invariant under A, starting with the trivial subspace 0 and ending with H, so that each one (after the first trivial one) includes its predecessor, and so that each one is of dimension just 1 greater than that predecessor. Project e into each of these subspaces, apply A to the results, and then form the modulus of the inner product with e. The resulting sequence need not be monotone in any sense, but it is a finite sequence of non-negative numbers beginning with 0 and ending with $|(Ae, e)|$. There is a last one among them less than or equal to $\frac{1}{2}|(Ae, e)|$ and the next is the first one greater than or equal to $\frac{1}{2}|(Ae, e)|$. Let P and Q be the projections onto the invariant subspaces corresponding to these stages.

For each n apply the lemma of the preceding paragraph to $E_n A E_n$ to find projections P_n and Q_n ($\leq E_n$) with the properties there described. Assume with no loss of generality (subsequences), that the sequences $\{P_n\}$ and $\{Q_n\}$ are weakly convergent, to P and Q, say; these limiting operators may, of course, fail to be projections. The compactness of A implies that $P_n A P_n \to PAP$ and $Q_n A Q_n \to QAQ$ (weakly); since $AP_n \to AP$ and $AQ_n \to AQ$ (strongly, but weak is all that is needed), it follows that

(i) $$AP - PAP = AQ - QAQ = 0.$$

(This is the main use made of the compactness of A.) Since the order relation between operators is defined in terms of inner products, and is therefore weakly continuous, and since rank is weakly lower semicontinuous, it follows that

(ii) $$P \leq Q, \qquad \operatorname{rank}(Q - P) \leq 1.$$

The numerical inequalities are obviously preserved under passage to the limit; they become

(iii) $$|(APe, e)| \leq \tfrac{1}{2}|(Ae, e)| \leq |(AQe, e)|.$$

The construction of a non-trivial invariant subspace for A is now accessible: it turns out that one or the other of $\ker(1 - P)$ and $\ker(1 - Q)$ will do. Clearly both are subspaces; let P^+ and Q^+ be their projections. It is to be proved that both P^+ and Q^+ are invariant under A, that $P^+ \neq 1$ and $Q^+ \neq 0$, and, finally, that either $P^+ \neq 0$ or $Q^+ \neq 1$.

If $f \in \ker(1 - P)$, so that $Pf = f$, then $P_n f \to f$ (strongly). (This is a well known and elementary result; $P_n f \to f$ weakly and $\|P_n f\|^2 = (P_n f, f) \to (f, f) = \|f\|^2$.) Hence $AP_n f \to Af$ (strongly) and therefore $P_n A P_n f \to PAf$ (weakly). It follows that

$$(AP_n - P_n A P_n)f \to Af - PAf = (1 - P)Af.$$

Since $\|AP_n - P_n A P_n\| \to 0$ (this is an easy consequence of $P_n \leq E_n$ and $\|AE_n - E_n A E_n\| \to 0$), it follows that $(1 - P)Af = 0$, i.e., that $Af \in \ker(1 - P)$. In other words, $\ker(1 - P)$ is invariant under A. The proof for Q is the same.

Since $|(APe, e)| \leq \tfrac{1}{2}|(Ae, e)|$ and $(Ae, e) \neq 0$, it follows that $Pe \neq e$, and hence that $P^+ \neq 1$.

Similarly, since $\tfrac{1}{2}|(Ae, e)| \leq |(AQe, e)|$ and $(Ae, e) \neq 0$, it follows that $AQe \neq 0$; since $(1 - Q)AQe = (AQ - QAQ)e = 0$ [by (i)], so that $AQe \in \ker(1 - Q)$, it follows that $Q^+ \neq 0$.

It remains to prove that either $P^+ \neq 0$ or $Q^+ \neq 1$, or, in other words, that if $Q^+ = 1$, then $P^+ \neq 0$. Suppose, accordingly, that $Q^+ = 1$, so that $\ker(1 - Q) = H$, i.e., $Q = 1$. Since $\operatorname{rank}((1 - P) - (1 - Q)) = \operatorname{rank}(Q - P) \leq 1$ [by (ii)], it follows that $\text{rank}(1 - P) \leqq 1$. Since rank is dimension of (closure of) range, and, therefore, codimension of kernel, it follows that $\ker(1 - P)$ is at least a hyperplane, and hence (since $\dim H \neq 1$) that $P^+ \neq 0$. The proof of the Aronszajn–Smith theorem is complete.

The last thing to be mentioned in this bird's-eye-view summary of the principal known facts about invariant subspaces has to do with compactness also. The easiest statement is this: if $A = B + iC$, where B and C are Hermitian and C has finite rank, then A is not transitive. The result can be improved by requiring merely that C be a Hilbert–Schmidt operator, or, still more generally, that C belong to the Schatten–von Neumann class of index p, $1 \leq p < \infty$ (18). A recent generalization gets the same intransitivity conclusion for operators of the form $B + C$, where B is normal and has a "thin" spectrum and C is as above (12). The most desirable result along these lines would require just that C be compact, but that desirable result is not known. "Polar" forms of these "Cartesian" results (and questions) exist: they concern contractions ($\|A\| \leq 1$), with the smallness condition on $1 - A^*A$. That is, if $1 - A^*A$ has finite rank, or belongs to a Schatten–von Neumann class of index p with $1 \leq p < \infty$, then A is not transitive; if $1 - A^*A$ is merely compact, the result is not known. The proofs of all known results along these lines are delicate and analytically difficult. For further details and references, especially to the Russian work, see the book of Nagy and Foiaş (14).

3. SHIFTS

What can be said about the existence of invariant subspaces for an operator A that satisfies no compactness conditions? There is clearly no loss of generality in assuming that A is a contraction, and it is convenient to do so. If neither $\{A^n\}$ nor $\{A^{*n}\}$ tends strongly to 0, then the Nagy–Foiaş theorem guarantees intransitivity; hence, from the present point of view, there is no loss of generality in assuming that $A^n \to 0$ strongly. Much recent work has concentrated attention on contractions whose powers tend strongly to 0. The approach is first to find certain "universal" operators of that kind, and then to try to get the necessary information about those universal operators.

Given an operator A on H and a vector f in H, it is natural to consider the orbit of f under S, i.e., the sequence $\langle f, Af, A^2f, \dots \rangle$. If f is replaced by Af, the sequence is shifted back by one step, i.e., it is replaced by $\langle Af, A^2f, A^3f, \dots \rangle$. This suggests that A can be interpreted to act on the direct sum $H \oplus H \oplus \cdots$, or on at least a part of it. The suggestion seems, however, to be technically quite inadequate. There is no reason why the sequence $\langle f, Af, A^2f, \dots \rangle$ should belong to the direct sum (the series $\Sigma_n \|A^n f\|^2$ need not converge), and, even if it does, the correspondence between f and $\langle f, Af, A^2f, \dots \rangle$ may be far from a unitary equivalence (even if $\Sigma_n \|A^n f\|^2$ converges, its sum will be equal to $\|f\|^2$ only in case $Af = 0$).

The inspiration that removes these difficulties is to apply to each term of the sequence $\langle f, Af, A^2f, \dots \rangle$ an operator T, so that the resulting sequence of square norms converges to $\|f\|^2$ the easy way, by telescoping. That is, replace $\langle f, Af, A^2f, \ldots \rangle$ by $\langle Tf, TAf, TA^2f, \ldots \rangle$, so that

$$\|Tf\|^2 = \|f\|^2 - \|Af\|^2,$$

$$\|TAf\|^2 = \|Af\|^2 - \|A^2f\|^2,$$

$$\|TA^2f\|^2 = \|A^2f\|^2 - \|A^3f\|^2, \qquad \text{etc.}$$

The first of these equations alone, if required to hold for all f, implies that $T^*T = 1 - A^*A$, and, conversely, if $T^*T = 1 - A^*A$, then all the equations hold.

The preceding paragraphs were intended as motivation. For a rigorous attack, proceed as follows. Since A is a contraction, $1 - A^*A$ is positive. Let T be its positive square root, and let R be the closure of the range of T. Form the direct sum $K = R \oplus R \oplus \cdots$. If $f \in H$, then $TA^nf \in R$ for all n, and

$$\sum_{n=0}^{k} \|TA^nf\|^2 = \sum_{n=0}^{k} ((1 - A^*A)A^nf, A^nf)$$

$$= \sum_{n=0}^{k} (\|A^nf\|^2 - \|A^{n+1}f\|^2)$$

$$= \|f\|^2 - \|A^{k+1}f\|^2.$$

Since $\|A^{k+1}f\| \to 0$ by assumption, it follows that if $f \in H$, and if a mapping V is defined by $Vf = \langle Tf, TAf, TA^2f, \ldots \rangle$, then V is an isometric embedding of H into K.

Consider next the *shift* U on K defined by $U\langle f_0, f_1, f_2, \ldots \rangle = \langle 0, f_0, f_1, f_2, \ldots \rangle$. It is easy to verify that $U^*\langle f_0, f_1, f_2, \ldots \rangle = \langle f_1, f_2, f_3, \ldots \rangle$. It follows that $VAf = U^*Vf$ for all f. Since the image of H under V is a subspace of K that is invariant under U^*, the conclusion is that A is unitarily equivalent to a part of U^*.

Shifts such as U, and the corresponding backward shifts, such as U^*, occur in many parts of Hilbert space theory. From the present point of view, the latter are the universal operators promised above, and the program of the immediate sequel is to study shifts and their adjoints. If it were known that every invariant subspace of U^*, of dimension greater than 1, had a non-trivial invariant subsubspace, the invariant subspace problem would be solved. Nothing that strong can be said, of course, but the study of the invariant subspaces of U^* (or U—it comes to the same thing) is still a promising enterprise.

The simplest U is called the *simple* unilateral shift; it is defined on l^2 (the direct sum of 1-dimensional Hilbert spaces) by $U\langle \xi_0, \xi_1, \xi_2, \ldots \rangle = \langle 0, \xi_0, \xi_1, \xi_2, \ldots \rangle$. In the present lectures, this U only will be studied in some detail; properties of the others will be briefly hinted at as generalizations of properties of this one.

Observe to begin with that the unilateral shift is an isometry. The range of U is not l^2 but a proper subspace of l^2, the subspace of vectors with vanishing first coordinate. The existence of an isometry whose range is not the whole space is characteristic of infinite-dimensional spaces.

If e_n is the vector $\langle \xi_0, \xi_1, \xi_2, \ldots \rangle$ for which $\xi_n = 1$ and $\xi_i = 0$ whenever $i \neq n$ ($n = 0, 1, 2, \ldots$), then the e_n's form an orthonormal basis for l^2. The effect of U on this basis is described by

$$Ue_n = e_{n+1} \ (n = 0, 1, 2, \ldots).$$

These equations uniquely determine U, and in most of the study of U they may be taken as its definition.

Except for size (dimension) one Hilbert space is very like another. To make a Hilbert space more interesting than its neighbors, it is necessary to enrich it by the addition of some external structure. One of the most important Hilbert spaces, known as H^2 ("H" is for Hardy this time), endowed with some structure not usually found in a Hilbert space, is defined as follows.

As once before, let C be the perimeter of the unit circle in the complex plane, with normalized Lebesgue measure μ, and write $e_n(z) = z^n$ ($n = 0, \pm 1, \pm 2, \ldots$). (Finite linear combinations of the e_n's are called trigonometric polynomials.) The space H^2 is, by definition, the subspace of L^2 spanned by the e_n's with $n \geq 0$; equivalently H^2 is the orthogonal complement in L^2 of $\{e_{-1}, e_{-2}, e_{-3}, \ldots\}$.

Fourier expansions with respect to the orthonormal basis $\{e_n : n = 0, \pm 1, \pm 2, \ldots\}$ are formally similar to the Laurent expansions that occur in analytic function theory. The analogy motivates calling the functions in H^2 the *analytic* elements of L^2. A subset of H^2 (a linear manifold but not a subspace) of considerable technical significance is the set H^∞ of bounded functions in H^2; equivalently, H^∞ is the set of all those functions f in L^∞ for which $\int f e_n \, d\mu = 0$ ($n = 1, 2, 3, \ldots$). Similarly H^1 is the set of all those elements of L^1 for which these same equations hold. What gives these Hardy spaces their special flavor is the structure of the semigroup of non-negative integers within the additive group of all integers.

It is customary to speak of the elements of spaces such as H^1, H^2, and H^∞ as functions. The custom is not likely to lead its user astray, so long as the qualification "almost everywhere" is kept in mind at all times. Thus "bounded" means "essentially bounded", and, similarly, all statements such as "$f = 0$" or "f is real" or "$|f| = 1$" are to be interpreted, when asserted, as holding almost everywhere.

Some authors define the Hardy spaces so as to make them honest function spaces (consisting of functions analytic on the unit disc). In that approach the almost everywhere difficulties are still present, but they are pushed elsewhere; they appear in questions (which must be asked and answered) about the limiting behavior of the functions on the boundary.

The connection between H^2 and analytic functions is most easily seen as follows. If $f \in H^2$, with Fourier expansion $f = \sum_{n=0}^\infty \alpha_n e_n$, then $\sum_{n=0}^\infty |\alpha_n|^2 < \infty$, and therefore the radius of convergence of the power series $\sum_{n=0}^\infty \alpha_n^2 z^n$ is greater than or equal to 1. It follows from the usual expression for the radius of convergence in terms of the coefficients that the power series $\sum_{n=0}^\infty \alpha_n z^n$ defines an analytic function $\hat{f}$ in the open unit disc D. The mapping $f \to \hat{f}$ (obviously linear) establishes a one-to-one correspondence between H^2 and the set of those functions analytic in D whose series of Taylor coefficients is square summable. The important properties of this mapping can be deduced from the fact that if φ is an analytic function in the open unit disc, $\varphi(z) = \sum_{n=0}^{\infty} \alpha_n z^n$, and if $\varphi_r(z) = \varphi(rz)$ for $0 < r < 1$ and $|z| = 1$, then $\varphi_r \in H^2$ for each r; the series $\sum_{n=0}^{\infty} |\alpha_n|^2$ converges if and only if the norms $|\varphi_r|$ are bounded. Many authors define H^2 to be the set of analytic functions in the unit disc with square-summable Taylor series, or, equivalently, with bounded concentric L^2 norms.

The algebraic properties of the Hardy spaces are a curious mixture of the properties of the corresponding Lebesgue spaces and the properties of analytic functions. Thus, for instance, if f is a real function in H^2, then f is a constant. The product of two functions in H^2 is in H^1, and, conversely, every function in H^1 is the product of two functions in H^2. If a function f in H^2 vanishes on a set of positive measure, then $f = 0$. Some of these assertions lie on the surface and others are much deeper; the last one, in particular, is a non-trivial part of the celebrated theorem of the brothers F. and M. Riesz.

In the space H^2 the effect of shifting the natural basis $\{e_0, e_1, e_2, \ldots\}$ forward by one index is the same as the effect of multiplication by e_1. In other words, the (simple) unilateral shift is the same as the multiplication operator on H^2 defined by $(Uf)(z) = zf(z)$. To say that it is the "same", and, in fact, to speak of "the" unilateral shift is a slight abuse of language. Properly speaking, the unilateral shift is a unitary equivalence class of operators, but no confusion will result from regarding it as one operator with many different manifestations.

A close relative of the unilateral shift is the *bilateral* shift. To define it, let H be the Hilbert space of all two-way (bilateral) square-summable sequences. The elements of H are most conveniently written in the form

$$\langle \ldots, \xi_{-2}, \xi_{-1}, (\xi_0), \xi_1, \xi_2, \ldots \rangle;$$

the term in parentheses indicates the one corresponding to the index 0. The bilateral shift is the operator W on H defined by

$$W\langle \ldots, \xi_{-2}, \xi_{-1}, (\xi_0), \xi_1, \xi_2, \ldots \rangle = \langle \ldots, \xi_{-3}, \xi_{-2}, (\xi_{-1}), \xi_0, \xi_1, \ldots \rangle.$$

Linearity is obvious, and boundedness is true with room to spare; the bilateral shift, like the unilateral one, is an isometry. Since the range of the bilateral shift is the entire space H, it is even unitary.

If e_n is the vector $\langle \ldots, \xi_{-1}, (\xi_0), \xi_1, \ldots \rangle$ for which $\xi_n = 1$ and $\xi_i = 0$ whenever $i \neq n$ ($n = 0, \pm 1, \pm 2, \ldots$), then the e_n's form an orthonormal basis for H. The effect of W on this basis is described by

$$We_n = e_{n+1} \quad (n = 0, \pm 1, \pm 2, \ldots).$$

The bilateral shift is a unitary operator; the unilateral shift is a non-unitary isometry. Since a direct sum (finite or infinite) of isometries is an isometry, the direct sum of a unitary operator and a number of copies (finite or infinite) of the unilateral shift is an isometry. (There is no point in forming direct sums of unitary operators—they are no more unitary than the summands.) The useful theorem along these lines is that that is the only way to get isometries.

Precisely, every isometry is either unitary, or a direct sum of one or more copies of the unilateral shift, or a direct sum of a unitary operator and some copies of the unilateral shift. It follows that the unilateral shift is more than just an example of an isometry, with interesting and peculiar properties; it is in fact one of the fundamental building blocks out of which all isometries are constructed. An isometry for which the unitary direct summand is absent is called *pure*. According to the fundamental decomposition theorem just stated, the pure isometries are exactly the direct sums of copies of the unilateral shift.

If U is an isometry on a Hilbert space H, and if there exists a unit vector e_0 in H such that the vectors $e_0, Ue_0, U^2e_0, \ldots$ form an orthonormal basis for H, then (obviously) U is unitarily equivalent to the unilateral shift, or, by the accepted abuse of language, U is the unilateral shift. This characterization of the unilateral shift can be reformulated as follows: U is an isometry on a Hilbert space H for which there exists a 1-dimensional subspace N such that the subspaces $N, UN, U^2N, \ldots$ are pairwise orthogonal and span H. If there is such a subspace N, then it must be equal to the *co-range* $(UH)^\perp$. In view of this comment, another slight reformulation is possible: the unilateral shift is an isometry U of *co-rank* 1 on a Hilbert space H such that the subspaces $(UH)^\perp, U(UH)^\perp, U^2(UH)^\perp, \ldots$ span H. (Since U is an isometry, it follows that they must be pairwise orthogonal.)

A generalization lies near at hand. Consider an isometry U on a Hilbert space H such that the subspaces $(UH)^\perp, U(UH)^\perp, \ldots$ are pairwise orthogonal and span H, but make no demands on the value of the co-rank. Every such isometry is a shift (a unilateral shift). The co-rank of a shift (also called its multiplicity) constitutes a complete set of unitary invariants for it; the original unilateral shift is determined (to within unitary equivalence) as the shift of multiplicity 1. The universality theorem discussed before says that every contraction A whose powers tend strongly to 0 is a part of the adjoint of a shift whose multiplicity is the rank of $1 - A^*A$.

4. THE BEURLING THEORY

At this point the study of shifts and their invariant subspaces acquires three branches, characterized by the multiplicity. The easiest (and still far from trivial) branch is the one for which the multiplicity is 1. In that case (i.e., if U is the simple unilateral shift) the best possible result is true: every invariant subspace of U^* (of dimension greater than 1) has a non-trivial invariant subsubspace. The proof of this will be outlined below. Another branch is characterized by the values of the multiplicity strictly between 1 and $\aleph_0$. The desired result is true here too, but it is much harder. The names associated with the work are Potapov and Helson:, most of Helson's book is devoted to this topic. The third branch corresponds to the multiplicity $\aleph_0$; in this case, the principal problem is still unsolved.

A subspace N is called a *wandering* subspace for U (the simple unilateral shift) if $N \perp U^kN$ for $k = 1, 2, 3, \ldots$. Since U is an isometry, it follows that

$U^n N \perp U^k N$ whenever h and k are distinct non-negative integers. Wandering subspaces are important because they are connected with invariant subspaces, in this sense: there is a natural one-to-one correspondence between wandering subspaces N and invariant subspaces M, given by setting $M = \bigvee_{k=0}^{\infty} U^k N$.

Observe that $UM = \bigvee_{k=1}^{\infty} U^k N$, so that $N = M \cap (UM)^{\perp}$; this proves that N is determined by M, and hence that the correspondence is one-to-one. The proof that the correspondence is onto all invariant subspaces is a slightly more detailed analysis of the equations that define M and N in terms of one another.

One thing that is not obvious from the definition is that every non-zero wandering subspace of U has dimension 1. To prove that, it is convenient to regard the unilateral shift U as the restriction to H^2 of the bilateral shift W on the larger space L^2. If f and g are orthogonal unit vectors in a wandering subspace N, then the set of all vectors of either of the forms $W^n f$ or $W^m g$ ($n, m = 0, +1, +2, \ldots$) is an orthonormal set in L^2. It follows that

$$2 = \|f\|^2 + \|g\|^2 = \sum_n |(f, e_n)|^2 + \sum_m |(g, e_m)|^2$$

$$= \sum_n |(f, W^n e_0)|^2 + \sum_m |(g, W^m e_0)|^2$$

$$= \sum_n |(W^{*n} f, e_0)|^2 + \sum_m |(W^{*m} g, e_0)|^2 \leq \|e_0\|^2 = 1.$$

(The inequality is Bessel's.) The absurdity shows that f and g cannot coexist.

The principal result about invariant subspaces of the shift is due to Beurling. It asserts that a non-zero subspace M of H^2 is invariant under U if and only if there exists a function φ in H^{∞}, of constant modulus 1 almost everywhere, such that M is the range of the restriction to H^2 of the multiplication induced by φ. In more informal language, M can be described as the set of all *multiples* of φ (multiples by functions in H^2, that is). Correspondingly it is suggestive to write $M = \varphi H^2$. For no very compelling reasons functions such as φ (functions in H^{∞}, of constant modulus 1) are called *inner* functions.

To prove Beurling's theorem, suppose first that $M = \varphi H^2$. Then $UM = e_1 M = e_1 \varphi H^2 = \varphi e_1 H^2 \subset \varphi H^2 = M$:, this proves the "if". To prove "only if", suppose that M is invariant under U and represent M in the form $\bigvee_{n=0}^{\infty} U^n N$, where N is a wandering subspace for U. Take a unit vector φ in N. Since, by assumption, $(U^n \varphi, \varphi) = 0$ when $n > 0$, or $\int e_n |\varphi|^2 \, d\mu = 0$ when $n > 0$, it follows (by the formation of complex conjugates) that $\int e_n |\varphi|^2 \, d\mu = 0$ when $n < 0$, and hence that $|\varphi|^2$ is a function in L^1 such that all its Fourier coefficients with non-zero index vanish. Conclusion: φ is constant almost everywhere, and, since $\int |\varphi|^2 \, d\mu = 1$, the constant modulus of φ must be 1. Since N has dimension 1, so that φ by itself spans N, the functions φe_n ($n = 0, 1, 2, \ldots$) span M. Equivalently, the set of all functions of the form φp, where p is a polynomial, spans M. Since multiplication by φ (restricted to H^2) is an isometry, its range is closed; since M is the span of the image under that isometry of a dense set, it follows that M is in fact equal to the range of that isometry, and hence that $M = \varphi H^2$. The proof of Beurling's theorem is complete.

There is a useful relation between the ordering of invariant subspaces and the divisibility properties of inner functions. The assertion is that if φ and ψ are inner functions such that $\varphi H^2 \subset \psi H^2$, then φ is divisible by ψ, in the sense that there exists an inner function θ such that $\varphi = \psi \theta$; if, conversely, φ is divisible by ψ, then $\varphi H^2 \subset \psi H^2$. Indeed, if $\varphi H^2 \subset \psi H^2$, then $\varphi = \varphi e_0 = \psi f$ for some f in H^2; since $f = \varphi \psi^*$, it follows then $|f| = 1$, so that f is an inner function. The converse is trivial. If, moreover, $\varphi H^2 = \psi H^2$, then φ and ψ are constant multiples of one another, by constants of modulus 1. To prove this, it is sufficient to prove that if both θ and θ^* are inner functions, then θ is a constant. Indeed, then both $\operatorname{Re} \theta$ and $\operatorname{Im} \theta$ are real functions in H^2, and therefore both $\operatorname{Re} \theta$ and $\operatorname{Im} \theta$ are constants.

Beurling's theorem answers all invariant subspace questions about the unilateral shift in the mathematician's peculiar sense: the questions are all reduced to questions about inner functions. At this point, fortunately, classical complex function theory comes to the rescue: a satisfactory characterization of inner functions exists.

To begin with, it is not at all obvious that there are many inner functions. A trivial example is $\varphi(z) = 1$ (i.e., $\varphi = e_0$), and other examples, not much more inspired, are the e_n's, $n = 1, 2, 3, \dots$. There are still other examples, somewhat less trivial, namely, the functions φ defined by

$$\varphi(z) = \frac{z - \alpha}{1 - \alpha^* z},$$

where $|\alpha| < 1$. Since the domain of the functions under consideration is the perimeter of the unit circle ($|z| = 1$), the denominator causes no trouble; since, moreover, $|1 - \alpha^* z| = |z(z^* - \alpha^*)| = |z - \alpha|$, it is clear that $|\varphi| = 1$. The expansion (Fourier or Taylor)

$$\frac{1}{1 - \alpha^* z} = 1 + \alpha^* z + \alpha^{*2} z^2 + \cdots$$

implies that $\varphi \in H^2$, and hence, indeed, φ is an inner function.

The product of two (or, for that matter, of any finite number of) inner functions is again an inner function, and with a modicum of attention to the courtesies of convergence, the same is true of infinite products. This remark, together with the preceding paragraph, yields an infinitely large supply of inner functions; the ones so obtained are called *Blaschke products*.

Not every inner function is a Blaschke product. There is an analytically complicated procedure that yields many others; it goes as follows. Let μ be a singular positive measure on the unit circle, and write

$$\varphi(z) = \exp\left(-\int \frac{\lambda + z}{\lambda - z} d\mu(\lambda)\right).$$

The proof that such a *singular* function is an inner function is a relatively easy piece of classical analysis; what is harder and deeper is the assertion that every inner function is a product of the form $\beta \sigma$, where β is a Blaschke product and σ is a singular function (and, of course, one or another of the two factors may be absent).

With the aid of the factorization theorem in the preceding paragraph, the solution of the invariant subspace problem for parts of the adjoint of the simple unilateral shift becomes easy to settle. The assertion that every invariant subspace of U^*, of dimension greater than 1, has a non-trivial invariant subsubspace, translates, in terms of U, to the assertion that every invariant subspace of U, of co-dimension greater than 1, has a non-trivial invariant supersubspace. In terms of inner functions this becomes: every inner function, other than a single Blaschke factor, has a non-trivial factorization into further inner functions, or, in other words, the Blaschke factors are the only primes. The proof is obvious. If φ is an inner function, $\varphi = \beta\sigma$, and if both factors are present, nothing more needs to be said. If only β is present, factorization is still possible, except only in the excluded case where β consists of a single factor. If, finally, only σ is present, then factorizability is trivial; the singular function determined by the singular measure μ is, for instance, the square of the singular function determined by $\frac{1}{2}\mu$.

5. THE DEFINITION OF QUASITRIANGULAR OPERATORS

The considerations that follow were originally motivated by the Aronszajn–Smith technique of treating compact operators. Begin with the observation that every square matrix with complex entries is unitarily equivalent to a triangular one. In other words, if A is an operator on a finite-dimensional Hilbert space H, then there exists an increasing sequence $\{M_n\}$ of subspaces such that dim $M_n = n$ $(n = 0, \ldots, \dim H)$, and such that each M_n is invariant under A. On a Hilbert space of dimension $\aleph_0$, the appropriate definition is this: A is *triangular* if there exists an increasing sequence $\{M_n\}$ of finite-dimensional subspaces whose union spans H such that each M_n is invariant under A. It is easy, but not obviously desirable, to fill in the dimension gaps, and hence to justify the added assumption that dim $M_n = n$ $(n = 0, 1, 2, \ldots)$.

In many considerations of invariant subspaces $(AM \subset M)$, it is convenient to treat their projections instead $(AE = EAE)$. In terms of projections, a necessary and sufficient condition that an operator A on a separable Hilbert space H be triangular is that

(Δ) there exists an increasing sequence $\{E_n\}$ of projections of finite rank such that $E_n \to 1$ (strong topology) and such that $AE_n - E_nAE_n = 0$ for all n.

This formulation suggests an asymptotic generalization of itself. An operator A is *quasitriangular* if

(Δ_1) there exists an increasing sequence $\{E_n\}$ of projections of finite rank such that $E_n \to 1$ (strong topology) and such that $\|AE_n - E_nAE_n\| \to 0$. (Informally, E_n is approximately invariant under A.)

The concept (but not the name) has been seen before; it plays a central role in the proofs of the Aronszajn–Smith theorem on the existence of invariant subspaces for compact operators, and in the proofs of its various known generalizations. Quasitriangular operators were officially introduced in (8).

It is interesting and useful to examine a variant of the condition (Δ_1); the variant requires that

(Δ_2) there exists a sequence $\{E_n\}$ of projections of finite rank such that $E_n \to 1$ (strong topology) and such that $\|AE_n - E_n AE_n\| \to 0$.

The only difference between Δ_1 and Δ_2 is that the latter does not require the sequence $\{E_n\}$ to be increasing.

There is still another pertinent condition. The set of all projections of finite rank, ordered by range inclusion, is a directed set. Since $E \mapsto \|AE - EAE\|$ is a net on that directed set, it makes sense to say that

(Δ_0)
$$\liminf_{E \to 1} \|AE - EAE\| = 0.$$

What it means is that for every positive number ε and for every projection E_0 of finite rank there exists a projection E of finite rank such that $E_0 \leqq E$ and $\|AE - EAE\| < \varepsilon$.

Sequences of approximately invariant projections that are not required to be "large" always exist. A precise statement is this: for each operator A, there exists a sequence $\{E_n\}$ of non-zero projections of finite rank such that $\|AE_n - E_n AE_n\| \to 0$; in fact, the E_n's can be chosen to have rank 1. The proof is immediate from the existence of approximate eigenvalues and eigenvectors. Let λ be a scalar and $\{e_n\}$ a sequence of unit vectors such that $\|Ae_n - \lambda e_n\| \to 0$. If the projections E_n are defined by $E_n f = (f, e_n)e_n$, then

$$(AE_n - E_n AE_n)f = (f, e_n)(Ae_n - (Ae_n, e_n)e_n).$$

Since $(Ae_n, e_n) \to \lambda$, it follows that $\|AE_n - E_n AE_n\| \to 0$.

Since every operator has approximately invariant projections of rank 1, it is tempting to conclude, via the formation of finite spans, that every operator on an infinite-dimensional Hilbert space has approximately invariant projections of arbitrarily large finite ranks. The theory of approximate invariance turns out, however, to be surprisingly delicate. It is, for instance, not true that the span of two approximate eigenvectors is approximately invariant. More precisely, there exists a 3×3 matrix A and there exist two projections F and G of rank 1 such that F is invariant under A, G is nearly invariant under A, but if $E = F \vee G$, then $\|AE - EAE\| = 1$. In detail, set

$$A = \begin{pmatrix} 0 & 0 & 0 \\ 0 & 0 & 0 \\ 1 & 0 & 0 \end{pmatrix},$$

let F be the projection onto $\langle 0, 1, 0 \rangle$, and let G be the projection onto $\langle a, b, 0 \rangle$, where $|a|^2 + |b|^2 = 1$ and a is "small" (but not 0). It is easy to verify that

$$F = \begin{pmatrix} 0 & 0 & 0 \\ 0 & 1 & 0 \\ 0 & 0 & 0 \end{pmatrix}, \qquad G = \begin{pmatrix} |a|^2 & ab^* & 0 \\ a^*b & |b|^2 & 0 \\ 0 & 0 & 0 \end{pmatrix}, \qquad E = \begin{pmatrix} 1 & 0 & 0 \\ 0 & 1 & 0 \\ 0 & 0 & 0 \end{pmatrix},$$

$\|AF - FAF\| = 0$, $\|AG - GAG\| = |a|$, and $\|AE - EAE\| = 1$.

This example, informal in its interpretation of "nearly invariant", can be used to construct an example of two sequences of approximately invariant projections, in the precise technical sense, such that the sequence of their spans is not approximately invariant, as follows. Let H be the direct sum $H_1 \oplus H_2 \oplus \ldots$ of 3-dimensional spaces such as played a role in the preceding paragraph, and let the operator A on H be the direct sum $A_1 \oplus A_2 \oplus \ldots$ of the corresponding operators. Let F_n be the direct sum projection whose summand with index n is the previous F and whose other summands are 0; let G_n be the direct sum projection whose summand with index n is the previous G with $a = 1/n$ and whose other summands are 0. It follows that $\|AF_n - F_n AF_n\| = 0$ for all n, $\|AG_n - G_n AG_n\| \to 0$, and, if $E_n = F_n \vee G_n$, then $\|AE_n - E_n AE_n\| = 1$.

It is slightly surprising that despite the negative evidence of the preceding example, approximately invariant projections of arbitrarily large ranks always do exist. If A is an operator on an infinite-dimensional Hilbert space. ε is a positive number, and n is a positive integer, then there exists a projection E of rank n such that $\|AE - EAE\| < \varepsilon$.

For $n = 1$, the result was derived from the existence of approximate eigenvectors. The idea of the inductive proof that follows is that although near invariance is not preserved by the formation of spans, it is preserved by the formation of orthogonal spans. Given ε and n, assume the result for n, and let F be a projection of rank n such that $\|AF - FAF\| < \varepsilon/2$. Since the compression of A to $\mathrm{ran}(1 - F)$ (i.e., the restriction of $(1 - F)A(1 - F)$ to $\mathrm{ran}\,(1 - F)$) has approximately invariant projections of rank 1, it follows that there exists a projection G of rank 1 such that $G \perp F$ and

$$\|(1 - F)A(1 - F)G - G(1 - F)A(1 - F)G\| < \frac{\varepsilon}{2}.$$

(Find G on $\mathrm{ran}(1 - F)$ first and then extend it by defining it to be 0 on $\mathrm{ran}\,F$.) Since $G(1 - F) = (1 - F)G = G$, the last inequality is equivalent to

$$\|(1 - F)(1 - G)AG\| < \varepsilon/2.$$

If $E = F + G$, then E is a projection of rank $n + 1$ and

$$\|AE - EAE\| = \|(1 - E)AE\| = \|(1 - F)(1 - G)A(F + G)\|$$
$$= \|(1 - G)(1 - F)AF + (1 - F)(1 - G)AG\|$$
$$\le \|(1 - F)AF\| + \|(1 - F(1 - G)AG\| < \varepsilon.$$

A diagram may help to clarify this proof; the idea of the diagram is to think of A as a matrix.

	0	
$\dfrac{\varepsilon}{2}$	$\dfrac{\varepsilon}{2}$	

Note that the non-increasing character of the projections obtained is emphasized by the proof: to obtain a projection for a smaller ε, it is necessary to keep backtracking.

It is trivial that the definition of quasitriangularity (Δ_1) implies the weakened form (Δ_2) (obtained from Δ_1 by omitting the word "increasingly"). It is also quite easy to prove that if, on a separable Hilbert space, $\liminf_{E \to 1} \|AE - EAE\| = 0$ (Δ_0), then A is quasitriangular (Δ_1). Indeed, let $\{e_1, e_2, \ldots\}$ be an orthonormal basis for the space. By Δ_0, there exists a projection E_1 of finite rank such that $e_1 \in \operatorname{ran} E_1$ and $\|AE_1 - E_1AE_1\| < 1$. Again, by Δ_0, there exists a projection E_2 of finite rank such that $E_1 \leq E_2$, $e_2 \in \operatorname{ran} E_2$, and $\|AE_2 - E_2AE_2\| < \frac{1}{2}$. In general, inductively, use Δ_0 to get a projection E_{n+1} of finite rank such that $E_n \leq E_{n+1}$, $e_{n+1} \in \operatorname{ran} E_{n+1}$, and $\|AE_{n+1} - E_{n+1}AE_{n+1}\| < 1/(n+1)$. Conclusion: $\{E_n\}$ is an increasing sequence of projections of finite rank such that $E_n \to 1$ and such that $\|AE_n - E_nAE_n\| \to 0$; in other words, A is quasitriangular, as promised.

The non-trivial implication along these lines is the one from Δ_2 to Δ_0. The proof depends on a lemma according to which if two projections have the same finite rank and are near, then there is a "small" unitary operator that transforms one onto the other. (For unitary operators "small" means "near to 1".) A possible quantitative formulation goes as follows: if E and F are projections of the same finite rank such that $\|E - F\| = \varepsilon < 1$, then the infimum of $\|1 - W\|$, extended over all unitary operators W such that $W^*EW = F$, is not more than $2\sqrt{\varepsilon}$. The lemma can be improved but the improvement takes considerably more work and for present purposes it is not needed. A trivial improvement is to drop the assumption that E and F have the same rank and recapture it from the known result that the inequality $\|E - F\| < 1$ implies rank $E = $ rank F. Another qualitative improvement is to drop the assumption that the ranks are finite and pay for it by introducing partial isometries instead of unitary operators. The best kind of improvement is quantitative; the estimate $2\sqrt{\varepsilon}$ can be sharpened to $\sqrt{2}(1 - \sqrt{1 - \varepsilon^2})^{1/2}$. For a discussion of such results and references to related earlier work, see (5). Conjecturally the sharpened estimate is best possible, but the proof of that does not seem to be in the literature. The proof of the lemma is computational and not illuminating.

The ground is now prepared for the proof (or at least a sketch of the proof) of the principal result about quasitriangular operators. If $\{E_n\}$ is a sequence of projections of finite rank such that $E_n \to 1$ (strong topology) and such that $\|AE_n - E_nAE_n\| \to 0$, then $\liminf_{E \to 1} \|AE - EAE\| = 0$.

It is to be proved that given $\varepsilon > 0$ and a projection E_0 of finite rank, there exists a projection E of finite rank, $E \geq E_0$, such that $\|AE - EAE\| < \varepsilon$. Choose an orthonormal basis $\{e_1, \ldots, e_k\}$ for ran E_0. If n is sufficiently large, then $\|e_j - E_ne_j\|$ is small (since $E_n \to 1$ strongly) and $\|AE_n - E_nAE_n\|$ is small. The vectors $E_ne_1, \ldots, E_ne_k$ are linearly independent; let F_0 be the projection onto their span (of dimension k). Clearly $F_0 \leq E_n$. Then $\|E_0 - F_0E_0\|$ is small (because $E_0e_j - F_0E_0e_j = e_j - E_ne_j$). A small technical strengthening of the argument shows that $\|E_0 - F_0\|$ is small. Now apply the lemma: find W so that

$\|1 - W\|$ is small and $F_0 = W^*E_0W$. Transform E_n ($\geq F_0$) by W in the reverse direction, i.e., form WE_nW^* ($\geq E_0$); because W is nearly 1, the result is just as approximately invariant as E_n.

6. PROPERTIES OF QUASITRIANGULAR OPERATORS

The first definition of quasitriangularity (Δ_1) is quite difficult to disprove; how does one show that there does not exist a sequence with the required properties? The theorem just proved makes the job easier. For an example, suppose that $\{e_0, e_1, \ldots\}$ is an orthonormal basis and let U be the corresponding unilateral shift. The properties of U that will be needed are that it is an isometry ($U^*U = 1$) whose adjoint has a non-trivial kernel ($U^*e_0 = 0$). Assertion: the unilateral shift is not quasitriangular. Indeed, if E is a projection of finite rank such that $e_0 \in \text{ran } E$, then $\|UE - EUE\| = 1$. The way to prove this is to consider $D = UE - EUE$; then $D^*D = E - (EUE)^*(EUE)$. The operator D acts on the finite-dimensional space ran E; if T is EUE on ran E, then D on ran E is $1 - T^*T$. Since T^* has a non-trivial kernel (consider e_0), finite-dimensionality implies that T must have, and therefore so must T^*T. It follows that $\|1 - T^*T\| \geq 1$. q.e.d.

There are several good positive results about quasitriangular operators that are "closure" theorems: they assert that the set of quasitriangular operators is closed under certain operations. Here are some of them.

If A is quasitriangular, so is every polynomial in A.

If A is quasitriangular, so is every operator similar to A (Arveson).

The direct sum of countably many quasitriangular operators is quasitriangular. (Use the $E_n \to 1$ definition; use "long" sequences of projections "approximately" invariant for "many" direct summands.)

The set of quasitriangular operators is norm closed.

Every normal operator is quasitriangular. (Approximate by diagonal operators.)

Every compact operator is quasitriangular. (Approximate by operators of finite rank.)

Better: if A is compact and if $\{E_n\}$ is a sequence of projections of finite rank with $E_n \to 1$ (strong), then $\|AE_n - E_nAE_n\| \to 0$. Hence:

The sum of a quasitriangular operator and a compact operator (and, in particular, of a normal operator and a compact operator) is quasitriangular.

The theory of quasitriangular operators is by no means complete. R. G. Douglas and C. M. Pearcy have several results (not yet published) that indicate further directions in which the work could proceed. Some samples follow. (1) If a bounded (infinite) matrix with operator entries is upper triangular and has quasitriangular entries on the main diagonal, then it is quasitriangular. (This is a strong generalization of the theorem about countable direct sums of quasitriangular operators.) (2) Each quasinilpotent operator and each polynomially compact operator is quasitriangular. (3) Every operator with a finite spectrum is quasitriangular. The principal unsolved question, of course, is whether every quasitriangular operator is intransitive. Experience with compact and poly-nomially compact operators suggests that the answer is yes. On the other hand, if the answer is yes, then it follows that every quasinilpotent operator is intransitive, and hence (?!?) every operator; in view of this the compact and polynomially compact experience becomes suspicious.

Peter Rosenthal has suggested a more concrete way of connecting the principal question with quasinilpotent operators. It is quite a reasonable conjecture that the spectrum of every unicellular operator is a singleton. (An operator is called unicellular if its lattice of invariant subspaces is a chain.) Every transitive operator is obviously unicellular. The truth of the conjecture would imply therefore that, except for an additive scalar, every transitive operator is quasinilpotent, and hence, once again, an affirmative answer to the question would imply an affirmative answer to the general invariant subspace question.

The theory of quasitriangular operators makes close and interesting contact with some other parts of operator theory. The contact rests on a new char-acterization of quasitriangularity: an operator is quasitriangular if and only if it is the sum of a triangular operator and a compact one. The "if" is old. To prove "only if", suppose $E_n \uparrow 1$ and $\|AE_n - E_n AE_n\| \to 0$. Choose a subse-quence of $\{E_n\}$ if necessary, and, hence, assume with no loss of generality that $\Sigma_n \|AE_n - E_n AE_n\| < \infty$. Write T for the upper triangular part of the corre-sponding block matrix, i.e., $T = \Sigma_{i \leq j} (E_i - E_{i-1}) A (E_j - E_{j-1})$, and write C for the subdiagonal remainder, i.e., $\hat{C} = \Sigma_{i > j} (E_i - E_{i-1}) A (E_j - E_{j-1})$. It is not obvious by inspection that T and C are operators, and even less that C is compact, but the argument is easy. Indeed, if $C_n = E_{n+1}(AE_n - E_n AE_n)$, then $\|C_n\| \leq \|AE_n - E_n AE_n\|$, and therefore $\Sigma_n C_n$ is norm convergent. Since each C_n has finite rank, the sum is compact. Since $C_n = E_{n+1} AE_n - E_n AE_n$ (recall that $E_{n+1} \geq E_n) = (E_{n+1} - E_n) AE_n = (E_{n+1} - E_n) A \Sigma (E_j - E_{j-1})$ (with $E_0 = 0$), it is clear that $\Sigma_n C_n = C$; q.e.d.

The theory of quasitriangular operators can be paralleled by a theory of *quasidiagonal* operators. For a definition, modify the definition of quasitrian-gularity to demand $E_n \uparrow 1$ and $\|AE_n - E_n A\| \to 0$. It is easy analysis to prove that the set of quasidiagonal operators is closed. Corollary: every normal operator is quasidiagonal.

Next, every quasidiagonal A is a sum $D + C$, where D is "block diagonal" (i.e., a direct sum of finite matrices) and C is compact. For the proof, imitate the quasitriangular theory as follows. Put

$$C_n = E_{n+1}(AE_n - E_n A) - E_n(AE_n - E_n A) E_{n+1};$$

clearly, $\|C_n\| \leq 2\|AE_n - E_n A\|$, and therefore $\Sigma_n C_n$ converges in the norm to a compact C. Since

$$C_n = (E_{n+1} - E_n) AE_n + E_n A(E_{n+1} - E_n)$$

$$= (E_{n+1} - E_n) A \sum_{j=1}^{n} (E_n - E_{j-1}) + \sum_{i=1}^{n} (E_i - E_{i-1}) A(E_{n+1} - E_n),$$

therefore C is exactly the off-diagonal part of the block matrix A; q.e.d.

A famous theorem of H. Weyl is an immediate corollary: every Hermitian operator is the sum of a diagonal operator and a compact one (20). Indeed, all that is needed is to note, in the preceding paragraph, that if A is Hermitian, then each $(E_i - E_{i-1})A(E_i - E_{i-1})$ is diagonable. Weyl's theorem was sharpened by von Neumann: C can be made Hilbert–Schmidt, with arbitrarily small Hilbert–Schmidt norm (19). The quasidiagonal technique does not seem to yield the sharpened version. Open question: is the Weyl theorem generalizable to normal operators? (The Hermitian character of A was used in the proof in an essential, apparently unavoidable manner, in diagonalizing the blocks.) What about the von Neumann theorem?

7. THE GALOIS THEORY OF INVARIANT SUBSPACES

Invariant subspace theory for individual operators can and should be embedded in a larger context.

There is a Galois theory for operators on Hilbert space and their invariant subspaces. A subspace M of a complex Hilbert space H is invariant under a set $\mathcal{C}$ of operators on H if $AM \subset M$ for each A in $\mathcal{C}$. The intersection of two subspaces invariant under $\mathcal{C}$ is another such subspace, and the same is true of their span; the subspaces 0 and H are invariant under every $\mathcal{C}$. In other words, the set of all subspaces invariant under $\mathcal{C}$ is always a lattice containing 0 and H; it will be denoted in what follows by Lat $\mathcal{C}$. If, dually, $\mathcal{L}$ is a set of subspaces of H, an operator A on H leaves $\mathcal{L}$ invariant if $AM \subset M$ for each M in $\mathcal{L}$. A linear combination of two operators that leave $\mathcal{L}$ invariant is another such operator, and the same is true of their product; the identity operator 1 leaves every $\mathcal{L}$ invariant. In other words, the set of all operators that leave $\mathcal{L}$ invariant is always an algebra containing 1; it will be denoted in what follows by Alg $\mathcal{L}$.

The lattice $\mathcal{L}_{\max}$ of all subspaces of H and the algebra $\mathcal{C}_{\max}$ of all operators on H possess, in addition to their natural algebraic structures, natural topological structures also. The weak operator topology is what turns out to be useful in both cases. For $\mathcal{C}_{\max}$, the meaning of the phrase is clear. Since, for each set $\mathcal{L}$ of subspaces, weak limits of operators that leave $\mathcal{L}$ invariant continue to have that property, it follows that Alg $\mathcal{L}$ is always a closed subset of $\mathcal{C}_{\max}$. For brevity in what follows, an unmodified use of the word "algebra" will always refer to a weakly closed subalgebra of operators containing 1.

The weak topology for $\mathcal{L}_{\max}$ is most efficiently defined by associating with each subspace M the projection pro M with range M and transplanting the weak operator topology for the set of all projections to the set $\mathcal{L}_{\max}$ of all subspaces. Assertion: for each set $\mathcal{C}$ of operators, Lat $\mathcal{C}$ is always a closed subset of $\mathcal{L}_{\max}$. Suppose indeed that $\{P_n\}$ is a net of projections weakly convergent to a projection P and that ran $P_n \in$ Lat $\mathcal{C}$ for each n; it is to be proved that ran P also belongs to Lat $\mathcal{C}$. The assumption means that $AP_n = P_n AP_n$ for each n; the desired conclusion is that $AP = PAP$. The only delicate point in the proof is that since multiplication is not jointly continuous in its

268

two arguments, it is not immediately clear that $P_n A P_n \to PAP$ weakly. What saves the day is that if a net of projections converges weakly *to a projection*, then it converges strongly. For bounded nets, multiplication is jointly strongly continuous, and that proves the assertion. Caution: what was just proved is that the set of projections corresponding to the subspaces in Lat $\mathcal{C}$ is a weakly closed subset of the set of all projections. This does not imply that it is a weakly closed subset of the set of all operators, or equivalently, of the weakly compact unit ball in operator space, and hence, for instance, it is not legitimate to infer that Lat $\mathcal{C}$ is always weakly compact. Counterexample: if $\mathcal{C}$ consists of 0 alone, then Lat $\mathcal{C}$ is $\mathcal{L}_{max}$, and, indeed, the set of *all* projections is not weakly compact. The trouble is, of course, that the set of all projections is not weakly closed in the unit ball. In any event, Lat $\mathcal{C}$ is always weakly closed within $\mathcal{L}_{max}$ and, for brevity in what follows, an unmodified use of the word "lattice" will always refer to such a weakly closed lattice containing 0 and H.

The topological property of weak closure implies the lattice-theoretic property of completeness. That is, if a set $\mathcal{L}$ of subspaces is an algebraic lattice (assumed closed under the formation of only finite intersections and spans) and if $\mathcal{L}$ is weakly closed in $\mathcal{L}_{max}$, then $\mathcal{L}$ is complete (closed under the formation of arbitrary intersections and spans). Suppose indeed that $\mathcal{L}_0$ is a subset of $\mathcal{L}$, and consider the directed set of all finite subsets of $\mathcal{L}_0$ ordered by inclusion. If n is such a finite subset, let M_n be the intersection of the subspaces in n. Since $n \mapsto M_n$ is a decreasing net of subspaces, the corresponding net of projections is weakly (in fact, strongly) convergent to the projection whose range is $\cap \mathcal{L}_0$; this proves that $\cap \mathcal{L}_0 \in \mathcal{L}$. The proof for spans is the same. Caution: the converse is not true; a complete lattice need not be weakly closed in $\mathcal{L}_{max}$. Counterexample: if H is 2-dimensional, then 0 and H, together with an arbitrary set of 1-dimensional subspaces of H, constitute a complete lattice; if the set of 1-dimensional subspaces is not closed, e.g., if it consists of every 1-dimensional subspace except one, then the lattice is not closed.

Observe that if an algebra $\mathcal{C}$ is generated by a set $\mathcal{C}_0$, then Lat $\mathcal{C}_0 =$ Lat $\mathcal{C}$, and if a lattice $\mathcal{L}$ is generated by a set $\mathcal{L}_0$, then Alg $\mathcal{L}_0 =$ Alg $\mathcal{L}$. There is, consequently, no point in considering sets of operators that are not algebras or sets of subspaces that are not lattices.

The basic properties of Lat and Alg are easy and shallow parts of universal algebra. Both are order reversing:

$$\text{if } \mathcal{C}_1 \subset \mathcal{C}_2, \text{ then Lat } \mathcal{C}_2 \subset \text{ Lat } \mathcal{C}_1$$

and

$$\text{if } \mathcal{L}_1 \subset \mathcal{L}_2, \text{ then Alg } \mathcal{L}_2 \subset \text{ Alg } \mathcal{L}_1.$$

If $\mathcal{C}$ is an algebra, then

$$\mathcal{C} \subset \text{Alg Lat } \mathcal{C};$$

if, dually, $\mathcal{L}$ is a lattice, then

$$\mathcal{L} \subset \text{Lat Alg } \mathcal{L}.$$

Apply Lat to the first of these inclusions and apply the second inclusion to

Lat; the two inclusions so obtained imply that

$$\text{Lat } \mathcal{Q} = \text{Lat Alg Lat } \mathcal{Q}$$

for every algebra $\mathcal{Q}$; the dual argument proves that

$$\text{Alg } \mathcal{L} = \text{Alg Lat Alg } \mathcal{L}$$

for every lattice $\mathcal{L}$.

A fundamental problem of operator theory is to characterize the cases in which one or another of the four inclusions

$$\mathcal{Q} \subset \text{Alg Lat } \mathcal{Q} \subset \mathcal{Q}_{max}$$

and

$$\mathcal{L} \subset \text{Lat Alg } \mathcal{L} \subset \mathcal{L}_{max}$$

degenerates to equality. When, for instance, is $\text{Alg Lat } \mathcal{Q} = \mathcal{Q}_{max}$? That is, when is each subspace in $\text{Lat } \mathcal{Q}$ left invariant by every operator? Since the only subspaces left invariant by every operator are 0 and H, it follows that if $\text{Alg Lat } \mathcal{Q} = \mathcal{Q}_{max}$, then $\text{Lat } \mathcal{Q} = \mathcal{L}_{min}$ (where $\mathcal{L}_{min}$ is the lattice that consists of 0 and H only). The algebras $\mathcal{Q}$ for which $\text{Lat } \mathcal{Q} = \mathcal{L}_{min}$ are called *transitive*. The result just proved is that if $\text{Alg Lat } \mathcal{Q} = \mathcal{Q}_{max}$, then $\mathcal{Q}$ is transitive; the converse is trivial.

Not much is known about transitive algebras. If H is small (finite-dimensional), then Burnside's classical theorem applies [(10), p. 276] and says that the only transitive algebra is $\mathcal{Q}_{max}$. If H is very large (non-separable), then an easy cardinality argument shows that no countably generated algebra is transitive. The cases between these two extremes remain shrouded in mystery.

Consider next the possibility $\mathcal{Q} = \text{Alg Lat } \mathcal{Q}$. Analogy with linear algebra (when is an object equal to its own second dual?) suggests that an algebra satisfying this condition be called *reflexive*. The reflexive algebras are exactly the ones that can be obtained as the Alg of some lattice. Indeed, by definition, if $\mathcal{Q}$ is reflexive, then $\mathcal{Q} = \text{Alg } \mathcal{L}$ with $\mathcal{L} = \text{Lat } \mathcal{Q}$; if, conversely, $\mathcal{Q} = \text{Alg } \mathcal{L}$ for some $\mathcal{L}$, then $\text{Alg Lat } \mathcal{Q} = \text{Alg Lat Alg } \mathcal{L} = \text{Alg } \mathcal{L} = \mathcal{Q}$.

The theory of reflexive algebras is relatively new and, judging from the analytic techniques that it has been using, quite deep. A typical result is Sarason's (17): a commutative algebra of normal operators is reflexive. Example: the set of all matrices of the form $\begin{pmatrix} a & 0 \\ 0 & b \end{pmatrix}$. In a different direction, Radjavi and Rosenthal (15) generalized a theorem of Arveson's (2) as follows: if $\mathcal{Q}$ includes a maximal abelian self-adjoint algebra and $\text{Lat } \mathcal{Q}$ is a chain (totally ordered), then $\mathcal{Q}$ is reflexive. Example: the set of all matrices of the form $\begin{pmatrix} a & 0 \\ b & c \end{pmatrix}$. A typical example of a non-reflexive algebra is the set of all matrices of the form $\begin{pmatrix} a & 0 \\ b & a \end{pmatrix}$.

So much for algebras. For lattices, the same questions can be asked. When, for instance, is it true that $\text{Lat Alg } \mathcal{L} = \mathcal{L}_{max}$? Answer: if and only if every subspace is invariant under $\text{Alg } \mathcal{L}$. Since an easy and well known argument proves that the only operators that leave every subspace invariant are the scalars, it follows that if $\text{Lat Alg} \, \mathcal{L} = \mathcal{L}_{max}$, then $\text{Alg} \, \mathcal{L} = \mathcal{Q}_{min}$ (where $\mathcal{Q}_{min}$ is the algebra that consists of the scalar multiples of the identity only). The lattices $\mathcal{L}$ for which $\text{Alg} \, \mathcal{L} = \mathcal{Q}_{min}$ are called *transitive*. The result just proved is that if $\text{Lat Alg} \, \mathcal{L} = \mathcal{L}_{max}$, then $\mathcal{L}$ is transitive; the converse is trivial.

The lattice $\mathcal{L}_{max}$ is always transitive; if the underlying Hilbert space has dimension greater than 2, no other examples of transitive lattices are immediately obvious. The following interesting (unpublished) comment of J. E. McLaughlin shows that they exist nevertheless. Coordinatize the space, i.e., represent it as the L^2 of some measure space, and consider the set $\mathcal{L}$ of all those subspaces that are invariant under the formation of complex conjugates. It is easy to prove that $\mathcal{L}$ is a lattice. Essentially the same argument as is needed to prove that $\text{Alg} \, \mathcal{L}_{max} = \mathcal{Q}_{min}$ proves that the lattice $\mathcal{L}$ is transitive. Conjecture: every non-trivial transitive lattice on a space of dimension greater than 2 can be obtained this way.

8. REFLEXIVE LATTICES

The last problem in this circle of ideas is the study of lattices $\mathcal{L}$ such that $\mathcal{L} = \text{Lat Alg} \, \mathcal{L}$. Such a lattice is called *reflexive*. The reflexive lattices are exactly those that can be obtained as the Lat of some algebra; the proof is the obvious dual of the corresponding proof for reflexive algebras.

Trivial examples of both reflexive and non-reflexive lattices are easy to obtain in a space of dimension 2. The first non-trivial theorem on the subject is a sharpening of some related work of Kadison and Singer (11); the statement, proved in complete generality by Ringrose (16), is that every chain is reflexive. Here "chain" means a lattice whose order (in general possibly partial) is in fact total. The convention that a lattice is necessarily weakly closed (and therefore complete) is still in force, but it is pertinent to observe that here it is sufficient to assume completeness only: a complete chain of subspaces (projections) is necessarily weakly closed.

The Ringrose theorem shows that the action of Lat on algebras is much easier to study than its action on singletons. The existence of an operator A such that Lat A is isomorphic to the lattice of positive integers (with $+\infty$) takes some proving, and the existence of an A such that Lat A is isomorphic to the closed unit interval is a deep theorem in analysis. [For detailed references, see (7).] For most chains, attainability by singletons seems to be out of reach so far; as challenging special cases, the set of all integers (with $\pm\infty$) and the Cantor set deserve special mention.

The smallest lattice that is not a chain has the Hasse diagram $\lozenge$. Consider any realization of this lattice in H; i.e., consider a set $\mathcal{L}$ consisting of four subspaces 0, M, N, and H, with $M \cap N = 0$ and $M \vee N = H$. Assertion: $\mathcal{L}$ is a reflexive lattice. (Emphasis: this is true for all M and N satisfying the stated conditions.)

For the proof, consider the projections P and Q with ranges M and N respectively, and let $\mathcal{Q}$ be the algebra generated by the set of all operators of either of the forms $PA(1-Q)$ and $QA(1-P)$. It is easy to verify that $\mathcal{L} \subset \text{Lat } \mathcal{C}$; the assertion will be proved by establishing the reverse inclusion.

It is useful first to prove the following lemma: if $K \in \text{Lat } \mathcal{C}$ and $K \not\subset M$, then $N \subset K$ (whence, by symmetry, if $K \in \text{Lat } \mathcal{C}$ and $K \not\subset N$, then $M \subset K$). To prove the lemma, let u be a vector in K with $u \notin M$, so that $(1-P)u \neq 0$. Given an arbitrary vector v in N, find an operator A such that $A(1-P)u = v$. Then $QA(1-P)u = v$, and it follows, as desired, that $N \subset K$.

Now, to prove the original assertion, suppose that $K \in \text{Lat } \mathcal{C}$, $K \neq 0, M, N$; it is to be proved that $K = H$. Since $M \cap N = 0$, therefore either $K \not\subset M$ or $K \not\subset N$; it follows from the lemma that either $N \subset K$ or $M \subset K$. If $N \subset K$, then $K \not\subset N$ (because $K \neq N$), and therefore $M \subset K$, whence $K = H$; similarly, if $M \subset K$, then $N \subset K$, and $K = H$, and the proof is complete.

The lattices $\mathcal{L}$ just discussed are the simplest non-trivial Boolean algebras. Conjecturally, every complete Boolean algebra of projections is reflexive, but that result is not yet known. What a suitable refinement of the proof just given does show is that every complete *atomic* Boolean algebra of subspaces is reflexive.

The preceding discussion disposes of the reflexivity of all lattices with four or fewer elements. The smallest non-reflexive lattice (with five elements) has the Hasse diagram $\langle\!\!\mid\!\!\rangle$; a simple realization of it consists of three 1-dimensional subspaces in a 2-dimensional space H, together with 0 and H, of course. This example can be generalized. Suppose that H is a finite-dimensional Hilbert space, and suppose that M, N, and L are subspaces with $M \cap N = N \cap L = L \cap M = 0$ and $M \vee N = N \vee L = L \vee M = H$, and write $\mathcal{L} = \langle 0, M, N, L, H \rangle$. Assertion: the lattice $\mathcal{L}$ is not reflexive.

A simple special case is obtained by assuming that H is a direct sum $K \oplus K$, and letting M, N, and L, respectively, be the sets of all vectors of the forms $\langle f, 0 \rangle$, $\langle 0, f \rangle$, and $\langle f, f \rangle$. In this case the assertion is easy to prove, even for infinite-dimensional spaces. This provides the only known example of a non-reflexive lattice in an infinite-dimensional space. Whether every lattice algebraically isomorphic to this one is non-reflexive is unknown.

The next simplest case to consider is that of lattices whose Hasse diagram looks like $\langle\!\!\diamond\!\!\rangle$. The fact that every (finite) matrix is similar to its transpose implies that if A is an operator on a finite-dimensional space, then Lat A is algebraically isomorphic to its own dual. Since the five-element lattice under consideration does not have that property, it cannot be the lattice of any *single* operator. For algebras the situation is different. The pertinent general theorem is that the ordinal sum of two reflexive lattices is reflexive. (Incidentally, the same is true for direct sums.)

A curious role is played by lattices whose Hasse diagram is the pentagon $\langle\!\!\pentagon\!\!\rangle$. Because such lattices are prototypically non-modular, and because every lattice of subspaces of a finite-dimensional space is modular, the pentagon is not realizable in a finite-dimensional space. In infinite-dimensional spaces it is, and, in the one realization that has been studied so far, it turns out to be reflexive. Whether all its realizations are reflexive is not known.

BIBLIOGRAPHY

(1) C. Apostol, A theorem on invariant subspaces, *Bull. Acad. Polon. Sci.* **16** (1968), 181–183.

(2) W. B. Arveson, A density theorem for operator algebras, *Duke Math. J.* **34** (1967), 635–647.

(3) W. B. Arveson and J. Feldman, A note on invariant subspaces, *Mich. Math. J.* **15** (1968), 61–64.

(4) A. R. Bernstein, Invariant subspaces of polynomially compact operators, *Pac. J. Math.* **21** (1967), 445–464.

(5) H. O. Cordes, On a class of C^* algebras, *Math. Ann.* **170** (1967), 283–313.

(6) T. A. Gillespie, An invariant subspace theorem of J. Feldman, *Pac. J. Math.* **26** (1968), 67–72.

(7) P. R. Halmos, *A Hilbert Space Problem Book*, Van Nostrand, Princeton, NJ, 1967.

(8) P. R. Halmos, Quasitriangular operators, *Acta Szeged* **29** (1968) 283–293.

(9) N. H. Hsu, Invariant subspaces of polynomially compact operator in Banach spaces, *Yokohama Math. J.* **15** (1967), 11–15.

(10) N. Jacobson, *Lectures in Abstract Algebra*, Vol. II., Van Nostrand, Princeton, NJ, 1953.

(11) R. V. Kadison and I. M. Singer, Triangular operator algebras, *Amer. J. Math.* **82** (1960), 227–259.

(12) K. Kitano, Invariant subspaces of some non-selfadjoint operators, *Tohoku Math. J.* **20** (1968), 313–322.

(13) P. Meyer-Nieberg, Invariante Unterräume von polynomkompakten Operatoren, *Arch. Math.* **19** (1968), 180–182.

(14) B. Sz.-Nagy and C. Foiaş, *Analyse Harmonique des Opérateurs de l'Espace de Hilbert*, Adadémiai Kiadó, Budapest, 1967.

(15) H. Radjavi and P. Rosenthal, On invariant subspaces and weakly closed algebras, *Bull. A.M.S.* **74** (1968), 1013–1014.

(16) J. R. Ringrose, On some algebras of operators, *Proc. L.M.S.* **15** (1965), 61–83.

(17) D. E. Sarason, Invariant subspaces and unstarred operator algebras, *Pac. J. Math.* **17** (1966), 511–517.

(18) J. T. Schwartz, Subdiagonalization of operators in Hilbert space with compact imaginary part, *Comm. Pure Appl. Math.* **15** (1962), 159–172.

(19) J. von Neumann, *Charakterisierung des Spektrums eines Integraloperators*, Hermann, Paris, 1935.

(20) H. Weyl, Über beschränkte quadratische Formen deren Differenz vollstätig ist, *Rend. Circ. Mat. Palermo* **27** (1909), 373–392.

Reprinted from the
TRANSACTIONS OF THE AMERICAN MATHEMATICAL SOCIETY
Vol. 144, pp. 381-389, Oct. 1969

TWO SUBSPACES

BY

P. R. HALMOS

In the study of pairs of subspaces M and N in a Hilbert space H there are four thoroughly uninteresting cases, the ones in which both M and N are either 0 or H. In the most general case H is the direct sum of five subspaces:

$$M \cap N, \qquad M \cap N^\perp, \qquad M^\perp \cap N, \qquad M^\perp \cap N^\perp,$$

and the rest. The parts of M and N in the first four are "thoroughly uninteresting". In "the rest", the orthogonal complement of the span of the first four, M and N are in *generic position* ("position p" in [2]), in the sense that all four of the special intersections listed above are equal to 0. The purpose of this paper is to use graphs of linear transformations to represent pairs of subspaces in generic position. The results arose in the study of the invariant subspace lattices of operators and promise to be useful there. A more immediate by-product, described below, is a reasonably transparent new proof of Dixmier's theorem on the unitary equivalence of pairs of subspaces. Even aside from such external applications, however, the results answer at least one natural question and may be considered to be of geometric interest in their own right. Specialization to the finite-dimensional case makes neither the conclusions more obvious nor the proofs substantially simpler.

Axis-graph. Suppose that T is a closed but not necessarily bounded linear transformation on a dense subset of a Hilbert space K, with zero kernel and dense range. Write $H = K \oplus K$, let M be the "horizontal axis" consisting of all vectors of the form $\langle f, 0 \rangle$ in H, and let N be the graph of T, i.e., the set of all vectors of the form $\langle f, Tf \rangle$ in H. Assertion: M and N are in generic position. The first step of the proof is to show that $M \cap N = 0$. Indeed, how can an $\langle f, 0 \rangle$ be equal to a $\langle g, Tg \rangle$? Answer: only if $Tg = 0$, whence $g = 0$ (because ker $T = 0$), and therefore $f = 0$. To prove the rest of the assertion, it is necessary to know $M^\perp$ (trivial: all $\langle 0, f \rangle$) and $N^\perp$ (easy and standard computation: all $\langle -T^*f, f \rangle$). From this it is easy to deduce that $M^\perp \cap N^\perp = 0$: since ran T is dense, it follows that ker $T^* = 0$, and the proof just given applies again. The equations $M \cap N^\perp = 0$ and $M^\perp \cap N = 0$ are trivial.

The construction of the preceding paragraph is not new; it has been used, for instance, to exhibit pairs of subspaces whose vector sum is different from their closed span [4, p. 110], [5, p. 26]. The first result of the present paper is that this way of constructing pairs of subspaces in generic position is, to within unitary equivalence, the only way. To say that a pair $\langle M_1, N_1 \rangle$ of subspaces is unitarily

---

Presented to the Society, August 29, 1969; received by the editors April 7, 1969.

381

equivalent to a pair $\langle M_2, N_2 \rangle$ means, of course, that there exists a unitary operator U such that $UM_1 = M_2$ and $UN_1 = N_2$. In case M_1 and N_1 are in one Hilbert space H_1 and M_2 and N_2 in another Hilbert space H_2, then the requirement that U be unitary is to be interpreted to mean that U is an isometry from H_1 onto H_2.

THEOREM 1. *If M and N are subspaces in generic position in a Hilbert space H, then there exists a Hilbert space K, and there exists a closed linear transformation T on K with zero kernel and dense range, such that the pair $\langle K \oplus 0, \text{graph } T \rangle$ is unitarily equivalent to the pair $\langle M, N \rangle$.*

Proof. Let P be the projection with range M. Assertion: the restriction $P|N$ of P to N maps N one-to-one onto a linear manifold dense in M. Suppose, indeed, that $Pg = 0$ for some g in N. It follows that $g \in M^\perp \cap N$, and hence (generic position) that $g = 0$: this proves that the kernel of $P|N$ is 0. To prove that PN is dense in M, suppose that $f \in M$ and $f \perp PN$. This means that if $g \in N$, then $0 = (f, Pg) = (Pf, g) = (f, g)$, so that $f \in M \cap N^\perp$. It follows (generic position) that $f = 0$; the proof of the assertion is complete.

The existence of a transformation with the properties just proved for $P|N$ implies [5, p. 27] that M and N have the same dimension. Since M and $M^\perp$ on the one hand and N and $N^\perp$ on the other hand enter the hypotheses with perfect symmetry, it follows that all four of these subspaces have the same dimension. Since this applies to M and $M^\perp$ in particular, there exists an isometric mapping I from M onto $M^\perp$.

Everything is now prepared for the necessary definitions. Write

$$K = M,$$

define T on the dense subset PN of M by

$$TPg = I^{-1}(1-P)g \qquad (g \in N),$$

and, if both f and g are in $K (= M)$, write

$$U\langle f, g \rangle = f + Ig.$$

The definition of T may look artificial at first, but it is not. Here is what it says. To represent a vector g in N in the form $\langle f, Tf \rangle$, recall that the components of g with respect to the decomposition $H = M \oplus M^\perp$ are Pg and $(1-P)g$ and, there-fore, let f be Pg and define T so that Tf is $(1-P)g$. Since T is to be a transformation on $K (= M)$, the last part does not quite make sense; the closest Tf can come to being $(1-P)g$ is to be $I^{-1}(1-P)g$, the vector in M that is identified with the vector $(1-P)g$ in M. That the definition of T is unambiguous is implied by the one-to-one character of $P|N$, proved above.

It remains to prove that K, T, and U have the required properties. Since I^{-1} is an isometry, the assertion that $\ker T = 0$ comes down to this: the restriction to N of the projection with range $M^\perp$ has kernel 0. That assertion is one of eight (go from M, or $M^\perp$, to N, or $N^\perp$, or back); it differs from what was shown in the first paragraph of this proof in notation only. The density of ran T in K is proved similarly.

Next, why is T closed? It is to be proved that if $f_n = Pg_n$ (with $g_n \in N$), $f_n \to f$, and $I^{-1}(1-P)g_n \to h$, then $f = Pg$ for some g in N, and $h = I^{-1}(1-P)g$. Apply I to $I^{-1}(1-P)g_n \to h$ and infer $(1-P)g_n \to Ih$. Combine this with $Pg_n \ (=f_n) \to f$ to get

$$g_n = Pg_n + (1-P)g_n \to f + Ih.$$

In view of this, the obviously indicated thing to do is to write $g = f + Ih$. Since $g_n \in N$ and N is closed, it follows that $g \in N$. Since $I^{-1}(1-P)g_n \in M$ and M is closed, it follows that $h \in M$; hence $Ih \in M^\perp$ and $(1-P)Ih = Ih$. Since, finally, $f \in M$, it follows that

$$(1-P)g = (1-P)(f+Ih) = (1-P)Ih = Ih,$$

and hence that $I^{-1}(1-P)g = h$. Consequence: T is closed.

Since I maps M onto $M^\perp$, the transformation U maps $K \oplus K$ onto H. The isometric character of U follows from the computation:

$$\|U\langle f, g\rangle\|^2 = \|f+Ig\|^2 = \|f\|^2 + \|Ig\|^2 = \|f\|^2 + \|g\|^2 = \|\langle f, g\rangle\|^2.$$

It is trivial that U maps $K \oplus 0$ onto M; how does U map the graph of T? The graph of T is the same as the set of all $\langle Pg, I^{-1}(1-P)g\rangle$, with g in N; the image of $\langle Pg, I^{-1}(1-P)g\rangle$ under U is $Pg + II^{-1}(1-P)g = g$, and therefore the image of the graph is exactly N. The proof of the theorem is complete.

COROLLARY. *The transformation T can be chosen selfadjoint and positive, and, if it is so chosen, then it is unique, in the sense that if the pair $\langle K \oplus 0, \text{graph } T_1\rangle$ is unitarily equivalent to the pair $\langle K \oplus 0, \text{graph } T_2\rangle$, then T_1 is unitarily equivalent to T_2.*

Proof. To make T selfadjoint, consider its polar decomposition $T = WA$, where W is a partial isometry and A is selfadjoint and positive [3, p. 1249]. The conditions on T (zero kernel, dense range) imply that the partially isometric factor W is unitary. Apply the unitary operator $1 \oplus W^*$ to both axis and graph; the axis remains invariant, and the graph becomes the graph of the positive selfadjoint transformation A.

To prove uniqueness, observe first that if a unitary operator on $K \oplus K$ maps the axis $K \oplus 0$ onto itself, then it is a direct sum $U_1 \oplus U_2$ of two unitary operators on K. (A subspace mapped onto itself by a unitary operator is invariant under both the operator and its inverse, and therefore reduces the operator.) The assumption that $U_1 \oplus U_2$ maps graph T_1 onto graph T_2 implies that if $f \in \text{dom } T_1$, then $U_1 f \in \text{dom } T_2$ and $T_2 U_1 f = U_2 T_1 f$; in other words $T_2 U_1 = U_2 T_1$. By adjunction $U_1^* T_2 = T_1 U_2^*$ (because T_1 and T_2 are selfadjoint); by multiplication $T_1^2 = T_1 U_2^* U_2 T_1 = U_1^* T_2^2 U_1$, and therefore $T_1 = U_1^* T_2 U_1$ (because T_1 and T_2 are positive).

Projection matrix. It is an elementary exercise in analytic geometry to calculate the matrix of a projection of rank 1 acting on a space of dimension 2. The projection whose range is a line of inclination θ turns out to be

$$\begin{pmatrix} \cos^2 \theta & \cos \theta \sin \theta \\ \cos \theta \sin \theta & \sin^2 \theta \end{pmatrix}.$$

The graph of a linear transformation on a Hilbert space is very much like a line in a plane. Since the elements of the graph of T are ordered pairs $\langle f, Tf \rangle$, it follows, purely formally, that the ratio of the second coordinate to the first is always T; the transformation T plays the role of the slope of the line, i.e., of the tangent of the inclination θ. The following result is the operator version of the elementary exercise in analytic geometry mentioned above. The boundedness of *sin* and *cos* (as opposed to the unboundedness of *tan*) are reflected in that only bounded linear transformations need to be mentioned.

THEOREM 2. *If M and N are subspaces in generic position in a Hilbert space H, with respective projections P and Q, then there exists a Hilbert space K, and there exist positive contractions S and C on K, with $S^2 + C^2 = 1$ and ker $S =$ ker $C = 0$, such that P and Q are unitarily equivalent to*

$$\begin{pmatrix} 1 & 0 \\ 0 & 0 \end{pmatrix} \quad and \quad \begin{pmatrix} C^2 & CS \\ CS & S^2 \end{pmatrix}$$

respectively.

Proof. Identify H with a direct sum $K \oplus K$ just as in the proof of Theorem 1. In this identification M and $M^\perp$ are the axes $K \oplus 0$ and $0 \oplus K$, and therefore

$$P = \begin{pmatrix} 1 & 0 \\ 0 & 0 \end{pmatrix}.$$

Since Q is a positive contraction, it follows that

$$Q = \begin{pmatrix} D & E \\ E^* & F \end{pmatrix},$$

where D $(= PQP | M)$ and F $(= (1 - P)Q(1 - P) | M^\perp)$ are positive contractions. Since $E = PQ(1 - P) | M^\perp$ and $E^* = (1 - P)QP | M$, it follows easily, just as in the first paragraph of the proof of Theorem 1, that ker $E =$ ker $E^* = 0$. This implies that in the polar decomposition $E = WA$ of E, the partially isometric factor W is unitary. Transform both P and Q by

$$\begin{pmatrix} W & 0 \\ 0 & 1 \end{pmatrix}$$

(i.e., multiply by

$$\begin{pmatrix} W^* & 0 \\ 0 & 1 \end{pmatrix}$$

on the left and

$$\begin{pmatrix} W & 0 \\ 0 & 1 \end{pmatrix}$$

on the right); the former remains invariant and the latter transforms to the same form as before except that now the off-diagonal terms are equal and positive. Assume therefore, with no loss of generality, that they were that way in the first place. (The identification of H with $K \oplus K$ was achieved by writing $K = M$ and choosing an arbitrary isometric correspondence between M and $M^\perp$. In these terms the change just made amounts to choosing a different, more convenient isometric correspondence.)

Let C and S be the positive square roots of D and F respectively, so that

$$Q = \begin{pmatrix} C^2 & E \\ E & S^2 \end{pmatrix},$$

where C, S, and E are positive contractions. The idempotence of Q says exactly that

$$C^2 - C^4 = S^2 - S^4 = E^2 \quad \text{and} \quad C^2 E + E S^2 = E.$$

Since, by the first two equations, C and S commute with E, the last equation says that $E(C^2 + S^2 - 1) = 0$. Since ker $E = 0$, it follows that $S^2 = 1 - C^2$; since $E^2 = C^2 - C^4 = C^2(1 - C^2) = C^2 S^2$, therefore $E = CS$. The last assertion, together with ker $E = 0$, implies that ker $C = \text{ker } S = 0$; the proof is complete.

Some of the known theory of pairs of subspaces in generic position can be recaptured from Theorem 2. Consider, as a sample, the problem of determining the commutant of the pair of projections P and Q. The commutant of P alone is easy to compute: it consists of all matrices of the form

$$\begin{pmatrix} X & 0 \\ 0 & Y \end{pmatrix}.$$

When does such a matrix commute with

$$\begin{pmatrix} C^2 & CS \\ CS & S^2 \end{pmatrix}?$$

Answer: if and only if both X and Y commute with C^2 (and hence with C, and S^2 and S), and also $CSX = CSY$. Since ker $CS = 0$, the last condition implies $X = Y$. Conclusion: the simultaneous commutant of P and Q consists of all

$$\begin{pmatrix} X & 0 \\ 0 & X \end{pmatrix}$$

where X commutes with C.

The mere statement that the commutant of $\{P, Q\}$ is nontrivial is trivial: $(P+Q-1)^2$ commutes with both P and Q. N.B.:

$$(P+Q-1)^2 = \begin{pmatrix} C^2 & 0 \\ 0 & C^2 \end{pmatrix}.$$

It follows easily that the von Neumann algebra generated by two projections (on a Hilbert space of dimension greater than 2) is never the algebra of all operators; the cases in which the ranges are not in generic position are easily disposed of. Equivalently: if P and Q are projections on a Hilbert space H of dimension greater than 2, then there exists a nontrivial subspace of H invariant under both P and Q. (Note that since P and Q are Hermitian, invariance is the same as reduction.) Results such as these are almost explicitly contained in the earlier work on pairs of subspaces; cf. in particular [1] and [2].

Reflected graphs. Theorems 1 and 2 were proved separately above, but there are other (not necessarily simpler) approaches to the theory. The same results can be obtained by proving only one of them directly (either one) and then deriving the other one from it.

The way to get Theorem 1 from Theorem 2, for instance, is to note first that the range of

$$\begin{pmatrix} C^2 & CS \\ CS & S^2 \end{pmatrix}$$

consists of all vectors of the form $\langle Cf, Sf \rangle$. (It really consists of all vectors of the form $\langle C(Cf+Sg), S(Cf+Sg) \rangle$. Since, however, the positive operator $C+S$ is invertible, because the function $x \to x+(1-x^2)^{1/2}$ is bounded away from zero in $[0, 1]$, the simpler description is valid.) To recognize that set as a graph, it is necessary to define a linear transformation T such that $TC=S$. (Formally: $\tan \theta = \sin \theta / \cos \theta$.) There is no conceptual difficulty here. The operator C has an inverse (not necessarily bounded), and the functional calculus for selfadjoint transformations can be invoked to justify the definition $T=C^{-1}(1-C^2)^{1/2}$.

To go in the other direction, from Theorem 1 to Theorem 2, it is necessary first to express in terms of T the projection whose range is the graph of T. Formally this too is an exercise in analytic geometry; the projection whose range is a line of slope $\tan \theta$ turns out to be

$$\begin{pmatrix} (1+\tan^2 \theta)^{-1} & \tan \theta(1+\tan^2 \theta)^{-1} \\ \tan \theta(1+\tan^2 \theta)^{-1} & \tan^2 \theta(1+\tan^2 \theta)^{-1} \end{pmatrix}.$$

(Alternatively, obtain this matrix from the one in terms of sines and cosines by trigonometric identities.) It is easy to derive the precise version of the formalism from the basic definitions; for an explicit treatment see [6]. Once the projection Q is recognized as

$$\begin{pmatrix} (1+T^2)^{-1} & T(1+T^2)^{-1} \\ T(1+T^2)^{-1} & T^2(1+T^2)^{-1} \end{pmatrix},$$

then the functional calculus for selfadjoint transformations can be invoked again; the result is that $C^2 = (1 + T^2)^{-1}$ and $S^2 = T^2(1 + T^2)^{-1}$, and all the conditions of Theorem 2 are satisfied.

The purpose of this section is to exploit the trigonometric analogy once more. The underlying geometric fact is that the diagram

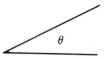

when rotated downward through $\theta/2$ becomes

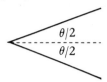

The operator version of this geometric fact is the following somewhat surprising assertion.

THEOREM 3. *If M and N are subspaces in generic position in a Hilbert space H, then there exists a Hilbert space K, and there exists a positive contraction T_0 on K, with* $\ker T_0 = \ker (1 - T_0) = 0$, *such that the pair $\langle M, N \rangle$ is unitarily equivalent to the pair $\langle \text{graph } T_0, \text{graph } (-T_0) \rangle$.*

Proof. The rotation that maps a line of inclination θ onto one of inclination $\theta/2$ and, at the same time, maps the horizontal axis onto the line of inclination $-\theta/2$ has the matrix

$$\begin{pmatrix} \cos (\theta/2) & \sin (\theta/2) \\ -\sin (\theta/2) & \cos (\theta/2) \end{pmatrix}.$$

This suggests that Theorem 3 could be derived from Theorem 1 by the operator analogue of that half-angle rotation. Formally this program is a simple computation, and precisely it is a straightforward argument about selfadjoint transformations. Since, however, the argument involves unbounded transformations, and since the same result can be achieved with bounded operators only, it is preferable to base the proof on Theorem 2. The boundedness of T_0 is, by the way, one of the mildly surprising features of Theorem 3. The heuristic reason for it is that if T corresponds to $\tan \theta$, $0 < \theta < \pi/2$, then T_0 corresponds to $\tan (\theta/2)$. As θ varies over the domain $(0, \pi/2)$, its tangent varies unboundedly, but, since $\theta/2$ varies over $(0, \pi/4)$, the tangent of the half-angle remains bounded.

By virtue of Theorem 2 there is no loss of generality in assuming that H is already represented as $K \oplus K$ in such a way that M is the axis $K \oplus 0$ and N is the range of the projection

$$Q = \begin{pmatrix} C^2 & CS \\ CS & S^2 \end{pmatrix}.$$

The problem is to find T_0 so that $\langle K \oplus 0, \operatorname{ran} Q \rangle$ is unitarily equivalent to $\langle \operatorname{graph} T_0, \operatorname{graph}(-T_0) \rangle$. Since the half-angle formula for the tangent is

$$\tan \frac{\theta}{2} = \left(\frac{1 - \cos \theta}{1 + \cos \theta} \right)^{1/2},$$

it is formally natural to write

$$T_0 = ((1 - C)(1 + C)^{-1})^{1/2}.$$

This makes sense. Since C is positive, $1 + C$ is invertible, and since C is a positive contraction, the same is true of $1 - C$. That T_0 and $1 - T_0$ have zero kernel is a consequence of the same properties of $1 - C$ and C, respectively.

The natural candidate for T_0 has been found; the next problem is to find a unitary operator that transforms M and N the desired way. The formalism suggests that too; since

$$\cos \frac{\theta}{2} = \left(\frac{1 + \cos \theta}{2} \right)^{1/2} \quad \text{and} \quad \sin \frac{\theta}{2} = \left(\frac{1 - \cos \theta}{2} \right)^{1/2},$$

it is natural to write

$$U = \begin{pmatrix} C_0 & S_0 \\ -S_0 & C_0 \end{pmatrix},$$

where

$$C_0 = \left(\frac{1 + C}{2} \right)^{1/2} \quad \text{and} \quad S_0 = \left(\frac{1 - C}{2} \right)^{1/2}.$$

All this makes sense; the verification that U is unitary is trivial.

The rest of the proof is computational: it verifies that as f varies over K, the images

$$\begin{pmatrix} C_0 & S_0 \\ -S_0 & C_0 \end{pmatrix} \begin{pmatrix} Cf \\ Sf \end{pmatrix} \quad \text{and} \quad \begin{pmatrix} C_0 & S_0 \\ -S_0 & C_0 \end{pmatrix} \begin{pmatrix} f \\ 0 \end{pmatrix}$$

fill out exactly the graph of T_0 and the graph of $-T_0$ respectively. It is to be verified, in other words, that
   (i) $\operatorname{ran}(C_0 C + S_0 S) = K$,
   (ii) $T_0(C_0 C + S_0 S) = -S_0 C + C_0 S$,
   (iii) $\operatorname{ran} C_0 = K$,
   (iv) $(-T_0)C_0 = -S_0$.
For (i): in fact $C_0 C + S_0 S = C_0$; since C_0 is invertible, this settles (i) and (iii). The verifications of (ii) and (iv) are mechanical.

COROLLARY. *If M and N are subspaces in generic position in a Hilbert space H, with respective projections P and Q, then a complete set of unitary invariants of the pair $\langle M, N \rangle$ is the unitary equivalence class of the Hermitian operator $P + Q$.*

   Proof. The statement means that $\langle M_1, N_1 \rangle$ is unitarily equivalent to $\langle M_2, N_2 \rangle$ if and only if $P_1 + Q_1$ is unitarily equivalent to $P_2 + Q_2$. Since the problem of unitary equivalence for Hermitian operators is in principle solved, this is a solution of the problem of unitary equivalence for pairs of subspaces. The result is due to Dixmier [2].

To prove the corollary, use Theorem 3 to justify the assumption that M and N are the graphs of T_0 and $-T_0$ respectively. In that case the matrices of the projections P and Q are given by

$$P = \begin{pmatrix} (1+T_0^2)^{-1} & T_0(1+T_0^2)^{-1} \\ T_0(1+T_0^2)^{-1} & T_0^2(1+T_0^2)^{-1} \end{pmatrix}$$

and

$$Q = \begin{pmatrix} (1+T_0^2)^{-1} & -T_0(1+T_0^2)^{-1} \\ -T_0(1+T_0^2)^{-1} & T_0^2(1+T_0^2)^{-1} \end{pmatrix},$$

or, equivalently, by

$$P = \begin{pmatrix} C_0^2 & C_0S_0 \\ C_0S_0 & S_0^2 \end{pmatrix} \quad \text{and} \quad Q = \begin{pmatrix} C_0^2 & -C_0S_0 \\ -C_0S_0 & S_0^2 \end{pmatrix}.$$

The rest of the proof is the same as Dixmier's. In view of the form of $P+Q$, it is natural to consider the Hermitian operator

$$R = P+Q-1 = \begin{pmatrix} C & 0 \\ 0 & -C \end{pmatrix}.$$

Since the positive and negative parts of a Hermitian operator are unitary invariants of it, it follows that the unitary equivalence class of R uniquely determines that of C. (Here it is important that ker $C=0$.) Since C, in turn, uniquely determines C_0, and thence P and Q, it follows that the unitary equivalence class of R uniquely determines that of the pair P, Q. The opposite direction is trivial; the proof of the corollary is complete. (The difference between $P+Q$ and R is a technical triviality. For R the parts that matter are the ones above and below 0, whereas for $P+Q$ they are the ones above and below 1.)

REFERENCES

1. Arlen Brown, *The unitary equivalence of binormal operators*, Amer. J. Math. **76** (1954), 414–434.

2. J. Dixmier, *Position relative de deux variétés linéaires fermées dans un espace de Hilbert*, Rev. Sci. **86** (1948), 387–399.

3. N. Dunford and J. T. Schwartz, *Linear operators*, Vol. II, Interscience, New York, 1963.

4. P. R. Halmos, *Introduction to Hilbert space and the theory of spectral multiplicity*, Chelsea, New York, 1951.

5. ———, *A Hilbert space problem book*, Van Nostrand, Princeton, N. J., 1967.

6. M. H. Stone, *On unbounded operators in Hilbert space*, J. Indian Math. Soc. **15** (1951), 155–192.

UNIVERSITY OF HAWAII,
HONOLULU, HAWAII

Reprinted from the
BULLETIN OF THE AMERICAN MATHEMATICAL SOCIETY
Vol. 76, No. 5, pp. 887-933, Sept. 1970

TEN PROBLEMS IN HILBERT SPACE[1]

BY P. R. HALMOS

Dedicated to my teacher and friend Joseph Leo Doob with admiration and affection.

TABLE OF CONTENTS

PREFACE

Machines can write theorems and proofs, and read them. The purpose of mathematical exposition for people is to communicate ideas, not theorems and proofs. Experience shows that almost always the best way to communicate a mathematical idea is to talk about concrete examples and unsolved problems. In what follows I try to communicate some of the basic ideas of Hilbert space theory by discussing a few of its problems.

Nobody, except topologists, is interested in problems about Hilbert space; the people who work in Hilbert space are interested in problems about operators. The problems below are about operators. I do not know the solution of any of them. I guess, however, that

AMS 1969 subject classifications. Primary 4702.
Key words and phrases. Hilbert space, operators, cyclic vectors, weighted shifts, invariant subspaces, compact operators, dilations, contractions, quasinilpotence, reducibility, reflexive lattices, transitive lattices.
[1] Lectures delivered at a Regional Conference in the Mathematical Sciences, sponsored by the National Science Foundation and arranged by the Conference Board of the Mathematical Sciences, at Texas Christian University, Fort Worth, Texas, May 25–29, 1970.

887

they vary a lot in depth. Some have been around for many years and are known to be both difficult and important; others are untested and may turn out to be trivial.

The problems are formulated as yes-or-no questions. That, I believe, is the only clear way to formulate any problem. What a mathematician usually wants to do is something vague, like study, determine, or characterize a class of objects, but until he can ask a clear-cut test question, the chances are he does not understand the problem, let alone the solution. The purpose of formulating the yes-or-no question, however, is not only to elicit the answer; its main purpose is to point to an interesting area of ignorance.

A discussion of unsolved problems is more ephemeral than an exposition of known facts. Deciding to make a virtue out of necessity, I chose to focus on a narrow segment of the present, rather than on a broad view of history. As often as possible the references are to recent publications, and the proofs presented in detail are more from current folklore than from standard texts. Consequently some results (with appropriate credit lines to their discoverers) are published here for the first time. Most of the results and proofs with no assigned credit are "well known"; only a small number are mine. If a result is attributed to someone but is not accompanied by a reference to the bibliography, that means I learned it through personal communication, and, as far as I could determine, it is not available in print.

The chief prerequisite for understanding the exposition is a knowledge of the standard theory of operators on Hilbert spaces. In some cases that means that not even the spectral theorem is needed, and in other cases every available ounce of multiplicity theory would help to understand things clearly. In any case, the presentation is not cumulative; some readers who find §1 obscure may find §10 obvious.

As an expository experiment I have put some exercises into each section. This is unusual in papers (as opposed to books). The reason I did it is that originally each section was a lecture, and lectures (as opposed to papers) can stand every bit of elasticity that anyone can think of to put in. The audience is trapped and cannot impatiently riffle the pages forward and back, and the lecturer can use something that is easily omitted or inserted without disturbing the continuity.

As for terminology: *Hilbert space* means a complete, complex, inner-product space; *subspace* means a closed linear manifold; *operator* means a bounded linear transformation. The Hilbert spaces of principal interest are the ones that are neither too large nor too small, i.e., they are separable but infinite-dimensional. The problems are stated without any such explicit size restrictions, but in some cases they become trivial for either the large spaces, or the small ones, or both.

I thank Peter Fillmore, Pratibha Gajendragadkar, and Eric Nordgren for reading the first written version of these lectures and for making many helpful suggestions.

1. CONVERGENT

Problem 1. *Does the set of cyclic operators have a non-empty interior?*

A vector f in a Hilbert space H is cyclic for an operator A on H in case the smallest subspace of H that contains f and is invariant under A is H itself. An operator is cyclic in case it has a cyclic vector.

In the finite-dimensional case, A is cyclic if and only if its minimal polynomial is equal to its characteristic polynomial. (Equivalently: each eigenvalue occurs in only one Jordan block, or the space of eigenvectors corresponding to each eigenvalue has dimension 1 [24, p. 69 et seq]; cf. also [33].) It follows easily that in the finite-dimensional case the set $\mathcal{C}$ of cyclic operators is open. If the dimension is n, then the operators with n distinct eigenvalues constitute a dense set; it follows that $\mathcal{C}$ is dense. (Consequence: $\mathcal{C}$ is not closed; for instance, $\mathrm{diag}\langle 1, 1+1/n\rangle \to \mathrm{diag}\langle 1, 1\rangle$.)

There is only one reasonable topology for the operators on a finite-dimensional space; in the infinite-dimensional case there are several. Problem 1 refers to the norm (or uniform) topology, the one according to which $A_n \to A$ in case $\|A_n - A\| \to 0$.

If f is a cyclic vector for A, then the countable set $\{f, Af, A^2f, \cdots\}$ spans H; it follows that in the non-separable case $\mathcal{C}$ is empty.

In the separable infinite-dimensional case what little is known is negative. The set $\mathcal{C}$ is surely *not* open; the cyclic operator $\mathrm{diag}\langle 1, \frac{1}{2}, \frac{1}{3}, \cdots\rangle$ is the limit of the non-cyclic operators $\mathrm{diag}\langle 1, \frac{1}{2}, \cdots, 1/n, 0, 0, 0, \cdots\rangle$. It is, of course, possible that the interior of $\mathcal{C}$ is non-empty; that is what Problem 1 is about. A reasonable candidate for an operator in the interior of $\mathcal{C}$ is the unilateral shift U. (According to the most easily accessible definition U is the operator on l^2 that sends $\langle \xi_0, \xi_1, \xi_2, \cdots \rangle$ onto $\langle 0, \xi_0, \xi_1, \xi_2, \cdots \rangle$.) That candidate happens to fail (J. G. Stampfli). The set $\mathcal{C}$ is also *not* dense; in face if V is the direct sum of two copies of U and $\|V - A\| < 1$, then A is not cyclic. Indeed:

$$\|1 - A^*V\| = \|V^*V - A^*V\| \leq \|V^* - A^*\| < 1,$$

so that A^*V is invertible; it follows that V^*A is invertible and hence that $\mathrm{ran}\, A \cap \ker V^* = 0$. Since, thus, $\mathrm{ran}\, A$ is disjoint from the 2-dimensional space $\ker V^*$, the co-dimension of $\mathrm{ran}\, A$ must be at least

2. (The proof so far was needed to get this inequality only. An alternative way to get it is to note that it is a special case of a small part of the theory of Fredholm operators [35].) Now it is clear that A cannot be cyclic; for each f, the span of $\{f, Af, A^2f, \cdots\}$ has co-dimension at least 1.

The point of the preceding discussion is to call attention to the existence of topological problems of analytic importance in operator theory. There are many; Problem 1 is just a sample. Other examples are Problems 7 and 8, about the closures of certain sets of operators, or, in other words, about approximation theory. Here is a well known example of the same kind: is the set of all invertible operators dense? The answer is yes for finite-dimensional spaces, and furnishes an occasionally useful proof technique; to prove something for all matrices, prove it for the non-singular ones first and then approximate. For infinite-dimensional spaces the answer is no [20, Problem 109]. It is slightly less well known that the answer remains no even for a somewhat larger set, the set of kernel-free operators. (If A is invertible, then A is kernel-free, i.e., ker $A = 0$, but the converse is not true.) In other words, there exists a non-empty open set every element of which has a non-trivial kernel. Indeed, the open ball $\{A: \|U^* - A\| < 1\}$ is such a set (where U is, as before, the unilateral shift). For the proof, apply the (Fredholm) argument that disproved the density of the cyclic operators to infer that, for each A in that ball, ran A^* has positive co-dimension, and then conclude that ker $A \neq 0$.

Here is a classically important question: is the set of invertible operators connected? The answer is yes [20, Problem 110]; the proof is an application of the polar decomposition of operators. A much less important question that takes most people a few seconds longer to answer than it should is this (**Exercise** 1): is the set of non-scalar normal operators connected?

All the questions so far concerned the norm topology; questions about the other operator topologies are likely to be (a) easier and (b) more pathological. The pathology of topology can be fun, however, and, although the ground has been pretty well worked over, new and even useful facts do turn up from time to time.

The strong operator topology is the one according to which $A_n \to A$ in case $\|A_nf - Af\| \to 0$ for each vector f. (The indices, here and elsewhere, need not and should not be restricted to integers; any directed set will do.) A famous and sometimes bothersome misbehavior of the strong topology is the discontinuity of the adjoint. Indeed, if $A_k = U^{*k}$ (the unilateral shift again), $k = 1, 2, 3, \cdots$, then $A_k \to 0$ strongly, but $\{A_k^*\}$ is not strongly convergent to anything [20, Problem 90]. This negative result is old; the following interesting positive observation of Kadison's is quite recent [26]: the restriction of the adjoint to the set of normal operators is strongly continuous. (That is: if $A_n \to A$ strongly, where A and each A_n is normal, then $A_n^* \to A^*$ strongly.) The proof is a trick, but a simple trick:

$$\|(A_n^* - A^*)f\|^2 = \|A_n f\|^2 - \|Af\|^2 + (f, (A - A_n)A^*f) + ((A - A_n)A^*f, f)$$

$$\leq \|(A - A_n)f\| (\|Af\| + \|A_n f\|) + 2\|(A - A_n)A^*f\| \cdot \|f\|.$$

Approximation techniques that rely on the strong topology do not seem to have been used much, although there are some interesting questions whose known and surprising answers might turn out to be useful. Example: what is the strong closure of the set of normal operators? Answer: the set of subnormal operators [4]. The proof is not trivial. Another example (**Exercise 2**): what is the strong closure of the set of unitary operators?

The weak operator topology is the one according to which $A_n \to A$ in case $(A_n f, g) \to (Af, g)$ for each f and g. A useful recent result is that rank (dimension of the closure of the range) is weakly lower semicontinuous; that is, if $A_n \to A$ weakly, then lim inf $_n$ rank $A_n \geq$ rank A [21]. The proof is not immediately obvious, but it is certainly not deep. Warning: the possible values of rank in this context are the nonnegative integers, together with ∞, no distinction must be made among different infinite cardinals. Were such a distinction to be made, the result would become false. Suppose indeed that the underlying Hilbert space has an uncountable orthonormal basis $\{e_j : j \in J\}$. Let D be the set of all countable subsets of J, ordered by inclusion; for each n in D, write A_n for the projection onto the span of $\{e_j : j \in n\}$. Since for each vector f there exists an n_0 in D such that $f \perp e_j$ whenever j is not in n_0, it follows that $A_n \to 1$ (not only weakly, but, in fact, strongly). Since rank $A_n = \aleph_0$ and rank $1 > \aleph_0$, the cardinal version of semicontinuity is false.

What is the weak closure of the set of normal operators? Answer: in the finite-dimensional case it is the set of normal operators [21]; in the infinite-dimensional case it is the set of all operators [18]. What is the weak closure of the set of projections? Answer: in the finite-dimensional case it is the set of projections; in the infinite-dimensional case it is the set of Hermitian operators A with $0 \leq A \leq 1$ [18]. The proofs of the infinite-dimensional statements are not difficult, but they depend on a little dilation theory. The results show why weak approximation theory is not likely to be fruitful; it is too good to be any good. **Exercise 3**: what is the weak closure of the set of scalars?

A simple example that illustrates many of the curious properties of the weak operator topology can be described in terms of matrices. Let A_n be the infinite matrix (or, by a slight abuse of language, the operator) whose entries at the positions $\langle 1, 1 \rangle$, $\langle 1, n \rangle$, $\langle n, 1 \rangle$, and $\langle n, n \rangle$ are $\frac{1}{2}$, and whose remaining entries are all 0 ($n = 2, 3, 4, \cdots$). It is easy to verify that A_n is idempotent, and it is plausible (and true, and easy to verify) that if A is the matrix whose $\langle 1, 1 \rangle$ entry is $\frac{1}{2}$ and whose remaining entries are all 0, then $A_n \to A$ weakly. The example shows how projections can converge to a non-projection, it proves that multiplication (in fact squaring) is not weakly continuous, and, incidentally, it implies that the strong and the weak topologies are distinct. (Here is an unimportant but unsolved teaser: does squaring have any points of weak continuity?)

ANSWERS TO THE EXERCISES. (1) The non-scalar *Hermitian* operators form a (norm) connected set. For a 1-dimensional space this is trivial. In any other case, the set of Hermitian operators is a real vector space of dimension 3 or more. If, therefore, A and B are non-scalar Hermitian operators, then there always exists a non-scalar Hermitian C such that $tA + (1-t)C$ and $tB + (1-t)C$ are non-scalar Hermitian operators whenever $0 \leq t \leq 1$. The *normal* case can be joined to the Hermitian one. Indeed, if $A + iB$ (with A and B Hermitian) is a non-scalar normal operator, then assume, with no loss of generality, that A is not a scalar, and consider $A + itB$, $0 \leq t \leq 1$. (This solution is due to J. J. Schäffer.)

(2) The strong closure of the set of unitary operators is the set of isometries. Since the set of isometries is easily seen to be strongly closed, and since every isometry is a direct sum of a unitary operator and a number of copies of the unilateral shift [20, Problem 118], it is sufficient to prove that U, the unilateral shift, is a strong limit of unitary operators. That is easy: If

$$U_n \langle \xi_0, \xi_1, \xi_2, \cdots \rangle = \langle \xi_n, \xi_0, \xi_1, \cdots, \xi_{n-1}, \xi_{n+1}, \xi_{n+2}, \cdots \rangle$$

for $n = 1, 2, 3, \ldots$, then each U_n is unitary and $U_n \to U$ strongly.

(3) The set of scalars is weakly closed. Indeed, if $\lambda_n(f, g) \to (Sf, g)$ for each f and g, then $\{\lambda_n\}$ is a Cauchy net of complex numbers, and therefore $\lambda_n \to \lambda$ for some complex number λ; it follows that $\lambda_n(f, g) \to \lambda(f, g)$ and hence that $Sf = \lambda f$ for each f.

2. WEIGHTED

Problem 2. *Is every part of a weighted shift similar to a weighted shift?*

If U is the unilateral shift and if e_n is the sequence $\langle \xi_0, \xi_1, \xi_2, \cdots \rangle$

with $\xi_n = 1$ and $\xi_i = 0$ when $i \neq n$, then $\{e_0, e_1, e_2, \cdots\}$ is an orthonormal basis for l^2 and $Ue_n = e_{n+1}$, $n = 0, 1, 2, \cdots$. If $\langle \alpha_0, \alpha_1, \alpha_2, \cdots \rangle$ is a bounded sequence of positive numbers, then the equations $Ae_n = \alpha_n e_{n+1}$, $n = 0, 1, 2, \cdots$, unambiguously define an operator A. Such operators are called *weighted shifts*; they are of interest because they can be used to construct examples of many different kinds of operatorial behavior. (To eliminate easy but dull case distinctions, the number 0 is not allowed as a weight.)

For any operator A on any Hilbert space H, a *part* of A is, by definition, the restriction of A to a subspace of H invariant under A. One of the main reasons for the success of the Beurling treatment of the unilateral shift is that every non-trivial part of U is unitarily equivalent to U [20, Problem 123]. Here "non-trivial" excludes the most trivial case only, the restriction to the subspace 0. That should, of course, be excluded in the formulation of Problem 2 also, and in all related contexts; that exclusion is hereby made, retroactively and forward to eternity.

What sense does it make to say that a certain operator "is" a weighted shift? If shifts are defined on the concrete Hilbert space l^2 (as above), then the best that can be hoped about a part of one is that it is "abstractly identical" with or "isomorphic" to another; the correct technical phrase is "unitarily equivalent". If shifts are defined with respect to arbitrary orthonormal bases $\{e_0, e_1, e_2, \cdots\}$ in arbitrary separable infinite-dimensional Hilbert spaces (the best way), then "is" means "is". There is a third interpretation, which may seem artificial but has some merit; it is obtained by demanding isomorphism with respect to the linear and the topological structures, but ignoring the metric structure. The correct technical word is "similar"; two operators A and B, on Hilbert spaces H and K, are similar in case there is a bounded linear transformation S from H onto K, with a bounded inverse, such that $A = S^{-1}BS$.

There are two reasons for the insistence on positive weights in the definition of weighted shifts: (a) it seems more natural, and (b) nothing is gained by allowing arbitrary complex weights. The proof of (b) is the following assertion (**Exercise 1**): if A and B are weighted shifts, with complex weight sequences $\langle \alpha_n \rangle$ and $\langle \beta_n \rangle$, then a necessary and sufficient condition that A and B be unitarily equivalent is that $|\alpha_n| = |\beta_n|$ for $n = 0, 1, 2, \cdots$. Consequence: if A is a weighted shift and if λ is a complex number of modulus 1, then A is unitarily equivalent to λA. This kind of "circular symmetry" is an unusual but useful property for an operator to have. More along the same lines cannot be asked; modulus 1 is in the nature of things. That is (**Exercise 2**): if a non-nilpotent operator A is similar to λA, then $|\lambda| = 1$.

The unitary equivalence theory of weighted shifts is easy; their similarity theory is slightly more complicated, but still near the surface. In the notation of the preceding paragraph, A is similar to B if and only if the sequence of quotients

$$\left| \frac{\alpha_0 \cdots \alpha_n}{\beta_0 \cdots \beta_n} \right|$$

is bounded away from 0 and from ∞ [20, Problem 76]. It is thus easy to tell when two weighted shifts are similar, but the circle of ideas centered at Problem 2 includes a question of a much more difficult type: when is an operator obtained by certain non-transparent constructions similar to a weighted shift? The formation of a part of a weighted shift is such a construction; there are others. Consider, for instance, the (Cesaro) operator C_0 on l^2 defined by

$$C_0 \langle \xi_0, \xi_1, \xi_2, \cdots \rangle = \langle \eta_0, \eta_1, \eta_2, \cdots \rangle,$$

where

$$\eta_n = \frac{1}{n+1} \sum_{i=0}^{n} \xi_i.$$

The known spectral properties of C_0 [7] suggest the question: is $1 - C_0$ similar to a weighted shift? The answer is not known. Another interesting operator (Volterra) is defined on $L^2(0, 1)$ by $Vf(x) = \int_0^x f(y)dy$; i.e., Vf is, for each f, the indefinite integral of f. (**Exercise** 3): is V similar to a weighted shift?

What are weighted shifts good for? The answer is that they can be used for examples and counterexamples to illustrate many properties of operators. Among such properties are the existence of square roots, spectral behavior, irreducibility, the structure of invariant subspaces, and subnormality.

Assertion: no weighted shift has a square root. Suppose, indeed, that A is a weighted shift, and suppose that $B^2 = A^*$. (The adjoint is easier to treat, and the result obviously comes to the same thing.) Since $\ker B \subset \ker A^*$, and since $\ker A^* \neq 0$, it follows that $\ker B = \ker A^*$. Since $\ker A^* \subset \operatorname{ran} A$, it follows that $\ker A^* \subset \operatorname{ran} B$. Hence if e is in $\ker A^*$, $e \neq 0$, then there exists a vector f such that $Bf = e$, and it is easy to deduce (from $\ker B = \ker A^*$) that e and f are linearly independent. Conclusion: the dimension of $\ker B^2$ ($= \ker A^*$) must be at least 2, which is absurd [29], [20, Problem 114].

A weighted shift provides the easiest example of a kernel-free operator whose spectrum consists of 0 only [20, Problem 80], and the more delicate spectral behavior of weighted shifts has also received quite a bit of attention [16], [29], [34], [41].

Weighted shifts are irreducible. That is, if A is a weighted shift and if M is a subspace that reduces A, then M is either 0 or the entire space; equivalently, if P is a projection that commutes with A, then $P = 0$ or $P = 1$. To prove this, all that is needed is that the kernel of A^* is spanned by a single (non-zero) vector e which happens to be cyclic for A. Indeed: $A^*Pe = PA^*e = 0$ implies (since e spans ker A^*) that $Pe = \lambda e$, and hence (since P is a projection) that either $Pe = 0$ or $Pe = e$. If $Pe = 0$, then $A^n e = A^n(1 - P)e = (1 - P)A^n e$, which belongs to ker P for all n, and therefore (since e is cyclic for A) $P = 0$; if $Pe = e$, then $A^n e = A^n Pe = PA^n e$, which belongs to ran P for all n, and therefore $P = 1$. (This proof is due to K. J. Harrison.) A somewhat stronger result is known: every operator similar to a weighted shift is irreducible [29], [51].

To what extent can the structure of the invariant subspaces of an operator be prescribed? Is it, for instance, possible that an operator has exactly one invariant subspace of each dimension between 0 and $\aleph_0$ inclusive? The answer is yes, and the standard example (due to Donoghue) is given by the adjoint of a suitably weighted shift [20, Problem 151]. (The unilateral shift itself, unweighted, won't do.)

There are many generalizations of the concept of normality; two important ones, in decreasing order of generality and increasing order of importance, are hyponormality and subnormality. An operator A is *hyponormal* in case $A^*A - AA^* \geq 0$; an operator is *subnormal* in case it is part of a normal operator. (The verificaton that every subnormal operator is hyponormal is the easiest part of this otherwise tricky subject.) Which weighted shifts are normal, which are hyponormal, and which are subnormal? The first question is settled just by looking at kernels; there is no weight sequence that can make a weighted shift normal. The answer to the second question is also easy [20, Problem 160]: the weight sequence must be monotone increasing. The answer to the third question is not easy. One formulation was offered by Stampfli [49]; the hitherto unpublished formulation below (due to C. Berger) is very different. Berger's condition is elegant and easy to state: a weighted shift is subnormal if and only if the squares of the partial products of the weights constitute the moment sequence of a probability measure in the unit interval.

For the proof, suppose first that A is a subnormal weighted shift, and assume, with no loss of generality, that $\|A\| = 1$. Let B (on a Hilbert space K) be the minimal normal extension of A; then $\|B\| = 1$. Since e_0 is "bicyclic" for B (i.e., by minimality, there is no proper subspace of K that contains e_0 and reduces B), the spectral theorem implies that B is unitarily equivalent to the position operator $(f(z) \rightarrow zf(z))$ on L^2 of the closed unit disc D with respect to some probability measure μ. The standard unitary equivalence, moreover, carries the vector e_0 onto the constant function 1. Identify B with its unitarily equivalent transform on $L^2(\mu)$, and, correspondingly, identify A with the restriction of B to the span of (the images of) the vectors $e_0\, e_1,\, e_2\, \cdot\, \cdot\, \cdot\,$.

Let ν be the "marginal" measure induced by μ in the closed unit interval I. That is: if $S:D \rightarrow I$ is the mapping defined by $Sz = |z|$, then $\nu(E) = \mu(S^{-1}E)$ for each Borel subset E of I. It follows [17, p. 163] that $\int_I f d\nu = \int_D^2 fS \cdot d\mu$ whenever either side of the equation makes sense. If, in particular, $f(\rho) = \rho^{2n}$, $n = 0, 1, 2, \cdot\, \cdot\, \cdot\,$, then

$$\int_I \rho^{2n} d\nu(\rho) = \int_D |z|^{2n} d\mu(z) = \int_D |z^n \cdot 1|^2 d\mu(z)$$

$$= \|B^n e_0\|^2 = \|A^n e_0\|^2.$$

The last term is easy to compute: if $p_0 = 1$ and $p_{n+1} = \alpha_n p_n$, then $A^n e_0 = p_n e_n$, and therefore

$$\int_I \rho^{2n} d\nu(\rho) = p_n^2, \qquad n = 0, 1, 2, \cdot\, \cdot\, \cdot\,.$$

The proof of the necessity of Berger's condition is complete; p_n^2 is the nth moment of $d\nu(\sqrt{\rho})$.

To prove sufficiency, start with a probability measure ν in I, and write $\int_I \rho^{2n} d\nu(\rho) = p_n$. If λ is normalized Lebesgue measure in the perimeter C of the unit circle, then $\nu \times \lambda$ in $I \times C$ induces (it is tempting to say "is") a measure μ in D. That is: if $T: I \times C \rightarrow D$ is the mapping defined by $\langle \rho,\, u \rangle \rightarrow \rho u$, then $\mu(E) = (\nu \times \lambda)(T^{-1}E)$ for each Borel subset E of D. Let B be the position operator on $L^2(\mu)$, and let A be the restriction of B to the subspace $H^2(\mu)$ spanned by $\{f_0, f_1, f_2, \ldots\}$, where $f_n(z) = z^n$, $n = 0, 1, 2, \ldots$. Since B is normal, it follows that A is subnormal.

A straightforward computation shows that

$$(f_n, f_m) = \int_I \rho^{n+m} d\nu(\rho) \cdot \int_C u^{n-m} d\lambda(u), \qquad n, m = 0, 1, 2, \cdot\, \cdot\, \cdot\,.$$

If $n \neq m$, the last factor vanishes, i.e., the f's are pairwise orthogonal. If $n = m$, the last factor is 1, and therefore

$$\|f_n\|^2 = p_n^2, \qquad n = 0, 1, 2, \cdot\, \cdot\, \cdot\,.$$

If, therefore, $e_n = f_n/p_n$, $n = 0, 1, 2, \cdot\, \cdot\, \cdot\,$, then the e's form an orthonormal basis for $H^2(\mu)$. Since, finally,

$$Ae_n = A\left(\frac{1}{p_n}f_n\right) = \frac{1}{p_n}f_{n+1} = \frac{p_{n+1}}{p_n}e_{n+1}, \qquad n = 0, 1, 2, \cdots,$$

it follows that A is (unitarily equivalent to) the weighted shift whose weight sequence has exactly the p's for its partial products. The proof of Berger's theorem is complete.

The preceding report on weighted shifts is a representative but not exhaustive summary of what is known. There is much that remains to be done, and, in particular, there are at least three directions of generalization that promise to be fruitful.

(a) Two-way shifts ($e_n \to \alpha_n e_{n+1}$ for *all* integers n, not the non-negative ones only) have many properties in common with the one-way kind, but not all. (Sample question with a known but not completely trivial answer: is the Volterra operator similar to a two-way weighted shift?)

(b) If the shift $n \to n+1$ on the integers is replaced by a measure-preserving transformation on a more interesting measure space, the theory of weighted shifts becomes the theory of "weighted translations"; although a few things about them are known [36], their theory is still largely undeveloped.

(c) The space l^2 is the direct sum of copies of a 1-dimensional Hilbert space; a generalization in the direction of higher multiplicities is obtained via the formation of direct sums of more general Hilbert spaces. In that case the role of the numerical weights is played by operator weights. This is essentially virgin territory; the idea has been used to construct counterexamples [20, Problem 164], but a general theory does not yet exist.

ANSWERS TO THE EXERCISES. (1) Given the α's and the β's, put $\delta_0 = 1$ and determine δ_n recursively from $\alpha_n \delta_n = \beta_n \delta_{n+1}$, $n = 0, 1, 2, \ldots$; if always $|\alpha_n| = |\beta_n|$, then always $|\delta_n| = 1$, and the "diagonal operator" determined by the δ's transforms A onto B. This proves sufficiency. To prove necessity, suppose that $A = W^*BW$, where W is unitary, so that $A^n = W^*B^nW$, and also $A^* = W^*B^*W$. It follows that W^* carries e_0 (in the kernel of B^*) onto a unit vector in the kernel of A^*; assume with no loss of generality that $W^*e_0 = e_0$. It follows that $W^*B^ne_0 = A^ne_0$, and hence (form norms and use induction) that $|\beta_n| = |\alpha_n|$ for $n = 0, 1, 2, \ldots$ [20, Problem 75; 28].

(2) Since

$$\frac{1}{\|S\|}\|A^n\|\frac{1}{\|S^{-1}\|} \leqq \|S^{-1}A^nS\| \leqq \|S^{-1}\| \cdot \|A^n\| \cdot \|S\|,$$

it follows that $\|S^{-1}A^nS\|^{1/n}/\|A^n\|^{1/n} \to 1$. If $S^{-1}AS = \lambda A$, then this implies that $|\lambda| = 1$ [28]. Note that if A is nilpotent, it is quite possible to have A be similar to λA for *every* non-zero complex number λ; an example is $A = \begin{pmatrix} 0 & 1 \\ 0 & 0 \end{pmatrix}$.

(3) The Volterra operator is not similar to a weighted shift. Reason: if V is similar to A, then V^* is similar to A^*; if A is a weighted shift, then ker $A^* \neq 0$, but an easy calculation shows that ker $V^* = 0$. (This proof is due to J. G. Stampfli.)

3. Invariant

Problem 3. *If an intransitive operator has an inverse, is its inverse also intransitive?*

An operator is called *intransitive* if it leaves invariant some subspace other than 0 or the whole space; in the contrary case it is transitive. The hope that non-trivial invariant subspaces always exist (i.e., that, except on a 1-dimensional space, every operator is intransitive) is perhaps still alive in the hearts of some. That existence theorem would be the first step toward a detailed structure theory for operators (possibly a generalization of the theory of the Jordan form in the finite-dimensional case). Repeated failure has convinced most of those who tried that the truth lies in the other direction, but the proof of that is, so far, just as elusive.

There are many special classes of operators that have been proved intransitive. Problem 3 (due to R. G. Douglas) seems to be the simplest problem of its kind (derive the intransitivity of something from that of something else). For the strong kind of invariance (that is, for reduction) the problem is easy: if a reducible operator has an inverse, then the inverse is also reducible. Indeed, more is true: if a subspace M reduces an invertible operator A, then the same M reduces A^{-1}. (Proof: if a projection commutes with A, then it commutes with A^{-1}.) For plain invariance the more is false; the invariance of a subspace under an invertible operator A does not imply its invariance under A^{-1}. (Consider the bilateral shift.)

A related question concerns square roots instead of inverses. Squaring preserves invariance (and reduction); what about the formation of square roots? If A^2 is reducible, is A reducible? The answer is no (consider weighted shifts). If A^2 is intransitive, is A intransitive? The answer is not known. Equally unknown is the answer to a mixed question (proposed by C. Pearcy): if A^2 is reducible, is A at least intransitive? The only thing along these lines that can be said has an undesirably special hypothesis (**Exercise 1**): if A^2 is normal, then A is intransitive.

If even the relatively elementary algebraic questions are unanswered, there is no immediate hope for the much deeper questions of perturbation (in imitation of the theory of spectral perturbation), but they deserve to be put on record. What is known is that if A and B are Hermitian and B has rank 1, then $A + iB$ is intransitive. The presently hoped-for generalization is obtained by replacing "rank 1" by "compact". Although large classes of compact operators are admissible here [47], not all are known to be. The mildest unproved perturbation statement appears to be this: if A is normal and B has rank 1, then $A + B$ is intransitive. The generalization of this perturbation problem to arbitrary intransitive operators in place of normal ones is not going to be easy to settle. Indeed: if A is an arbitrary operator, and if P is a projection of rank 1, then $A = AP + A(1 - P)$; the first summand has rank 1, and the second has a non-trivial kernel.

For normal operators the spectral theorem yields many invariant subspaces. The step from normal to subnormal is, however, large: it is not known whether every subnormal operator is intransitive.

Replacement of the algebraic condition of normality by the analytic condition of compactness yields a famous and useful intransitivity theorem: every compact operator (on a space of dimension greater than 1, of course) has a non-trivial invariant subspace [2]. Do two commutative compact operators always have a non-trivial invariant subspace in common? The answer is not known.

Two more invariant subspace theorems deserve mention; they are elementary, but they are sufficiently different from the preceding ones and from each other that they might suggest a new idea to someone.

(i) The strictly algebraic version of the invariant subspace problem has a positive solution (**Exercise** 2): every linear transformation (on a vector space of dimension greater than 1) has a non-trivial invariant linear manifold [46].

(ii) For every operator A there exists a hyperplane M (subspace of co-dimension 1) such that the compression of A to M has an eigenvector. (The compression of A to M is the operator B on M defined as follows: if P is the projection with range M, then $Bf = PAf$ for each f in M.) In fact more is true: for each non-zero vector f, there exists a hyperplane M containing f such that the compression of A to M has f as an eigenvector. Proof: if f is an eigenvector for A, the assertion is trivial. If f is not an eigenvector for A, then f and Af span a 2-dimensional space; let g be a non-zero vector in that space orthogonal to f, and let M be the orthogonal complement of g. Clearly f belongs to the hyperplane M. Since Af is a linear combination of f and g, the projection of Af into M is a scalar multiple of f. This proof is due to L. J.

Wallen. The result was suggested by the following theorem of C. Apostol's: if A is an operator such that 0 is in the spectrum of A but ker A = ker $A^* = 0$, then there exists an infinite-dimensional subspace M such that the compression of A to M is compact.

All the results reported on so far concern intransitivity; they assert that under certain conditions certain subspaces are invariant. The last two results to be mentioned in this context go in the direction of transitivity: certain subspaces are not invariant.

(iii) **Exercise 3**: for each countable set of non-trivial subspaces, there exists an operator that leaves none of them invariant.

(iv) There exists a linear transformation on a Hilbert space H that leaves no non-trivial subspace of H invariant. Warning: this does not pretend to be a solution of the invariant subspace problem. "Subspace" here means closed linear manifold, as always, but "linear transformation" does not mean operator, i.e., the linear transformation whose existence is asserted is not necessarily bounded. (There is, however, no fudging about domains: the assertion is about linear transformations that act on the entire space H.) The result is due to A. L. Shields [48]; the proof goes as follows.

Let ψ be the smallest ordinal number with cardinal number c (continuum). If H is a separable infinite-dimensional Hilbert space, then the set of all infinite-dimensional proper subspaces has cardinal number c, and is therefore in one-to-one correspondence $\alpha \rightarrow M_\alpha$ with the predecessors of ψ.

Let f_0 be a non-zero vector in M_0, and let g_0 be a vector *not* in M_0. Suppose that f_β and g_β have been defined whenever $\beta < \alpha < \psi$ so that f_β is in M_β, g_β is not in M_β, and the set of all f's and g's is linearly independent. The linear (not necessarily closed) manifold V_α spanned by the f's and g's cannot include M_α (because the cardinal number of a Hamel basis of M_α is c, and the cardinal number of the predecessors of α is less than c). Let f_α be a vector in M_α that is not in V_α. The linear manifold W_α spanned by V_α and f_α is not H (same argument), and, therefore, it cannot include a non-empty open set (such as the set-theoretic complement of M_α). It follows that there exists a vector g_α that belongs to neither M_α nor W_α. The inductive process so described defines a linearly independent set $E = \{f_\alpha, g_\alpha : \alpha < \psi\}$. Extend E to a Hamel basis for H; i.e., let F be a set disjoint from E such that $E \cup F$ is a Hamel basis.

Let A be the transformation on E defined by $Af_\alpha = g_\alpha$ and $Ag_\alpha = f_{\alpha+1}$. If F is empty, stop here; if F is not empty but finite, $F = \{h_1, \cdots, h_n\}$, write $Ah_i = h_{i+1}$, $i = 1, \cdots, n-1$, and $Ah_n = f_0$; and if, finally, F is infinite, let A on F be any permutation of F that has no finite orbits. In all cases, A is a transformation on the Hamel basis $E \cup F$, and, therefore, A has a unique extension to a linear transformation on H.

The linear transformation A can have no invariant infinite-dimensional proper subspace (because Af_α is never in M_α), and it can have no eigenvector (because it maps each finite linear combination of the vectors in the basis $E \cup F$ onto a linear combination that involves at least one new vector).

ANSWERS TO THE EXERCISES. (1) Since A commutes with A^2 it follows that A commutes with all the spectral projections of A^2. The only way this comment can fail to yield an invariant (in fact reducing) subspace for A is if A^2 is a scalar. That case has to be examined separately; the examination is straightforward and the desired result follows easily.

(2) Assume that the linear transformation A is such that, for some vector f, the set $\{f, Af, A^2f, \cdots\}$ is a linear basis; in all other cases the conclusion is obvious. Since every vector is a (finitely non-zero) linear combination of the form $\sum_{n=0}^{\infty} \alpha_n A^n f$, the mapping $\sum_{n=0}^{\infty} \alpha_n A^n f \to \sum_{n=0}^{\infty} \alpha_n$ is a linear functional; the kernel of that linear functional is an invariant linear manifold.

(3) If M is a non-trivial subspace, then so is $M^\perp$, and, therefore, both M and $M^\perp$ are nowhere dense sets. It follows that the set-theoretic union of a countable set of non-trivial subspaces and their orthogonal complements is a set of the first category (meager set), and that, consequently, there exists a vector that does not belong to that union. The projection onto the 1-dimensional space spanned by that vector cannot leave invariant any of the originally given subspaces.

4. TRIANGULAR

Problem 4. *Is every normal operator the sum of a diagonal operator and a compact one?*

An operator A on a Hilbert space H is *diagonal* if H has an orthonormal basis each element of which is an eigenvector of A. On a finite-dimensional space the answer to Problem 4 is trivially yes, because every operator is compact. **Exercise 1**: on a non-separable space the answer is no, even for Hermitian operators.

It is a remarkable theorem of H. Weyl [53] that every *Hermitian* operator on a separable space is the sum of a diagonal operator and a compact one. What follows is first the description of a context within which it is possible to give a proof of Weyl's theorem (different from the classical one), and then a natural generalization that has already been found useful in other parts of operator theory; it is reasonable to hope that this circle of ideas will have further applications.

On a separable space (the only kind to be considered in this section from now on) if an operator A is diagonal, then there exists an increasing sequence $\{E_n\}$ of projections of finite rank such that $E_n \to 1$ strongly and $AE_n = E_n A$ for each $n = 1, 2, 3, \cdots$. (The converse is not true. What the condition implies is that A is "block-diagonal". That is: the space is the direct sum of finite-dimensional reducing subspaces, which, however, need not be 1-dimensional.) A surprisingly rich class of operators is obtained if commutativity is replaced by asymptotic commutativity. Call an operator A *quasi-diagonal* in case there exists an increasing sequence $\{E_n\}$ of projections of finite rank such that $E_n \to 1$ strongly and $\|AE_n - E_n A\| \to 0$.

There are three kinds of basic results about quasidiagonal operators: characterization, closure, and inclusion. Characterization theorems are equivalences; they replace the definition by something else that is sometimes easier to apply. Closure theorems assert that certain operations on quasidiagonal operators lead to other operators of the same kind. Inclusion theorems assert that certain other, more familiar, classes of operators are included among the quasidiagonal ones.

The most elegant characterization theorem replaces the unnatural existence requirement in the definition of quasidiagonality by a simple equation. To describe that equation, note first that the set of all projections of finite rank is partially ordered by range inclusion, and, endowed with that order, it becomes a directed set. If ϕ is a real net (i.e., a real-valued function) on that directed set, then the convergence of ϕ to a limit α, denoted by

$$\phi(E) \to \alpha \quad \text{or} \quad \lim_{E \to 1} \phi(E) = \alpha,$$

means that for every positive number ϵ there exists a projection E_0 (of finite rank, of course) such that $|\phi(E) - \alpha| < \epsilon$ whenever $E_0 \leqq E$. Similarly, the assertion that

$$\liminf_{E \to 1} \phi(E) = \alpha,$$

means that α is the smallest number such that for every positive number ϵ and for every E_0 there exists a projection E with $E_0 \leqq E$ and $|\phi(E) - \alpha| < \epsilon$. Theorem: an operator A is quasidiagonal if and only if $\liminf_{E \to 1} \|AE - EA\| = 0$.

As for closure theorems, there are several. The most important one takes the word "closure" seriously: the set of quasidiagonal operators is (norm) closed. There are also algebraic closure theorems; e.g., every polynomial in a quasidiagonal operator, the adjoint of a quasidiagonal operator, and countable direct sums of quasidiagonal operators are again quasidiagonal.

The most trivial inclusion theorem is that diagonal operators are quasidiagonal; together with norm closure (and the spectral theorem) this yields the valuable inclusion theorem that normal operators are quasidiagonal. Another trivial inclusion theorem is that operators of finite rank are quasidiagonal; together with the characterization of compact operators as the limits of operators of finite rank [20, Problem 137] this yields the result that compact operators are quasidiagonal. The latter result can be strengthened; in fact, if A is compact, then

$$\lim_{E \to 1} \| AE - EA \| = 0.$$

(That is: lim inf can be replaced by lim.) This, in turn, implies that a compact operator will never interfere with the lim inf that establishes the quasidiagonality of some other operator; precisely, if A is quasi-diagonal and C is compact, then $A + C$ is quasidiagonal. Noteworthy special case: the sum of a normal operator and a compact one is always quasidiagonal. Is the converse true? That is (**Exercise 2**): is every quasidiagonal operator the sum of a normal operator and a compact one?

One of the facts that make quasidiagonal operators interesting is that every quasidiagonal operator is the sum of a block diagonal operator and a compact one. A slightly stronger statement is true: every quasidiagonal operator has a matrix such that, for a suitable way of dividing it into finite blocks, the result of replacing the blocks on the diagonal by 0's is compact. To prove this, suppose that $\{E_n\}$ is an increasing sequence of projections of finite rank such that $E_n \to 1$ strongly and $\| AE_n - E_n A \| \to 0$. Choose a subsequence, if necessary, to justify the assumption that $\sum_n \| AE_n - E_n A \| < \infty$. Write D for the matrix formed of the diagonal blocks $(E_n - E_{n-1}) A (E_n - E_{n-1})$ (where $E_0 = 0$) and write $C = A - D$. It is not obvious by inspection that D and C are operators, and it is even less obvious that C is compact, but that is what the proof is about to show. Indeed: put

$$C_n = E_{n+1}(AE_n - E_n A)E_n - E_n(AE_n - E_n A)E_{n+1}.$$

Clearly $\| C_n \| \leq 2 \| AE_n - E_n A \|$, and therefore $\sum_n C_n$ converges in the norm to a compact operator; an obvious computation shows that that operator is equal to C.

Weyl's theorem (the affirmative answer to Problem 4 for Hermitian operators) is an immediate consequence. Reason: if A is Hermitian, then so is each diagonal block in any matrix of A; it follows that the block diagonal summand that the preceding result yields is, in fact, diagonal. The proof shows also what blocks the proof in the normal case; a diagonal block in a normal matrix may fail to be normal. For unitary operators J. A. Dyer (simplifying and generalizing an argument of I. D. Berg) proved that Weyl's theorem remains true; the trick is to represent a unitary operator as e^{iA}, with A Hermitian, and apply the Hermitian result to A.

Quasidiagonal operators are a special case of a concept that has received some attention in the literature [10], [12], [22]; the more general operators are called quasitriangular. By definition an operator A is *quasitriangular* in case there exists an increasing sequence $\{E_n\}$ of projections of finite rank such that $E_n \to 1$ strongly and such that $\|AE_n - E_nAE_n\| \to 0$. (Motivation: replace asymptotic reduction by asymptotic invariance. Note that the appropriate definition of a *triangular* operator requires the existence of a sequence $\{E_n\}$ of projections of finite rank such that $AE_n = E_nAE_n$.)

Most of the characterization, closure, and inclusion theorems stated above for quasidiagonal operators are true for quasitriangular operators also [22]; the proofs for quasidiagonal operators are easy adaptations of the proofs in [22]. A case in point is the characterization theorem: A is quasitriangular if and only if $\lim\inf_{E \to 1}\|AE - EAE\| = 0$. A notable exception is the closure theorem: the class of quasidiagonal operators is closed under the formation of adjoints. For quasitriangular operators that is not true. In order to support the latter (negative) statement, it is, of course, necessary to have a technique for proving that something is *not* quasitriangular. The tool in [22] is the assertion that if $A^*A = 1$ and $\ker A^* \neq 0$, then A is not quasitriangular. A slightly sharper version of this is true [12]: if $A^*A \geq 1$ and $\ker A^* \neq 0$, then A is not quasitriangular. Suppose, indeed, that $e \neq 0$ and $A^*e = 0$; let E_0 be the projection onto the (1-dimensional) span of e. If E is a projection of finite rank with $E_0 \leq E$, then the restriction of EA^*E to ran E has e in its kernel; the finite-dimensionality of ran E implies the existence of a unit vector f in ran E such that $EAEf = 0$. It follows that

$$\|(AE - EAE)f\| = \|AEf\| = \|Af\| \geq \|f\|,$$

whence $\|AE - EAE\| \geq 1$; this proves that A cannot be quasitriangular. Very small consequence: since the adjoint of the unilateral shift is quasitriangular (in fact, triangular), the adjoint of a quasitri-angular operator can fail to be one.

For all that has been said so far it could be true that, for every operator A, either A or A^* is quasitriangular. A natural candidate for a counterexample is $U \oplus U^*$ (where U, of course, is the unilateral shift), but the candidate fails; $U \oplus U^*$ is, in fact, quasidiagonal. (Proof: it differs from the bilateral shift, which is normal, by an operator of rank 1.) **Exercise 3**: is there an operator A such that neither A nor A^* is quasitriangular?

The sum theorem for quasidiagonal operators (block diagonal plus compact) is not the specialization of a theorem in [22], but it might as well be; a slight complication of the proof serves to show that every quasitriangular operator is the sum of a triangular operator and a compact one.

Two further facts deserve at least brief mention. (i) Douglas and Pearcy [12] have proved that every operator with a finite spectrum is quasitriangular, and they conjectured that the same is true for operators with a countable spectrum. (ii) What happens if the lim inf in the definition of quasitriangularity is replaced by lim? Answer: a necessary and sufficient condition that $\lim_{E \to 1} \|AE - EAE\| = 0$ is that A be the sum of a compact operator and a scalar [11].

ANSWERS TO THE EXERCISES. (1) For every compact operator C there exists a separable reducing subspace whose orthogonal com-plement is included in ker C. This fact is usually stated for compact *normal* operators only [20, Problem 133]; the general case follows by applying the usual statement to C^*C and forming the smallest sub-space that reduces C and includes the separable part. Another ap-plication of the same technique shows that if $A = D + C$, where C is compact, then there exists a separable subspace that reduces both D and C and whose orthogonal complement is included in ker C. If the space is non-separable and D is diagonal, then it follows that A has many eigenvectors. If, therefore, A has no point spectrum, then A admits no sum representation of the required kind.

(2) The operators of the form "normal plus compact" do not ex-haust all quasidiagonal operators. For an example, consider the direct sum A of infinitely many copies of the operator on C^2 with matrix $\begin{pmatrix} 0 & 0 \\ 1 & 0 \end{pmatrix}$; since A is block diagonal, it is quasidiagonal. Then commutator $A^*A - AA^*$ is the direct sum of infinitely many copies of $\begin{pmatrix} 1 & 0 \\ 0 & -1 \end{pmatrix}$, which is about as far as possible from being compact; if, on the other hand, N is normal and C is compact, then $(N+C)^*(N+C) - (N+C)(N+C)^*$ is compact.

(3) If U is the unilateral shift and $A = (U-i) \oplus (U^*+i)$, the

neither A nor A^* is quasitriangular. For the proof, consider $A + i$ $= U \oplus (U^* + 2i)$ and note that it is bounded from below by 1 but its adjoint has a non-zero kernel. This answer to a question in [22] appears in [12].

5. DILATED

Problem 5. *Is every subnormal Toeplitz operator either analytic or normal?*

Toeplitz operators are the best known and analytically most important examples of compressions; they are compressions of the simplest "continuous" generalizations of diagonal matrices. If C is the unit circle ($\{z : |z| = 1\}$) in the complex plane, endowed with normalized Lebesgue measure, then the functions e_n, defined by $e_n(z) = z^n$, $n = 0, \pm 1, \pm 2, \cdots$, form an orthonormal basis for L^2 ($= L^2(C)$); let H^2 be the subspace of L^2 spanned by $\{e_0, e_1, e_2, \cdots\}$. If ϕ is a bounded measurable function on C, then the *Toeplitz operator* T_ϕ induced by ϕ is the compression to H^2 of the multiplication operator induced by ϕ on L^2. Explicitly, if P is the projection with domain L^2 and range H^2, then $T_\phi f = P(\phi \cdot f)$ for each f in H^2. The Toeplitz operator T_ϕ is *analytic* in case ϕ itself belongs to H^2. For the basic facts about Toeplitz operators see [6].

One of the first things noticed about Toeplitz operators was that their matrices (with respect to the basis $\{e_0, e_1, e_2, \cdots\}$) have an exceptionally simple form. Indeed, if $i, j = 0, 1, 2, \cdots$, then

$$(T_\phi e_j, e_i) = (P(\phi \cdot e_j), e_i) = (\phi \cdot e_j, e_i);$$

since the multiplication operator induced by e_1 is the unilateral shift U, and hence isometric, it follows that

$$(T_\phi e_j, e_i) = (U(\phi \cdot e_j), U e_i) = (\phi \cdot e_{j+1}, e_{i+1})$$
$$= (P(\phi \cdot e_{j+1}), e_{i+1}) = (T_\phi e_{j+1}, e_{i+1}).$$

The conclusion is that the diagonals of the matrix of T_ϕ are constants; the entries in row number 0 and column number 0 determine all others. Those entries are, in fact, the Fourier coefficients of ϕ (the ones with non-positive and non-negative indices respectively). From this observation it is easy to prove that not only does every Toeplitz operator have a Toeplitz matrix (i.e., one with constant diagonals), but the converse is true also; for each (bounded) Toeplitz matrix, the ϕ that induces it is given by the Fourier series whose coefficients appear in row 0 and column 0. (An algebraic way of expressing the same conclusion is this: an operator A on H^2 is a Toeplitz operator if and only if $U^*AU=A$.)

If a Toeplitz operator T_ϕ is analytic (which, by the way, happens if and only if its matrix is lower triangular), then $\phi \cdot f$ is in H^2 whenever f is in H^2; the projection P is not needed. Consequence: if T_ϕ is analytic, then it is a part of the multiplication operator induced by ϕ on L^2, and hence it is subnormal.

An easy way to get a normal Toeplitz operator is to take a Hermitian one (i.e., to take ϕ real). It is somewhat surprising but true [6] that every normal Toeplitz operator has the form $\alpha+\beta T_\phi$, where ϕ is real.

These two trivial ways of getting subnormal Toeplitz operators (analytic and normal) are the only ones known; Problem 5 asks if they are the only ones that exist. There are many other problems about Toeplitz operators, big and small, solved and unsolved; here are two small ones, one old, one new. **Exercise 1**: is there a nonzero compact Toeplitz operator? **Exercise 2**: is every operator the sum of a Toeplitz operator and a compact one?

The opposite of compression is dilation. If, that is, A and B are operators on Hilbert spaces H and K respectively, where $H \subset K$, if P is the projection from K onto H, and if $Af=PBf$ for each f in H, then A is the compression of B to H, and B is a *dilation* of A to K. One reason for the analytic importance of Toeplitz operators is that they have easily treatable dilations.

Compressions and dilations can be usefully described in terms of matrices. If K is decomposed into H and $H^\perp$, and, correspondingly, operators on K are written as two-by-two matrices (whose entries are operators on H, operators on $H^\perp$, and transformations between the two), then a necessary and sufficient condition that B be a dilation of A is that the matrix of B have the form $\left(\begin{smallmatrix} A & X \\ Y & Z \end{smallmatrix}\right)$.

The purpose of dilation theory in general is to get information about difficult operators by finding their easy dilations. The program has been spectacularly successful. Unitary operators are among the easiest to deal with, and it turns out that, except for an easily adjusted normalization, every operator has a unitary dilation. Some normalization is clearly necessary: if B is unitary, then $\|Bf\| = \|f\|$ for every vector f, and it follows that $\|Af\| \le \|f\|$ for every vector f; in other words, if A has a unitary dilation, then A must be a contraction. That much normalization is sufficient: *every contraction has a unitary dilation*. The proof is explicit; given A on H, let B on $H \oplus H$ be defined by the matrix

$$\begin{pmatrix} A & S \\ T & -A^* \end{pmatrix},$$

where S and T are the positive square roots off $-AA^*$ and $1-A^*A$ respectively. The verification that B is unitary is trivial but not obvious [20, Problem 177].

Projections, that is to say perpendicular projections, or projections onto a subspace along its orthogonal complement, are the right things to look at in Hilbert space, but, from a geometrical point of view, skew projections are just as good. In algebraic language skew projections are idempotent operators; perpendicular projections are the ones among them that happen to have unit norm. It is this normalization that forces every compression of a unitary operator to be a contraction; what happens if skew projections are allowed? **Exercise 3:** is it true that for *every* operator A on a Hilbert space H there exists a unitary operator B on a larger Hilbert space K and a skew projection P from K onto H such that $Af = PBf$ for each f in H?

What makes the theory of unitary dilations useful is Nagy's theorem [20, Problem 178] on power dilations. The assertion is that for every contraction A on H there exists a unitary operator B on a larger K such that B^n is a dilation of A^n simultaneously for every positive integer n.

One of the most spectacular applications of unitary dilation theory is the proof of von Neumann's beautiful and powerful analytic theorem about contractions [20, Problem 180]. The assertion is that if $\|A\| \leq 1$ and if p is a polynomial such that $|p(z)| \leq 1$ whenever $|z| = 1$, then $\|p(A)\| \leq 1$. For the proof, let B be a unitary power dilation of A to, say, K, let P be the (perpendicular) projection from K onto H, and note that, for each f in H,

$$\|p(A)f\| = \|Pp(U)f\| \leq \|p(U)\| \cdot \|f\| \leq \|f\|.$$

The crucial inequality, the second one, is an elementary consequence of the functional calculus for normal operators; all that is needed is the observation that, since U is unitary, the spectrum of U is on the unit circle, where $|p|$ is, by assumption, bounded by 1.

The success of this "one-variable" theory made it tempting to look for possible "several-variable" extensions. The first step was taken by Andô [1]; he proved that if A_1 and A_2 are commutative contractions on H, then there exist commutative unitary operators B_1 and B_2 on a larger K such that $B_1^m B_2^n$ is a dilation of $A_1^m A_2^n$ for every pair of positive integers m and n.

Much to everyone's surprise, Andô's theorem turned out to be a characterization of the integer 2; Parrott [37] proved that for three or more contractions the corresponding statement is false. In fact even something much weaker is false: there exist three commutative contractions A_0, A_1, A_2 such that if B_0, B_1, B_2 respectively are isometric dilations (not necessarily unitary and not necessarily power dilations), then the set $\{B_0, B_1, B_2\}$ cannot be commutative. The following presentation of Parrott's idea is due to Chandler Davis.

Suppose that C_0, C_1, C_2 are isometries on some Hilbert space; write

$$A_i = \begin{pmatrix} 0 & 0 \\ C_i & 0 \end{pmatrix}.$$

It is clear that the A's are contractions, and, since the product of any two of them is 0, it is clear that they commute. If B_0, B_1, and B_2 are dilations of A_0, A_1, and A_2 respectively, to the same enlarged space, then they can be written in the form

$$B_i = \begin{pmatrix} 0 & 0 & * \\ C_i & 0 & * \\ D_i & E_i & * \end{pmatrix},$$

where the entries indicated by $*$ need not be known. If each B_i is an isometry, so that

$$\|B_i\langle f, 0, 0\rangle\|^2 = \|\langle f, 0, 0\rangle\|^2$$

for all f, then it follows that

$$\|C_i f\|^2 + \|D_i f\|^2 = \|f\|^2$$

for all f; since C_i is an isometry, it follows that D_i must be 0. A similar glance at $\langle 0, f, 0 \rangle$ shows that E_i must be an isometry.

It is to be proved that the C's can be chosen so that the B's do not commute. Assertion: if $C_0 = 1$ and C_1 and C_2 do not commute (so that the Hilbert space they act on has dimension at least 2), then the desired result is achieved. Suppose, on the contrary, that the B's do commute. Since the entry in row 3 and column 1 of the product $B_i B_j$ is $E_i C_j$, it follows that

$$E_1 C_2 = E_2 C_1, \quad E_0 C_1 = E_1, \quad \text{and} \quad E_0 C_2 = E_2.$$

This implies that

$$E_0 C_1 C_2 = E_1 C_2 = E_2 C_1 = E_0 C_2 C_1,$$

and therefore, since E_0 is an isometry, that $C_1 C_2 = C_2 C_1$; the contradiction has arrived and the proof is complete.

ANSWERS TO THE EXERCISES. (1) The only compact Toeplitz operator is 0. Indeed, if ϕ is a bounded measurable function on C, and if n and $n+k$ are non-negative integers, then $(\phi, e_k) = (T_\phi e_n, e_{n+k})$. If

T_ϕ is compact, then $\|T_\phi e_n\| \to 0$ (since $e_n \to 0$ weakly); it follows that $(\phi, e_k) = 0$ for all k (positive, negative, or zero), and hence that $\phi = 0$.

(2) If $A = T + C$, where T is a Toeplitz operator and C is compact, then $U^*A U = T + U^*CU$, where U is the unilateral shift, and therefore $U^*A U = A + K$, where K is compact. This implies that not every A has the form $T + C$. For an example, let A be the projection onto the span of $\{e_0, e_2, e_4, \cdots\}$, so that $Ae_n = e_n$ when n is even and $Ae_n = 0$ when n is odd. If n is even, then $U^*A Ue_n = 0$; if n is odd, then $U^*A Ue_n = e_n$. Conclusion: $A - U^*A U$ is not compact.

(3) If skew projections are allowed, then every operator has a unitary dilation. Indeed, given A on H, write $K = H \oplus H$, identify H, as usual, with $H \oplus 0$, let B on K be given by $\begin{pmatrix} 0 & 1 \\ 1 & 0 \end{pmatrix}$, and let P the skew projection $\begin{pmatrix} 1 & A \\ 0 & 0 \end{pmatrix}$ from K onto H. Since PB has the matrix $\begin{pmatrix} A & 1 \\ 0 & 0 \end{pmatrix}$, the result follows. A generalization of this technique can be used to produce, for each positive integer n, a unitary operator B such that B^k is a skew dilation of A^k for each $k = 1, \cdots, n$. For the typical case $n = 2$, write $K = H \oplus H \oplus H$,

$$B = \begin{pmatrix} 0 & 0 & 1 \\ 1 & 0 & 0 \\ 0 & 1 & 0 \end{pmatrix} \quad \text{and} \quad P = \begin{pmatrix} 1 & A & A^2 \\ 0 & 0 & 0 \\ 0 & 0 & 0 \end{pmatrix}.$$

These results are due to L. J. Wallen and J. S. Johnson.

6. SIMILAR

Problem 6. *Is every polynomially bounded operator similar to a contraction?*

There is a sense in which questions of similarity are not "natural" in Hilbert space, but the few elegant results and interesting examples already known make it hard to resist the temptation to continue looking for an adequate general theory.

As is often true, the easiest operators to begin with are the unitary ones. If U is unitary and $A = S^{-1}US$, then, of course, $A^n = S^{-1}U^nS$, and therefore $\|A^n\| \leq c$ for every positive integer n (where $c = \|S^{-1}\| \cdot \|S\|$); in other words, each operator similar to a unitary operator is *power bounded*. Since A is invertible, the inequality makes sense and is true for negative exponents also. One of the earliest non-trivial results about similarity is Nagy's converse [31]: if A is invertible and both A and A^{-1} are power bounded, then A is similar to a unitary operator. (Caution: small diagonal matrices show that the inverse of a power bounded invertible operator may fail to be power bounded.) This is a satisfactory state of affairs; for similarity to a unitary operator there is a simple and usable necessary and sufficient condition.

Exercise 1: is every idempotent operator similar to a projection? What happens if the invertibility part of Nagy's condition is dropped? In other words: is there a simply describable class of operators such that power boundedness is a necessary and sufficient condition for being similar to one of them? Superficially it might seem that the class of isometries is the answer (drop the invertibility condition from the definition of unitary operators), but that is not right. (A quick way to see that it is not right is to recall that on finite-dimensional spaces all isometries are unitary.) A reasonable second guess might lead to the question: is every power bounded operator similar to a contraction? For some classes of operators the answer is yes. It is, for instance, an easy exercise in analysis to prove that every power bounded weighted shift is similar to a contraction.

It is clear that every eigenvalue of a power bounded operator must have modulus less than or equal to 1. A finite Jordan block with eigenvalue of modulus 1 (and size greater than 1) is not power bounded; if the eigenvalue has modulus less than 1, then the block is power bounded and it is easily seen to be similar to a contraction. These considerations imply that in the finite-dimensional case every power bounded operator is similar to a contraction, and it follows, of course, that the same is true for operators of finite rank on spaces of arbitrary dimension. Nagy [32] extended the result to compact operators. For not necessarily compact operators the answer is no; a counterexample was constructed by Foguel [15], [19]. **Exercise 2**: every power bounded subnormal operator is similar to a contraction.

At this point it is reasonable to begin to think that the whole approach is wrong. The preceding two paragraphs encouraged a search for a class of operators whose conjugates (i.e., transforms by similarities) were characterized by a prescribed condition (power boundedness). It might be better to prescribe an interesting class of operators (e.g., contractions) and search for a characterization of their conjugates.

With hindsight it might be said that two-sided power boundedness is a natural characterization of the conjugates of unitary operators in that it is a condition on the *group* of invertible operators. To characterize the conjugates of contractions (which may, of course, fail to be invertible) it seems reasonable to put a condition on operators considered as elements of an *algebra*. To find a good candidate for a characteristic condition, invert the problem: what can be said about the elements of the algebra generated by an operator similar to a contraction? In other words: if C is a contraction, $A = S^{-1}CS$, and p is a polynomial, what can be said about $p(A)$? Since $p(A) = S^{-1}p(C)S$, and since (by the von Neumann theorem mentioned in §5) $\|p(C)\| \leqq \|p\|_\infty$ ($= \sup\{|p(z)| : |z| \leqq 1\}$), one answer is the inequality $\|p(A)\| \leqq c\|p\|_\infty$ (where $c = \|S^{-1}\| \cdot \|S\|$). Operators satisfying this condition for all p are called *polynomially bounded*; Problem 6 asks whether they are exactly the conjugates of contractions.

Foguel's work provides an operator that is power bounded but not similar to a contraction; if that operator were polynomially bounded also, Problem 6 would be solved (in the negative). Lebow [30] examined Foguel's operator and proved that it is not polynomially bounded; Problem 6 is still unsolved.

Some interesting special cases of Problem 6 are solved.

(i) If $r(A) < 1$ (r is the spectral radius), then A is similar to a contraction. The result is due to Rota [20, Problem 122]. It takes more than a casual glance to see that it is a special case of Problem 6. The easiest proof uses the result itself. That is: if $r(A) < 1$, then A is polynomially bounded because it is similar to a contraction. There is an elegant quantitative version of Rota's theorem; it asserts that $r(A)$ is always equal to the infimum of the norms of all conjugates of A.

(ii) If A^2 is a contraction, then A is similar to a contraction. (This was proved in collaboration by A. L. Shields, L. J. Wallen, and myself; simultaneously and independently it was proved by J. P. Williams.) For the proof, define a new inner product by $((f, g)) = (f, g) + (Af, Ag)$. Since $(f, f) \leqq ((f, f)) \leqq (1 + \|A\|^2)(f, f)$, the new inner product is equivalent to the old one. The assumption that A^2 is a contraction with respect to the old inner product implies that $((Af, Af)) \leqq ((f, f))$. This means that A is a contraction with respect to the new inner product, and hence that A was similar to a contraction in the first place. For an alternative approach to essentially the same proof, let S be the positive square root of $1 + A^*A$ and verify that SAS^{-1} is a contraction. Generalization: if A^2 is similar to a contraction, then so is A. Proof: if $A^2 = S^{-1}CS$, where C is a contraction, then write $B = SAS^{-1}$. Since $B^2 = SA^2S^{-1} = C$, the result just proved implies that B is similar to a contraction and hence so is A. Both the result and the generalization extend automatically to A^n in place of A^2.

What does (ii) have to do with Problem 6? Answer: if A^2 is a contraction, then A is polynomially bounded. For the proof, given a polynomial p, express it in terms of its even and odd parts, i.e., let

q^+ and q^- be the polynomials defined by

$$q^+(z^2) = \frac{1}{2}(p(z) + p(-z)) \quad \text{and} \quad q^-(z^2) = \frac{1}{2z}(p(z) - p(-z)),$$

so that $p(z) = q^+(z^2) + zq^-(z^2)$. Since A^2 is a contraction, the von Neumann theorem on contractions applies and proves that

$$\|q^\pm(A^2)\| \leqq \|q^\pm\|_\infty.$$

Consequence:

$$\|p(A)\| \leqq \|q^+\|_\infty + \|A\| \cdot \|q^-\|_\infty \leqq (1 + \|A\|)\|p\|_\infty.$$

(The last inequality follows from the definitions of q^+ and q^-.) What the argument really proves is that if A^2 is polynomially bounded, then so is A. The extension to A^n is routine.

A final small comment about (ii) must be made before the subject can be abandoned: it is not the general case. That is: there exists a polynomially bounded operator A such that no power of A is a contraction. Example (A. L. Shields): a unilateral weighted shift with weights $1 + 1/2^n$, $n = 0, 1, 2, \cdots$. The norm of A^n is $\prod_{k=0}^{n}(1 + 1/2^k)$, which is always strictly greater than 1; to prove that A is polynomially bounded, apply the general theory of similarity for weighted shifts [20, Problem 76] and infer that A is similar to the (unweighted) unilateral shift.

(iii) **Exercise 3**: every polynomially bounded operator is the limit (in the norm) of operators that are similar to contractions. That is: granted that equality is not yet attainable, at least asymptotic equality can be achieved.

Unitary operators are too special and contractions are too general, there are several interesting classes between. What, for instance, can be said about the conjugates of isometries? Answer: A is similar to an isometry if and only if the powers of A are uniformly bounded from both above and below. To say that the powers of A are uniformly bounded from above means, of course, that the supremum of the set of numbers

$$\{\|A^n f\| : n = 0, 1, 2, \cdots, \|f\| = 1\}$$

is finite (i.e., that A is power bounded); the corresponding condition from below is that the infimum of the same set of numbers is (strictly) positive. The proof of the assertion is a routine imitation of Nagy's proof for the unitary case.

Although this characterization of the conjugates of isometries looks as if it might be awkward to apply, it does have some pleasant consequences. These consequences are minor, and each one can be proved directly, but it is more efficient to derive them from a common source. Sample: every square root of an isometry is similar to an isometry. (Isometries can have non-isometric square roots. A trivial example is $\begin{pmatrix} 1 & 1 \\ 0 & -1 \end{pmatrix}$; a non-trivial infinite-dimensional example is a unilateral weighted shift with alternating weights 2 and $\frac{1}{2}$.) For the proof, observe that if A^2 is an isometry, then A has a left inverse and, consequently, A is bounded from below; since A^n is alternately an isometry and A times an isometry, the desired result follows. The extension to higher powers is obvious.

Another sample: if a normal operator A is similar to an isometry, then A *is* an isometry (and hence A is unitary). Proof: use the power characterization of the conjugates of isometries and the spectral theorem. Here is a small surprise: for subnormal operators the conclusion is false; a counterexample was constructed by Sarason [20, Problem 156].

The problem of characterizing the conjugates of partial isometries has apparently not been studied.

ANSWERS TO THE EXERCISES. (1) If $E^2 = E$, form $A = 2E - 1$, note that $A^n = 1$ or A according as n is even or odd, and conclude that A is similar to a unitary operator U. Since $U^2 = 1$ (because $A^2 = 1$), it follows that U is Hermitian, and hence so is $\frac{1}{2}(U+1)$. Conclusion: $\frac{1}{2}(U+1)$ is a projection similar to E ($=\frac{1}{2}(A+1)$). This is an elegant application of Nagy's similarity theorem, but it has a flaw; the trouble is that the result is easier to prove directly. Indeed: decompose the space into ran$E$ ($=\ker(1-E)$) and its orthogonal complement; the matrix corresponding to E then takes the form $\begin{pmatrix} 1 & A \\ 0 & 0 \end{pmatrix}$. If $S = \begin{pmatrix} 1 & A \\ 0 & 1 \end{pmatrix}$, then S is invertible and $S^{-1}ES = \begin{pmatrix} 1 & 0 \\ 0 & 0 \end{pmatrix}$. No harm done: the conclusion is worth mentioning in any discussion of similarity. The second proof is due to J. G. Stampfli.

(2) Suppose that A is subnormal and power bounded, and let B be its minimal normal extension. If Λ denotes spectrum, then $\Lambda(B^n)$ $= (\Lambda(B))^n$ (spectral mapping theorem) $\subset (\Lambda(A))^n$ (spectral inclusion theorem for subnormal operators) $= \Lambda(A^n)$. If r denotes spectral radius, then it follows that

$$\|B^n\| = r(B^n) \leq r(A^n) \leq \|A^n\|.$$

Consequence: B is power bounded. Since the only power bounded complex numbers are the ones of modulus less than or equal to 1, the spectral theorem implies that B must be a contraction, and it follows that so is A. Conclusion: a power bounded subnormal operator is not only similar to a contraction, but actually is one.

(3) Every operator A with $r(A) \leq 1$ is the limit of operators similar to contractions. (Note that if A is polynomially bounded, then it is power bounded, and therefore $r(A) \leq 1$.) Indeed: if $A_n = (1 - 1/n)A$, then $r(A_n) < 1$, so that Rota's theorem (i) applies, and, clearly, $A_n \rightarrow A$. Since there are easy examples of operators A with $r(A) \leq 1$ that are not polynomially bounded (a hard example is the Foguel operator), this shows incidentally that the set of polynomially bounded operators is not closed.

7. Nilpotent

Problem 7. *Is every quasinilpotent operator the norm limit of nilpotent ones?*

The similarity theory of linear transformations on finite-dimensional spaces reduces to that of nilpotent ones, and the theory of nilpotent transformations turns out to be algebraically tractable. In the infinite-dimensional case nilpotence no longer plays the same central role, and the theory of the appropriately generalized concept is both algebraically and topologically refractory.

If A is nilpotent, say $A^n = 0$, then, by the spectral mapping theorem, the spectrum of A consists of 0 alone. In the finite-dimensional case the converse is true; in the infinite-dimensional case it is not. The standard example [20, Problem 80] is any unilateral weighted shift whose weights tend to 0. If A is such a shift, then A is not nilpotent (in fact if $A^n f = 0$ for some positive integer n, then $f = 0$). The operator A is, however, *quasinilpotent* in the sense that $\|A^n\|^{1/n} \rightarrow 0$. Since $\lim_n \|A^n\|^{1/n}$ is always equal to the spectral radius of A [20, Problem 74], the spectrum of a quasinilpotent operator is the singleton $\{0\}$.

Problem 7 is important because it calls for the discovery of new techniques; the question itself is "wrong". What is wrong is that the condition whose necessity is in question is already known to be *not* sufficient; a limit of nilpotent operators may fail to be quasinilpotent. The basic example is due to Kakutani [20, Problem 87]. It is a unilateral weighted shift A whose weight sequence $\{\alpha_0, \alpha_1, \alpha_2, \cdots\}$ is obtained as follows: every second α is equal to 1, (i.e., $\alpha_0 = 1$, $\alpha_2 = 1$, $\alpha_4 = 1$, $\cdots$); every second one of the remaining α's is equal to $\frac{1}{2}$ ($\alpha_1 = \frac{1}{2}$, $\alpha_5 = \frac{1}{2}$, $\alpha_9 = \frac{1}{2}$, $\cdots$); every second one of the remaining α's is $\frac{1}{4}$, etc., etc. If A_n is the "weighted shift" obtained from A by replacing $1/2^n$ by 0, then A_n is nilpotent of index 2^{n+1} and $A_n \rightarrow A$. The spectral radius of A can be obtained by a mildly onerous computation; the result is that $r(A) = 1$. Problem 7 is specific but misdirected. The more honest formulation has to be more vague; it should be something like "what is the closure of the set of nilpotent operators?". As long as the subject has been raised: what is the closure of the set of quasinilpotent operators?

There is an elegant partial solution of Problem 7 due to R. G. Douglas: every compact quasinilpotent operator is the limit of nilpotent ones. The main tool in the proof is the upper semicontinuity of the spectrum [20, Problem 86]. According to that result if A is quasinilpotent, then corresponding to each positive integer n there exists a positive number ϵ_n such that every operator within ϵ_n of A has a spectrum smaller than $1/n$. Assume, with no loss of generality, that $\epsilon_n \to 0$. If now A is compact (as well as quasinilpotent), then for each n there exists an operator B_n of finite rank such that $\|A - B_n\| < \epsilon_n$. Since all the eigenvalues of B_n are less than $1/n$ in modulus, it follows (triangular form) that B_n has the form $C_n + D_n$, where C_n is nilpotent and D_n is diagonal, with $\|D_n\| < 1/n$. Conclusion: $C_n \to A$.

Exercise 1: is every quasinilpotent operator either nilpotent or compact? (It is clear from the context that the answer must be no; otherwise Problem 7 would be solved in the affirmative.)

The general structure of quasinilpotent operators is not known at all. There are some large questions, and there are a few fragmentary theorems. Two questions have to do with the invariant subspace problem. (i) Does every quasinilpotent operator have nontrivial invariant subspaces? What makes the question interesting is not only that the answer is not known, but also that, possibly, it is equivalent to the general invariant subspace problem. No convincing reduction is known, but meditation on the finite-dimensional case, and on the connectedness of the spectrum of a transitive operator, has led some people sometimes to hope that one exists. (ii) An operator is called *unicellular* in case all its invariant subspaces are comparable (i.e., if M and N are invariant subspaces, then $M \subset N$ or $N \subset M$). The terminology comes from the finite-dimensional case: there the phenomenon occurs if and only if the Jordan form has just one block (cell). Since one block has just one eigenvalue, it is natural to raise the general question [44]: is every unicellular operator equal to the sum of a quasinilpotent operator and a scalar? For compact operators the answer is yes.

What follows is a discussion of some of the fragmentary theorems about quasinilpotence that are known. The first two are more fragmentary than the rest; their discussion is left to the reader.

Exercise 2: An analytic quasinilpotent operator is nilpotent. (Here A is called "analytic" in case there exists a function f analytic in a neighborhood of 0 such that $f(A) = 0$.)

Exercise 3: If U is the unilateral shift and A is quasinilpotent, then $\|U - A\| \geqq 1$.

Perhaps the most important place where quasinilpotent operators enter functional analysis is in Dunford's theory of spectral operators [13]. The concept belongs to Banach spaces, but in Hilbert spaces, by virtue of an elegant theorem of Wermer [52], it has an especially simple characterization. An operator A (on a Hilbert space) is *spectral* if and only if it can be represented as a sum, $A = S + Q$, where S is similar to a normal operator, Q is quasinilpotent, and S and Q commute; the representation is unique. The theory of the classical Jordan form shows that on a finite-dimensional Hilbert space every operator is spectral.

A basic result about spectral operators is that if $A = S + Q$, as above, then the spectra of A and S are the same. To prove this, observe first that if T is an operator that commutes with Q, then QT is quasinilpotent (because $\|(QT)^n\| = \|Q^n T^n\|$), and it follows that if, moreover, T is invertible, then $T + Q$ is invertible (because $T + Q = T(1 + T^{-1}Q)$). Suppose now that λ is not in the spectrum of S and write $T = S - \lambda$; by what was just said, $T + Q$ is invertible, so that λ is not in the spectrum of $S + Q$. Since S and $S + Q$ play symmetric roles, it follows that, indeed, their spectra are the same. This theorem has an ingenious generalization due to Colojoară and Foiaş [8]. They define two operators (which might as well be called A and S) to be *quasinilpotent equivalent* in case

$$\left\| \sum_{k=0}^{n} (-1)^k \binom{n}{k} A^k S^{n-k} \right\|^{1/n} \to 0$$

and in case the same is true with A and S interchanged. The point is that if A and S happen to commute (which happens exactly when $A - S$ and S commute), then the sum inside the norm is equal to $(A - S)^n$. In the commutative case quasinilpotent equivalence reduces to having a quasinilpotent difference; the non-commutative case is a proper generalization. The Colojoară-Foiaş theorem is that any two operators that are quasinilpotent equivalent have equal spectra.

If the imaginary part of a quasinilpotent operator is compact, does it follow that the operator itself is compact? The answer is yes, and the result plays an important role in the theory of invariant subspaces. The fact was discovered almost simultaneously by Ringrose [42] and Schwartz [47]. Schwartz's proof is sophisticated but short; it goes as follows. Consider the algebra $\mathfrak{G}$ of all operators, the ideal $\mathcal{C}$ of compact operators, the quotient algebra $\mathcal{B}/\mathcal{C}$, and the canonical homomorphism π from $\mathcal{B}$ to $\mathcal{B}/\mathcal{C}$. If Q is quasinilpotent (in $\mathcal{B}$), then $\pi(Q)$ is quasinilpotent (in $\mathcal{B}/\mathcal{C}$); if Im Q ($=(1/2i)$ $(Q-Q^*)$) is compact, then Im $\pi(Q)=0$. Since $\mathcal{B}/\mathcal{C}$ is representable as an operator algebra, and since the image of Q in such a representation has now been proved to be quasinilpotent and Hermitian, it follows that $\pi(Q)=0$, so that Q is compact.

A nilpotent operator is *locally nilpotent* also; that is, if $A^n=0$, then $A^nf=0$ for each f. The converse is true: if for each f there is an n ($=n(f)$) such that $A^nf=0$, then there is an n (independent of f) such that $A^n=0$. Indeed, if A is locally nilpotent, then the space is the union of the subspaces kerA^n, $n=1, 2, 3, \cdots$. The Baire category theorem implies that kerA^n has a non-empty interior for at least one n, and a subspace with a non-empty interior equals the whole space. Note: local nilpotence on a dense set is not enough to guarantee nilpotence.

The preceding paragraph extends elegantly to quasinilpotence, as follows. A quasinilpotent operator is *locally quasinilpotent* also; that is, if $\|A^n\|^{1/n}\to 0$, then $\|A^nf\|^{1/n}\to 0$ for each f. To prove the converse, observe that if $\|A^nf\|<\epsilon^n$ whenever n is sufficiently large (for each fixed f and each fixed positive number ϵ), then $\{\|A^nf/\epsilon^n\|:n =1, 2, 3, \cdots\}$ is bounded. The principle of uniform boundedness [20, Problem 40] implies that $\{\|A^n/\epsilon^n\|:n=1, 2, 3, \cdots\}$ is bounded, so that there exists a constant c such that $\|A^n\|\leq c\epsilon^n$ for all n and all ϵ. Conclusion: $\|A^n\|^{1/n}\to 0$, as promised. This smooth proof is due to Colojoară and Foias [9]. Local quasinilpotence on a dense set is not enough to guarantee quasinilpotence. Indeed, if U is the unilateral shift, and if f is a polynomial (i.e., a finite linear combination of the basis vectors being shifted), then $U^{*n}f=0$ for n sufficiently large; U^*, however, is obviously not quasinilpotent.

ANSWERS TO THE EXERCISES. (1) Begin with a quasinilpotent operator A with ker$A=0$, and put $B=\begin{pmatrix}0&A\\1&0\end{pmatrix}$. It follows that ker$B =0$, and hence that B is not nilpotent. Since $B^2=\begin{pmatrix}A&0\\0&A\end{pmatrix}$, it follows that B^2 is quasinilpotent, and therefore so is B. A simple special case is a unilateral weighted shift with every other weight equal to 1 and the in-between ones tending to 0. For another example, start with an operator that is quasinilpotent and not nilpotent, but possibly compact, and form the direct sum of infinitely many copies of it.

(2) Write $f(z)=z^ng(z)$, where g is analytic in a neighborhood of 0 and $g(0)\neq 0$. The function g has an analytic reciprocal; i.e., there exists a function h analytic in a neighborhood of 0 such that $g(z)h(z) =1$. It follows that $0=f(A)=f(A)h(A)=A^ng(A)h(A)=A^n$. (This

proof is due to D. E. Sarason.)

(3) If $\|U-A\|<1$, then $\|1-U^*A\|<1$, so that U^*A is invertible, and therefore A is left invertible. But if $BA=1$, then $B^nA^n=1$ (discard BA's from the middle), and therefore $1\leq\|B^n\|^{1/n}\|A^n\|^{1/n}$ for all n. It follows that $1\leq r(B)r(A)$, which cannot happen when A is quasinilpotent. Corollary: the set of quasinilpotent operators is not dense. As far as the corollary is concerned, however, Exercise 3 is not needed. For that purpose it is enough to note that if A is quasinilpotent, then $\|1-A\|\geq1$, and that is obvious: a quasinilpotent operator surely cannot be invertible. The corollary should be contrasted with the result that the set of all nilpotent operators of index 2 is strongly dense [20, Problem 91]. The statement of Exercise 3 is due to R. A. Hirschfeld.

8. Reducible

Problem 8. *Is every operator the norm limit of reducible ones?*

Recall that a subspace M of a Hilbert space H *reduces* an operator A on H in case both M and $M^\perp$ are invariant under A; equivalently, M reduces A in case the projection with range M commutes with A. The operator A is reducible in case it has a non-trivial reducing subspace.

On a finite-dimensional Hilbert space the answer to the question is no: the set $\mathfrak{R}$ of reducible operators is closed. Suppose, indeed, that A_n is reducible and $A_n\to A$, and, for each n, let P_n be a non-trivial projection that commutes with A_n. Finite-dimensionality implies the compactness of the unit ball in the space of operators. There is, therefore, no loss of generality in assuming that $P_n\to P$, where, of course, P is a projection, and, clearly, $AP=PA$. If $\dim H=K$, then $1\leq$ trace $P_n\leq k-1$; since trace is continuous, it follows that $P\neq0,1$. This proves that $\mathfrak{R}$ is closed; since not every operator is reducible (witness $\left(\begin{smallmatrix}0&0\\1&0\end{smallmatrix}\right)$), it follows that $\mathfrak{R}$ cannot be dense.

Exercise 1: on a non-separable Hilbert space every operator is reducible.

In view of the preceding comments, the solution of Problem 8 is negative for small spaces and trivially affirmative for large ones. The scope of the problem has been reduced to medium-sized Hilbert spaces (infinite-dimensional but separable), and there it is unsolved. Irreducible operators on such spaces certainly do exist; the unilateral shift is one of them. The unilateral shift, however, does not yield a negative solution of Problem 8; it can be approximated by reducible operators.

Another pertinent comment concerns bilateral weighted shifts. Such a shift is reducible if and only if its weight sequence is periodic [20, Problem 129]. If the weight sequence is not periodic but is the uniform limit of periodic ones, then, of course, the corresponding shift is the norm limit of reducible ones. If, however, the weight sequence is not the uniform limit of periodic ones (if, for instance, all the weights are 1 except the one with index 0, which is 2), then it does not follow that the shift is not reducibly approximable; all that follows is that the obvious approximation breaks down. The facts are not known.

Irreducible operators not only exist, they exist in profusion: the set $\mathfrak{I}$ of irreducible operators (on a separable Hilbert space) is dense. To prove this, consider an arbitrary operator A and write $A = B + iC$ with B and C Hermitian. Represent B as a multiplication on L^2 over a finite measure space. (This is one place where separability comes in.) The (real) multiplier can be uniformly approximated by (real) simple functions. Multiplication by a real simple function is the direct sum of a finite set of real scalars, and consequently it is a diagonal operator; a Hermitian diagonal operator can obviously be approximated by one with no repeated eigenvalues. Call such an approximant B_0, and consider the matrix of C with respect to a basis formed by the eigenvectors of B_0. That matrix might have some entries equal to 0, but in any event it can be approximated by a (Hermitian) matrix with all non-zero entries; let C_0 be the operator corresponding to such a matrix. The operator $A_0 = B_0 + iC_0$ approximates A; it remains to prove that A_0 is irreducible. If a subspace M reduces A_0, then M is invariant under both B_0 and C_0. A subspace invariant under B_0 is spanned by the subset of the eigenvectors of B_0 that it contains; this is a consequence of the distinctness of the eigenvalues of B_0 and is proved by a standard and elementary computation. Such a subspace, however, cannot be invariant under C_0, unless it is either 0 or the whole space; this is a consequence of the non-zeroness of the matrix entries of C_0. Conclusion: A_0 is irreducible, and the proof of the density of $\mathfrak{I}$ is complete. The first appearance of the theorem is in [21]; the proof here presented is due to Radjavi and Rosenthal [39].

The proof works for all separable spaces, and, in particular, for the finite-dimensional ones. Since in the finite-dimensional case $\mathfrak{I}$ is open (because its complement $\mathfrak{R}$ is closed), and since the complement of a dense open set is nowhere dense, it follows that in the finite-dimensional case $\mathfrak{R}$ is, in the sense of topological size, very small indeed. It is of interest to note that for separable spaces $\mathfrak{R}$ is always topologically small (meager, set of the first category). Explicitly: if the space is separable, then $\mathfrak{R}$ is an F_σ [21].

The proof that $\mathfrak{R}$ is an F_σ is tricky. Let $\mathcal{P}$ be the set of all those Hermitian operators P for which $0 \leq P \leq 1$. Recall that $\mathcal{P}$ is exactly the weak closure of the set of projections. Let $\mathcal{P}_0$ be the subset of those elements of $\mathcal{P}$ that are *not* scalars. Since $\mathcal{P}$ is a weakly closed subset of the unit ball, it is weakly compact, and hence the weak topology for $\mathcal{P}$ is metrizable. Since the set of scalars is weakly closed, it follows that $\mathcal{P}_0$ is weakly locally compact. Since the weak topology for $\mathcal{P}$ has a countable base, the same is true for $\mathcal{P}_0$, and therefore $\mathcal{P}_0$ is weakly σ-compact. Let $\mathcal{P}_1$, $\mathcal{P}_2$, $\cdots$ be weakly compact subsets of $\mathcal{P}_0$ whose union is $\mathcal{P}_0$, and, for each n, let $\hat{\mathcal{P}}_n$ be the set of all those operators A that commute with some element of $\mathcal{P}_n$. The spectral theorem implies that $\bigcup_{n=1}^{\infty} \hat{\mathcal{P}}_n = \mathfrak{R}$.

The proof can be completed by showing that each $\hat{\mathcal{P}}_n$ is norm closed. Suppose therefore that A^k is in $\hat{\mathcal{P}}_n$ and that $\|A_k - A\| \to 0$. For each k, find P_k in $\mathcal{P}_n$ so that it commutes with A_k. Since $\mathcal{P}_n$ is weakly compact and metrizable, there is no loss of generality in assuming that the sequence $\{P_k\}$ is weakly convergent to P, say. (This is the point where it is advantageous to consider all the operators in $\mathcal{P}$, and not just the projections; there is no guarantee that P is a projection even if the P_k's are. Note that P is in $\mathcal{P}_n$, so that, in particular, P is not a scalar.)

Since $A_k \to A$ (norm) and $P_k \to P$ (weak), it follows that $A_k P_k \to A P$ and $P_k A_k \to PA$ (weak). (The proof of this continuity assertion is elementary; all that it needs is that the sequence $\{P_k\}$ is bounded.) Conclusion: A commutes with P, so that A is in $\hat{\mathcal{P}}_n$.

Category arguments are sometimes used for existence proofs. That is certainly not the point here; the assertion that $\mathfrak{I}$ is not empty was obvious long before the proof that $\mathfrak{I}$ is a dense G_δ. Consider, however, invariance instead of reducibility. That is, let $\mathfrak{I}$ be the set of transitive operators; one way to try to prove that $\mathfrak{I}$ is not empty might be to prove that it is (or includes) a dense G_δ. As it stands, this is doomed to failure; since every transitive operator is cyclic, $\mathfrak{I}$ cannot be dense (cf. §1).

How far is $\mathfrak{I}$ from being topologically large? Could it be that it is so small that its complement is (or includes) a dense G_δ? The answer is not known, but the evidence is toward the affirmative: the complement of $\mathfrak{I}$ is at least dense. A much stronger statement is true (**Exercise 2**): the set of all operators with an eigenvalue is dense.

There is not much more to the theory of irreducible operators, but there is a little; it would be good to have more. Here, in conclusion for now, are two known results. (i) The set $\mathfrak{R}_1$ of all operators (on a separable infinite-dimensional Hilbert space) that have a reducing subspace of dimension 1 is not dense, but (**Exercise** 3): its closure contains every isometry. (ii) Every operator (on a separable Hilbert space) is the sum of two irreducible ones [14], [38].

ANSWERS TO THE EXERCISES. (1) If A is an operator on H and f is a non-zero vector in H, then the smallest subspace of H that contains f and reduces A is spanned by f together with the set of all vectors of the form $A_1 \cdots A_n f$, where n is an arbitrary positive integer and each A_j is either A or A^*. It follows that that (non-zero) subspace is separable; if H is not separable, then the construction has yielded a non-trivial subspace that reduces A.

(2) Given A, find an approximate eigenvector for A, use it as the first element of a basis, form the matrix of A, and then find an approximant by replacing by 0 all but the first entry in the first column. Conclusion: every operator can be approximated by operators with eigenvalues.

(3) It is to be proved that if U is an isometry, then U can be approximated by operators with reducing eigenvectors. The proof is similar to the one in (2). Let λ be an approximate eigenvalue of U with modulus 1. It follows, by definition, that corresponding to each positive number ϵ there exists a unit vector e such that $\| Ue - \lambda e \| < \epsilon$, and hence such that $\| U^*e - \lambda^*e \| = \| -\lambda^*U^*(Ue - \lambda e) \| < \epsilon$. Use e as the first element of a basis, form the matrix of U, and then find an approximant by replacing by 0 all but the first entry in both the first row and the first column.

9. REFLEXIVE

Problem 9. *Is every complete Boolean algebra reflexive?*

The invariant subspace problem asks whether the set of all invariant subspaces of an operator can consist of the two extremes only, but what people would really like to know is what the set of all invariant subspaces of an operator, or, for that matter, of any set of operators can look like. A few necessary conditions are easy to obtain. If, for instance, $\mathcal{Q}$ is a set of operators (on H) and $\mathcal{L}$ is the set of all those subspaces (of H) that are invariant under every operator in $\mathcal{Q}$, then it is clear that $\mathcal{L}$ is a lattice (i.e., that $\mathcal{L}$ is closed under the formation of intersections and spans), and it is even clearer that $\mathcal{L}$ contains 0 and H. In addition to these algebraic conditions, $\mathcal{L}$ satisfies a somewhat less obvious topological condition, namely that it is strongly closed. (Explanation: the set of all projections whose ranges are in $\mathcal{L}$ is strongly closed. Proof: if A is in $\mathcal{Q}$ and $\{P_n\}$ is a net of projections strongly convergent to P and satisfying $AP_n = P_n AP_n$ for each n, then, because multiplication is jointly continuous on bounded sets, $AP = PAP$.) The topological condition has algebraic reverberations. **Exercise 1**: a strongly closed lattice is complete. (Caution: the converse is false. Example: discard a 1-dimensional space from the lattice of all subspaces of a 2-dimensional space; the remainder is a complete but non-closed lattice.) It is easy to show that the necessary conditions listed so far are nowhere near sufficient; the characterization problem for the set of all invariant subspaces of a set of operators is still open.

There is a kind of Galois theory connecting sets of operators and sets of subspaces. Half of it was just described: to each set $\mathcal{Q}$ of operators there corresponds the set of those subspaces that are invariant under the elements of $\mathcal{Q}$; call that set Lat$\mathcal{Q}$. The other half goes backwards: to each set $\mathcal{L}$ of subspaces there corresponds the set Alg$\mathcal{L}$ of those operators that leave invariant each element of $\mathcal{L}$. It is clear that Alg$\mathcal{L}$ is always an algebra (closed under the formation of sums, products, and scalar multiples); it is even clearer that Alg$\mathcal{L}$ contains 1; and the one-sided weak continuity of multiplication implies that Alg$\mathcal{L}$ is weakly closed.

The basic properties of Alg and Lat are easy; their proofs belong to universal algebra. The facts are that both Alg and Lat are order-reversing, and that if $\mathcal{L}$ is any set of subspaces and $\mathcal{Q}$ is any set of operators, then

$$\mathcal{L} \subset \text{Lat Alg}\,\mathcal{L} \quad \text{and} \quad \mathcal{Q} \subset \text{Alg Lat}\,\mathcal{Q}.$$

(**Exercise 2**: Alg $\mathcal{L}$ = Alg Lat Alg $\mathcal{L}$ and Lat $\mathcal{Q}$ = Lat Alg Lat $\mathcal{Q}$. Analogy with linear algebra (when is an object equal to its own second dual?) suggests that if $\mathcal{L}$ = Lat Alg$\mathcal{L}$, then the lattice $\mathcal{L}$ be called *reflexive*; the same word is used for an algebra $\mathcal{Q}$ such that $\mathcal{Q}$ = Alg Lat $\mathcal{Q}$. Word-saving convention: every lattice contains 0 and H and is strongly closed, every algebra contains 1 and is weakly closed.

The theory of reflexive algebras is relatively new and, judging from the analytic techniques it has been using, quite deep. A typical result is Sarason's [45]: a commutative algebra of normal operators is reflexive. Example: the set of all matrices of the form $\left(\begin{smallmatrix} a & 0 \\ 0 & b \end{smallmatrix}\right)$. In a different direction, Radjavi and Rosenthal [40] generalized a theorem of Arveson's [3] as follows: if $\mathcal{Q}$ includes a maximal abelian self-adjoint algebra and if Lat$\mathcal{Q}$ is a chain (totally ordered), then $\mathcal{Q}$ is reflexive. Example: the set of all matrices of the form $\left(\begin{smallmatrix} a & 0 \\ b & c \end{smallmatrix}\right)$. A typical example of a non-reflexive algebra is the set of all matrices of the form $\left(\begin{smallmatrix} a & 0 \\ 0 & a \end{smallmatrix}\right)$. A complete characterization of reflexive algebras is unknown even for finite-dimensional spaces.

To ask about invariant subspace lattices of sets of operators is the same as to ask about reflexive lattices. Here are two trivial but illuminating examples.

(i) If H is 2-dimensional and $\mathcal{L}$ consists of 0, H, and two distinct 1-dimensional subspaces of H, then $\mathcal{L}$ is reflexive. To say of an operator that it leaves invariant each subspace in $\mathcal{L}$ is to say just that it has two prescribed eigenvectors, and hence that its matrix with respect to the basis they form is diagonal. Since the only subspaces simultaneously invariant under all such diagonal operators are the ones in $\mathcal{L}$, the lattice $\mathcal{L}$ is reflexive. (ii) Suppose again that H is 2-dimensional, and let $\mathcal{L}$ consist of 0, H, and three distinct 1-dimensional subspaces. It is very hard for an operator on H to have three distinct eigenvectors; the only operators that can do it are the scalars. Scalars, on the other hand, leave invariant every subspace of H; the lattice $\mathcal{L}$ is not reflexive.

The first non-trivial theorem about reflexive lattices is a sharpening of some related work of Kadison and Singer [27]; the statement, proved in complete generality by Ringrose [43], is that every complete chain is reflexive. Here again "chain" means a lattice whose order (in general partial) is in fact total. No topological assumption is needed: it is easy to prove that a complete chain of subspaces is necessarily strongly closed.

The smallest lattice that is not a chain is a 4-element Boolean algebra, i.e., a lattice of the form $\mathcal{L} = \{0, M, N, H\}$, where the subspaces M and N are such that $M \cap N = 0$ and $M \vee N = H$. It is true but not obvious that each such $\mathcal{L}$ is reflexive. To prove this it is sufficient (and necessary) to exhibit a set $\mathcal{a}$ of operators such that Lat $\mathcal{a} = \mathcal{L}$. Let P and Q be the projections onto M and N respectively and let $\mathcal{a}$ be the set of all operators that are either of the form $PA(1-Q)$ or else of the form $QA(1-P)$. Since $PA(1-Q)$ annihilates N and maps everything into M, and, similarly, $QA(1-P)$ annihilates M and maps everything into N, it follows that $\mathcal{L} \subset \text{Lat}\,\mathcal{a}$. It remains to prove the reverse inclusion.

A useful lemma is that if K is in Lat$\mathcal{a}$ and $K \not\subset M$, then $K \supset N$. (Similarly, of course, if K is in Lat$\mathcal{a}$ and $K \not\subset N$, then $K \supset M$.) For the proof, take a vector f in K but not in M, and take an arbitrary vector g in N. Since f is not in M, it follows that $(1-P)f \neq 0$, and hence that there exists an operator A such that $A(1-P)f = g$. Since $QA(1-P)f$ is in K (because K is in Lat$\mathcal{a}$) and $QA(1-P)f = g$ (because g is in N), it follows that g is in K; this proves the lemma.

Suppose now that K is in Lat $\mathcal{C}$, $K \neq 0$, M, N; it is to be proved that $K = H$. The subspace K cannot be included in both M and N (because $K \neq 0$). If, say, $K \not\subset M$, then, by the lemma, $K \supset N$. Since, however, $K \neq N$, it is not the case that $K \subset N$; the lemma applies again and yields $K \supset M$. The conclusion follows from the assumption that $M \vee N = H$.

The result just proved is contained in a much more general theorem: every complete atomic Boolean algebra is reflexive. The complementation implicit in the phrase "Boolean algebra" is not, of course, assumed to be orthogonal; compare the $\mathcal{L}$ just treated. To say of a Boolean algebra of subspaces that it is atomic means that every subspace in the algebra is the span of all the atoms it includes; an *atom* is a non-zero subspace in the algebra that includes no subspace in the algebra other than 0 and itself. Problem 9 asks whether the assumption of atomicity can be omitted. The atomic special case was first announced in [23].

There are at least two ways of making new reflexive lattices out of old ones: direct sums and ordinal sums. The latter is more useful. A lattice $\mathcal{L}$ of subspaces of a Hilbert space H is the ordinal sum of two lattices in case it contains a comparable element H_0. That means that, for each M in $\mathcal{L}$, either $M \subset H_0$ or $H_0 \subset M$. If that is the case, then there are, indeed, two other lattices associated with $\mathcal{L}$; (i) the lattice $\mathcal{L}^-$ of all those subspaces of H_0 that happen to be in $\mathcal{L}$, and (ii) the lattice $\mathcal{L}^+$ of all those subspaces of $H_0^\perp$ that are of the form $M \cap H_0^\perp$ for some M in $\mathcal{L}$. (This ad hoc definition is not elegant, but it avoids a long digression, and suggests the correct picture.) Theorem: if both $\mathcal{L}^-$ and $\mathcal{L}^+$ are reflexive, then so is $\mathcal{L}$. Sample application: if dim $H = 3$ and $\mathcal{L}$ (the "pendulum") consists of 0, two 1-dimensional spaces, their span, and H, then $\mathcal{L}$ is reflexive. This is of interest because (**Exercise 3**) $\mathcal{L}$ is not the Lat of any single operator.

The theorems and techniques mentioned so far yield many reflexive lattices and a few non-reflexive ones, and they are sufficient to decide the status of all sufficiently small lattices. The following comments exhaust all that is known, and show therefore how much remains to be done.

If a lattice has four elements, or fewer, then it is reflexive. There are five isomorphism types of lattices with five elements: the chain, the pendulum, the pendulum upside down, the non-modular pentagon, and the double triangle. Every lattice of subspaces that belongs to one of the first three types is reflexive. The pentagon ($= \{0, L, M, N, H\}$, with $M \subset N$, $L \cap M = L \cap N = 0$ and $L \vee M = L \vee N = H$) is not realizable as a lattice of subspaces of a finite-dimensional space. If dim H is infinite, the pentagon can occur, and in the one manifestation that has been studied it turns out to be reflexive. (Can it have a non-reflexive occurrence? The answer is not known.) As for the double triangle ($= \{0, L, M, N, H\}$, with $L \cap M = L \cap N = M \cap N = 0$ and $L \vee M = L \vee N = M \vee N = H$), all its known manifestations are non-reflexive. The known manifestations include all possible finite-dimensional ones; the unknown cases are all infinite-dimensional. The finite-dimensional facts are covered by a theorem of R. E. Johnson [25]: a finite lattice of subspaces of a finite-dimensional space is reflexive if and only if it is distributive. The "if" remains true for infinite-dimensional spaces (K. J. Harrison); the reflexive realizability of the pentagon shows that the "only if" is false for them.

ANSWERS TO THE EXERCISES. (1) Suppose that $\mathcal{L}$ is a strongly closed lattice and that $\mathcal{L}_0$ is a subset of $\mathcal{L}$. Consider the directed set of all finite subsets of $\mathcal{L}_0$ ordered by inclusion. If n is such a finite subset, let M_n be the intersection of the subspaces in n. Since $n \to M_n$ is a decreasing net of subspaces, the corresponding net of projections is strongly convergent to the projection whose range is $\cap \mathcal{L}_0$; this proves that $\cap \mathcal{L}_0$ is in $\mathcal{L}$. The proof for spans is similar.

(2) Apply Lat to the relation $\mathcal{Q} \subset \mathrm{Alg\ Lat}\mathcal{Q}$, and apply the relation $\mathcal{L} \subset \mathrm{Lat\ Alg}\mathcal{L}$ to $\mathrm{Lat}\mathcal{Q}$ in place of $\mathcal{L}$. This proves that $\mathrm{Lat}\mathcal{Q} = \mathrm{Lat\ Alg\ Lat}\ \mathcal{Q}$; the equation $\mathrm{Alg}\mathcal{L} = \mathrm{Alg\ Lat\ Alg}\mathcal{L}$ is proved dually. Corollary: every lattice that is the Lat of something is the Lat of its own Alg.

(3) If dim $H < \infty$, and if $\mathcal{L} = \mathrm{Lat}\ A$ for some operator A on H, then $\mathcal{L}$ is self-dual, in the sense that $\mathcal{L}$ is isomorphic to a lattice in which all the order relations of $\mathcal{L}$ are reversed. Proof: the formation of orthogonal complements proves that $\mathrm{Lat}\ A$ is anti-isomorphic to $\mathrm{Lat}\ A^*$; since every matrix is similar to its transpose, and similar operators have isomorphic lattices, the proof can be completed by recalling that the matrix of A^* is the conjugate transpose of that of A [5].

10. TRANSITIVE

Problem 10. *Is every non-trivial strongly closed transitive atomic lattice either medial or self-conjugate?*

This is the most awkward of the ten problems here proposed; the reason is that it points to the darkest area of ignorance.

The crucial word is "transitive". A set $\mathcal{Q}$ of operators is called transitive in case $\mathrm{Lat}\mathcal{Q} = \mathcal{L}_{\min}$ ($=$ the smallest lattice, the one consisting of 0 and H only); a set $\mathcal{L}$ of subspaces is transitive in case $\mathrm{Alg}\mathcal{L} = \mathcal{Q}_{\min}$ ($=$ the smallest algebra, the one consisting of scalars only).

If a set $\mathcal{C}$ of operators is transitive, then so is every subset of $\mathcal{C}$ that generates the same weakly closed algebra. It follows that the search for transitive sets might as well be restricted to weakly closed algebras. (Word-saving convention: every algebra contains 1, every lattice contains 0 and H.) Problem 10 has a dual, a problem for algebras, that is easier to state: is every weakly closed transitive algebra equal to $\mathcal{C}_{max}$ (=the largest algebra, the one consisting of all operators)? It is implicit in the question that $\mathcal{C}_{max}$ itself is transitive; if a subspace M of H is invariant under every operator on H, then $M = 0$ or $M = H$. Conjecture: the solution of this dual of Problem 10 is negative. In fact, there probably exists a commutative transitive algebra. This conjecture, in turn, is a special case of one already on record (in §3): if a transitive operator exists (in present language an operator A is transitive in case Lat $A = \mathcal{L}_{min}$), then the weakly closed algebra it generates is a transitive algebra. All these conjectures seem to be out of reach at present. If H is small (finite-dimensional), then Burnside's classical theorem applies [24, p. 276] and says that the only transitive algebra is $\mathcal{C}_{max}$. If H is very large (non-separable), then an easy cardinality argument shows that no countably generated algebra can be transitive. The cases between these two extremes remain shrouded in mystery.

The corresponding questions for lattices are newer, and, although the principal characterization problem is still unsolved, they promise to be somewhat more accessible.

The first thing to settle is that the definition is not vacuous: are there any transitive lattices? Since a lattice larger than a transitive one is also transitive, the question reduces to a known projective geometric fact (**Exercise 1**): if $\mathcal{L} = \mathcal{L}_{max}$ (=the largest lattice, the one consisting of all subspaces), then $\mathcal{L}$ is transitive.

Once that's settled, it is natural to regard $\mathcal{L}_{max}$ as a trivial example and to ask whether there exist any non-trivial strongly closed transitive lattices. The answer is yes (J. E. McLaughlin), but that's not obvious. To get an interesting class of examples, consider an arbitrary conjugation J on a Hilbert space H. (A *conjugation* is an involutory semilinear isometry, i.e., a mapping J of H into itself such that $J^2 = 1$, $J(\alpha f + \beta g) = \alpha^* J f + \beta^* J g$, and $\|Jf\| = \|f\|$. If H is an L^2 space, then an example of a conjugation can be obtained by writing $Jf = f^*$, and, conversely, every conjugation can be represented in this way. For a detailed discussion see [50, p. 357].) Call a subspace M of H self-conjugate (with respect to J) in case Jf is in M whenever f is in M, and let $\mathcal{L}$ ($= \mathcal{L}_J$) be the set of all self-conjugate subspaces. It is easy to verify that $\mathcal{L}$ is a strongly closed transitive lattice. (The proof of transitivity uses the same technique as is needed to prove the transitivity of $\mathcal{L}_{\max}$.) Call every $\mathcal{L}$ obtained in this way a *self-conjugate* lattice; there are as many of them as there are J's.

Once that's settled, it is again natural to ask whether every transitive lattice has been found by now. In other words, do there exist non-trivial strongly closed transitive lattices other than the self-conjugate ones? The answer is yes again, but that's even less obvious than anything that led up to it.

To construct an example, let K be a Hilbert space, write $H = K \oplus K$, and let $\mathcal{L}$ be the 7-element lattice of subspaces of H consisting of 0, H, the two axes ($K \oplus 0$ and $0 \oplus K$), the diagonal (the set of all $\langle f, f \rangle$'s with f in K), and the graphs of two operators S and T on K. Assertion: for suitable choice of S and T, the lattice $\mathcal{L}$ is transitive.

Since H is given as a direct sum of two copies of K, every operator on H is a two-by-two matrix of operators on K. The assumption that $\mathcal{L}$ contains the two axes implies that every operator in Alg $\mathcal{L}$ has the form $\left(\begin{smallmatrix} z & 0 \\ 0 & y \end{smallmatrix}\right)$; the assumption that $\mathcal{L}$ contains the diagonal implies that $Y = X$.

The operators S and T are to be chosen so that their graphs are disjoint from one another and from the other non-trivial elements of $\mathcal{L}$ (i.e., from the axes and the diagonal). When can $\langle f, Sf \rangle$ be equal to $\langle g, Tg \rangle$? Answer: exactly when $f = g$ and $(S - T)f = 0$. Consequence: graph $S \cap$ graph $T = 0$ if and only if ker $(S - T) = 0$. Since the horizontal axis ($K \oplus 0$) and the diagonal are graphs, and since every graph is disjoint from the vertical axis, it follows that the five non-trivial elements of $\mathcal{L}$ are pairwise disjoint if and only if

$$\ker S = \ker T = \ker(1 - S) = \ker(1 - T) = \ker(S - T) = 0.$$

The consideration of the orthogonal complements of graphs shows that the five non-trivial elements of $\mathcal{L}$ pairwise span H if and only if their adjoints satisfy the vanishing kernel conditions just found. (If K is finite-dimensional, the resulting conditions are equivalent to the original ones; in the infinite-dimensional case they are not. Note that the orthogonal complement of the graph of, say, S is the "co-graph" consisting of all vectors of the form $\langle -S^*f, f \rangle$.)

When does $\left(\begin{smallmatrix} X & 0 \\ 0 & X \end{smallmatrix}\right)$ leave invariant the graph of S? An obvious computation shows that a necessary and sufficient condition is that $SX = XS$. Consequence: $\mathcal{L}$ is transitive if and only if

$$\text{Com } S \cap \text{Com } T = \mathcal{C}_{\min},$$

where "Com" means commutant.

The problem of finding a transitive lattice of the kind promised above comes down to this: find two operators S and T satisfying the commutant condition just stated and such that both they and their adjoints satisfy the vanishing kernel condition. That is not difficult, but it does seem to lead to some matrix computation. One example (with dim $K < \infty$) is obtained by letting S be the sum of a nilpotent Jordan block and a scalar, say, 2 (what is important about 2 is that it is neither 0 nor 1), and letting T be a diagonal matrix none of whose eigenvalues is 0, 1, or 2. The only condition whose verification is not obvious is the one involving ker $(S-T)$. To prove it, compute $T^{-1}S$ and note that it is a lower triangular matrix with all diagonal entries distinct from 1; it follows that $T^{-1}S-1$ is invertible, and hence so is $S-T$.

A similar example can be constructed on infinite-dimensional spaces. Suppose, for instance, that K is L^2 of the circle, let S be the bilateral shift ($Se_n = e_{n+1}$ for all n), and let T be a diagonal operator ($Te_n = \lambda_n e_n$ for all n). If the λ's are bounded, if none of them is 0 or 1, and if they are such that

$$\sum_{n=1}^{\infty} |1/\lambda_1 \cdots \lambda_n|^2 < \infty \quad \text{and} \quad \sum_{n=1}^{\infty} |1/\lambda_{-1} \cdots \lambda_{-n}|^2 < \infty,$$

then S and T satisfy all the requirements. The proof is straightforward.

The lattices $\mathcal{L}$ just constructed have the property that any two of their non-trivial elements are complements. Since in the lattice diagram of such an $\mathcal{L}$ all the non-trivial elements occur on the same level, halfway between 0 and H, the word *medial* might serve as an adequate description. P. Rosenthal reports that in an infinite-dimensional space a transitive medial lattice can be constructed with only four non-trivial elements. The construction uses, as above, the two axes and the diagonal, but only one graph, namely the graph of a suitably chosen unbounded transformation. It is not known whether three non-trivial elements can do the job. In finite-dimensional spaces (of dimension greater than 2) a medial lattice with four (or fewer) non-trivial elements can never be transitive; the proof proceeds by observing that any such lattice can be represented, as above, by two axes and two (or fewer) graphs, and then proving that there are too many degrees of freedom to make transitivity possible. **Exercise 2**: is every medial lattice with five (or more) non-trivial elements transitive?

Each new construction brings with it the question whether it's the last. Do there exist non-trivial strongly closed transitive lattices other than the medial or the self-conjugate ones? Once again the answer is yes, but that yes is at the boundary of what is known today. K. J. Harrison has constructed a transitive lattice of 16 non-trivial elements (the dimension of the space is at least 8), of which 5 are atoms, 5 are co-atoms, and 6 are in the middle. It seems hard to describe a general class of which Harrison's example is an instance and which, together with the classes described before, has a chance of catching all transitive lattices. A reasonable guess is that atoms play a crucial role. Recall that a lattice is called *atomic* in case every element is the span of all the atoms it includes. (**Exercise 3**: if $2 \leq \dim H < \infty$ and if a lattice $\mathcal{L}$ of subspaces of H has exactly one atom, then $\mathcal{L}$ is not transitive.) The trouble with Harrison's 18-element lattice is that it is not atomic. A fussy examination of low-dimensional cases indicates that transitive lattices want to be atomic. In the infinite-dimensional case atomicity still makes sense but it may be too special to play an important role. In any event, since no intelligent guess at the structure of all (strongly closed) transitive lattices is at hand, the best that can be done is to grasp at the atomic straw; that is what Problem 10 does.

ANSWERS TO THE EXERCISES. (1) It is to be proved that if an operator A on H leaves invariant every subspace of H, then A is a scalar. The assumption implies that to every non-zero vector f there corresponds a unique scalar $\lambda(f)$ such that $Af = \lambda(f)f$. Since $A(f+g) = Af + Ag$, it follows that $(\lambda(f) - \lambda(f+g))f + (\lambda(g) - \lambda(f+g))g = 0$; this implies that if f and g are linearly independent, then $\lambda(f) = \lambda(g)$ $(= \lambda(f+g))$. If, on the other hand, $f = \alpha g$ (with $\alpha \neq 0$), then $Af = \alpha Ag$ and therefore, again, $\lambda(f) = \lambda(g)$, i.e., the function λ is a constant.

(2) There exist intransitive medial lattices with exactly five non-trivial elements. For a simple example in $H = K \oplus K$ (where K is separable and $\dim K \geq 2$), let $\mathcal{L}$ consist of the two axes, the diagonal, and the graphs of two operators S and T. If, for instance, both S and T are diagonal matrices, such that the set of all diagonal entries in both of them together is a set of distinct numbers all distinct from 0 and 1, then the vanishing kernel conditions are satisfied. It follows that $\mathcal{L}$ is indeed a medial lattice. Since both Com S and Com T consist of all diagonal matrices, it follows that Alg$\mathcal{L}$ consists of all $\begin{pmatrix} X & 0 \\ 0 & X \end{pmatrix}$, where X is diagonal; since $\dim K \geq 2$, this implies that Alg$\mathcal{L}$ $\neq \mathcal{C}_{\min}$.

(3) If $\mathcal{L}$ has exactly one atom M_0, then, since finite-dimensionality implies that every element of $\mathcal{L}$ includes at least one atom, it follows that every element of $\mathcal{L}$ includes M_0. If $M_0 = H$, then $\mathcal{L}$ is not transitive (because $\dim H \geq 2$). If $M_0 \neq H$, then there exists a non-scalar

operator A on H with range M_0. Such an A leaves invariant every M in $\mathfrak{L}$; indeed $A M \subset A H = M_0 \subset M$.

BIBLIOGRAPHY

1. T. Andô, *On a pair of commutative contractions*, Acta Sci. Math. (Szeged) **24** (1963), 88–90. MR **27** #5132.

2. N. Aronszajn and K. T. Smith, *Invariant subspaces of completely continuous operators*, Ann of Math. (2) **60** (1954), 345–350. MR **16**, 488.

3. W. B. Arveson, *A density theorem for operator algebras*, Duke Math. J. **34** (1967), 635–647. MR **36** #4345.

4. E. Bishop, *Spectral theory for operators on a Banach space*, Trans. Amer. Math. Soc. **86** (1957), 414–445. MR **20** #7217.

5. L. Brickman and P. A. Fillmore, *The invariant subspace lattice of a linear transformation*, Canad. J. Math. **19** (1967), 810–822. MR **35** #4242.

6. A. Brown and P. R. Halmos, *Algebraic properties of Toeplitz operators*, J. Reine Angew. Math. **213** (1963/64), 89–102. MR **28** #3350; errata MR **30**, 1205.

7. A. Brown, P. R. Halmos, and A. L. Shields, *Cesàro operators*, Acta Sci. Math. (Szeged) **26** (1965), 125–137. MR **32** #4539.

8. I. Colojoară and C. Foiaş, *Quasi-nilpotent equivalence of not necessarily commuting operators*, J. Math Mech. **15** (1966), 521–540. MR **33** #570.

9. ———, *Theory of generalized spectral operators*, Mathematics and its Applications, vol. 9, Gordon and Breach, New York, 1968.

10. D. Deckard, R. G. Douglas, and C. Pearcy, *On invariant subspaces of quasi-triangular operators*, Amer. J. Math. **91** (1969), 637–647.

11. R. G. Douglas and C. Pearcy, *A characterization of thin operators*, Acta Sci. Math. (Szeged) **29** (1968), 295–297. MR **38** #2628.

12. ———, *A note on quasitriangular operators*, Duke Math. J. **37** (1970), 177–188.

13. N. Dunford, *A survey of the theory of spectral operators*, Bull. Amer. Math. Soc. **64** (1958), 217–274. MR **21** #3616.

14. P. A. Fillmore and D. M. Topping, *Sums of irreducible operators*, Proc. Amer. Math. Soc. **20** (1969), 131–133. MR **38** #1549.

15. S. Foguel, *A counterexample to a problem of Sz.-Nagy*, Proc. Amer. Math. Soc. **15** (1964), 788–790. MR **29** #2646.

16. R. Gellar, *Cyclic vectors and parts of the spectrum of a weighted shift*, Trans. Amer. Math. Soc. **146** (1969), 69–85.

17. P. R. Halmos, *Measure theory*, Van Nostrand, Princeton, N. J., 1950. MR **11**, 504.

18. ———, *Normal dilations and extensions of operators*, Summa Brasil. Math. **2** (1950), 125–134. MR **13**, 359.

19. ———, *On Foguel's answer to Nagy's question*, Proc. Amer. Math. Soc. **15** (1964), 791–793. MR **29** #2647.

20. ———, *A Hilbert space problem book*, Van Nostrand, Princeton, N. J., 1967. MR **34** #8178.

21. ———, *Irreducible operators*, Michigan Math. J. **15** (1968), 215–223. MR **37** #6788.

22. ———, *Quasitriangular operators*, Acta Sci. Math. (Szeged) **29** (1968), 283–293. MR **38** #2627.

23. ———, *Invariant subspaces*, Abstract Spaces and Approximation, Proc. M. R. I. Oberwolfach, Birkhäuser, Basel, 1968, pp. 26–30.

24. N. Jacobson, *Lectures in abstract algebra.* Vol. II: *Linear algebra*, Van Nostrand, Princeton, N. J., 1953. MR **14**, 837.

25. R. E. Johnson, *Distinguished rings of linear transformations*, Trans. Amer. Math. Soc. **111** (1964), 400–412. MR **28** #5088.

26. R. V. Kadison, *Strong continuity of operator functions*, Pacific J. Math. **26** (1968), 121–129. MR **37** #6766.

27. R. V. Kadison and I. M. Singer, *Triangular opeartor algebras. Fundamentals and hyperreducible theory*, Amer. J. Math. **82** (1960), 227–259. MR **22** #12409.

28. G. K. Kalisch, *On similarity, reducing manifolds, and unitary equivalence of certain Volterra operators*, Ann. of Math. (2) **66** (1957), 481–494. MR **19**, 970.

29. R. L. Kelley, *Weighted shifts on Hilbert space*, Dissertation, University of Michigan, Ann Arbor, Mich., 1966.

30. A. Lebow, *A power-bounded operator that is not polynomially bounded*, Michigan Math. J. **15** (1968), 397–399. MR **38** #5047.

31. B. Sz.-Nagy, *On uniformly bounded linear transformations in Hilbert space*, Acta Sci. Math. (Szeged) **11** (1947), 152–157. MR **9**, 191.

32. ———, *Completely continuous operators with uniformly bounded iterates*, Magyar Tud. Akad. Mat. Kutató Int. Közl. **4** (1959), 89–93. MR **21** #7436.

33. B. Sz.-Nagy and C. Foiaş, *Opérateurs sans multiplicité*, Acta Sci. Math. (Szeged) **30** (1969), 1–18.

34. N. K. Nikol'skiĭ, *Invariant subspaces of weighted shift operators*, Mat. Sb. **74** (**116**) (1967), 171–190-Math. USSR Sb. **3** (1967), 159–176. MR **37** #4659.

35. R. S. Palais, *Seminar on the Atiyah-Singer index theorem*, Ann. of Math. Studies, no. 57, Princeton Univ. Press, Princeton, N. J., 1965. MR **33** #6649.

36. S. K. Parrott, *Weighted translation operators*, Dissertation, University of Michigan, Ann Arbor, Mich., 1965.

37. ———, *Unitary dilations for commuting contractions*, Pacific J. Math. (to appear).

38. H. Radjavi. *Every operator is the sum of two irreducible ones*, Proc. Amer. Math. Soc. **21** (1969), 251–252. MR **38** #6388.

39. H. Radjavi and P. M. Rosenthal, *The set of irreducible operators is dense*, Proc. Amer. Math. Soc. **21** (1969), 256. MR **38** #5042.

40. ———, *On invariant subspaces and reflexive algebras*, Amer. J. Math. **91** (1969), 683–692.

41. W. C. Ridge, *Approximate point spectrum of a weighted shift*, Trans. Amer. Math. Soc. **147** (1970), 349–356.

42. J. R. Ringrose, *On the triangular representation of integral operators*, Proc. London. Math. Soc. (3) **12** (1962), 385–399. MR **25** #3372.

43. ———, *On some algebras of operators*, Proc. London Math. Soc. (3) **15** (1965), 61–83. MR **30** #1405.

44. P. M. Rosenthal, *On lattices of invariant subspaces*, Dissertation, University of Michigan, Ann Arbor, Mich., 1967.

45. D. E. Sarason, *Invariant subspaces and unstarred operator algebras*, Pacific J. Math. **17** (1966), 511–517. MR **33** #590.

46. H. H. Schaefer, *Eine Bemerkung zur Existenz invarianter Teilräume linearer Abbildungen*, Math. Z. **82** (1963), 90. MR **27** #4815.

47. J. Schwartz, *Subdiagonalization of operators in Hilbert space with compact imaginary part*, Comm. Pure Appl. Math. **15** (1962), 159–172. MR **26** #1759.

48. A. L. Shields, *A note on invariant subspaces*, Michigan Math. J. (to appear).

49. J. G. Stampfli, *Which weighted shifts are subnormal?* Pacific J. Math. **17** (1966), 367–379. MR **33** #1740.

50. M. H. Stone, *Linear transformations in Hilbert space*, Amer. Math. Soc. Colloq. Publ., vol. 15, Amer. Math. Soc., Providence, R. I., 1932.

51. N. Suzuki, *On the irreducibility of weighted shifts*, Proc. Amer. Math. Soc. **22** (1969), 579–581.

52. J. Wermer, *Commuting spectral measures on Hilbert space*, Pacific J. Math. **4** (1954), 355–361. MR **16**, 143.

53. H. Weyl, *Über beschränkte quadratische Formen deren Differenz vollstetig ist*, Rend. Circ. Mat. Palermo **27** (1909), 373–392.

INDIANA UNIVERSITY, BLOOMINGTON, INDIANA 47401

NOTE ADDED IN PROOF. One of the consequences of the preprint system of scientific communication is that some part of almost everything that is printed is superseded by the time it is published. The present paper is no exception apparently. Solutions have been reported to two of the ten problems above: Problem 1, negative, J. G. Stampfli; Problem 4, affirmative, I. D. Berg.

Reprinted from the
INDIANA UNIVERSITY MATHEMATICS JOURNAL
Vol. 20, No. 9, pp. 855-863, Mar. 1971

Capacity in Banach Algebras

P. R. HALMOS*

An element A of an algebra $\mathbf{A}$ over a field $\mathbf{F}$ is *algebraic* if there exists a monic polynomial p with coefficients in $\mathbf{F}$ such that $p(A) = 0$. ("Monic" means that the leading coefficient is 1.) The main purpose of what follows is to introduce and characterize an analytic version of this concept that is suitable for use in Banach algebras. (To avoid extraneous complications, the field of scalars is assumed to be the field of complex numbers throughout.)

The first section is devoted to the (algebraic) special case that the others generalize. Section 2 introduces the new concept of quasialgebraic element and stresses its relation to the potential-theoretic notion of capacity; Section 3 shows that the relation is more than an analogy. Section 4 connects the concept of quasialgebraic operator with the slightly less new concept of quasitriangularity, and, incidentally, solves a problem of Douglas and Pearcy about quasitriangular operators. The final section asks a couple of questions.

1. Algebraic elements. Are sums and products of algebraic elements algebraic? In the familiar theory of field extensions the answer is yes; for algebras in general the answer is no. For a concrete example, let $\mathbf{A}$ be the algebra of all operators (bounded linear transformations) on a Hilbert space H, let $\{e_0, e_1, e_2, \cdots\}$ be an orthonormal basis for H, and let A and B be the elements of $\mathbf{A}$ such that $Ae_n = e_{n+1}$ or 0 according as n is even or odd and $Be_n = e_{n+1}$ or 0 according as n is odd or even. Clearly $A^2 = B^2 = 0$, so that both A and B are algebraic. Since, however, $(A + B)e_n = e_{n+1}$ for all n, i.e., $A + B$ is the unilateral shift, the sum $A + B$ is not algebraic. For products consider $A + 1$ and $B + 1$; both are algebraic, but their product is not.

Despite the fiasco described in the preceding paragraph, the class of algebraic elements does have at least two reasonable closure properties. (1) If A is algebraic and if g is a monic polynomial such that $A = g(B)$, then B is algebraic. This is trivial; if $p(A) = 0$, then $p(g(B)) = 0$. (2) If A is algebraic, f is a polynomial, and $B = f(A)$, then B is algebraic. This is easy, but not obvious. For the proof, suppose that $p(A) = 0$, where $p(z) = (z - \lambda_1) \cdots (z - \lambda_n)$, and consider the polynomial q defined by $q(z) = (z - f(\lambda_1)) \cdots (z - f(\lambda_n))$. Since $f(z) - f(\lambda_j)$ is divisible by $z - \lambda_j$, $j = 1, \cdots, n$, it follows that $q(f(z))$ is divisible by $p(z)$. Consequence: $q(B) = q(f(A)) = 0$.

* Research supported in part by a grant from the National Science Foundation.

855

What was just proved is that if A is algebraic and either A is a polynomial in B or B is a polynomial in A, then B is algebraic. A natural generalization might go this way: if A is algebraic and B is an algebraic function of A, then B is algebraic. What does it mean to say that B is an algebraic function of A? One possible interpretation is this: A and B commute and there exists a polynomial h in two variables such that $h(A, B) = 0$. In this interpretation the generalization is false. Suppose indeed that $h(u, v) = uv$, let $\mathbf{A}$ be the algebra of 2×2 matrices over an algebra $\mathbf{B}$, and suppose that C and D are elements of $\mathbf{B}$ such that C is algebraic and D is not. If

$$A = \begin{bmatrix} C & 0 \\ 0 & 0 \end{bmatrix} \quad \text{and} \quad B = \begin{bmatrix} 0 & 0 \\ 0 & D \end{bmatrix},$$

then A and B are in $\mathbf{A}$, the matrices A and B commute, and $h(A, B) = 0$. All the assumptions are satisfied, A is algebraic, but B is not; the generalization of the preceding paragraph to all algebraic functions is false. Less sweeping generalizations are, however, accessible. Here is a sample: if A is algebraic and if f and g are non-zero polynomials such that $f(A) = g(B)$, then B is algebraic. Condensed proof: if $p(A) = 0$, where $p(z) = (z - \lambda_1) \cdots (z - \lambda_n)$, then $r(B) = 0$, where $r(z) = (g(z) - f(\lambda_1)) \cdots (g(z) - f(\lambda_n))$.

More is true about algebraic operators than about algebraic elements of general algebras. The first additional result is that algebraic operators have a special kind of matrix.

Theorem 1. *A necessary and sufficient condition that an operator A on a Hilbert space H be algebraic is that H be a finite (orthogonal) direct sum of subspaces such that the operator matrix of A corresponding to that direct sum is upper triangular and has scalars for its diagonal entries.*

Proof. Suppose that p is the minimal polynomial of A, $p(z) = (z - \lambda_1) \cdots (z - \lambda_n)$, and write $M_k = \ker (A - \lambda_1) \cdots (A - \lambda_k)$, $k = 1, \cdots, n$. The subspaces M_k increase with k, and $M_n = H$. (The minimality of p even implies that the increase is strict.) Since each M_k is invariant under A, the matrix of A corresponding to the decomposition

$$H = M_1 \oplus (M_2 \cap M_1^\perp) \oplus \cdots \oplus (M_n \cap M_{n-1}^\perp)$$

is upper triangular. If $f \in M_k \cap M_{k-1}^\perp$, $k = 1, \cdots, n$ (where $M_0 = 0$), then $(A - \lambda_k)f \in M_{k-1}$, which implies that the component of Af in $M_k \cap M_{k-1}^\perp$ is $\lambda_k f$.

This proves the necessity of the condition; the proof of sufficiency is a straightforward calculation.

Another fact about operators, which cannot even be formulated in general algebras, has to do with local algebraic behavior. If an operator A on H is algebraic, then it is locally algebraic, in the sense that for each f in H there exists a monic polynomial $p^{(f)}$ such that $p^{(f)}(A)f = 0$. The central fact about locally algebraic operators on Banach spaces is that they are algebraic; see [10, p. 40].

2. Quasialgebraic elements. The role of nilpotent elements in general algebras is played by the quasinilpotent elements of Banach algebras. (Recall that A is quasinilpotent if $||A^n||^{1/n} \to 0$. Every Banach algebra in the sequel is assumed to have a unit.) What, in this sense, is the appropriate Banach algebraic version of an algebraic element? Here is a possible answer: an element A of a Banach algebra $\mathbf{A}$ is *quasialgebraic* in case there exists a sequence $\{p_n\}$ of monic polynomials, with degree $p_n = d(n)$, such that $||p_n(A)||^{1/d(n)} \to 0$ as $n \to \infty$. Clearly every algebraic element is quasialgebraic, and so is every quasinilpotent element.

It is plausible, and true, that if the sequence $\{d(n)\}$ of degrees has a bounded infinite subsequence, then A is algebraic. Indeed, in that case it may be assumed (consider a suitable subsequence) that there is a positive integer k such that $d(n) = k$ for all n. If A is not algebraic, then $||p||_A = ||p(A)||$ defines a norm on the finite-dimensional vector space of all polynomials of degree not greater than k. Since $||p_n(A)|| \to 0$, it follows that the sequence $\{p_n\}$ tends to 0 coefficientwise; since, however, each p_n is monic, that is impossible.

The definition of quasialgebraic elements can be formulated more elegantly. The reformulation puts the concept into a larger context and makes contact with some interesting known facts.

For each positive integer n, how close does A come to being algebraic of degree n? The quantitative answer to this question is given by

$$\mathrm{cap}_n A = \inf \{||p(A)||: p \, \varepsilon \, \mathbf{P}_n^1\},$$

where $\mathbf{P}_n^1$ is the set of all monic polynomials of degree n or less. (The reason for the abbreviation "cap" will be explained presently.) A standard text-book proof in approximation theory (see for instance [3, pp. 137–139]) shows that for each positive integer n there exists a polynomial t_n in $\mathbf{P}_n^1$ such that $\mathrm{cap}_n A = ||t_n(A)||$; any such t_n may be called a Tchebycheff polynomial (of degree n) for A. If $p \, \varepsilon \, \mathbf{P}_n^1$, $q \, \varepsilon \, \mathbf{P}_m^1$, and $r = pq$, then $r \, \varepsilon \, \mathbf{P}_{n+m}^1$, and

$$\mathrm{cap}_{n+m} A \leq ||r(A)|| \leq ||p(A)|| \cdot ||q(A)||;$$

since p and q are arbitrary (in $\mathbf{P}_n^1$ and $\mathbf{P}_m^1$ respectively), it follows that

$$\mathrm{cap}_{n+m} A \leq (\mathrm{cap}_n A)(\mathrm{cap}_m A).$$

This implies [12, p. 171] that the sequence $\{(\mathrm{cap}_n A)^{1/n}\}$ has a limit; write

$$\mathrm{cap} \, A = \lim_n (\mathrm{cap}_n A)^{1/n}.$$

The considerations of the preceding paragraph apply in particular to the following special case: X is a non-empty compact subset of the complex plane, $\mathbf{A} = C(X)$ is the usual algebra of all continuous functions on X, and A is the identity function (i.e., $A(z) = z$ for all z in X). In this case cap A (which might as well be denoted by cap X) is known by several names (Tchebycheff's constant, Robin's constant, transfinite diameter, exterior mapping radius, capacity), or, better said, there are several definitions that arose historically in different contexts but turned out to lead to the same number. The abbreviation "cap" stands, obviously, for the shortest name on the list; cap A will be called the *capacity* of A, or of X. A classical result (to be used below) asserts that if X is infinite, then for each positive integer n there is only one Tchebycheff polynomial of degree n for A (for X); the requirement that $||t_n||_X = \text{cap}_n X$ uniquely determines t_n (in $\mathbf{P}_n^1$). The proof is standard, often reproduced. A simple version of it is in [1]. All the zeros of the Tchebycheff polynomials for X lie in the convex hull of X [5].

The terminology (capacity) and notation (cap) are used below for elements of arbitrary Banach algebras, not for $C(X)$ only.

Theorem 2. *An element A of a Banach algebra is quasialgebriac if and only if it has capacity 0.*

Proof. If cap $A = 0$, let t_n be a Tchebycheff polynomial of degree n for A, $n = 1, 2, 3, \cdots$. Since $||t_n(A)||^{1/n} = (\text{cap}_n A)^{1/n} \to 0$, it follows that A is quasialgebraic. Suppose, conversely, that A is quasialgebraic, and let $\{p_n\}$ be a sequence of monic polynomials, with degree $p_n = d(n)$, such that $||p_n(A)||^{1/d(n)} \to 0$. If the sequence $\{d(n)\}$ of degrees has a bounded infinite subsequence, then A is algebraic, and therefore quasialgebraic. If, on the other hand, $d(n) \to \infty$, then, since $||p_n(A)|| \geq \text{cap}_{d(n)} A$, it follows that a subsequence of $\{(\text{cap}_n A)^{1/n}\}$ tends to 0. Since, however, the entire sequence is always convergent, it follows that the limit of the entire sequence must be 0, and hence that A has capacity 0.

Some of the properties of algebraic elements are easy to prove for quasialgebraic ones also. Thus, for instance, if A is quasialgebraic and g is a monic polynomial such that $A = g(B)$, then B is quasialgebraic. Proof: if p_n is a monic polynomial of degree n and $||p_n(A)||^{1/n} \to 0$, then $||p_n(g(B))||^{1/kn} \to 0$, where $k = \text{degree } g$.

One more simple general observation: if the spectrum of A is finite, then A is quasialgebraic. Proof: find a monic polynomial g that annihilates the spectrum of A; then $g(A)$ is quasinilpotent, hence quasialgebraic, and the preceding paragraph applies.

3. Spectral capacity. If A is an element with spectrum X, in an arbitrary Banach algebra, then it is possible to associate two numbers with A: the capacity, cap A, and *spectral capacity*, cap X. (The symbol "cap X" was defined above, just before Theorem 2.)

Theorem 3. *Spectral capacity is equal to capacity.*

Proof. If p is a (monic) polynomial, then

$$||p||_X = \sup \{|p(z)|: z \, \varepsilon \, X\} = r(p(A)) \leq ||p(A)||$$

(where r denotes spectral radius). It follows that for each n and for each p_0 in $\mathbf{P}_n^1$

$$\text{cap}_n X = \inf \{||p||_X : p \,\varepsilon\, \mathbf{P}_n^1\} \leq ||p_0(A)||,$$

and therefore

$$\text{cap}_n X \leq \inf \{||p(A)|| : p \,\varepsilon\, \mathbf{P}_n^1\} = \text{cap}_n A.$$

It follows immediately that $\text{cap } X \leq \text{cap } A$.

To prove the reverse inequality, it is convenient to use the Tchebycheff polynomials s_k (of degree k) for X; they have the property that

$$||s_k||_X = \text{cap}_k X,$$

and hence

$$(||s_k||_X)^{1/k} \to \text{cap } X.$$

Since

$$||s_k||_X = r(s_k(A)),$$

it follows that

$$||(s_k(A))^n||^{1/n} \to ||s_k||_X$$

as $n \to \infty$. If $||s_k||_X = 0$, then X is finite and A is quasialgebraic, so that $\text{cap } X = 0$ and $\text{cap } A = 0$. If $||s_k||_X > 0$, then, corresponding to each such fixed k, there exists an $n = n(k)$ such that

$$||(s_k(A))^n||^{1/n} \leq 2 ||s_k||_X .$$

(This uses the definition of spectral radius in terms of norm.) It follows that

$$||(s_k(A))^n||^{1/kn} \leq 2^{1/k}(||s_k||_X)^{1/k}.$$

The right term tends to $\text{cap } X$ as $k \to \infty$, and therefore

$$\limsup_k ||(s_k(A))^{n(k)}||^{1/kn(k)} \leq \text{cap } X.$$

On the other hand, by definition,

$$(\text{cap}_{kn(k)} A)^{1/kn(k)} \leq ||(s_k(A))^{n(k)}||^{1/kn(k)},$$

and therefore

$$\text{cap } A \leq \limsup_k ||(s_k(A))^{n(k)}||^{1/kn(k)}.$$

The proof is complete.

Theorem 3 has many corollaries. One of them is that every element with a finite spectrum is quasialgebraic, but, of course, that is not fair: the corollary (proved earlier by a special technique) was used in the proof. Another corollary is that every compact operator is quasialgebraic. The point here is that if X is a compact set with a single cluster point, then cap $X = 0$. This is easy to prove directly from the definition; it is also a special case of the pertinent part of capacity theory [13, p. 57]. A corollary of the second corollary is that every polynomially compact operator is quasialgebraic. Indeed: if $A = g(B)$, where g is a monic polynomial and A is compact, then A is quasialgebraic, and therefore so is B.

If a quasialgebraic element A is "analytic", in the sense that there exists a non-zero function f analytic in a neighborhood of X (the spectrum of A) such that $f(A) = 0$, then A is algebraic. Proof: the function f maps X onto 0. Since f is not identically zero, X must be finite. Let p be a polynomial such that $f = pg$, where g is analytic and does not vanish on X. Then $g(A)$ is invertible; since $f(A) = 0$, it follows that $p(A) = 0$.

A weighted shift (in the algebra of operators on a Hilbert space) can be quasialgebraic only if it is quasinilpotent. Proof: the spectrum is circularly symmetric about 0; if it consists of more than 0, then it contains a continuum, and therefore [13, p. 56] it cannot have capacity 0.

Neither the set of algebraic operators nor the set of quasialgebraic operators is closed. Indeed: since every normal operator is the norm limit of operators with finite spectra, algebraic operators can converge to one that is not quasialgebraic.

The set of quasialgebraic operators is not dense. In fact, if U is the unilateral shift and $||U - A|| < 1$, then A is not quasialgebraic. For the proof, suppose that $||U - A|| = \rho < 1$ and $|\lambda| < 1 - \rho$. Then $||U - (A - \lambda)|| \leq ||U - A|| + |\lambda| = \rho + |\lambda| < 1$, and therefore $A - \lambda$ is not invertible [7, p. 266]. Consequence: the spectrum of A has positive measure, and therefore positive capacity [13, p. 58].

The examples in Section 1 of algebraic elements whose sums and products are not algebraic are in fact such that the sums and products are not even quasialgebraic. After this comment only one of the preliminary results of Section 1 still remains to be extended to quasialgebraic elements; that extension goes as follows.

Theorem 4. *Every polynomial in a quasialgebraic element is quasialgebraic.*

Proof. Suppose that A is quasialgebraic, with spectrum X, and let f be an arbitrary polynomial; it is to be proved that if $B = f(A)$, then cap $B = 0$, or, equivalently, in view of Theorem 3, that cap $f(X) = 0$.

Let s_n be the Tchebycheff polynomial of degree n for X; the assumption cap $X = 0$ implies that $(||s_n||_X)^{1/n} \to 0$. If $s_n(z) = (z - \lambda_1^{(n)}) \cdots (z - \lambda_n^{(n)})$, write

$$q_n(z) = (z - f(\lambda_1^{(n)})) \cdots (z - f(\lambda_n^{(n)})).$$

Since $f(z) - f(\lambda_i^{(n)})$ is divisible by $z - \lambda_i^{(n)}$, the composite $q_n(f(z))$ is divisible by $s_n(z)$; let r_n be a polynomial such that

$$q_n(f(z)) = r_n(z)s_n(z).$$

If λ is a zero of r_n, then $f(\lambda) = f(\lambda_j^{(n)})$ for some j, i.e., $\lambda \in f^{-1}f(\lambda_j^{(n)})$. If X^+ is the convex hull of X, then $\lambda_j^{(n)} \in X^+$ for all j and all n; this implies that all the zeros of r_n are in $f^{-1}f(X^+)$. Since X^+ is compact and f is continuous, it follows that $f^{-1}f(X^+)$ is compact. If d is the largest distance from the origin to a point of $f^{-1}f(X^+)$, if α is the leading coefficient of f, and if k is the degree of f, then

$$||r_n||_X \leq (2\,|\alpha|\,d)^{(n-1)k}.$$

Reason: factor r_n, note that its leading coefficient is α, and, for each factor of the form $(z - \lambda)$, where λ is a zero of r_n, use the estimate $|z| + |\lambda| \leq 2d$. Note that the degree of r_n is $nk - n$. Consequence:

$$(||q_n \circ f||_X)^{1/nk} \leq ((||s_n||_X)^{1/n})^{1/k} \cdot (2\,|\alpha|\,d)^{(n-1)/n},$$

and therefore

$$(||q_n \circ f||_X)^{1/nk} \to 0.$$

This, in turn, implies that $(||q_n||_{f(X)})^{1/n} \to 0$, and hence, by Theorem 2, that cap $f(X) = 0$.

4. Quasialgebraic operators. Quasialgebraic operators (i.e., quasialgebraic elements of the algebra of operators on a Hilbert space) and quasitriangular operators [8] were introduced in different contexts for different purposes; the following generalization of Theorem 1 is sheer terminological serendipity.

Theorem 5. *Every quasialgebraic operator (on a separable Hilbert space) is quasitriangular.*

Proof. There are several possible definitions of quasitriangularity. A mildly artificial but quickly stable sufficient condition is this: A is quasitriangular in case it has a matrix (a_{ij}) (with respect to an orthonormal basis) such that every diagonal two steps or more below the main one vanishes ($a_{ij} = 0$ unless $i \leq j + 1$) and some subsequence of the diagonal just below the main one tends to 0 ($a_{k(n)+1,k(n)} \to 0$).

The proof of the theorem is almost the same as the proof that every polynomially compact operator is quasitriangular [6]. Suppose that A is quasialgebraic and suppose, to begin with, that it is cyclic. Let p_n be a monic polynomial of degree n such that $||p_n(A)||^{1/n} \to 0$, and let e be a cyclic vector. Assume (with no loss of generality) that the underlying Hilbert space is infinite-dimensional. Orthonormalize $\{e, Ae, A^2e, \cdots\}$ to get $\{e_1, e_2, e_3, \cdots\}$; the matrix (a_{ij}) of A has the property that $a_{ij} = 0$ unless $i \leq j + 1$. If $(a_{ij}^{(n)})$ is the matrix of A^n, then $a_{ij}^{(n)} = 0$ unless $i \leq j + n$, and

$$a_{j+n,j}^{(n)} = \prod_{1 \leq k \leq n} a_{j+k,j+k-1}.$$

It follows that if p is a monic polynomial of degree n, and if the matrix of $p(A)$

is $(a_{ij}^{(p)})$, then $a_{ij}^{(p)} = 0$ unless $i \leq j + n$ and

$$a_{j+n,i}^{(p)} = a_{j+n,i}^{(n)} \; .$$

This result with $p = p_n$ implies the existence of an increasing sequence $\{k(n)\}$ of positive integers such that the corresponding subdiagonal terms $a_{k(n)+1,k(n)}$ tend to 0 as n tends to ∞.

The general (not necessarily cyclic) case is now immediate from the work of Douglas and Pearcy. They prove (for separable Hilbert spaces) that every operator has an upper triangular operator matrix whose diagonal entries are cyclic [4]. Assertion: if A is so represented, if, say,

$$A = \begin{bmatrix} X & Y \\ 0 & Z \end{bmatrix} ,$$

and if A is quasialgebraic, then both X and Z are quasialgebraic. Indeed: if p is a polynomial, then

$$p(A) = \begin{bmatrix} p(X) & * \\ 0 & p(Z) \end{bmatrix} ,$$

and therefore both $||p(X)||$ and $||p(Z)||$ are dominated by $||p(A)||$. Consequence: every quasialgebraic operator has an upper triangular operator matrix whose diagonal entries are quasitriangular. Another result of Douglas and Pearcy [4] then becomes applicable: exactly under the stated conditions it follows that the operator in question is quasitriangular.

The result yields a solution of a problem that Douglas and Pearcy raised but did not settle.

Corollary. *On a separable Hilbert space, every operator with a countable spectrum is quasitriangular.*

Proof. Every countable compact set has capacity 0 [13, p. 57].

5. Locally quasialgebraic operators. Is every locally quasialgebraic operator quasialgebraic? One possible way to define a locally quasialgebraic operator is to put

$$\mathrm{cap}_n \,(A, f) = \inf \,\{||p(A)f|| : p \,\varepsilon\, \mathbf{P}_n^1\}$$

and require that

$$\lim_n \,(\mathrm{cap}_n \,(A, f))^{1/n} = 0$$

for every f. Another, perhaps nearer the surface, is to assume, for each vector f, the existence of a sequence $\{p_n^{(f)}\}$ of monic polynomials, with degree $p_n^{(f)} = d(n, f)$, such that $||p_n^{(f)}(A)f||^{1/d(n,f)} \to 0$ as $n \to \infty$. A word of caution may not be amiss: although the limit that defines cap A always exists, the analogous local assertion is false. Even for locally quasinilpotent operators ($\lim_n ||A^n f||^{1/n} = 0$

for each f) the assumption should be formulated this way: for each f, the limit exists and is equal to 0. It is easy to construct examples (weighted shifts will do) such that for certain vectors f the limit fails to exist.

Kaplansky's theorem (locally algebraic operators are algebraic) belongs to a small group of results that have been receiving some attention in recent times. The question of the preceding paragraph can be expressed this way: is the "quasi" version of Kaplansky's theorem true? Kaplansky's theorem is a generalization of the assertion that every locally nilpotent operator is nilpotent [9], and hence a cousin of the theorem that every locally quasinilpotent operator is quasinilpotent [2, p. 28]. Another related result is that if an operator is "locally inner-analytic" (of class C_0), then it is globally so [11]. Is there a grandfather theorem that includes all these results as special cases and includes the "quasi" version of Kaplansky's theorem also?

References

[1] J. C. Burkill, *Lectures on approximation by polynomials*, Tata Institute of Fundamental Research, Bombay, (1959).

[2] I. Colojoară & C. Foiaş, *Theory of generalized spectral operators*, Gordon and Breach, New York, (1968).

[3] P. J. Davis, *Interpolation and approximation*, Blaisdell, New York, (1963).

[4] R. G. Douglas & C. Pearcy, A note on quasitriangular operators, *Duke Math. J.*, **37** (1970) 177–188.

[5] L. Fejér, Über die Lage der Nullstellen von Polynomen, die aus Minimalforderungen gewisser Art entspringen, *Math. Ann.*, **85** (1922) 41–48.

[6] P. R. Halmos, Invariant subspaces of polynomially compact operators, *Pac. J. Math.*, **16** (1966) 433–437.

[7] P. R. Halmos, *A Hilbert space problem book*, Van Nostrand, Princeton, (1967).

[8] P. R. Halmos, Quasitriangular operators, *Acta Szeged*, **29** (1968) 283–293.

[9] P. R. Halmos, Ten problems in Hilbert space, *Bull. A. M. S.*, **76** (1970) 887–933.

[10] I. Kaplansky, *Infinite abelian groups*, University of Michigan Press, Ann Arbor, (1954).

[11] B. Sz.-Nagy & C. Foiaş, Local characterization of operators of class C_0, to appear.

[12] G. Pólya & G. Szegö, *Aufgaben und Lehrsätze aus der Analysis, vol. 1*, Dover, New York, (1945).

[13] M. Tsuji, *Potential theory in modern function theory*, Maruzen, Tokyo, (1959).

Indiana University
Date communicated: October 5, 1970

PROCEEDINGS OF THE
AMERICAN MATHEMATICAL SOCIETY
Volume 31, No. 1, January 1972

CONTINUOUS FUNCTIONS OF HERMITIAN OPERATORS[1]

P. R. HALMOS

ABSTRACT. Theorem: every normal operator is a continuous function of a Hermitian one. Corollary: every normal operator on a separable Hilbert space is the sum of a diagonal operator and a compact one.

THEOREM. *Every normal operator is a continuous function of a Hermitian one.*

COROLLARY. *Every normal operator on a separable Hilbert space is the sum of a diagonal operator and a compact one.*

Any two commutative Hermitian operators are Borel functions of some Hermitian operator; the version of the theorem in which "continuous" is replaced by "Borel" is old. Both the Borel and the continuous versions extend to countable sets of commutative Hermitian operators, with no additional effort. The present version is an easy consequence of well known Hilbert space techniques, together with an elegant fact of general topology. The result, with a similar but different proof, is implicit in a construction of Schwartz [6, pp. 15–16]. Much is known about normal operators, but not everything. The theorem is presented here in the hope that it may yield some new information. The basis of the hope is the derivation of the corollary from the theorem.

The version of the corollary in which "normal" is replaced by "Hermitian" is the Weyl-von Neumann theorem; till quite recently the extension to the normal case was an open problem [4]. The first solution is due to I. D. Berg [1]. The proof below is shorter, and it may perhaps be considered more translucent.

PROOF OF THE THEOREM. Every compact metric space is a continuous image of the Cantor set [5, §41, VI]. If, in particular, Λ is a non-empty compact subset of the complex plane and Γ is the Cantor set in the unit interval, then there exists a continuous function φ from Γ onto Λ. The elegant fact from general topology enters here: it is the existence of a Borel cross section. That is: there exists a Borel function ψ from Λ into Γ such that the composition $\varphi \circ \psi$ is the identity on Λ [5, §43, IX]. (Short

Received by the editors February 3, 1971.

AMS 1970 subject classifications. Primary 47B15.

[1] Research supported in part by a grant from the National Science Foundation.

130

proof for the case at hand: for each z in Λ, let $\psi(z)$ be the infimum of the numbers x in Γ for which $\varphi(x) = z$. If $z_n \to z$, then the continuity of φ implies that the limit of each convergent subsequence of $\{\psi(z_n)\}$ is greater than or equal to $\psi(z)$. It follows that $\psi(z) \leqq \lim \inf_n \psi(z_n)$, so that ψ is lower semicontinuous, and therefore of Baire class 1.)

Suppose that μ is a positive finite Borel measure with support in Λ, and write $\nu = \mu \circ \psi^{-1}$ for the induced measure in Γ. The mapping T from $L^2(\nu)$ into $L^2(\mu)$ defined by $Tg = g \circ \psi$ is an isometry. (This is standard; all that is needed to verify it is to look at the expressions for $\|g\|^2$ and $\|Tg\|^2$ and to recall the way to change variables in Lebesgue integrals [2, §39].) The isometry T maps $L^2(\nu)$ onto $L^2(\mu)$; indeed the inverse of T, which in this situation is the same as the adjoint of T, is defined by $T^*f = f \circ \varphi$. Since $\varphi \circ \psi$ is the identity on Λ, the equation $TT^*f = f$ is obvious. Since $\psi \circ \varphi$ is, in general, not the identity on Γ, the equation $T^*Tg = g$ is true for a different reason; it follows from the isometric character of T.

The position operator on $L^2(\mu)$ $(f(z) \mapsto zf(z))$ is the transform by T^* of multiplication by φ on $L^2(\nu)$. Proof: $(T(\varphi \cdot T^*f))(z) = (T(\varphi \cdot (f \circ \varphi)))(z)$ $= ((\varphi \cdot (f \circ \varphi)) \circ \psi)(z) = \varphi(\psi(z)) \cdot f(\varphi(\psi(z)))$.

The spectral theorem says that every normal operator is the direct sum of position operators [3]. More precisely: to within unitary equivalence every normal operator A can be obtained as follows. Fix a non-empty compact set Λ in the plane, let μ vary over an arbitrary set of positive finite Borel measures in Λ, and form the direct sum of the corresponding position operators. As μ varies, ν varies. The direct sum of the corresponding position operators $(g(x) \mapsto xg(x))$ on the $L^2(\nu)$'s is a Hermitian operator B (because Γ is a subset of the real line). Since $A = \varphi(B)$, the proof of the theorem is complete.

PROOF OF THE COROLLARY. In view of the theorem, it is sufficient to prove that if B is a Hermitian operator on a separable Hilbert space, and if φ is a complex-valued continuous function on the spectrum of B, then $\varphi(B)$ is the sum of a diagonal operator and a compact one.

By the Weyl-von Neumann theorem, $B = D + C$, where D is diagonal and C is compact. Extend φ to a continuous function (into the complex plane) on a compact set that includes the spectra of both B and D. The Weierstrass approximation theorem implies that there exists a sequence $\{p_n\}$ of (complex) polynomials that converges uniformly to (the extended) φ; it follows that $p_n(B) \to \varphi(B)$ and $p_n(D) \to \varphi(D)$ in the norm. If $C_n = p_n(B) - p_n(D)$, then each C_n is compact, and the sequence $\{C_n\}$ converges in the norm to a (necessarily compact) operator K. Since $\varphi(D)$ is diagonal (any orthonormal basis that diagonalizes D does the same for each $p_n(D)$), and since $\varphi(B) = \varphi(D) + K$, the proof of the corollary is complete.

REMARK.   Is the theorem true for arbitrary C^* algebras? (Question via J. P. Williams.) The answer is no. Indeed, if Λ is a non-empty compact subset of the complex plane, then the identity mapping $(z \mapsto z)$ on Λ is an (obviously normal) element of $C(\Lambda)$. The theorem for $C(\Lambda)$ would imply the existence of a Hermitian (i.e., real-valued) continuous function ψ on Λ and the existence of a continuous function φ on $\psi(\Lambda)$ such that $\varphi(\psi(z)) = z$ for all z in Λ. Under these circumstances ψ is one-to-one, and, consequently, Λ is homeomorphic to a subset of the real line. If Λ is chosen to make this impossible (e.g., if Λ is the perimeter of a circle), then no such ψ can exist.

REFERENCES

1. I. D. Berg, *An extension of the Weyl-von Neumann theorem to normal operators,* Trans. Amer. Math. Soc. **160** (1971), 365–371.

2. P. R. Halmos, *Measure theory,* van Nostrand, New York, 1950. MR **11**, 504.

3. ——, *What does the spectral theorem say?,* Amer. Math. Monthly **70** (1963), 241–247. MR **27** #595.

4. ——, *Ten problems in Hilbert space,* Bull. Amer. Math. Soc. **76** (1970), 887–933.

5. K. Kuratowski, Topology. Vol. II, Academic Press, New York, 1968. MR **41** #4467.

6. J. T. Schwartz, *W*-algebras,* Gordon and Breach, New York, 1967. MR **38** #547.

DEPARTMENT OF MATHEMATICS, INDIANA UNIVERSITY, BLOOMINGTON, INDIANA 47401

Reprinted from the
INDIANA UNIVERSITY MATHEMATICS JOURNAL
Vol. 21, No. 10, pp. 951-960, April 1972

Positive Approximants of Operators

P. R. HALMOS

Dedicated to Eberhard Hopf on his seventieth birthday,
April 17, 1972.

Introduction. How near can a positive matrix come to $\left(\begin{smallmatrix} 0 & 0 \\ 1 & 0 \end{smallmatrix}\right)$? ("Positive" means non-negative semidefinite.) Unless the question is approached just right, it can become very exasperating, and, even after it is answered, the problem of putting the answer into a suitable context, so as to understand why it is what it is, can still remain a non-trivial challenge. It is quite possible, for instance, to find the answer, and still not have any technique for answering the same question about $\left(\begin{smallmatrix} i & 1 \\ 1 & 0 \end{smallmatrix}\right)$, which is much harder. The main purpose of this paper is to prove a theorem that answers these two questions, and, in a sense, solves the general problem of positive approximation.

By way of introduction to such ideas, consider the problem of real approximation. Each complex number has a unique real approximant, namely its real part. If, that is, $\alpha = \beta + i\gamma$, with β and γ real, then

$$|\alpha - \beta| \leqq |\alpha - \rho|$$

whenever ρ is real, and if equality holds, then $\rho = \beta$. [Comment on terminology. It is customary to speak of "best" approximation to a point in a metric space by the points in a prescribed set. Since the unmodified noun is not a meaningful technical term in its own right, it seems simpler to define things so that the modifier is never needed. The word "approximant" will be used below, as it was used just above, in this sense: to say that β is a real *approximant* of α means that β is real and that the distance from α to β is as small as possible.]

In classical approximation theory the objects of central interest are not numbers but functions; the problem is to get information about bad functions via approximation by good ones. A non-typically trivial problem in classical, functional, approximation theory is to approximate each bounded complex-valued function α uniformly as closely as possible by a real one. There are usually many solutions. If $\alpha = \beta + i\gamma$, with β and γ real, then one solution is the real part, β. The distance from α to this approximant is the norm $||\gamma||$ of the imaginary part. Each real-valued function ρ such that

951

$$|\alpha(x) - \rho(x)| \leqq ||\gamma||$$

for all x is another real approximant.

Functions can often be regarded as operators on vector spaces. If, for instance, α is a bounded measurable function on a measure space with measure μ, then multiplication by α is a normal operator on the complex Hilbert space $L^2 (\mu)$. From this point of view it is natural to regard classical analysis as a special case of non-commutative analysis, *i.e.*, of the theory in which the objects of central interest are not functions but possibly non-normal (but always bounded) operators. The first one to formulate and study such a program (even for unbounded operators) was I. E. Segal [3]. Analysis as a whole ranges from its most trivial aspect (functions on finite sets) to its deepest (operators on infinite-dimensional spaces), through the two intermediate and incomparable stages of functions on infinite sets and matrices of finite size. In some parts of non-commutative analysis, and, in particular, in non-commutative approximation theory, the conceptual depth of the problem is completely visible in the finite-dimensional case.

Few things about non-commutative approximation theory are easy; the basic fact about the analogue of the problem of real approximation is one of those few. If A is an operator on a Hilbert space, $A = B + iC$ with B and C Hermitian, then B is a Hermitian approximant of A [1]. That is:

$$||A - B|| \leqq ||A - R||$$

whenever R is Hermitian. Proof: $||A - R|| = ||(B - R) + iC|| \geqq ||C||$.

The problem of positive approximation is similar to the problem of real approximation, but harder. Every real number β has a unique positive approximant, namely β or 0 according as $\beta \geqq 0$ or $\beta \leqq 0$. (The word "positive" is used, here and throughout, to mean non-negative.) For a complex number α ($= \beta + i\gamma$, with β and γ real), the unique positive approximant is the positive approximant of β.

The problem of uniform positive approximation to a bounded real function usually has many solutions. One of them is the positive part β^+ of the given function β ($\beta^+(x) = \beta(x)$ or 0 according as $\beta(x) \geqq 0$ or $\beta(x) \leqq 0$). The distance from β to this approximant is the norm $||\beta^-||$ of the negative part β^- of β. Each positive function π such that

$$|\beta(x) - \pi(x)| \leqq ||\beta^-||$$

for all x is another positive approximant. Equivalently: a function π is a positive approximant of β if and only if

$$(\beta - ||\beta^-||)^+ \leqq \pi \leqq \beta + ||\beta^-||.$$

The problem of positive approximation to bounded complex-valued functions has a small additional geometric complication. The problem is, given α, to find a positive π so as to minimize the radius of the smallest closed disc with center at the origin that includes the range of $\alpha - \pi$. To subtract a positive number from a complex number α means to push α horizontally to the left. It follows that if $\alpha(x) = \beta(x) + i\gamma(x)$, with β and γ real, then the range of $\alpha - \pi$, with π positive, can never be covered by a disc with center 0 and smaller radius than

$$r^- = \sup \{|\alpha(x)| : \beta(x) \leqq 0\},$$

nor by a disc with center 0 and smaller radius than

$$r^+ = \sup \{|\gamma(x)| : \beta(x) \geqq 0\}.$$

If $r = \max \{r^-, r^+\}$ and if $\pi = \beta + (r^2 - \gamma^2)^{1/2}$, then $\|\alpha - \pi\| = r$, so that π is a positive approximant of α, clearly the largest one. More generally, a function π is a positive approximant of α if and only if

$$(\beta - (r^2 - \gamma^2)^{1/2})^+ \leqq \pi \leqq \beta + (r^2 - \gamma^2)^{1/2}.$$

Even the easy parts of the non-commutative analogue of the problem of positive approximation are hard; as the examples $\begin{pmatrix} 0 & 0 \\ 1 & 0 \end{pmatrix}$ and $\begin{pmatrix} i & 1 \\ 1 & 0 \end{pmatrix}$ show, even very special cases of the finite-dimensional special case are far from transparent.

Main theorem. In what follows *operators* are bounded linear transformations on a complex Hilbert space. The customary operator norm defined by

$$\|A\| = \sup \{\|Af\| : \|f\| = 1\}$$

induces a metric on the set of all operators. Write

$$\delta(A) = \inf \{\|A - P\| : P \geqq 0\};$$

the problem is to determine $\delta(A)$ for each operator A. The numerical discussion of the preceding section makes it seem reasonable to consider operators of the form

$$B + (r^2 - C^2)^{1/2},$$

where B and C are the real and imaginary parts of A and r is a positive number. Write

$$\eta(A) = \inf \{r : B + (r^2 - C^2)^{1/2} \geqq 0\}.$$

Theorem 1. $\delta(A) = \eta(A)$.

Proof. The positive numbers of which $\eta(A)$ is the infimum must satisfy two conditions: (i) $r^2 - C^2 \geqq 0$ (so that the positive square root of $r^2 - C^2$ can be formed), and (ii) $B + (r^2 - C^2)^{1/2} \geqq 0$. It should be observed that numbers satisfying these two conditions always exist. Indeed, if $r^2 = \|B\|^2 + \|C\|^2$, then $r^2 - C^2 \geqq r^2 - \|C\|^2 = \|B\|^2$, and therefore

$$B + (r^2 - C^2)^{1/2} \geqq B + \|B\| \geqq 0.$$

Suppose now that r is a positive number such that $r^2 - C^2 \geqq 0$ and such that if $P = B + (r^2 - C^2)^{1/2}$, then $P \geqq 0$. Since

$$A - P = -(r^2 - C^2)^{1/2} + iC$$

is normal, it follows that

$$||A - P||^2 = ||(A - P)^*(A - P)|| = ||(r^2 - C^2) + C^2|| = r^2.$$

Since $\delta(A) \leqq ||A - P||$, it follows that $\delta(A) \leqq r$ and hence (take the infimum over r) that $\delta(A) \leqq \eta(A)$.

Suppose next that P is an arbitrary positive operator. Since P is Hermitian, and since B is a Hermitian approximant of A, it follows that

$$||A - P|| \geqq ||A - B|| = ||C||.$$

Consequence: if $r = ||A - P||$, then $r^2 - C^2 \geqq 0$. The proof of the theorem can be completed by showing that $B + (r^2 - C^2)^{1/2} \geqq 0$; indeed if that is true, then $\eta(A) \leqq r = ||A - P||$, and therefore (take the infimum over P) $\eta(A) \leqq \delta(A)$.

The first step in the proof that $B + (r^2 - C^2) \geqq 0$ is the general observation that if X and Y are Hermitian operators, then

$$||X^2 + Y^2|| \leqq ||X + iY||^2.$$

To see that, set $Z = X + iY$, so that $X = \frac{1}{2}(Z + Z^*)$ and $Y = (1/2i)(Z - Z^*)$; it follows that $X^2 + Y^2 = \frac{1}{2}(Z^*Z + ZZ^*)$, whereas $||X + iY||^2 = ||Z||^2 = ||Z^*Z|| = ||ZZ^*||$.

The result of the preceding paragraph applied to $B - P$ and C yields

$$||A - P||^2 \geqq (B - P)^2 + C^2,$$

and hence

$$(r^2 - C^2)^{1/2} \geqq |B - P|.$$

(The absolute value of a Hermitian operator is, as usual, the positive square root of its square. Note the second application of the order-preserving character of square root.)

The last step is to prove the general lemma that whenever B is Hermitian and P is positive, then

$$B + |B - P| \geqq 0.$$

(Since this inequality is true when B is a real number and P a positive number, it is plausible that it should be true here. It is well known, however, that in this subject such plausibility considerations are often misleading. Thus, for instance, if $|B| \leqq P$, then $-P \leqq B \leqq P$, plausibly enough, but, rather implausibly, the converse implication is false. Counterexample:

$$B = \begin{pmatrix} 1 & 0 \\ 0 & -1 \end{pmatrix}, P = \begin{pmatrix} 2 & \sqrt{2} \\ \sqrt{2} & 2 \end{pmatrix}.)$$

It is to be proved that $|B - P| \geqq -B$, or, equivalently, that

$$|P - B| \geqq -P + (P - B).$$

This is the same as

$$(P - B)^+ + (P - B)^- \geqq -P + (P - B)^+ - (P - B)^-.$$

(Here the superscripts "+" and "−" indicate positive and negative parts, both of which are positive operators.) That is, the inequality to be proved reduces to

$$(P - B)^- \geqq -P - (P - B)^-,$$

or

$$P + 2(P - B)^- \geqq 0,$$

which is obvious. The proof of the theorem is complete.

(Donald Rogers pointed out that the proof proves a curiously strengthened version of the statement: not only is it true that the infima of the two sets

$$\{\|A - P\| : P \geqq 0\} \quad \text{and} \quad \{r : B + (r^2 - C^2)^{1/2} \geqq 0\}$$

are equal, but, in fact, the sets are the same.)

Consequences. Theorem 1 can be viewed as a solution of the problem of positive approximation. All it does, to be sure, is replace one infimum by another, but the new infimum is very much easier to evaluate. The value of $\delta(A)$ depends on all positive operators; for $\eta(A)$ only positive numbers need to be looked at. The technical difficulties involved in determining $(r^2 - C^2)^{1/2}$ and in deciding whether $B + (r^2 - C^2)^{1/2}$ is positive are of a familiar kind; they need the spectral theorem, or less, applied first to C and then to $B + (r^2 - C^2)^{1/2}$.

Consider, as an application, the case of $A = \left(\begin{smallmatrix} 0 & 0 \\ 1 & 0 \end{smallmatrix}\right)$. In this case $B = \frac{1}{2}(A + A^*) = \frac{1}{2}\left(\begin{smallmatrix} 0 & 1 \\ 1 & 0 \end{smallmatrix}\right)$ and $C = (1/2i)(A - A^*) = (1/2i)\left(\begin{smallmatrix} 0 & -1 \\ 1 & 0 \end{smallmatrix}\right)$. Since $C^2 = \frac{1}{4}\left(\begin{smallmatrix} 1 & 0 \\ 0 & 1 \end{smallmatrix}\right)$, it follows that $r^2 - C^2 \geqq 0$ if and only if $r \geqq \frac{1}{2}$; note that $r^2 - C^2$ is a scalar (i.e., a scalar multiple of the identity). Since $B^2 = \frac{1}{4}\left(\begin{smallmatrix} 1 & 0 \\ 0 & 1 \end{smallmatrix}\right)$, so that $|B| = \frac{1}{2}\left(\begin{smallmatrix} 1 & 0 \\ 0 & 1 \end{smallmatrix}\right)$, it follows that $B^- = \frac{1}{2}(|B| - B) = \frac{1}{4}\left(\begin{smallmatrix} 1 & -1 \\ -1 & 1 \end{smallmatrix}\right)$, and hence that $\|B^-\| = \frac{1}{2}$. That is: $\frac{1}{2}$ is the smallest scalar that can be added to B to produce a positive result, and hence, in particular, $B + (r^2 - C^2)^{1/2} \geqq 0$ if and only if $(r^2 - \frac{1}{4})^{1/2} \geqq \frac{1}{2}$. Conclusion: $\delta(A) = \sqrt{2}/2$, and $B + \frac{1}{2} = \frac{1}{2}\left(\begin{smallmatrix} 1 & 1 \\ 1 & 1 \end{smallmatrix}\right)$ is a positive approximant of A. In case $A = \left(\begin{smallmatrix} i & 1 \\ 1 & 0 \end{smallmatrix}\right)$, the evaluation of $\delta(A)$ is, if anything, easier, but numerically it is uglier; the result is that $\delta(A) = (\frac{1}{2}(1 + \sqrt{5}))^{1/2}$.

The truncated shifts U_n, $n = 1, 2, 3, \cdots$, furnish interesting special problems in positive approximation. The first one is the 1×1 matrix $U_1 = (0)$, with $\delta(U_1) = 0$; the second one is $U_2 = \left(\begin{smallmatrix} 0 & 0 \\ 1 & 0 \end{smallmatrix}\right)$, with $\delta(U_2) = \sqrt{2}/2$; the third one,

$$U_3 = \begin{bmatrix} 0 & 0 & 0 \\ 1 & 0 & 0 \\ 0 & 1 & 0 \end{bmatrix},$$

is a somewhat greater computational challenge, but still accessible; it turns out that $\delta(U_3) = \frac{1}{2}(1 + \sqrt{5})^{1/2}$. The computational difficulties seem to get out of hand after this. In the limit, however, all is well again: if U is the unilateral shift, then $\delta(U) = 1$. Proof: since $||U - 0|| = 1$, it follows that $\delta(U) \leq 1$; since $||U - A|| \geq 1$ whenever A is normal [2, Problem 113], it follows that $\delta(U) = 1$. Conjecture: $\delta(U_n) \uparrow 1$.

Corollary 1. *If $A = B + iC$, with B and C Hermitian, and if $P = B + ((\delta(A))^2 - C^2)^{1/2}$, then P is a positive approximant of A.*

Proof. Find $r_n \geq 0$, $n = 1, 2, 3, \cdots$, so that $r_n \downarrow \delta(A)$, and conclude that

$$B + (r_n^2 - C^2)^{1/2} \downarrow P;$$

since each term of the sequence is positive, so is the limit.

It is not *a priori* obvious that a positive approximant always exists, *i.e.*, that the infimum that defines $\delta(A)$ is always attained. Existence could have been proved before the theorem, as follows. Observe that (i) since $\delta(A) \leq ||A - 0||$, in the evaluation of $\delta(A)$ there is no loss of generality in restricting attention to the positive operators in the ball with center A and radius $||A||$, (ii) the positive operators form a weakly closed set, and (iii) the norm $||A - P||$ is a weakly lower semicontinuous function of P; from (i), (ii), and (iii), and the weak compactness of closed balls, it follows that $||A - P||$ attains its infimum. Corollary 1 says more; it explicitly describes a positive operator that does the job.

Corollary 1 implies (cf. the second paragraph of the proof of Theorem 1) that $A - P$ is a normal operator and has, moreover, the property that $(A - P)^*(A - P)$ is a scalar. It follows that every operator A can be written in the form

$$A = P + \delta U,$$

where P is positive, U is unitary, and $\delta \geq 0$. More can be said.

Corollary 2. *If A is an operator and if r is a real number, $r \geq \delta(A)$, then there exists a positive operator P and a unitary operator U such that the spectrum of $\mathrm{Re}\ U$ is negative and*

$$A = P + rU;$$

the positive operator P is unique, and, in case $r \neq 0$, so is U.

Proof. If $r = 0$, then $\delta(A) = 0$, so that A is positive. In this case, put $P = A$ and let U be an arbitrary unitary operator.

If $r \neq 0$, put $P = B + (r^2 - C^2)^{1/2}$ (where, as before, $B = \mathrm{Re}\ A$, $C = \mathrm{Im}\ A$) and $U = (1/r)(A - P)$. Clearly $P \geq 0$, and $U = -(1 - (C/r)^2)^{1/2} + i(C/r)$; this settles existence. As for uniqueness: if $P_1 + rU_1 = P_2 + rU_2$, then $U_1 - U_2 = (1/r)(P_2 - P_1)$, so that $U_1 - U_2$ is Hermitian, or $\mathrm{Im}\ U_1 = \mathrm{Im}\ U_2$. Since

U_1 and U_2 are unitary, so that

$$(\text{Re } U_1)^2 + (\text{Im } U_1)^2 = (\text{Re } U_2)^2 + (\text{Im } U_2)^2 = 1,$$

it follows that $(\text{Re } U_1)^2 = (\text{Re } U_2)^2$. The assumption that $\text{Re } U_1 \leq 0$ and $\text{Re } U_2 \leq 0$ implies that $U_1 = U_2$ and hence that $P_1 = P_2$.

The following special case of Corollary 2, which can be stated without anything but standard notation and terminology, would probably have been surprising if it had been met outside the present context.

Corollary 3. *Every contraction can be written uniquely as the sum of a positive operator and a unitary operator with negative real part.*

Proof. If A is a contraction, then $1 \geq \delta(A)$.

Corollary 3 is the operator extension of a simple geometric fact about complex numbers, in the same spirit as the Cartesian representation $(A = B + iC)$ extends the representation of complex numbers by their real and imaginary parts, and the polar decomposition extends the representation of complex numbers by polar coordinates. The geometric fact is that if z is a complex number of modulus not greater than 1, then z can be written uniquely in the form $z = p + u$, where p is on the positive real axis ($p \geq 0$) and u is on the left half of the perimeter of the unit circle ($|u| = 1$, $\text{Re } u \leq 0$). If, for instance, $z = 1$, then $p = 2$ and $u = -1$; no value of p greater than 2 ever occurs.

Normality. Can anything special be said about the problem of positive approximation for pleasant classes, such as the Hermitian operators, or, more generally, the normal ones? The answer is yes; in case A is normal, there is a simple formula for $\delta(A)$. The proof depends on a lemma of some independent interest.

Lemma. *If X, Y, and Z are Hermitian operators with $0 \leq X \leq Y$, then $w(X + iZ) \leq w(Y + iZ)$; if $X + iZ$ is normal, then $||X + iZ|| \leq ||Y + iZ||$.*

Proof. (Recall that w is the numerical radius; if T is an operator, then $w(T) = \sup \{|(Tf, f)| : ||f|| = 1\}$.) If $||f|| = 1$, then

$$|((X + iZ)f, f)|^2 = (Xf, f)^2 + (Zf, f)^2 \leq (Yf, f)^2 + (Zf, f)^2$$
$$= |((Y + iZ)f, f)|^2 \leq (w(Y + iZ))^2.$$

This settles the first assertion. The second one is an immediate consequence: $w(T) \leq ||T||$ for all T, and if T is normal, then equality holds. (In case $X + iZ$ is not normal, the norm inequality may fail. A counterexample is given by

$$X = \begin{pmatrix} 1 & 1 \\ 1 & 1 \end{pmatrix}, \; Y = \begin{pmatrix} 2 & 0 \\ 0 & 2 \end{pmatrix}, \; Z = \begin{pmatrix} 0 & i \\ -i & 0 \end{pmatrix}.)$$

Theorem 2. *If A is normal, $A = B + iC$ with B and C Hermitian, then*

$$\delta(A) = ||B^- + iC||;$$

in other words, the positive part B^+ of B is a positive approximant.

Proof. Decompose the Hilbert space so that B becomes

$$\begin{bmatrix} B_1 & 0 \\ 0 & -B_2 \end{bmatrix},$$

with B_1 and B_2 positive. The normality of A implies that this can be done so as to "commute" with C, in the sense that C becomes

$$\begin{bmatrix} C_1 & 0 \\ 0 & C_2 \end{bmatrix},$$

where C_1 and C_2 are Hermitian, of course, and each of them commutes with the corresponding part of B.

Suppose that P is an arbitrary positive operator; with respect to the decomposition of B it becomes a matrix

$$\begin{bmatrix} P_1 & Q \\ Q^* & P_2 \end{bmatrix},$$

with P_1 and P_2 positive. Since

$$||A - P|| = \left\| \begin{bmatrix} (B_1 - P_1) + iC_1 & Q \\ -Q^* & (-B_2 - P_2) + iC_2 \end{bmatrix} \right\|$$

$$\geq \max \{ ||(B_1 - P_1) + iC_1||, \ ||-(B_2 + P_2) + iC_2|| \},$$

it is sufficient to prove that $||(B_1 - P_1) + iC_1|| \geq ||C_1||$ and $||-(B_2 + P_2) + iC_2|| \geq ||-B_2 + iC_2||$. (Indeed it will then follow that

$$||A - P|| \geq \max \{ ||C_1||, \ ||-B_2 + iC_2|| \} = \left\| \begin{pmatrix} iC_1 & 0 \\ 0 & -B_2 + iC_2 \end{pmatrix} \right\|$$

$$= \left\| \begin{bmatrix} B_1 + iC_1 & 0 \\ 0 & -B_2 + iC_2 \end{bmatrix} - \begin{bmatrix} B_1 & 0 \\ 0 & 0 \end{bmatrix} \right\| = ||A - B^+||.)$$

The first inequality is a consequence of the elementary facts about Hermitian approximation: since P is Hermitian, it cannot be a strictly better Hermitian approximation to $B_1 + iC_1$ than B_1. The second inequality follows from the lemma: put $X = B_2$, $Y = B_2 + P_2$, and $Z = -C_2$.

For each operator A, let $\mathbf{P}(A)$ be the set of all positive approximants of A, i.e., the set of all those positive operators P for which $||A - P|| = \delta(A)$. For Hermitian A, the description of $\mathbf{P}(A)$ is near the surface.

Corollary 4. *If A is Hermitian, then $\mathbf{P}(A)$ consists of all positive operators P such that*

$$A - ||A^-|| \leq P \leq A + ||A^-||.$$

Proof. The inequalities in the statement can also be expressed in the form $||A - P|| \leq ||A^-||$. If P is a positive approximant of A, then $||A - P|| \leq ||A^-||$ $(= ||A - A^+||)$ because $A^+ \geq 0$. Suppose, conversely, that $P \geq 0$ and $||A - P|| \leq ||A^-||$; since A^+ is a positive approximant of A (by Theorem 2), and since P is positive, it follows that $||A^-|| \leq ||A - P||$.

A word of caution may be in order. Lattice-theoretic experience suggests that the two inequalities $A - ||A^-|| \leq P$ and $0 \leq P$ be combined into the single inequality

$$(A - ||A^-||)^+ \leq P.$$

The latter inequality clearly implies the two separate ones; the converse, however, is false. Counterexample:

$$A = \begin{bmatrix} 4 & 0 & 0 \\ 0 & 2 & 0 \\ 0 & 0 & -3 \end{bmatrix}, \qquad P = \begin{bmatrix} 2 & 2 & 0 \\ 2 & 3 & 0 \\ 0 & 0 & 0 \end{bmatrix}.$$

An operator is Hermitian if its imaginary part is 0; it is positive if both the imaginary part and the negative part of the real part are 0. If only the negative part of the real part is required to be 0, the operator is called *accretive*; in other words, A is accretive in case Re $A \geq 0$. The problem of accretive approximation is to determine

$$\alpha(A) = \inf \{||A - E||: \text{Re } E \geq 0\}.$$

That turns out to be easy.

Corollary 5. *If $A = B + iC$ with B and C Hermitian, then*

$$\alpha(A) = ||B^-||;$$

in other words, the "accretive part" $B^+ + iC$ of A is an accretive approximant.

Proof. If $E = F + iG$, with F and G Hermitian, is accretive, *i.e.*, $F \geq 0$, then

$$||A - E|| = ||(B - F) + i(C - G)|| \geq ||B - F||.$$

Since B^+ is a positive approximant of B (by Theorem 2) and since F is positive, it follows that $||B - F|| \geq ||B - B^+||$, and the proof is complete.

Observe incidentally that for accretive operators the problem of positive approximation is easy: if A is accretive, then Re A is a positive approximant of A, and, consequently, $\delta(A) = ||\text{Im } A||$. Proof: Re A is known to be a Hermitian approximant of A in any case; in case Re A is positive, no other positive (and a fortiori Hermitian) operator can come closer.

Problems. Non-commutative approximation theory is in its infancy. The number of not yet considered questions is so large that every answer raises more problems than it settles. Here are some of the unsettled ones.

(1) To solve the problem of positive approximation means more than just to devise a process for finding the distance from a given operator to the set of all positive operators, more than to prove that the infimum that defines the distance is attained, and more than to devise a process for exhibiting a positive approximant that attains it. It should also mean in some sense to describe the set $\mathbf{P}(A)$ of *all* positive approximants of A. Problems of this sort are vague till they become solved. One way to make them a little less vague is to point to some of the subproblems that the solution should also solve. Subproblem 1: describe the set of all positive approximants of a *normal* operator. Subproblem 2: describe the extreme points of the (convex) set of all positive approximants. Subproblem 3: devise a way to tell whether $\mathbf{P}(A)$ is a singleton. Thus, for instance, $\frac{1}{2}\left(\begin{smallmatrix}1&1\\1&1\end{smallmatrix}\right)$ turns out to be the unique positive approximant of $\left(\begin{smallmatrix}0&0\\1&0\end{smallmatrix}\right)$; but why? What about $\left(\begin{smallmatrix}i&1\\1&0\end{smallmatrix}\right)$?

The same questions can and should be asked about approximants of any kind: describe them all, especially in the normal case; find the extreme ones; and devise a test for uniqueness. This applies, in particular, to the problem of accretive approximation, and even to the presumably simpler problem of Hermitian approximation.

(2) Theorem 2 gives a simple formula for the distance $\delta(A)$ from A to the set of all positive operators in case A is normal. For non-normal operators that formula may fail. It fails, for instance, in case $A = \left(\begin{smallmatrix}0&0\\1&0\end{smallmatrix}\right)$. Recall now that $\|B^- + iC\| \geqq \|(B^-)^2 + C^2\|^{1/2}$ whenever B and C are Hermitian; in case $A = B + iC$ with A normal, then equality holds. Theorem 2 could therefore have been stated this way: if A is normal, then $\delta(A) = \|(B^-)^2 + C^2\|^{1/2}$. Could this formula work for all A? It does work for $\left(\begin{smallmatrix}0&0\\1&0\end{smallmatrix}\right)$, as well as for normal operators, and it works in some other special cases too: *e.g.*, when A is accretive, or when C^2 happens to be a scalar. It does not work all the time; one counterexample is $\left(\begin{smallmatrix}i&1\\1&0\end{smallmatrix}\right)$. What does work? Is there a simple formula for $\delta(A)$?

(3) Approximation answers depend on the norm used in the questions. What happens to the theory of Hermitian, and positive, and accretive approximation when "unitarily invariant" [1] norms are used different from the customary operator norm?

REFERENCES

1. K. FAN & A. J. HOFFMAN, *Some metric inequalities in the space of matrices*, Proc. A. M. S. 6 (1955), 111–116.
2. P. R. HALMOS, *A Hilbert space problem book*, Van Nostrand, New York (1967).
3. I. E. SEGAL, *Algebraic integration theory*, Bull. A. M. S. 71 (1965), 419–489.

This research was supported in part by a grant from the National Science Foundation.

Indiana University
Date communicated: DECEMBER 15, 1971

Reprinted from the
DUKE MATHEMATICAL JOURNAL
Vol. 39, No. 4, pp. 779-787, Dec. 1972

PRODUCTS OF SHIFTS

P. R. HALMOS

A *shift* on a Hilbert space H is an isometry S on H such that for some subspace K of H the subspaces K, SK, S^2K, $\cdots$ are pairwise orthogonal and span H. (This definition describes *unilateral* shifts only; the other kind, bilateral shifts, will be referred to by their full name.) The *multiplicity* of S is the dimension of K; since K is the co-range of S, the multiplicity of a shift is the same as its co-rank.

Every isometry on H is either a unitary operator, or a shift, or a direct sum of two operators of those two kinds. The set of all isometries on H is a semigroup with somewhat mysterious properties; the purpose of this paper is to illuminate a small corner of the algebraic theory of that semigroup.

Which operators on H have the form US, where U is unitary and S is a shift? All that is obvious is that every operator of that form is an isometry. Is it a shift? Which operators on H have the form S_1S_2, where S_1 and S_2 are shifts? Once again it is obvious that every operator of that form is an isometry. Is it a shift?

The answers are not deep, but they are somewhat surprising, and the techniques shed at least a little light on the chicanery of shifts. The heart of the matter is in the separable case, and that is treated first; afterward the general case is reduced to the separable one by considerations of cardinal arithmetic.

THEOREM 1. *On a separable Hilbert space, every isometry is either a unitary operator, or a shift, or a product of two operators of those two kinds.*

THEOREM 2. *On a separable Hilbert space, every isometry of co-rank at least 2 is a product of two shifts.*

Proof of Theorem 1. It is to be proved that if U is unitary, with $1 \leq$ size $U \leq \aleph_0$ (the *size* of an operator is the dimension of its domain), and if S is a shift, with $1 \leq$ mult $S \leq \aleph_0$ ("mult" stands for multiplicity), then the direct sum $U \oplus S$ is a product of a unitary operator and a shift.

It is sufficient to treat the case in which mult $S = 1$. Indeed, if mult $S > 1$, then express S as a direct sum of shifts of multiplicity 1, and apply the theorem to the direct sum of U and one of the direct summands of S. To obtain the result for $U \oplus S$, form the direct sum of the unitary factor that the theorem yields and an identity operator of size $\aleph_0$ for each unused summand of S, form the direct sum of the shift factor that the theorem yields and the unused

Received July 3, 1972. Research was supported in part by a grant from the National Science Foundation.

summands of S, and note that the product of the unitary direct sum and the shift direct sum is exactly $U \oplus S$.

The proof is most conveniently expressed in terms of matrices.

If size $U = n \; (< \aleph_0)$, then, with respect to a suitable orthonormal basis $\{f_1, \cdots, f_n, e_0, e_1, e_2, \cdots\}$, the operator $U \oplus S$ has a matrix of the form

$$
\begin{bmatrix}
u_{11} & \cdots & u_{1n} \\
\vdots & & \vdots \\
u_{n1} & \cdots & u_{nn} \\
& & & 0 & 0 & 0 & 0 \\
& & & 1 & 0 & 0 & 0 \\
& & & 0 & 1 & 0 & 0 \\
& & & 0 & 0 & 1 & 0 \\
& & & & & & & \ddots
\end{bmatrix}
$$

Let $\hat{U}$ be the operator whose matrix is obtained from that of $U \oplus S$ by adjoining a column at the left, the new column consisting of all 0's except for a single 1 in row $n + 1$. The operator $\hat{U}$ is unitary; its matrix is the direct sum of the unitary northwest corner (of size $n + 1$) and the identity in the southeast corner. (The effect of the adjoined left column is to pull the sub-diagonal 1's in $U \oplus S$ up to the main diagonal.) Let $\hat{S}$ be the unilateral shift of multiplicity 1 that shifts the basis $\{f_1, \cdots, f_n, e_0, e_1, e_2, \cdots\}$, so that the matrix of $\hat{S}$ consists of all 0's except for 1's on the diagonal just below the main one. Since the effect of multiplying a matrix by $\hat{S}$ on the right is to erase the leftmost column, it follows that the product $\hat{U}\hat{S}$ is equal to $U \oplus S$.

(Remark 1. By a benevolent interpretation of symbols such as "$\{f_1, \cdots, f_n\}$" in case $n = 0$, the proof covers the case in which size $U = 0$; it yields the factorization $S = 1 \cdot S$.)

If size $U = \aleph_0$, then identify the domains of U and S, and write

$$
\hat{U} = \begin{bmatrix} 0 & U \\ 1 & 0 \end{bmatrix}, \qquad \hat{S} = \begin{bmatrix} 0 & S \\ 1 & 0 \end{bmatrix}.
$$

Since

$$
\hat{U}\hat{S} = \begin{bmatrix} U & 0 \\ 0 & S \end{bmatrix} = U \oplus S,
$$

it remains only to verify that $\hat{U}$ is unitary and $\hat{S}$ is a shift. The first statement is obvious. The verification that $\hat{S}$ is a shift is easy; if S shifts the orthonormal basis $\{e_0, e_1, e_2, \cdots\}$ in H, say, then $\hat{S}$ shifts the basis $\{[e_0, 0], [0, e_0], [e_1, 0], [0, e_1], [e_2, 0], [0, e_2], \cdots\}$ in $H \oplus H$. The proof of Theorem 1 is complete.

(*Remark* 2. The proof yielded a product of the form "unitary times shift". To get "shift times unitary", multiply the result on the right first by the inverse of the unitary factor and then by the unitary factor itself.)

Proof of Theorem 2. It is to be proved that if U is unitary, with $0 \leq$ size $U \leq \aleph_0$, and if S is a shift, with $2 \leq$ mult $S \leq \aleph_0$, then the direct sum $U \oplus S$ is a product of two shifts. It is sufficient to treat the case in which mult $S < \aleph_0$. Indeed, if mult $S = \aleph_0$, then express S as a direct sum of shifts of multiplicity 2, and apply the theorem to the direct sum of U and one of the direct summands of S. To obtain the result for $U \oplus S$, express each unused direct summand of S as a product of two shifts (in fact as the square of a shift of multiplicity 1), form the direct sum of the first factor the theorem yields and all the other first factors, do the same for the second factors, and note that the product of the two direct sums is exactly $U \oplus S$.

If size $U = n \ (< \aleph_0)$ and mult $S = m \ (< \aleph_0)$, then, with respect to a suitable orthonormal basis $\{f_1, \cdots, f_n, e_0, e_1, e_2, \cdots\}$, the operator $U \oplus S$ has a matrix of the form

$$
m \left\{ \begin{array}{l}
\begin{pmatrix}
\lambda_1 & & & & & & \\
& \ddots & & & & & \\
& & \lambda_n & & & & \\
& & & 0 & 0 & 0 & \\
& & & 0 & 0 & 0 & \\
& & & \vdots & \vdots & \vdots & \\
& & & 0 & 0 & 0 & \\
& & & 1 & 0 & 0 & \\
& & & 0 & 1 & 0 & \\
& & & 0 & 0 & 1 & \\
& & & & & & \ddots
\end{pmatrix}
\end{array} \right.
$$

where $\lambda_1, \cdots, \lambda_n$ are complex numbers of modulus 1. (A unitary operator on a finite-dimensional space has an orthonormal basis of eigenvectors; corresponding to a shift S of multiplicity m there exists an orthonormal basis $\{e_0, e_1, e_2, \cdots\}$ such that $Se_k = e_{k+m}$ for all k.)

Let $\hat{S}_1$ be the operator whose matrix is obtained from that of $U \oplus S$ by adjoining a column at the left, the new column consisting of all 0's except for a single 1 in row $n + 2$. The operator $\hat{S}_1$ is a shift of multiplicity $m - 1$. To prove that, look at what $\hat{S}_1$ does to a suitable permutation of the basis:

$$\hat{S}_1 e_0 = \lambda_n f_n ,$$
$$\hat{S}_1 f_n = \lambda_{n-1} f_{n-1} ,$$
$$\cdots$$

$$\hat{S}_1 f_2 = \lambda_1 f_1 ,$$

$$\hat{S}_1 f_1 = e_1 ,$$

$$\hat{S}_1 e_k = e_{k+m-1} , \qquad\qquad k = 1, 2, 3, \cdots .$$

This proves that $\hat{S}_1$ is a weighted shift; since the "weights" are complex numbers of modulus 1, it follows that $\hat{S}_1$ is (unitarily equivalent to) a shift. If $\hat{S}_2$ is the unilateral shift of multiplicity 1 that shifts the basis $\{f_1 , \cdots , f_n , e_0 , e_1 , e_2 , \cdots \}$, then $U \oplus S = \hat{S}_1 \hat{S}_2$.

(*Remark* 3. Note, just as in Remark 1 in the proof of Theorem 1, that the proof covers the case in which size $U = 0$.)

If size $U = \aleph_0$, the proof requires more technical detail. The first step is to recall that U can be expressed as the product $W_1 W_2$ of two bilateral shifts of infinite multiplicity [1; Problem 112]. Next, express S as the direct sum $S_1 \oplus S_2$ of two (unilateral) shifts, and thus reduce the problem to that of expressing

$$V = \begin{bmatrix} W_1 W_2 & 0 & 0 \\ 0 & S_1 & 0 \\ 0 & 0 & S_2 \end{bmatrix}$$

as a product of two shifts.

Write

$$\hat{U} = \begin{bmatrix} 1 & 0 & 0 \\ 0 & U_1 & 0 \\ 0 & 0 & U_2 \end{bmatrix},$$

where U_1 and U_2 are unitary operators, and form $\hat{U}^* V \hat{U}$. The result is a matrix just like V except that the direct summands S_1 and S_2 are replaced by $U_1^* S_1 U_1$ and $U_2^* S_2 U_2$. That such a replacement is possible reduces the problem some more: once the W's are given, the S's can be adjusted, separately, so as to stand in as good a relation to the W's as unitarily equivalent versions of the S's can possibly maintain.

Observe now that if

$$\hat{S}_1 = \begin{bmatrix} 0 & 0 & W_1 \\ S_1 & 0 & 0 \\ 0 & 1 & 0 \end{bmatrix}, \qquad \hat{S}_2 = \begin{bmatrix} 0 & 1 & 0 \\ 0 & 0 & S_2 \\ W_2 & 0 & 0 \end{bmatrix},$$

then $V = \hat{S}_1 \hat{S}_2$; the problem now is to prove that $\hat{S}_1$ and $\hat{S}_2$ are shifts. If

$$T = \begin{bmatrix} 0 & 0 & 1 \\ 0 & 1 & 0 \\ 1 & 0 & 0 \end{bmatrix},$$

then T is unitary and, except for the subscripts on the W's and S's,

$$T^* \hat{S}_1 T = \hat{S}_2 .$$

In view of this fact, the problem is reduced to this: whenever W is a bilateral shift of multiplicity $\aleph_0$, and S is a (unilateral) shift of finite multiplicity, and

$$\hat{S} = \begin{bmatrix} 0 & 0 & W \\ S & 0 & 0 \\ 0 & 1 & 0 \end{bmatrix},$$

then $\hat{S}$ is a shift. The "adjustability" of S_1 and S_2 can be used to weaken this last implication without weakening its desired effect. Since any two shifts of the same multiplicity are unitarily equivalent, it is enough to prove that for some W and S the matrix $\hat{S}$ is a shift.

Since $\hat{S}[f, g, h] = [Wh, \hat{S}f, g]$, it is obvious that S is an isometry. The co-range of S is the set of all those $[u, v, w]$ for which

$$(Wh, u) + (Sf, v) + (g, w) = 0$$

identically in $[f, g, h]$; this implies that the co-range of $\hat{S}$ is the set of all $[0, v, 0]$ with v in the co-range of S. Equivalently: if the co-range of S is K, then the co-range of $\hat{S}$ is $\hat{K} = 0 \oplus K \oplus 0$. (Note in particular that co-rank $\hat{S}$ = co-rank S.)

Observe now that under repeated applications of $\hat{S}$, the co-range $\hat{K}$ transforms as follows:

$$\begin{bmatrix} 0 \\ K \\ 0 \end{bmatrix} \rightarrow \begin{bmatrix} 0 \\ 0 \\ K \end{bmatrix} \rightarrow \begin{bmatrix} WK \\ 0 \\ 0 \end{bmatrix} \rightarrow \begin{bmatrix} 0 \\ SWK \\ 0 \end{bmatrix} \rightarrow \begin{bmatrix} 0 \\ 0 \\ SWK \end{bmatrix} \rightarrow \begin{bmatrix} WSWK \\ 0 \\ 0 \end{bmatrix} \rightarrow \cdots .$$

It is therefore sufficient to prove that SW is a shift (with co-range K) and WS is a shift (with co-range WK). Since $WS = W(SW)W^*$, the second assertion follows from the first. Consequence: it is sufficient to prove that if W is a bilateral shift of multiplicity $\aleph_0$, then, for each m, with $1 \le m < \aleph_0$, there exists a (unilateral) shift S of multiplicity m such that SW is a shift.

The last reduction (before something is actually proved) is to the case $m = 1$. Indeed: if mult $S = 1$ and SW is a shift, then form the direct sum of m copies of S and the direct sum of m copies of W. The former is a shift of multiplicity m, the latter is a bilateral shift of multiplicity $\aleph_0$ (and therefore unitarily equivalent to any other bilateral shift of multiplicity $\aleph_0$); the temporarily assumed result for the case $m = 1$ implies that their product is a shift.

The final step begins with an orthonormal basis $\{e(i, j) : i = 0, \pm1, \pm2, \cdots ;$ $j = 1, 2, 3, \cdots\}$, and a corresponding bilateral shift W such that

$$We(i, j) = e(i + 1, j)$$

for all i and j. What is needed is a shift S of multiplicity 1 such that (i) S maps each $e(i, j)$ onto another, and (ii) SW is a shift. This is a purely combinatorial problem. The construction could be arithmetically coded, but such a coding would convey nothing except the facts; to understand it, the reader would have to decode it and draw a picture (the one that the coding came from). It is more efficient to present the construction pictorially and thus convey both facts and understanding.

In Figure 1 each boldface dot represents one of the e's. The effect of W is not indicated; it is understood to be a shift to the right by one unit. The effect of S is indicated by the arrows. In Figure 2 the arrows indicate the effect of SW.

The proof of Theorem 2 is complete.

(*Remark* 4. Which operators have the form W_1W_2, where W_1 and W_2 are bilateral shifts? The question is only formally similar to the one that Theorem 2 answers. The answer, which has been known for some time, and was used above in the proof of Theorem 2, is that on every Hilbert space of infinite dimension the operators so representable are exactly the unitary ones; not even separability is relevant.)

What can be said about the co-rank of an isometry in terms of the co-ranks of the factors described in Theorems 1 and 2? The answer is almost a part of standard Fredholm theory, except that there the additivity of the index is usually stated under the assumption that at least one of the factors is a Fredholm operator [2]. (An isometry is always a semi-Fredholm operator; its index is the negative of the co-rank.) The facts pertinent here are easy to prove directly, with no finiteness assumptions.

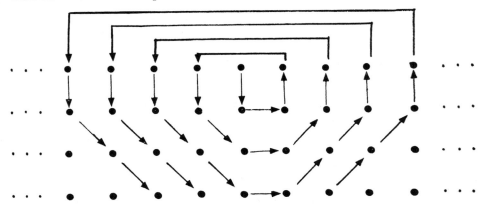

FIGURE 1

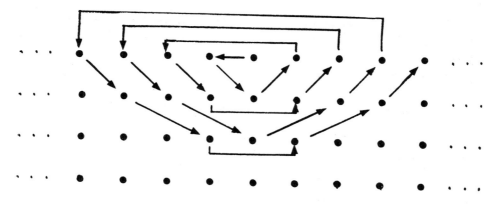

FIGURE 2

PROPOSITION. *If S and T are isometries, then*

$$\text{co-rank } (ST) = \text{co-rank } S + \text{co-rank } T.$$

Proof. Since co-rank X = null X^* for all operators X (where "null" is nullity, i.e., dimension of the kernel), the conclusion is equivalent to

$$\text{null } (T^*S^*) = \text{null } S^* + \text{null } T^*.$$

Observe, to begin with, that ker $S^* \subset$ ker T^*S^*; this observation reduces the problem to proving that

$$\dim (\text{ker } T^*S^* \cap \text{ker}^\perp S^*) = \dim \text{ker } T^*.$$

That, in turn, can be proved by exhibiting an isometry that maps ker T^* onto ker $T^*S^* \cap$ ran S.

An isometry that does the job is S itself. Suppose, in fact, that $f \varepsilon$ ker T^*; it follows that $T^*S^*(Sf) = T^*f = 0$, i.e., that $Sf \varepsilon$ ker T^*S^*, and hence that S maps ker T^* into ker $T^*S^* \cap$ ran S. To prove that S maps ker T^* *onto* ker $T^*S^* \cap$ ran S, consider g in ker $T^*S^* \cap$ ran S. Since $g = Sf$ for some f, therefore $f = S^*Sf = S^*g$; since $T^*f = T^*S^*g = 0$, it follows that $f \varepsilon$ ker T^*, and the proof of the proposition is complete.

The time has come to dispose of the non-separable case. The conclusions of Theorems 1 and 2 are not always true in that case; the facts, however, are easy to state.

THEOREM 3. *If U is unitary and S is a shift, then a necessary and sufficient condition that the direct sum $U \oplus S$ be a product of a unitary operator and a shift is that* size $U \leq$ size S.

THEOREM 4. *If U is unitary and S is a shift of multiplicity at least 2, then a necessary and sufficient condition that the direct sum $U \oplus S$ be a product of two shifts is that* size $U \leq$ size S.

Remark 5.   Since the size of a shift is always at least $\aleph_0$ (in fact size $S = \aleph_0 \cdot \text{mult } S$), the condition in Theorems 3 and 4 is always satisfied on separable spaces.

Proof of Theorem 3.   Suppose that $U \oplus S = \hat{U}\hat{S}$, where $\hat{U}$ is unitary and $\hat{S}$ is a shift.   Since

$$\text{mult } \hat{S} = \text{co-rank } \hat{S} = \text{co-rank } \hat{U}\hat{S} = \text{co-rank } (U \oplus S)$$

$$= \text{co-rank } S = \text{mult } S,$$

it follows that

$$\text{size } U \leq \text{size } (U \oplus S) = \text{size } \hat{S} = \aleph_0 \cdot \text{mult } \hat{S}$$

$$= \aleph_0 \cdot \text{mult } S = \text{size } S;$$

this proves the necessity of the condition.

Suppose now that size $U \leq$ size S. The argument is similar to one used for Theorem 1. That is: express U as a direct sum of unitary operators on separable spaces, and express S as a direct sum of shifts on separable spaces. Match each unitary direct summand with a shift direct summand, with, perhaps, some shifts left over. Apply Theorem 1 to each of the direct sums of the matched pairs. To obtain the result for $U \oplus S$, form the direct sum of the unitary factors that Theorem 1 yields and an identity operator of size $\aleph_0$ for each leftover shift, form the direct sum of the shift factors that Theorem 1 yields and the leftover shifts, and note that the product of the unitary direct sum and the shift direct sum is exactly $U \oplus S$. The proof of Theorem 3 is complete.

Proof of Theorem 4.   Suppose that $U \oplus S = \hat{S}_1 \cdots \hat{S}_n$, where $\hat{S}_1, \cdots, \hat{S}_n$ are shifts. As before

$$\text{mult } \hat{S}_j \leq \sum_{i=1}^{n} \text{mult } \hat{S}_i = \sum_{i=1}^{n} \text{co-rank } \hat{S}_i$$

$$= \text{co-rank } (U \oplus S) = \text{mult } S$$

for each j, and therefore

$$\text{size } U \leq \text{size } (U \oplus S) = \text{size } \hat{S}_i = \aleph_0 \cdot \text{mult } \hat{S}_i$$

$$\leq \aleph_0 \cdot \text{mult } S = \text{size } S.$$

This proves the necessity of the condition, and a little bit more.

Suppose now that size $U \leq$ size S. The case in which the multiplicity of S is finite (or, for that matter, $\aleph_0$) is covered by Theorem 2. In the remaining cases, express U as a direct sum of unitary operators on separable spaces, and express S as a direct sum of shifts, each of multiplicity at least 2, on separable spaces. Match each unitary direct summand with a shift direct summand, with, perhaps, some shifts left over; note that each leftover shift is a product of two shifts. Apply Theorem 2 to each of the direct sums of the matched pairs. To obtain the result for $U \oplus S$, form the direct sum of all the first factors that

Theorem 2 yields and all the first factors of the leftover shifts, do the same for the second factors, and note that the product of the first direct sum and the second direct sum is exactly $U \oplus S$. The proof of Theorem 4 is complete.

There are many open problems related to the ones solved above; they are the kind whose value is hard to assess till they are solved. Here are some examples. (1) What are all the different factorizations, of the kinds described in Theorems 1 and 2, that an isometry can have—i.e., what is the extent of non-uniqueness? (2) Which operators can be expressed as products of weighted shifts? (3) Does every contraction have a representation of the form $S_1^* S_2$, where S_1 and S_2 are shifts?

REFERENCES

1. P. R. HALMOS, *A Hilbert Space Problem Book*, Van Nostrand, 1967.
2. R. S. PALAIS, *Seminar on the Atiyah-Singer Index Theorem*, Princeton University Press, 1965.

DEPARTMENT OF MATHEMATICS, INDIANA UNIVERSITY, BLOOMINGTON, INDIANA 47401

Reprinted from the
ACTA SCIENTIARUM MATHEMATICARUM
Tomus 34, pp. 131-139, 1973

Limits of shifts

By P. R. HALMOS in Bloomington (Indiana, U.S.A.)

Dedicated to B. Sz.-Nagy on his sixtieth birthday, July 29, 1973

What is the closure of the unilateral shifts?

The question looks odd; by long-standing tradition the unilateral shift is regarded as one operator, not a set of operators. On the occasions when the plural is used it usually indicates multiplicities, or weights, but neither of those is what is meant here. A moment's thought reveals that the question makes unambiguous sense. An operator S on a Hilbert space H is a unilateral shift (of multiplicity 1) in case there exists an orthonormal basis $\{e_0, e_1, e_2, \ldots\}$ for H such that $Se_n = e_{n+1}, n = 0, 1, 2, \ldots$. From this point of view there are as many unilateral shifts of multiplicity 1 as there are orthonormal bases enumerated by the natural numbers. The problem is to determine the closure of the set of all such shifts with respect to the norm topology of operators.

The same question can be asked and the same comments can be made about bilateral shifts, which shift an orthonormal basis enumerated by all integers.

Unilateral shifts are isometric, and, therefore, so are their limits. (Reason: if $S_n \to T$, then $S_n^* S_n \to T^* T$.) If, moreover, all the terms of a convergent sequence of unilateral shifts have the same multiplicity, then the co-rank of the limit is equal to that common multiplicity. (Reason: for n large, the projections $1 - S_n S_n^*$ and $1 - TT^*$ are near, and, therefore, they have the same rank; the rank of $1 - S_n S_n^*$ is the multiplicity of S_n.)

Bilateral shifts are unitary, and, therefore, so are their limits. Since, moreover, the spectrum of every bilateral shift is the entire unit circle, it follows that the spectrum of a limit of bilateral shifts is also the entire unit circle. (Reason: the spectrum is upper semicontinuous, [6, Problem 86].)

The preceding two paragraphs describe some necessary conditions that limits of shifts must satisfy; it is natural to ask how near those conditions come to being sufficient. Can a limit of unilateral shifts of multiplicity 1, say, have a unitary direct

---

*) Research supported in part by a grant from the National Science Foundation.

summand? Can a limit of bilateral shifts of multiplicity 1 have anything other than an absolutely continuous spectral measure of uniform multiplicity? On first consideration both questions seem to call for a negative answer. It is remarkable, however, that the already stated necessary conditions turn out to be sufficient also. The facts are described in the following statement; the main purpose of the sequel is to prove it.

Theorem. *On a separable Hilbert space the norm closure of the set of unilateral shifts of multiplicity n* $(1 \leqq n \leqq \infty)$ *is the set of all isometries of co-rank n, and the norm closure of the set of bilateral shifts of multiplicity n* $(1 \leqq n \leqq \infty)$ *is the set of all unitary operators whose spectrum is the entire unit circle.*

The proof uses a slight sharpening of the proof of a result of R. G. DOUGLAS (which will be described later). That result became part of the oral tradition sometime in 1971. I learned the statement from P. A. FILLMORE and the proof of the central lemma (which appears as Lemma 2 below) from I. D. BERG. A treatment of the Douglas result in an extended context is to appear later [3]. The present sharpening is applied, along the way, to the proof of a theorem of von Neumann's (the so-called von Neumann converse of Weyl's theorem [11]). The result (Lemma 4) is a quantitative improvement of von Neumann's theorem for a large class of normal operators (the ones for which the spectrum coincides with the essential spectrum).

Lemma 1. *If A is a normal operator on a separable Hilbert space, then* $A = D + C$, *where D is diagonal, with its spectrum included in that of A, and C is compact, with its norm arbitrarily small.*

Except for the statement about the spectrum of D, this is the Berg extension [2] to normal operators of the Weyl—von Neumann theorem [11] for Hermitian ones. In my subsequent proof [9] no restriction was placed on the spectrum of D or on the size of C. There is perhaps some merit in knowing that the restrictions can be captured in the framework of that proof; the next two paragraphs show how that can be done.

As far as the size of C is concerned, the result in the Hermitian case goes back to von NEUMANN [11], who proved that the compact summand of a Hermitian operator could in fact be made a Hilbert—Schmidt operator with arbitrarily small Hilbert—Schmidt norm. (Cf. also [8, p. 904].) To extend the result to the normal case, use the fact that if A is normal, then $A = \varphi(A')$, where A' is Hermitian and φ is continuous [9]. Recall now that the mapping $X \mapsto \varphi(X)$, defined for each Hermitian operator X whose spectrum is in the domain of φ, is continuous in the norm topology. (This is an easy exercise whose proof uses nothing more than the Weierstrass polynomial approximation theorem and the norm continuity of the algebraic operations on operators. The statement is true for continuous functions of normal operators, as well as Hermitian ones; the only additional technique needed is the planar version of the Weierstrass theorem.) Consequence: if $A' = D' + C'$, with D' diagonal and C' compact, the norm of the (compact) operator $C = A - D(= \varphi(A') - \varphi(D'))$ can be made as small as desired by making $\|C'\|$ small enough. (Observe that because of the passage to a limit implied by the formation of a continuous function, the Hilbert—Schmidt character of the compact summand cannot automatically be asserted in the normal case. It is not known whether the reason is in the proof or in the facts.)

The problem of putting the spectrum of D into the spectrum of A can be handled as follows. For each positive number δ, there can be only finitely many eigenvalues of D farther than δ from the spectrum of A. (Reason: otherwise the eigenvalues of D would have a cluster point not in the spectrum of A, in contradiction to the fact that A and D have the same essential spectrum.) Suppose now that $A = D + C$, with D diagonal, C compact, and $\|C\|$ small enough for two purposes: (1) if the ultimate C is to have norm below ε, make the present one have norm below $\varepsilon/2$, and (2) use the upper semicontinuity of the spectrum [6, Problem 86] to guarantee that if $\|A - X\| < \|C\|$, then the spectrum of X is in the $\varepsilon/2$ neighborhood of the spectrum of A. Consider, successively, the values of δ equal to $\|C\|, \|C\|/2, \|C\|/3, \dots$, and, each time, replace the eigenvalues of D outside the δ neighborhood of the spectrum of A by numbers in the spectrum as near as possible. The total alteration is compact and has norm not more than $\varepsilon/2$. Absorb it in C, increasing $\|C\|$ thereby to ε at worst. In case C happened to have not only small norm but small Hilbert—Schmidt norm as well, the altered C will have the same property.

Lemma 2. *If S is a shift of multiplicity 1 (unilateral or bilateral), if $\{\lambda_1, \lambda_2, \lambda_3, \dots\}$ is a sequence of complex numbers of modulus 1, and if $\varepsilon > 0$, then there exist operators D and E such that D is diagonal, with eigenvalues $\lambda_1, \lambda_2, \lambda_3, \dots$, and such that $S = (D \oplus E) + C$, where C is a Hilbert—Schmidt operator with Hilbert—Schmidt norm not greater than ε.*

To prove the lemma, consider an orthonormal set $\{e_0, e_1, e_2, \dots\}$ that S shifts. If $|\lambda| = 1$, $m = 0, 1, 2, \dots$, and $n = 1, 2, 3, \dots$, write

$$f = (1/\sqrt{n})(e_m + e_{m+1}/\lambda + \cdots + e_{m+n-1}/\lambda^{n-1}).$$

Clearly $\|f\| = 1$. Since

$$Sf = (1/\sqrt{n})(e_{m+1} + e_{m+2}/\lambda + \cdots + e_{m+n}/\lambda^{n-1})$$

$$= \lambda(1/\sqrt{n})(e_{m+1}/\lambda + e_{m+2}/\lambda^2 + \cdots + e_{m+n}/\lambda^n),$$

it follows that

$$Sf - \lambda f = (1/\sqrt{n})(e_{m+n}/\lambda^{n-1} - \lambda e_m),$$

and hence that

$$\|Sf - \lambda f\|^2 = 2/n.$$

In other words, the vector $f = f(\lambda, m, n)$ is an approximate eigenvector for S, with approximate eigenvalue λ and degree of approximation $\sqrt{2/n}$. Since

$$S^*f - \bar{\lambda}f = -\bar{\lambda}S^*(Sf - \lambda f),$$

it follows that f is, at the same time, an approximate eigenvector for S^*, with approximate eigenvalue $\bar{\lambda}$ and degree of approximation $\sqrt{2/n}$.

The preceding construction can be applied to each of the given numbers λ_k. Choose n_k so that

$$\Sigma_k(2/n_k) \leq (\varepsilon/2)^2,$$

and choose m_k so that the index intervals $[m_k, m_k + n_k - 1]$ are pairwise disjoint. If $f_k = f(\lambda_k, m_k, n_k)$, then $\{f_1, f_2, f_3, \ldots\}$ is an orthonormal sequence such that $\Sigma_k \|Sf_k - \lambda_k f_k\|^2 \leq (\varepsilon/2)^2$, $\Sigma_k \|S^*f_k - \bar{\lambda}_k f_k\|^2 \leq (\varepsilon/2)^2$. Let M be the span of $\{f_1, f_2, f_3, \ldots\}$, let P be the projection with range M, and let D be the diagonal operator defined on M by

$$Df_k = \lambda_k f_k, \quad k = 1, 2, 3, \ldots.$$

Write the Hilbert space as $M \oplus M^\perp$, and, correspondingly, consider the matrices

$$S = \begin{pmatrix} X & Y \\ Z & E \end{pmatrix}, \quad P = \begin{pmatrix} 1 & 0 \\ 0 & 0 \end{pmatrix}, \quad Q = \begin{pmatrix} D & 0 \\ 0 & 0 \end{pmatrix}, \quad C = \begin{pmatrix} X-D & Y \\ Z & 0 \end{pmatrix}.$$

Assertion: $X - D$, Y, and Z are Hilbert—Schmidt operators, and the sum of the squares of their Hilbert—Schmidt norms is not more than ε^2. Indeed:

$$\|(X-D)f_k\|^2 + \|Zf_k\|^2 = \|(S-Q)f_k\|^2 = \|Sf_k - \lambda_k f_k\|^2 = 2/n_k,$$

and, similarly,

$$\|(X^* - D^*)f_k\|^2 + \|Y^*f_k\|^2 = 2/n_k;$$

the proof of Lemma 2 is complete.

Remark. The control that the proof gives over the differences $Sf_k - \lambda_k f_k$ is strong enough to make it possible to put C into the trace class. Indeed: choose the n_k's so large that $\Sigma_k \sqrt{2/n_k}$ is small, and apply the lemma of Dunford and Schwartz [4, p. 1116] according to which $\Sigma_k \|Tf_k\| < \infty$, for an orthonormal basis $\{f_1, f_2, f_3, \ldots\}$, implies that T is in the trace class. (The statement in [4] does not seem to be formulated for the right set of exponents. In any event, it is true and the proof is valid for the exponent $p = 1$.)

Lemma 3. *If $\{a_n\}$ and $\{b_n\}$ are sequences with the same cluster set C in a compact metric space, if $a_n \in C$ and $b_n \in C$ for all n, and if $\varepsilon > 0$, then there exists a permutation π of the natural numbers such that $\Sigma_n d(a_n, b_{\pi n}) < \varepsilon$.*

It is convenient to have a word to describe the sequences that occur in this statement: call a sequence recurrent if each of its terms is a cluster point of it. (The "cluster set" of a sequence is, of course, the set of all cluster points. In this language a sequence is recurrent if it is included in its own cluster set.) Lemma 3 is a sharpened version of the von Neumann permutation theorem [7], which is used in the proof of the von Neumann converse of Weyl's theorem. The original version does not assume that the given sequences are recurrent, and cannot conclude that, after the permutation, the sum of distances is small. If, for instance, $a_1 = 1$, $a_n = 0$ for $n > 1$, and $b_n = 0$ for all n, then, clearly, there is no permutation π such that $d(a_n, b_{\pi n}) \leqq 1/2$ for all n. The trouble is not that the ranges of the sequences are different; if $b_1 = = b_2 = 1$ and $b_n = 0$ for $n > 2$, the inequalities $d(a_n, b_{\pi n}) \leqq 1/2$ for all n can still not be achieved. The trouble is that the cluster sets (which, to be sure, are the same) do not contain all the terms; in the first example one of the sequences fails to be recurrent, and in the second example they both do.

Now for the proof of Lemma 3.

Write $\sigma(1) = 1$. Since $a_{\sigma(1)} \in C$, there exists an index $\tau(1)$ such that $d(a_{\sigma(1)}, b_{\tau(1)}) \leqq \leqq \varepsilon/2$. Let $\tau(2)$ be the smallest index distinct from $\tau(1)$ (so that, typically, $\tau(2)$ will be 1). Since $b_{\tau(2)} \in C$, there exists an index $\sigma(2)$ distinct from $\sigma(1)$ such that $d(a_{\sigma(2)}, b_{\tau(2)}) \leqq \varepsilon/4$. The preceding four sentences describe a two-step process that is now to be applied infinitely often. The second application will indicate how the general one is to be made. Let $\sigma(3)$ be the smallest index not contained in $\{\sigma(1), \sigma(2)\}$. Find $\tau(3)$ not contained in $\{\tau(1), \tau(2)\}$ so that $d(a_{\sigma(3)}, b_{\tau(3)}) \leqq \varepsilon/8$. Let $\tau(4)$ be the smallest index not contained in $\{\tau(1), \tau(2), \tau(3)\}$. Find $\sigma(4)$ not contained in $\{\sigma(1), \sigma(2), \sigma(3)\}$ so that $d(a_{\sigma(4)}, b_{\tau(4)}) \leqq \varepsilon/16$.

When, ultimately, $\sigma(n)$ and $\tau(n)$ are defined for all n, each of σ and τ is a permutation of the set of all natural numbers. Indeed: since the definition of $\sigma(n)$ guarantees that $\sigma(n)$ is not contained in $\{\sigma(1), ..., \sigma(n-1)\}$, $n > 1$, the mapping σ is one-to-one; the definition for odd values of n guarantees that every natural number is in the range of σ. The argument for τ is, of course, the same, except that "odd" has to be replaced by "even".

The result is a pair of permutations σ and τ such that $d(a_{\sigma(n)}, b_{\tau(n)}) \leqq \varepsilon/2^n$ for all n. If π is defined so that $\tau(n) = \pi(\sigma(n))$ for all n, i.e., if $\pi = \tau\sigma^{-1}$, then $\Sigma_n d(a_n, b_{\pi(n)}) = = \Sigma_n d(a_{\sigma(n)}, b_{\tau(n)}) \leqq \varepsilon$.

My original statement of Lemma 3 had "$d(a_n, b_{\pi(n)}) \to 0$ and $d(a_n, b_{\pi(n)}) \leqq \varepsilon$ for all n" instead of "$\Sigma_n d(a_n, b_{\pi(n)}) \leqq \varepsilon$", and my proof of it was longer; the simplification is due to J. G. STAMPFLI.

To apply Lemma 3, I introduce a new concept: a normal operator is *essential* if its spectrum is the same as its essential spectrum.

Lemma 4. *If A and B are essential normal operators with the same spectrum, on a separable Hilbert space, and if $\varepsilon > 0$, then there exists a unitary operator U and a compact operator K such that $A = U^* B U + K$ and $\|K\| \leq \varepsilon$.*

Lemma 4 is a sharpened version of the von Neumann converse of Weyl's theorem. The original version does not assume that the given operators are essential, and cannot conclude that, after the unitary equivalence, they are within ε of one another. VON NEUMANN [11] remarked that, in fact, if a single compact operator is excluded from the competition, the conclusion becomes false. The point of Lemma 4 is that for essential normal operators the compact operators that appear can be made to satisfy severe and useful size restrictions.

To prove Lemma 4, use Lemma 1 to write $A = D_A + C_A$ and $B = D_B + C_B$, where D_A and D_B are diagonal, with each diagonal entry in the common spectrum, and C_A and C_B are compact, with $\|C_A\| \leq \varepsilon/3$, $\|C_B\| \leq \varepsilon/3$. Since D_A and A have the same essential spectrum, and since the essential spectrum of D_A is the cluster set of the diagonal, it follows that that cluster set is the common spectrum of A and B. (This step uses the assumption that A is essential.) Similarly the cluster set of D_B is that common spectrum. By Lemma 3 there exists a unitary operator U (induced by a permutation) and a compact operator C such that $D_A = U^* D_B U + C$ and $\|C\| \leq \leq \varepsilon/3$. Consequence:

$$A = D_A + C_A = U^* D_B U + C + C_A = U^* (B - C_B) U + C + C_A =$$
$$= U^* B U - U^* C_B U + C + C_A;$$

since $- U^* C_B U + C + C_A$ is compact and has norm not greater than ε, the proof of Lemma 4 is complete.

Remark. In case A and B are such that C_A and C_B can be made to have small Hilbert—Schmidt norm (e.g., in case A and B are Hermitian or unitary), then K can be made to have small Hilbert—Schmidt norm; the perturbation C that Lemma 3 introduces belongs, in fact, to the trace class.

The statement of Lemma 4 does not include the von Neumann converse (for not necessarily essential operators) as a special case, but the proof of Lemma 4 is, in spirit, the same as that of the unmodified version; cf. [1], [10], [11]. The main difference is that the present proof uses the quantitative version (Lemma 3) of the von Neumann permutation theorem.

For some of the statements that follow it is convenient to introduce a shorthand notation: if A and B are operators and $\varepsilon > 0$, write

$$A \sim B \quad (\varepsilon)$$

in case there exists an operator B', unitarily equivalent to B, such that $A - B'$ has Hilbert—Schmidt norm not greater than ε. (The operators A and B need not even be defined on the same Hilbert space. The generality gained thereby is shallow but useful.)

Lemma 5. *If S is a shift of multiplicity n $(1 \leqq n \leqq \infty)$ (unilateral or bilateral), if U is a unitary operator on a separable Hilbert space, and if $\varepsilon > 0$, then*

$$S \sim U \oplus S \quad (\varepsilon).$$

For the proof, let $\{\lambda_1, \lambda_2, \lambda_3, \dots\}$ be a sequence (of complex numbers of modulus 1) whose closure is dense in the spectrum of U, in which each term occurs infinitely often. Apply Lemma 2 to write

(1) $$S \sim D \oplus E \quad (\varepsilon/4),$$

where D is a diagonal operator with eigenvalues $\lambda_1, \lambda_2, \lambda_3, \dots$. (The unitary equivalence is, in fact, effected by the identity operator in this case, but nothing is lost by forgetting that.) In Lemma 2, to be sure, the shift was assumed to have multiplicity 1. The lemma is, however, applicable to shifts of all non-zero multiplicities; all that needs to be done is to break off a shift of multiplicity 1 as a direct summand, apply Lemma 2 as is, and then glue the fracture together again.

Let U^∞ be the direct sum of countably infinitely many copies of U; observe that U^∞ is unitary (and hence, in particular, normal) and that U^∞ is essential. Since D and $U^\infty \oplus D$ are essential unitary operators with the same spectrum, the remark following the proof of Lemma 4 shows that

(2) $$D \sim U^\infty \oplus D \quad (\varepsilon/4).$$

Substitute (2) into (1) to get

$$S \sim U^\infty \oplus D \oplus E \quad (\varepsilon/2).$$

It follows that

(3) $$U \oplus S \sim U \oplus U^\infty \oplus D \oplus E \quad (\varepsilon/2).$$

Since $U \oplus U^\infty$ is unitarily equivalent to U^∞, (3) implies

$$U \oplus S \sim U^\infty \oplus D \oplus E \quad (\varepsilon/2)$$

and hence, by (2)

(4) $$U \oplus S \sim D \oplus E \quad (3\varepsilon/4).$$

Use (1) to replace the right side of (4) by S, and conclude that

$$U \oplus S \sim S \quad (\varepsilon).$$

The proof of Lemma 5 is complete.

Proof of the theorem. Suppose that V is an isometry of co-rank n $(1 \leqq n \leqq \infty)$ on a separable Hilbert space, and suppose that $\varepsilon > 0$. Write V as $U \oplus S$, where U

is unitary and S is a unilateral shift (with, of course, multiplicity n) [6, Problem 118]. By Lemma 5

$$V \sim S \quad (\varepsilon).$$

Since an operator unitarily equivalent to a unilateral shift is a unilateral shift, this proves that in every ε neighborhood of V there is a unilateral shift (necessarily of the same co-rank as V), and the first half of the theorem follows.

The second half is proved similarly. Suppose that U is a unitary operator whose spectrum is the entire unit circle, and suppose that $\varepsilon > 0$. By Lemma 5

$$[U \oplus S \sim S \quad (\varepsilon/2),$$

where S is a bilateral shift of multiplicity n. Since U and $U \oplus S$ are essential unitary operators with the same spectrum (here is where the hypothesis about the spectrum of U is used), it follows from the remark following the proof of Lemma 4 that

$$U \sim U \oplus S \quad (\varepsilon/2).$$

Consequence:

$$U \sim S \quad (\varepsilon),$$

and the proof is completed as in the unilateral case.

Scholium. *On a separable Hilbert space every isometry of non-zero co-rank is the sum of a pure isometry and an operator of arbitrarily small Hilbert—Schmidt norm.*

Except for the description of the size of the perturbation, this is the original version of the Douglas result mentioned after the statement of the theorem.

Experience shows that norm approximation theorems are likely to be difficult but worth the trouble; they give useful analytic insights into the behavior of operators. Strong and weak approximation theorems are usually easier to prove, but harder to find applications for. A comparison of the theorem proved above and the proposition below indicates that for approximation by shifts the customary situation prevails.

In what follows it is convenient to use the word "shift" ambiguously. A true statement and a valid proof result if it is interpreted consistently as either "unilateral shift" or "bilateral shift".

Proposition. *On a separable infinite-dimensional Hilbert space the strong closure of the set of shifts of multiplicity 1 is the set of all isometries; the weak closure of the set of shifts of multiplicity 1 is the set of all contractions.*

For the proof, consider first an arbitrary operator A on the given Hilbert space H, and a direct sum of the form $A \oplus B$ on $H \oplus H$. Assertion 1: if $f_1, \ldots, f_n$ are in H, then there exists an operator on H unitarily equivalent to $A \oplus B$ that agrees with

A on each f_j. To prove that, let V be an isometry from H onto $H \oplus H$ such that if f is in the (finite-dimensional) subspace spanned by $f_1, \ldots, f_n, Af_1, \ldots, Af_n$, then $Vf = [f, 0]$. (Here is where the infinite-dimensionality of H is used.) It follows that $V^*(A \oplus B)Vf_j = V^*(A \oplus B)[f_j, 0] = V^*[Af_j, 0] = Af_j$ for $j = 1, \ldots, n$.

Suppose now that U is an arbitrary unitary operator on H. Assertion 2: every strong neighborhood of U contains a shift of multiplicity 1. To prove this, consider a basic strong neighborhood of U, consisting of all operators T such that $\|Uf_j - Tf_j\| < \varepsilon$, $j = 1, \ldots, n$, where $f_1, \ldots, f_n$ are unit vectors in H and $\varepsilon > 0$. If S is a shift of multiplicity 1, then, by Assertion 1, there exists an operator unitarily equivalent to $U \oplus S$ that agrees with U on each f_j. Since, by Lemma 5, $U \oplus S \sim S$ (ε), it follows that some operator unitarily equivalent to S differs from U by less than ε on each f_j; this implies Assertion 2.

The preceding two paragraphs imply that the strong closure of the set of shifts of multiplicity 1 contains all unitary operators, and from this all else follows. Indeed, the *strong* closure of the set of unitary operators is known to be the set of all isometries [8, p. 892], and the *weak* closure of the set of unitary operators is known to be the set of all contractions [5, p. 128].

Problem. What are the answers to the corresponding questions for weighted shifts?

References

[1] S. K. BERBERIAN, The Weyl spectrum of an operator, *Indiana U. Math. J.*, **20** (1970), 529—544.

[2] I. D. BERG, An extension of the Weyl—von Neumann theorem to normal operators, *Trans. Amer. Math. Soc.*, **160** (1971), 365—371.

[3] L. G. BROWN, R. G. DOUGLAS, and P. A. FILLMORE. Unitary equivalence modulo compact operators and extensions of C^*-algebras. (Preprint, October 1972).

[4] N. DUNFORD and J. T. SCHWARTZ, *Linear operators*. Part II, Interscience (New York, 1963).

[5] P. R. HALMOS, Normal dilations and extensions of operators, *Summa Brasil. Math.*, **2** (1950), 125—134.

[6] P. R. HALMOS, *A Hilbert space problem book*, Van Nostrand (Princeton, 1967).

[7] P. R. HALMOS, Permutations of sequences and the Schröder—Bernstein theorem, *Proc. Amer. Math. Soc.*, **19** (1968), 509—510.

[8] P. R. HALMOS, Ten problems in Hilbert space, *Bull. Amer. Math. Soc.*, **76** (1970), 887—933.

[9] P. R. HALMOS, Continuous functions of Hermitian operators, *Proc. Amer. Math. Soc.*, **31** (1972), 130—132.

[10] W. SIKONIA, The von Neumann converse of Weyl's theorem, *Indiana U. Math. J.*, **21** (1971), 121—123.

[11] J. von NEUMANN, *Charakterisierung des Spektrums eines Integraloperators*, Hermann (Paris, 1935).

INDIANA UNIVERSITY
BLOOMINGTON, INDIANA 47401

(Received May 27, 1972)

Reprinted from PROCEEDINGS OF THE EDINBURGH MATHEMATICAL SOCIETY
Vol. 19 (Series II), Part 1, March 1974

SPECTRAL APPROXIMANTS OF NORMAL OPERATORS

by P. R. HALMOS †
(Received 19th February 1973)

Introduction

For each non-empty subset Λ of the complex plane, let $\mathscr{S}(\Lambda)$ be the set of all those operators (on a fixed Hilbert space H) whose spectrum is included in Λ. The problem of *spectral approximation* is to determine how closely each operator on H can be approximated (in the norm) by operators in $\mathscr{S}(\Lambda)$. The problem appears to be connected with the stability theory of certain differential equations. (Consider the case in which Λ is the right half plane.) In its general form the problem is extraordinarily difficult. Thus, for instance, even when Λ is the singleton $\{0\}$, so that $\mathscr{S}(\Lambda)$ is the set of quasinilpotent operators, the determination of the closure of $\mathscr{S}(\Lambda)$ has been an open problem for several years (**3**, Problem 7).

The problem of *normal spectral approximation* is the one in which $\mathscr{S}(\Lambda)$ is replaced by the set $\mathscr{N}(\Lambda)$ of all those normal operators whose spectrum is included in Λ. This problem too is important and difficult. Thus, for instance, even when Λ is the entire complex plane, for a long time it was not known whether $\mathscr{N}(\Lambda)$ approximants always exist. In other words: is it true that, for each operator A, the distance $\inf\{\|A-N\| : N \in \mathscr{N}(\Lambda)\}$ is always attained? An example proving that the answer is no was recently found by D. D. Rogers, but, even with that new information, the visible facts are greatly outnumbered by the submerged ones.

For some special sets Λ the theory of $\mathscr{N}(\Lambda)$ approximation is at least partly known. If, for instance, Λ is the real line, the problem becomes that of Hermitian approximation, and, at least as far as the existence of approximants is concerned, it is solved by the mapping $A \mapsto \frac{1}{2}(A+A^*)$. If Λ is the unit circle (perimeter), the problem is that of unitary approximation; this case has been studied by Fan and Hoffman (**1**) and van Riemsdijk (**7**). If Λ is the set of non-negative real numbers, the problem is that of positive approximation, and it can be regarded as solved (**4**). There are, however, many quite innocent-looking sets Λ for which almost nothing is known about $\mathscr{N}(\Lambda)$ approximation. An interesting example is the problem of projection approximation, corresponding to the case in which Λ is the pair $\{0, 1\}$.

It may be that the general problem of spectral approximation is, in some

† Research supported in part by grants from the National Science Foundation (U.S.A.) and the Science Research Council (U.K.).

sense, not the right thing to ask about. A piece of disturbing evidence is that
the set $\mathscr{S}(\Lambda)$ is not in general closed; witness the classical example of a sequence
of quasinilpotent operators whose limit is not quasinilpotent (**2**, Problem 87).
It is comforting to observe that the problem of *normal* spectral approximation
rests on a sounder basis: if Λ is a closed set of complex numbers, then $\mathscr{N}(\Lambda)$
is a closed set of operators. This comfort is a consequence of the continuity
of the spectrum for normal operators; see Newburgh (**6**).

The main purpose of this paper is to prove a theorem about normal spectral
approximation to normal operators. A special case of the theorem asserts
that if A is normal, then A has a projection approximant, and, moreover, it
has one that is a function of A (and that belongs, therefore, to the second
commutant of A).

In order to understand the point at issue, it is wise to consider the commutative analogue of the problem. How does one approximate a complex-
valued bounded measurable function ϕ (modulo sets of measure zero) by a
characteristic function? A reasonable answer goes as follows. Let F be the
function defined on the complex plane by

$$F(z) = \begin{cases} 0 & \text{if } \operatorname{Re} z \le \tfrac{1}{2}, \\ 1 & \text{if } \operatorname{Re} z > \tfrac{1}{2}. \end{cases}$$

(In other words: $F(z) = 0$ or 1 according as z is nearer to 0 or to 1. In case z
is equidistant from 0 and 1, i.e., in case $\operatorname{Re} z = \tfrac{1}{2}$, it doesn't much matter
how F decides between 0 and 1; the choice indicated in the formal definition
above is one of the two simplest ones to write down.) The function $F \circ \phi$ is a
characteristic function (and a function of ϕ at that), and it is, clearly, as near
to ϕ as any characteristic function can get. (Nearness is measured by the
supremum norm, of course; that is the appropriate commutative version of
the operator norm.)

Every normal operator can be represented as multiplication by a suitable
bounded measurable function ϕ on a suitable L^2 space. (This statement has
to be interpreted with some care for non-separable spaces, but there is no
essential loss of generality in restricting attention to the separable case.) Why
then doesn't the preceding paragraph solve the problem of projection approximation to normal operators? Given a normal A, use the F defined above and
write $P = F(A)$; surely the operator P is a projection that is as near to A as any
projection can get. Or is it?

Since P is a function of A, the functional facts described above imply that
P is as near to A as any projection that is a function of A can get. In non-
commutative approximation theory, however, the competition is keener than
in the commutative case. Nothing in the commutative situation inhibits the
existence of a projection that is *not* a function of A and that is nearer to A
than P is. The whole point of this paper is to show (not for projection approximation only, but for normal spectral approximation in general) that the keener
competition does not change winners to losers.

Two comments are in order.

(1) The enlarged competition may properly enlarge the list of those tied for first place. Consider, for example, the problem of projection approximation to the matrix

$$A = \begin{pmatrix} 0 & 0 & 0 \\ 0 & 1 & 0 \\ 0 & 0 & 2 \end{pmatrix}.$$

The functional approximation process described above yields the approximant

$$P = \begin{pmatrix} 0 & 0 & 0 \\ 0 & 1 & 0 \\ 0 & 0 & 1 \end{pmatrix}.$$

There are, however, many other projections that are as near to A as P is; the only thing that is " wrong " with them is that they are not functions of A. An example of such a near projection is given by

$$Q = \begin{pmatrix} \frac{1}{2} & \frac{1}{2} & 0 \\ \frac{1}{2} & \frac{1}{2} & 0 \\ 0 & 0 & 1 \end{pmatrix}.$$

Note that Q doesn't even commute with A.

(2) The solution of the problem of normal spectral approximation to normal operators does not necessarily yield a solution of the problem of general spectral approximation to normal operators. More precisely: it is not always true that if A is normal, then the distance from A to $\mathcal{N}(\Lambda)$ is the same as the distance from A to $\mathcal{S}(\Lambda)$; the latter may be strictly smaller. An example with $\Lambda = \{0\}$ is $A = \begin{pmatrix} 2 & 0 \\ 0 & 0 \end{pmatrix}$. The unique $\mathcal{N}(\Lambda)$ approximant to A is

$$0 \left(= \begin{pmatrix} 0 & 0 \\ 0 & 0 \end{pmatrix} \right);$$

the distance $\| A - 0 \|$ is 2. From the point of view of general spectral approximation, however, 0 is not a very good nilpotent approximation to A. The (non-normal) nilpotent operator $Q = \begin{pmatrix} 1 & -1 \\ 1 & -1 \end{pmatrix}$ is better; the distance $\| A - Q \|$ is $\sqrt{2}$.

Retraction

If, as before, Λ is a non-empty closed subset of the complex plane, a *distance-minimizing retraction* for Λ is a function F mapping the complex plane into Λ so that

$$| z - F(z)| \leqq | z - \lambda | \quad \text{for all } \lambda \text{ in } \Lambda.$$

The inequality explains the term " distance-minimizing "; the fact that if z is in Λ, then $| z - F(z)| \leqq 0$, so that $F(z) = z$, explains " retraction ". Since the only retractions to be considered in the sequel are distance-minimizing ones, the modifier can (and will) be omitted with no danger of confusion. The assumptions (non-empty and closed) imply that for each such Λ at least one retraction does exist.

The solution of the problem of projection approximation to normal operators depends on a retraction for the pair $\{0, 1\}$; the general case depends, similarly, on retractions for general closed sets. If Λ is convex, then it is easy to verify, and well known, that there is a *unique* retraction for Λ. It is of interest to observe that the converse is true: if Λ is a closed subset of the plane such that there is a unique retraction for Λ, then Λ is convex. (This converse is known as Motzkin's theorem; see (**8**, p. 94).)

Not much can be said about retractions in general. One easy observation that is worth recording is that every retraction is bounded on bounded sets. Proof: if $\rho(z)$ is the distance from z to Λ, then

$$| F(z)| \leqq | z - F(z)| + | z | \leqq \rho(z) + | z |$$

for all z, and the majorant is continuous in z.

Are retractions necessarily continuous, or, failing that, does at least one continuous retraction exist for each Λ? The answer is no. Retractions can refuse to be continuous; if $\Lambda = \{0, 1\}$, then there is no continuous retraction for Λ. It is, however, a part of the standard lore of the subject that if Λ is convex, then the unique retraction for Λ is necessarily continuous. In the converse direction, Motzkin's theorem says that uniqueness implies convexity; is there any sense in which continuity also implies convexity? The answer is yes: if Λ is such that there exists a continuous retraction F for Λ, then there can be only one retraction for Λ and therefore, by Motzkin's theorem, Λ is convex. Here is a simple proof (discovered by A. M. Davie). If some z_0 has two distinct nearest points in Λ, say α and β, then the open disc with centre z_0 and radius $| z_0 - \alpha |(= | z_0 - \beta |)$ can contain no point of Λ. It follows that each point z of the open segment (z_0, α) has α as its unique nearest point in Λ (i.e., $F(z) = \alpha$), and, similarly, each point z of the open segment (z_0, β) has β as its unique nearest point in Λ (i.e., $F(z) = \beta$). Since this contradicts the continuity of F at z_0, the proof is complete.

Retractions may even fail to be Borel measurable. If, for instance,

$$\Lambda = \{0, 1\},$$

then the value of a retraction for Λ is uniquely determined whenever $\mathrm{Re}\, z \neq \frac{1}{2}$. If, however, $\mathrm{Re}\, z = \frac{1}{2}$, then there is great freedom of choice; assign to each point on that line either 0 or 1, arbitrarily, and conclude, by a familiar cardinal number argument, that it is possible to do so in a manner that is not Borel measurable. In view of this example it is pleasant to learn that, although pathology can exist, it can always be avoided. This is, in fact, the most useful single assertion about retractions: they cannot refuse to be Borel measurable.

Lemma. *If Λ is a non-empty closed subset of the complex plane, then there exists a Borel measurable retraction for Λ.*

The assertion can be made to follow from the work of Kuratowski and Ryll-Nardzewski (5); what follows is an easy special proof for the special case at hand.

For each complex number z, write $\rho(z)$ for the distance from z to Λ, and consider the circle (perimeter) with centre z and radius $\rho(z)$. (The degenerate case $\rho(z) = 0$, i.e., the case in which z belongs to Λ, does not require special treatment.) By definition, $\rho(z)$ is the smallest radius with the property that the circle meets Λ; the intersection of the circle with Λ is, of course, a closed set. If λ is in that closed set, then $\lambda - z = \rho(z) \exp i\theta$, with $0 \leq \theta < 2\pi$; let $\theta(z)$ be the smallest value of θ (i.e. the smallest for which λ is in Λ), and write

$$F(z) = z + \rho(z) \exp i\theta(z).$$

That is: among the points λ in Λ that are at minimal distance from z, let $F(z)$ be the one with smallest argument as viewed from z.

It is to be proved that F is Borel measurable; it will, in fact, be proved that θ is lower semicontinuous. Since ρ is continuous, that is indeed more than enough.

Suppose $z_n \to z_0$. Write $\theta_n = \theta(z_n)$, and consider any convergent subsequence $\{\theta_{n_k}\}$ with $\theta_{n_k} \to \theta_0$, say. Since

$$z_{n_k} + \rho(z_{n_k}) \exp i\theta_{n_k} \to z_0 + \rho(z_0) \exp i\theta_0,$$

the point $z_0 + \rho(z_0) \exp i\theta_0$ is in Λ; since

$$| z_0 - (z_0 + \rho(z_0) \exp i\theta_0)| = \rho(z_0),$$

so that $z_0 + \rho(z_0) \exp i\theta_0$ is at minimal distance from z_0, it follows from the definition of the function θ that $\theta(z_0) \leq \theta_0$. That is: the limit of every convergent subsequence of $\{\theta(z_n)\}$ is greater than or equal to $\theta(z_0)$, which implies that

$$\liminf_n \theta(z_n) \geq \theta(z_0),$$

i.e., that θ is lower semicontinuous.

Conclusion

Theorem. *If F is a Borel measurable retraction for a non-empty closed subset Λ of the complex plane, and if A is an arbitrary normal operator, then* spect $F(A) \subset \Lambda$; *if N is any other normal operator with* spect $N \subset A$, *then*

$$\| A - F(A) \| \leq \| A - N \|.$$

Proof. To prove the spectral assertion, use the spectral theorem. Represent A as a multiplication by, say, ϕ on some L^2 space. Since ran $F \subset \Lambda$, therefore ran $F \circ \phi \subset \Lambda$. Since Λ is closed, it follows that the essential range of $F \circ \phi$ is included in Λ also, and the essential range of $F \circ \phi$ is exactly spect $F(A)$.

(Note that since F is bounded on bounded sets, the functional calculus is applicable, so that $F(A)$ makes sense.)

The main assertion of the theorem is the inequality. Its proof for convex Λ is somewhat easier than for the general case, and deserves at least to be outlined.

Proof for the convex case. If A has an eigenvalue α, with an eigenvector e of norm 1, so that $Ae = \alpha e$, then, of course, $F(A)e = F(\alpha)e$. Since

$$((A-N)e, e) = \alpha - (Ne, e) = \alpha - \lambda,$$

where λ belongs to the numerical range $W(N)$, and since

$$W(N) = \text{conv spect } N \subset \text{conv } \Lambda = \Lambda,$$

it follows that

$$\|(A - F(A))e\| = |\alpha - F(\alpha)| \leq |\alpha - \lambda|$$

$$= |((A-N)e, e)| \leq w(A-N) \leq \|A-N\|.$$

(Here w is the numerical radius.) If A happens to be diagonal (i.e., if A has an orthonormal basis full of eigenvectors), then the preceding conclusion implies that

$$\|A - F(A)\| \leq \|A - N\|.$$

By the spectral theorem an arbitrary normal operator is the limit (in the norm) of diagonal ones. Since the continuity of F implies that the mapping $X \mapsto F(X)$, defined for normal operators X, is continuous also, the proof in this case is complete.

Proof for the general case. Write, as before, $\rho(z)$ for the distance from z to Λ.

(a) If e is a unit vector, then $\rho(\alpha) \leq \|(\alpha - N)e\|$ for every complex number α.

For the proof, represent N as multiplication by ϕ on $L^2(\mu)$. Since the spectrum of N is the essential range of ϕ, it follows that the essential range of ϕ is included in Λ, and therefore (2, Problem 97) $\rho(\alpha) \leq |\alpha - \phi|$ almost everywhere. Hence

$$\|(\alpha - N)e\|^2 = \int |\alpha - \phi|^2 |e|^2 d\mu$$

$$\geq (\rho(\alpha))^2 \int |e|^2 d\mu = (\rho(\alpha))^2.$$

(b) If α is an eigenvalue of A, then $\rho(\alpha) \leq \|A - N\|$.

Indeed: if e is a unit vector such that $Ae = \alpha e$, then, by (a),

$$\rho(\alpha) \leq \|(\alpha - N)e\| = \|(A-N)e\| \leq \|A-N\|.$$

(c) If α is in spect A, then there exists a sequence $\{A_k\}$ of normal operators such that $\| A_k - A \| \to 0$ and α is an eigenvalue of each A_k.

Represent A as multiplication by ϕ on $L^2(\mu)$. Since α is in the essential range of ϕ, it follows that if D_k is the open disc with centre α and diameter $1/k$, then $\phi^{-1}(D_k)$ has positive measure. Write $\phi_k = \alpha$ in $\phi^{-1}(D_k)$ and $\phi_k = \phi$ outside $\phi^{-1}(D_k)$, and let A_k be multiplication by ϕ_k. Clearly A_k is normal and α is an eigenvalue of A_k for each k. Since $|\phi_k - \phi| < 1/k$ in $\phi^{-1}(D_k)$ and $|\phi_k - \phi| = 0$ outside $\phi^{-1}(D_k)$, it follows that $\phi_k \to \phi$ uniformly, and hence that $A_k \to A$ in the norm.

(d) If α is in spect A, then $\rho(\alpha) \leq \| A - N \|$.

By (c), there is a sequence $\{A_k\}$ of normal operators such that $A_k \to A$ and α is an eigenvalue of each A_k. By (b), $\rho(\alpha) \leq \| A_k - N \|$ for each k. Since $\| A_k - N \| \to \| A - N \|$ as $k \to \infty$, the assertion follows.

(e) The ground is now completely prepared for the final proof.

Represent A as multiplication by ϕ on L^2; then $A - F(A)$ is multiplication by $\phi - F \circ \phi$. Since almost every value of ϕ is in the spectrum of A, and since

$$| \phi - F \circ \phi | = \rho \circ \phi,$$

it follows from (d) that

$$| \phi - F \circ \phi | \leq \| A - N \|$$

almost everywhere. The stated inequality is an immediate consequence.

A special case of the theorem (in fact, a special case of the convex special case of the theorem) appears in (4); there Λ is the set of non-negative real numbers. The proof there is not as simple as the proof of the (more general) convex case above.

Non-convex special cases of the theorem, appearing outside the context established above, can be quite puzzling. A case in point is that of projection approximation ($\Lambda = \{0, 1\}$). It is tempting in each such special case to look for a proof appropriate to it, but the search, even if successful, is not likely to throw much light on the general problem.

The theorem implies that, for each Λ, each normal oprator has an $\mathcal{N}(\Lambda)$ approximant. A direct proof of even this mild existential statement does not seem to be on the surface.

The methods used above make heavy use of the normality of the operator to be approximated. The general problem of, for instance, projection approximation to non-normal operators is still open, and not even the existence of projection approximants is known. Sometimes, however, it happens that the results for normal operators are more widely applicable than they should be. Hermitian approximation is an example (it is easy to the point of triviality for all operators, not only for normal ones), and the problems of unitary approximation and contraction approximation are also easier than one would have dared to predict. Why?

REFERENCES

(1) K. FAN and A. J. HOFFMAN, Some metric inequalities in the space of matrices, *Proc. Amer. Math. Soc.* **6** (1955), 111-116.

(2) P. R. HALMOS, *A Hilbert space problem book* (Van Nostrand, Princeton, 1967).

(3) P. R. HALMOS, Ten problems in Hilbert space, *Bull. Amer. Math. Soc.* **76** (1970), 887-933.

(4) P. R. HALMOS, Positive approximants of operators, *Indiana Univ. Math. J.* **21** (1972), 951-960.

(5) K. KURATOWSKI and C. RYLL- NARDZEWSKI, A general theorem on selectors, *Bull. Acad. Polon. Sci.* **13** (1965), 397-403.

(6) J. D. NEWBURGH, The variation of spectra, *Duke Math. J.* **18** (1951), 165-176.

(7) D. J. VAN RIEMSDIJK, Some metric inequalities in the space of bounded linear operators on a separable Hilbert space, *Nieuw Arch. Wisk.* (3) **20** (1972), 216-230.

(8) F. A. VALENTINE, *Convex sets* (McGraw-Hill, New York, 1964).

INDIANA UNIVERSITY

AND

UNIVERSITY OF EDINBURGH

Proceedings of the Royal Society of Edinburgh, **76A**, 67–76, 1976

Some unsolved problems of unknown depth about operators on Hilbert space

P. R. Halmos†

Department of Mathematics, University of California (Santa Barbara)

(MS received 5 February 1976. Read 3 May 1976)

Synopsis

The paper presents a list of unsolved problems about operators on Hilbert space, accompanied by just enough definitions and general discussion to set the problems in a reasonable context. The subjects are: quasitriangular matrices, the resemblances between normal and Toeplitz operators, dilation theory, the algebra of shifts, some special invariant subspaces, the category (in the sense of Baire) of the set of non-cyclic operators, non-commutative (i.e. operator) approximation theory, infinitary operators, and the possibility of attacking invariance problems by compactness or convexity arguments.

Preface

The longer I study mathematics, the more (to my joy) I find questions to which no one knows the answers. The purpose of what follows is to list some of those questions. They are special questions, about the theory of bounded linear operators on complex infinite-dimensional Hilbert spaces. Some of them are vague, some of them may be trivial, and some of them may turn out to be equivalent to well-known problems that have for long been known to be deep and difficult. They are the kind of questions one may put to an aspiring Ph.D. candidate who is looking for something to work on, with the warning that he should regard them not as targets but as starting points. (Some of them I have, in fact, used just that way.) I hope that despite (or possibly because of) their vagueness and their dangerously unknown degree of difficulty, some specialists may find them interesting, and, who knows, may even answer them.

1. Quasitriangular Matrices

By a *triangular + 1 matrix* I shall mean a matrix of the form

$$\alpha = \begin{pmatrix} \alpha_{11} & \alpha_{12} & \alpha_{13} & \alpha_{14} \\ \alpha_{21} & \alpha_{22} & \alpha_{23} & \alpha_{24} \\ 0 & \alpha_{32} & \alpha_{33} & \alpha_{34} \\ 0 & 0 & \alpha_{43} & \alpha_{44} \\ & & & & \ddots \end{pmatrix},$$

i.e. one in which all the entries in every diagonal more than one step below the main one are equal to 0. (The terminology allows itself to be extended to triangular + n matrices for every n; a triangular + 0 matrix, in particular, is simply an upper triangular matrix.) A *quasitriangular matrix* is a triangular + 1 matrix whose entries in the diagonal exactly one step below the main one cluster at 0 (i.e., $\liminf_n |\alpha_{n+1,n}| = 0$).

† Work supported in part by a grant from the National Science Foundation.

The Aronszajn-Smith proof [2] of the existence of non-trivial invariant subspaces for compact operators has three steps. (1) Reduce the problem to one about quasitriangular matrices. (2) Apply a sequential algorithm that constructs invariant subspaces for each quasitriangular matrix. (3) Prove that at least one of the subspaces so constructed is non-trivial. Step (1) is elementary; step (2) uses quasitriangularity to approximate an infinite matrix by finite ones; step (3) makes essential use of compactness, both in the original version and in all subsequent extensions such as the Bernstein-Robinson proof [6] for polynomially compact operators.

If an operator A has a triangular $+1$ matrix and if any of the entries $\alpha_{n+1,n}$ in the subdiagonal is 0, then A has an obvious finite-dimensional subspace. If none of the entries in the subdiagonal is 0, then A is cyclic; the basis vector $(1, 0, 0, \ldots)$ is a cyclic vector for A. It is, in fact, easy to prove that an operator is representable by a triangular $+1$ matrix with no vanishing entries in the subdiagonal if and only if it is cyclic.

The general concept inspired by the Aronszajn-Smith technique is that of a (bounded) quasitriangular operator. Here is a quick definition in matrix terms: an operator is *quasitriangular* if it can be represented as the sum of a triangular matrix and a compact one. (This is not the original definition [13], but is equivalent to it [14].)

What is the relation between quasitriangular matrices and quasitriangular operators? Half the answer is easy: every quasitriangular matrix that defines an operator at all (i.e. every bounded quasitriangular matrix) defines a quasitriangular operator. Proof: find an infinite subsequence $\{\alpha_{n_k+1,n_k}\}$ of the subdiagonal that tends to 0, let β be the matrix obtained from α by replacing each α_{n_k+1,n_k} by 0, and let γ be the difference $\alpha - \beta$. Since β is 'block-triangular' (and therefore unitarily equivalent to a triangular matrix) and γ is compact, the operator induced by $\beta + \gamma$ is quasitriangular.

Knowing the result of the preceding paragraph, and anxious to communicate the concept of quasitriangularity as quickly as possible, in a lecture I once defined quasitriangular operators by saying that the cyclic ones among them have quasitriangular matrices. In other words, I assumed and blithely stated the converse of the preceding paragraph. On reflection I realised that that converse is not known, and I hereby propose it as a problem.

Does every cyclic quasitriangular operator have a quasitriangular matrix?

Even special cases of this problem are recalcitrant. Hermitian operators, for instance, constitute a small subclass of the class of quasitriangular operators, but the answer is not known for them. To be very specific: does the real part of the unilateral shift have a quasitriangular matrix? For typographical simplicity, multiply by 2 and rephrase the question as follows: is the matrix

$$\begin{pmatrix} 0 & 1 & 0 & 0 \\ 1 & 0 & 1 & 0 \\ 0 & 1 & 0 & 1 \\ 0 & 0 & 1 & 0 \\ & & & & \ddots \end{pmatrix}$$

unitarily equivalent to a quasitriangular one? (Note that the displayed matrix is triangular $+1$ with no vanishing entries in the subdiagonal, and therefore cyclic.)

It could be that the general question is dangerous—it could be, that is, that an affirmative answer to it implies an affirmative solution of the invariant subspace problem. The Aronszajn-Smith technique yields invariant subspaces for quasi-triangular matrices. If it could be shown that at least one of them is always non-trivial, then the implication chain would be complete: use the Rumanian results [1] on non-quasitriangular operators to reduce the invariant subspace problem to the one for quasitriangular ones, reduce that problem to the cyclic case, apply the (temporarily assumed) affirmative answer to the quasitriangular matrix question, and draw the desired conclusion from the Aronszajn-Smith technique. The last two links of the chain are weak, of course, but at least one of them is something new to think about.

The quasitriangular matrix question is a definite yes-or-no version of a vaguer but more general and perhaps more valuable question: what are the asymptotic properties of the subdiagonal generated by a vector? More precisely: given an operator A, consider a unit vector f; form the sequence $\{f, Af, A^2f, ...\}$; orthonormalise (excluding the trivial case in which A is algebraic at f); if $\{e_0, e_1, e_2, ...\}$ is the orthonormal sequence so obtained, let α_n be the coefficient of e_{n+1} in the expansion of Ae_n. Question: what are the asymptotic properties of the sequence $\{\alpha_0, \alpha_1, \alpha_2, ...\}$? When, in particular, is is true that $\lim\inf_n |\alpha_n| = 0$?

To say that the subdiagonal clusters at 0 means that α_n is small for many values of n. Since

$$\vee\{e_0, e_1, ..., e_n\} = \vee\{f, Af, ..., A^nf\},$$

that condition means that $\vee\{f, Af, ..., A^nf\}$ is nearly invariant under A for many values of n.

The problem is non-trivial even for one value of n, e.g. for $n = 1$. In that case what is required is that f be an approximate eigenvector. The problem is the push-me-pull-you kind. To say that A is not algebraic at f (in the best case, that f is a cyclic vector for A) is to say that f is 'large', i.e. that $\vee\{f, Af, A^2f, ...\}$ is large; to say that f is an eigenvector is to say that, in the same sense, f is 'small'. What is wanted here is an f that is large and, at the same time, nearly small.

2. Normal versus Toeplitz

There are several instances in which a pair of results is known: something is true for normal-like operators, and, for apparently different reasons, it is true for (some or all) Toeplitz operators too.

Perhaps the most useful single result about subnormal operators is the spectral inclusion theorem: if A is subnormal and if B is its minimal normal extension, then spec $B \subset$ spec A [12]. There is a Toeplitz analogue, and its statement is as near to the subnormal one as can be: if A is a Toeplitz operator and if B is the corresponding Laurent dilation, then spec $B \subset$ spec A [17].

The spectral inclusion theorem is about *sub*-normal operators. The second example of the normal-Toeplitz echo is about normal ones; both its parts are due to Sarason [20]. The first part says that if A is normal, and if B is an arbitrary operator on the same space such that Lat $A \subset$ Lat B, then $\mathfrak{W}(B) \subset \mathfrak{W}(A)$. Explanation: 'Lat' denotes the set of subspaces invariant under the indicated operator, and '$\mathfrak{W}$' denotes the weakly closed algebra generated by the operator and the identity. Remark: the theorem is like the spectral inclusion theorem in that it asserts that if A is smaller than B in a suitable sense, then A is larger than B in another sense. The Toeplitz version says that if A is an analytic Toeplitz operator, and if B is an arbitrary operator on the same space such that Lat $A \subset$ Lat B, then $\mathfrak{W}(B) \subset \mathfrak{W}(A)$.

Still another example concerns *hypo*normal operators; it is due to Coburn [8]. The first version is that if A is hyponormal, then the Weyl spectrum of A is obtained from its spectrum by discarding isolated eigenvalues of finite multiplicity. (The Weyl spectrum of A is the intersection of the spectra of all compact perturbations of A.) If A is a Toeplitz operator, then A has no isolated eigenvalues of finite multiplicity; the theorem in that case should be and is that the Weyl spectrum of A is the same as the spectrum of A.

I shall mention one more normal-Toeplitz pair, but not describe it in detail. The two results are due to Arveson [3]. Each of the two theorems concludes that a certain algebra is strongly dense in the algebra of all bounded operators. The hypotheses are the same except that a certain maximal abelian self-adjoint subalgebra in one (i.e. a certain L^{∞}) is replaced by the algebra of all multiplications by functions in H^{∞} in the other—which is very like replacing normal operators by analytic Toeplitz operators.

The facts demand the question: what is going on here? What is there about Toeplitz behaviour that produces the same effect as normal (subnormal, hyponormal) behaviour? A modest special case of the question is this:

Is there a spectral inclusion theorem that has both the one for subnormal operators and the one for Toeplitz operators as special cases?

One obvious generalisation of Toeplitz operators looks in the direction of dilations. Devinatz and Shinbrot [10] looked in that direction and found some analogues of the invertibility properties of Wiener-Hopf (= Toeplitz) operators. Their treatment appears, however, to be too general; it is not clear that it can ever make contact with normality or subnormality.

Perhaps subnormality is the best place to start: what is there about Toeplitz operators (only some of which are subnormal) that produces behaviour resembling that of subnormal operators? Are there perhaps special commutation relations between Laurent operators and the projection $P: L^2 \to H^2$?

Is there a clue in the behaviour of the operators that have both the properties that are being compared, i.e. in the behaviour of subnormal Toeplitz operators? The only non-trivial known examples of such operators are the analytic Toeplitz operators; whether others exist is an open problem [14]. Toeplitz operators in general, analytic or not, are in some sense reminiscent of analyticity; they are compressions of Laurent operators to the 'analytic half' of L^2. Subnormal operators are also analytic-like; the typical cyclic one is the restriction of a position operator (i.e. multiplication by the independent variable) to the closure of the set of polynomials.

Another possible clue is in the behaviour of Toeplitz operators reduced modulo compact operators: they are quite anxious to become normal. Sample: the Calkin image of a Toeplitz operator with continuous symbol is normal [9]. (It takes some effort, small, but some, to construct a Toeplitz operator whose Calkin image is not normal. Sample: let the symbol be an inner function with infinitely many zeros.)

Is there, finally, a possible generalise-specialise procedure, that might give insight into the problem? Here is an outline of what might work. Toeplitz operators are multiplication operators compressed to a subspace that is from the point of view of multiplication operators not 'natural'. What makes them natural is the intervention of the Fourier transform; what happens if that transform is replaced by other unitary operators? To be more specific: let X and Y be measure spaces (playing the roles of T, the circle, and $\mathbb{Z}$, the set of integers, respectively), and let U be a unitary operator from $L^2(X)$ onto $L^2(Y)$ (playing the role of the Fourier transform). For each subset Y_0 of positive measure in Y (playing the role of $\mathbb{Z}^+$), there is a generalised Toeplitz concept: take a multiplication operator M on $L^2(X)$, and compress the transform UMU^* (defined on $L^2(Y)$) to $L^2(Y_0)$. Programme: try to understand the classically important case by varying the parameters X, Y, Y_0, and U and looking for the easier problems that should be solved before the difficult one can be.

The preceding paragraph suggests the construction of many possibly interesting non-commutative pairs of operators, as follows. Let A be a multiplication operator on $L^2(X)$, let B be a multiplication operator on $L^2(Y)$, and consider, on $L^2(X)$, say, the operators A and $B' = U^*BU$. What can be said about the product AB'? The answer is not obvious, not even in the classical Fourier case. Consider, that is, the product of a multiplication operator on $L^2(T)$ by the Fourier transform of a diagonal operator on $L^2(\mathbb{Z})$. The so-called Bishop operator is a special case: the unitary operator on $L^2(T)$ induced by an irrational rotation of T is the Fourier transform of a diagonal operator on $L^2(\mathbb{Z})$. The other way around, consider the product of a multiplication operator on $L^2(\mathbb{Z})$, i.e. a diagonal operator, by the Fourier transform of a multiplication operator on $L^2(T)$. This process includes the Toeplitz construction (if the restriction step is ignored): transform a multiplication to $L^2(\mathbb{Z})$ and multiply it by the diagonal operator that projects $L^2(\mathbb{Z})$ onto $L^2(\mathbb{Z}^+)$.

If this generalised Fourier-Toeplitz notion is taken seriously, a possibly suggestive specialisation with which to start computing is the finite-dimensional case: let X and Y be finite sets (with the counting measure).

3. Minimal Dilations

Experience shows that sometimes bad operators can be studied via their good dilations. The connections between the properties of an operator and of its 'good' dilations are far from completely known; many challenging questions remain.

To establish notation, suppose that A and B are operators on Hilbert spaces H and K, with $H \subset K$, and let P be the projection from K onto H. To say that B is a dilation of A means that $Af = PBf$ for each f in H. Equivalently, B is a dilation of A if and only if B acts on K, in the representation of K as $H \oplus H^+$, as a matrix $\begin{pmatrix} A & X \\ Y & Z \end{pmatrix}$. (Emphasis: the top-left corner of B is A.)

Dilations can always be enlarged in an unhelpful way: replace B by $B \oplus C$ (orthogonal direct sum). Such an enlarged dilation has a non-trivial reducing subspace on which it is still a dilation. A dilation B of A is called *minimal* if it has no reducing subspaces strictly between the domain of B and the domain of A.

Dilations of various and sometimes surprising kinds exist. Thus, for instance, every operator has a nilpotent dilation (proof: $\begin{pmatrix} A & A \\ -A & -A \end{pmatrix}$), and every operator has an idempotent dilation (proof: $\begin{pmatrix} A & A \\ 1-A & 1-A \end{pmatrix}$). Is that useful? Still another: any two operators have commutative dilations (proof: $\begin{pmatrix} A & B \\ B & A \end{pmatrix}$ and $\begin{pmatrix} B & A \\ A & B \end{pmatrix}$.) I am inclined to believe that all these comments must be usable, but I do not as yet know of any applications.

The most useful dilations so far have been the normal (and, in particular, the unitary) ones. They have yielded a lot of information; how much can they yield? Do their spectra, for instance, determine the spectrum of the original undilated operator? That is, given A on H, consider the spectrum of each minimal normal dilation B of A; is there a way of reconstructing the spectrum of A from the set of dilation spectra so obtained? Similar question: do the spectra of all the minimal unitary dilations of a contraction determine its spectrum?

Speaking of spectra and of subnormal operators, when does equality occur in the spectral inclusion theorem? One example is $U \oplus A$, where U is the unilateral shift and A is a normal operator with spectrum equal to the unit disk. If T is an analytic Toeplitz operator, then T is subnormal and the spectrum of T is definitely *not* equal to the spectrum of its minimal normal extension. Associated with T, however, there is a natural normal operator, namely an analytic multiplication on the disc, whose spectrum is equal to the spectrum of T. Is there a general construction behind this?

4. ALGEBRA OF SHIFTS

Is the product of two commuting subnormal operators subnormal? This is an old question (which I learned from S. K. Berberian), and the answer is not known. In case one of the prescribed operators is normal, then it seems to be folklore that the answer is yes.

There is a not yet completely explored direction of investigation that may yield examples and counter-examples pertinent to such questions and may contain a bit of theory of interest in itself: the algebra of shifts. 'Shift' here means a unilateral shift of any multiplicity between 1 and $\aleph_0$ inclusive. The product of two shifts is obviously always an isometry; the less trivial fact is that every isometry (of co-rank at least 2) is the product of two shifts [16]. Open questions: when is a product of two shifts a shift?; when do two shifts commute? It is not clear what kind of answer would be acceptable; it could perhaps be one in terms of the geometry of the underlying bases that are being shifted.

There are some related questions in a slightly enlarged context, where adjunction is allowed. If U and V are shifts, then U^*V is a contraction; L. G. Brown [7] showed that every contraction has that form. Which contractions are products U^*V with commutative U^* and V?

5. SPECIAL INVARIANT SUBSPACES

The invariant subspace problem asks whether every (bounded linear) operator on an infinite-dimensional Hilbert space has a non-trivial invariant subspace. The answer is known in some special cases (compact, normal), and not known in others (quasinilpotent, subnormal). Some people have felt that the latter two (unsolved) cases are typical of the general case in more than a vague sense of resemblance. Specifically: can the general problem be reduced to either one of these two unsolved special problems?

Another set of special invariant subspace problems is in the Toeplitz direction. Does every Toeplitz operator have a non-trivial invariant subspace? That seems very hard; the answer could be no. What about the special case of a Toeplitz operator whose symbol is a trigonometric polynomial? The subject is delicate, to say the least; invariant subspaces are not plainly visible even for Hermitian operators. (A case in point: consider $U + U^*$, where U is the unilateral shift.)

Even when existence is not in question, the structure of the lattice of invariant subspaces for specific Toeplitz operators (even simple analytic ones such as $U + U^2$) would be of definite interest.

6. Non-cyclic Operators

One topological problem deserves mention; the answer to it would, at the very least, satisfy a natural curiosity. It is known [11] that the set of non-cyclic operators (on an infinite-dimensional separable Hilbert space) is dense (with respect to the norm topology). Is that set in fact a G_δ? If so, then we could at least say that 'almost' every operator has a non-trivial invariant subspace. What is known is that the set of non-cyclic operators is of second category; this is an unpublished comment of Norberto Salinas. Proof: the set of semi-Fredholm operators with index $-\infty$ is an open set of non-cyclic operators.

7. Operator Approximation

For each class of operators an approximation problem makes sense: what is the best way to approximate a prescribed operator by a member of the class? (Approximation is meant here in the sense of operator norm.) For some classes the problem turns out to be trivial (Hermitian), for some it turns out to be interesting (positive [15]), and for some it turns out to be impossible (normal [19]). In all cases, however, there are interesting questions nearby (uniqueness of best approximation, closure of the class, characterisation of all approximants), and new classes are always promising. Approximation problems that have not been studied include the following ones: algebraic (of fixed degree?), quasi-algebraic, reducible, completely reducible, nilpotent, quasinilpotent, and positive contractions.

A problem that seems deep and difficult but valuable is that of commutative approximation: given a pair $\langle A, B \rangle$ of operators, find a commutative pair $\langle A', B' \rangle$ as near as possible. (The distance between pairs can be measured by, for instance, $\max(\|A - A'\|, \|B - B'\|)$ or $\|A - A'\| + \|B - B'\|$.)

To make the problem specific, with a yes-or-no answer, consider the subproblem of the implication from almost commutativity to near commutativity. Thus: is it true that for every $\varepsilon > 0$ there is a $\delta > 0$ such that if

$$\|A\| \leq 1 \quad \text{and} \quad \|B\| \leq 1 \quad \text{and} \quad \|AB - BA\| < \delta,$$

then the distance from $\langle A, B \rangle$ to the set of commutative pairs is less than ε? On finite-dimensional spaces the answer is affirmative (this is an easy compactness argument), but, of course, δ may depend not only on ε but on the dimension as well. Is it in fact independent of the dimension?

[Caution: a nearby problem has an easy negative answer. Thus: is it true that for every $\varepsilon > 0$ there is a $\delta > 0$ such that if $\|AB - BA\| < \delta$, then the distance from B to the commutant of A is less than ε? Counterexample: $A = \text{diag}(1, \frac{1}{2}, \frac{1}{3}, \ldots)$.]

An important special case of the problem of commutative approximation concerns pairs of the form $\langle A, A^* \rangle$. Sample question: is there always a nearest commutative pair of the form $\langle B, B^* \rangle$?

A related problem is the implication from almost normality to near normality. Is it true that for every $\varepsilon > 0$ there is a $\delta > 0$ such that if $\| A \| \leq 1$ and $\| A^*A - AA^* \| < \delta$, then the distance from A to the set of normal operators is less than ε? On finite-dimensional spaces the answer is affirmative. Is δ independent of dimension?

[On infinite-dimensional spaces the answer is negative. Counterexample:

$$A_n = \text{shift } (\sqrt{1/n}, \sqrt{2/n}, \ldots, \sqrt{(n-1)/n}, 1, 1, 1, \ldots).$$

Then $A_n^*A_n - A_nA_n^* \to 0$, but distance $(A_n, \text{Normal}) \geq 1$.]

The proper infinite-dimensional formulation of the question is due to I. D. Berg [5]: he restricts the operator A by requiring that the Fredholm index of $A - \lambda$ be 0 (whenever it makes sense). In this formulation the answer is not known yet; it is unknown even for compact operators. Special cases (having to do with weighted shifts) have been conquered by Berg. Two other pertinent contributions are the work of Bastian and Harrison [4] and the work of Pearcy and Shields [18].

8. INFINITARY OPERATORS

Imitating the Dedekind definition of infinity, I propose to call a structure 'infinitary' if it is isomorphic to some proper substructure. Here 'structure' is to be given any reasonable definition in the sense of universal algebra: e.g. 'object in a category'. A set is infinitary in this sense if and only if it is infinite; a Hilbert space is infinitary if and only if it is infinite-dimensional.

Which groups are infinitary? In the group of all dyadic rational numbers modulo 1 every proper subgroup is finite, hence that group is 'finitary'. The weak direct product (finitely non-zero sequences) of the cyclic groups of prime orders is finitary.

Which topological spaces are finitary? The unit circle is, because a continuous image of it is connected, and the only connected subset of the circle that is homeomorphic to the circle is in the circle.

Which metric spaces are finitary? The compact ones are known to be.

What about operators on Hilbert space? Which ones are unitarily equivalent to a proper part? The identity is; unilateral shifts of all multiplicities are.

The purpose of the classification, apart from whatever intrinsic interest it may have, is possibly to identify the manageable objects in a category and isolate the difficult ones for separate study.

(There is, by the way, an alternative approach, a dual approach, to this concept, as to most others—the quotient approach. The quotient definition of an infinitary structure requires isomorphism with a proper quotient. Example: in the category of groups, $\mathbb{Z}$ is quotient finitary. Reason: there is no proper infinite quotient group.)

Suppose that A is an infinitary operator on a Hilbert space H, and suppose that M is a subspace of H that stands witness to the infinitary character of A. Precisely speaking that means that $M \neq H$, that M is invariant under A, and that $A \mid M$ is unitarily equivalent to A. That in turn means the existence of an isometry V from H onto M such that $V^*(A \mid M)V = A$. (Observe in particular that this can happen only if dim $H = \infty$.) The last equation can be compressed; since $VH = M$ and $Af \in M$ whenever f is in M, it says the same as $V^*AV = A$.

Since V is an isometry, $V^*V = 1$; the other product VV^* is the projection with range M. Multiply $V^*AV = A$ on the left by V; the invariance of M under A implies that the left term becomes AV. Conclusion: $AV = VA$. This necessary condition is sufficient also. If, that is, V is a proper isometry (i.e. a non-unitary isometry, an isometry with range distinct from H) such that $AV = VA$, then ran V is obviously invariant under A; since, moreover (multiply $AV = VA$ on the left by V^*) $V^*VA = A$, the restriction of A to ran V is unitarily equivalent to A itself. Summary: an operator A is infinitary (in the unitary sense) if and only if it commutes with some proper isometry. (Incidental question: is the size of an isometry with which A can commute uniquely determined?)

The preceding observation immediately yields a large supply of infinitary operators, namely all analytic Toeplitz operators. (Consider the unilateral shift in the role of V.) These are, however, by no means the only examples; the proper isometry in question may not be pure.

Question: could it be that here, as elsewhere, finite is much more amenable than infinite, and that, in particular, the usual hard questions (e.g. unitary equivalence, structure of invariant subspaces) are easy for the finitary operators?

9. INVARIANCE, COMPACTNESS AND CONVEXITY

I conclude by mentioning two problems that are rather vague, but that may capture someone's imagination. Both have to do with invariant subspaces.

The first asks for a re-examination of the Aronszajn-Smith technique. No matter who uses it, no matter how, the constructions are heavily sequential; they involve many suffixes. Modest hope: can the proof be rearranged so as to use honest compactness and continuity instead of their sequential versions? Is there, that is, a way of re-proving the existence of invariant subspaces for compact operators *not* approximating by finite-dimensional spaces, *not* arguing step by step, and *not* saying that because something is small at the beginning and large at the end therefore it must be medium-sized in the middle? Can all this be replaced just by saying that, for instance, a certain continuous function on a certain connected space has a connected image? If this modest hope is ever fulfilled, then I would brashly hope that not only could Lomonosov's result be recaptured similarly, but, possibly, the general invariant subspace problem could be solved, one way or the other.

The second problem, and last for now, asks if there is a connection between invariant subspaces and convexity. That is, for instance: is the set of projections onto the invariant subspaces of an operator the set of extreme points of an easily determinable convex set associated with the operator? The examination of this question for some obvious 2×2 matrices leads to some amusing pictures, but, so far in any event, it has not lead to any hint about the general case.

Is there perhaps a clue in the corresponding *reducing* question?

REFERENCES

1   C. Apostol, C. Foiaş and D. Voiculescu. Some results on non-quasitriangular operators IV. *Rev. Roumaine Math. Pures Appl.* **18** (1973), 487–514.
2   N. Aronszajn and K. T. Smith. Invariant subspaces of completely continuous operators. *Ann. of Math.* **60** (1954), 345–350.

3 W. B. Arveson. A density theorem for operator algebras. *Duke Math. J.* **34** (1967), 635–648.
4 J. J. Bastian and K. J. Harrison. Subnormal weighted shifts and asymptotic properties of normal operators. *Proc. Amer. Math. Soc.* **42** (1974), 475–479.
5 I. D. Berg. On approximation of normal operators by weighted shifts, to appear.
6 A. R. Bernstein and A. Robinson. Solution to an invariant subspace problem of K. T. Smith and P. R. Halmos. *Pacific J. Math.* **16** (1966), 421–431.
7 L. G. Brown. Almost every proper isometry is a shift. *Indiana Univ. Math. J.* **23** (1973), 429–431.
8 L. A. Coburn. Weyl's theorem for nonnormal operators. *Michigan Math. J.* **13** (1966), 285–288.
9 L. A. Coburn. The *C**-algebra generated by an isometry. II. *Trans. Amer. Math. Soc.* **137** (1969), 211–217.
10 A. Devinatz and M. Shinbrot. General Wiener-Hopf operators. *Trans. Amer. Math. Soc.* **145** (1969), 467–469.
11 P. A. Fillmore, J. G. Stampfli and J. P. Williams. On the essential numerical range, the essential spectrum, and a problem of Halmos. *Acta Sci. Math.* (*Szeged*), **33** (1972), 179–192.
12 P. R. Halmos. Spectra and spectral manifolds. *Ann. Soc. Polon. Math.* **25** (1952), 43–49.
13 P. R. Halmos. Quasitriangular operators. *Acta Sci. Math.* (*Szeged*), **29** (1968), 283–293.
14 P. R. Halmos. Ten problems in Hilbert space. *Bull. Amer. Math. Soc.* **76** (1970), 887–933
15 P. R. Halmos. Positive approximants of operators. *Indiana Univ. Math. J.* **21** (1972), 951–960.
16 P. R. Halmos. Products of shifts. *Duke Math. J.* **39** (1972), 779–787.
17 P. Hartman and A. Wintner. On the spectra of Toeplitz's matrices. *Amer. J. Math.* **76** (1954), 867–882.
18 C. M. Pearcy and A. L. Shields. Almost commuting versus commuting finite matrices, to appear.
19 D. D. Rogers. *Normal spectral approximation* (Indiana Univ. Dissertation, 1975).
20 D. E. Sarason. Invariant subspaces and unstarred operator algebras. *Pacific J. Math.* **17** (1966), 511–517.

(*Issued* 18 *November* 1976)

Integral Equations and Operator Theory, Vol. 2/4 1979
© **Birkhäuser Verlag, CH—4010 Basel (Switzerland), 1979**

TEN YEARS IN HILBERT SPACE [*]

P.R. Halmos

This is a report on progress in the theory of single operators in the 1970's. It is based for the most
part, but not exclusively, on ten problems in Hilbert
space posed in 1970 [21]; it reports which of those
problems have been solved and what the solutions are.
It reports some closely related results also, notably
those of Apostol, Foiaş, and Voiculescu on the spectral
characterization of non-quasitriangular operators,
Scott Brown on invariant subspaces of subnormal operators, Gambler on invariant subspaces of some Toeplitz
operators, Kriete and Trutt on the subnormality of the
Cesàro operator, and Lomonosov on hyperinvariant subspaces of compact operators.

§0.   Introduction

The 1970's were a remarkably fecund period in the

---

[*]
A compressed version of this survey was given as
a lecture to the Special Session on C* - algebras and
Operator Theory at the San Antonio meeting of the
American Mathematical Society in January 1980. The
writing of the paper was supported in part by a grant
from the National Science Foundation.

history of operator theory. The progress was so spectacular and so extensive that in one lecture there
isn't time to present more than an annotated table of
contents of the decade, and even that has to be trimmed
severely to remain within decent bounds. To achieve
that end I had to make subjective choices, based, una-
voidably, on my own interests and competence, and, in
addition, I decided to restrict the discussion to results that satisfy two somewhat vague and arbitrary conditions: (1) they should be about single operators (and
not about groups, semigroups, algebras, and the like),
and (2) they should solve problems that were explicitly
and publicly raised before. The first of these conditions is the harder one to maintain (or to defend).
Single operator theory and algebraic operator theory are
inextricably intertwined; the results and even the techniques of each are frequently used in the other.

A consequence of these exclusionist principles is
that, for example, I will discuss neither the Berger-
Shaw inequality for multicyclic operators [10], nor the
Cowen-Douglas study of the connection between complex
geometry and operator theory [15]. Forgive me please if
your favorite candidate isn't mentioned; some of mine
aren't either. A complete survey would have to be at
least twice as long as this one can be, and I had to be
ruthless even with such pet hobbies as Banach algebraic
capacity, the structure of integral operators, and non-
commutative approximation theory.

There are many possible ways of arranging and organizing a survey such as this. Since most of the subjects fall under an earlier organizational scheme, I chose to adopt that one, and to discuss a couple of subjects that do not fit it separately at the end.

What I am referring to is a set of ten problems in Hilbert space proposed at Texas Christian University ten years ago [21]. It is always risky to propose research problems; they can turn out to be impossibly difficult or utterly trivial. Here is how the TCU problems (all phrased as yes-or-no questions) look ten years later. Two of them (3, 6) are as unsolved now as they were then, and they seem to be out of reach for the foreseeable future. (They have to do with intransitive inverses and polynomial boundedness.) The answer to two of them (2, 10) is no, and knowledge of the subject is so slight that the counterexamples succeeded in showing only that the questions were not the right ones to ask. The answer to two others (1, 9) is no, but, in both cases, an interesting area of ignorance was revealed, and the lode is still promising. The answer to one of them (5) is "sometimes"; some very good mathematics has gone into that answer, and the future is hopeful. For three of them (4, 7, 8) the answers are yes, and the answers are deep new theorems. (Conjecture: when all the returns are in, the final score will be five yeses and five noes.)

Let me proceed to tell you some of the details.   In
what follows I shall assume that you know, or once knew,

some of the terminology of the subject.   The definitions
I insert from time to time are intended to remind the
person who has passed by this way once before, more than
to initiate the newcomer.   The statements of facts (as
opposed to the definitions of the words they use) are
intended to be clear and correct, but, except for occa-
sional "descriptions" of techniques, and equally occa-
sional fragments of connective tissue that lead from one
statement to the next, all proofs are omitted.

§1.   Shifts and lattices

Problem 2: is every part of a weighted shift similar to a weighted shift?   ("Part" means restriction to
an invariant subspace.)   Ralph Gellar [20] studied, for
each weighted shift   S , the lattice of all those subspaces that are invariant under   S   and have finite codimension.   The algebraic structure of such a lattice is
clearly a similarity invariant.   The examples Gellar constructs to get a negative answer to Problem 2 are such
that their lattices of invariant subspaces of finite codimension are not equal to the corresponding lattice for
any weighted shift.   The structure of weighted shifts,
and, in particular, of their full lattices of invariant
subspaces, is just about as unknown as ever.

Problem 10 is a complicated technicality about transitive lattices of subspaces. (A lattice is transitive if the only operators that leave all its elements

invariant are the scalars.) The question arose out of the desire to see some transitive lattices. A small handful of examples came to light, one by one, and each construction seemed to depend on some ad hoc trick. Problem 10 ruled out all the tricks known at the time, and asked whether an example could be constructed without them. Kenneth Harrison [24] constructed an example based on a new application of an old trick (quarternionic Hilbert spaces). The subject as a whole is still complicated and mysterious.

§2. Cyclic and reflexive

Problem 1 is the first of three outright topological ones: does the set of cyclic operators (on a Hilbert space of dimension $\aleph_0$) have a non-empty interior (in the sense of the norm topology)? (An operator A is cyclic if it has a cyclic vector; a vector f is cyclic for A if the vectors $f, Af, A^2f, \ldots$ span the whole space.) Soon after the preprint of [21] started circulating, Joseph Stampfli proved that the answer is no: the set of non-cyclic operators is everywhere dense [18].

The no answer is not just a counterexample but a strong positive result. If there were any known way of

getting useful invariant subspace information about an
operator by approximation, the Stampfli result might imply that every operator (on a Hilbert space of dimension
$\aleph_0$) has a non-trivial invariant subspace. The point is
that all non-cyclic operators have non-trivial invariant
subspaces. So also do all known cyclic operators, of
course (otherwise the invariant subspace problem
wouldn't exist), but for the non-cyclic ones the conclusion comes quickly. (The cyclic subspace generated by
any non-zero vector is a non-trivial invariant subspace.)
Perhaps (most likely?) there is no universally effective
way of getting invariant subspaces by approximation, but
there might be a way that works under some restrictions
on the limit operator. If so, then the negative solution
of Problem 1 might yield some special but usable invariant subspace theorems.

The idea of the solution of Problem 1 is simple. If
the operator A has a cyclic vector f , then the span
of the vectors $Af, A^2f, A^3f, \ldots$ (which are all in the
range of A) has codimension at most 1 (because together with f they span the whole space). Consequence: if
codim ran A $\geq$ 2 , then A is not cyclic. Since translation by a scalar preserves cyclicity, it follows that
if codim ran $(A - \lambda) \geq 2$ for some scalar λ , then A
is not cyclic. Expressed in terms of adjoints: if
dim ker $(A^* - \bar{\lambda}) \geq 2$ for some λ , then A is not cyclic. One way to prove the density of the set of non-
cyclic operators, therefore, is to prove that *some scalar*

translate of every operator is arbitrarily near to oper-
ators with large kernels.

There are straightforward ways of proving the ital-
icized assertion just above; the roundabout way via
matrices might make it easier for some people. If an
infinite matrix has columns of arbitrarily small ℓ^2
norm, then it is arbitrarily near to matrices with large
kernels (replace the small columns by zeroes). If an operator T has 0 in its left essential spectrum (meaning that there is no operator S such that $1 - ST$ has
finite rank), then the matrix of T (with respect to a
suitable orthonormal basis) has columns of arbitrarily
small norm. If T does not have 0 in its left essential spectrum, then replace T by a suitable scalar
translate of itself. (That is: use the fact that the
left essential spectrum is never empty.) The details
can be written down precisely in hardly any more space
than this condensation.

The only other problem with a known negative answer
is Problem 9: is every complete Boolean algebra (of subspaces of a Hilbert space) reflexive? Explanation: the
Boolean operations are intersection and span; it does
not follow, and it is not assumed, that Boolean complementation is orthogonal complementation. Completeness
is meant in the Boolean sense (every subset has a supremum). To say that the lattice (or, in particular,
Boolean algebra) $\mathcal{E}$ is reflexive means that $\mathcal{E}$ is the

set of all subspaces invariant under some set of operators; equivalently, $\mathcal{E}$ is reflexive if and only if
$\mathcal{E}$ = Lat Alg $\mathcal{E}$.

The answer to Problem 9 was found by John Conway
(Indiana, not England). Here is a brief summary of the
technique [13]. Let μ be a (positive) singular measure on the unit circle C (perimeter, not disc), and
for each Borel set E in C , let φ_E be the inner
function defined for $|z| < 1$ by

$$\varphi_E(z) = \exp \left(- \int_E \frac{\lambda + z}{\lambda - z} \, d\mu(\lambda) \right) .$$

If M_E is the orthogonal complement of $\varphi_C H^2$ in $\varphi_E H^2$
$(M_E = \varphi_E H^2 \cap \left(\varphi_C H^2 \right)^\perp)$, then M_E is a subspace of
$H^2 \cap \left(\varphi_C H^2 \right)^\perp$. The correspondence $E \longmapsto M_E$ is a
Boolean anti-isomorphism. Its range, that is, the set
of all M_E 's , is a complete Boolean algebra of subspaces, which, it turns out, is reflexive if and only if
the measure μ is purely atomic.

Conway makes two interesting comments. (1) The example is not strongly closed. Could there be one that
is? (2) The example (not reflexive) is lattice isomorphic to a reflexive Boolean algebra. Are they all like
that?

The intention of Problem 9 was to call attention to the theory of reflexive lattices. The subject is still very much open.

§3. Subnormal Toeplitz

Most of the principal problems of operator theory concern one of three concepts: similarity (usually unitary equivalence), invariance, and approximation. Problem 5 is different: it has the more special flavor of analytic function theory. The question is this: is every subnormal Toeplitz operator either analytic or normal? (Recall: a subnormal operator is one that has a normal extension. The Toeplitz operator with symbol φ, where $\varphi \in L^\infty(C)$, is defined on $H^2(C)$ as multiplication by φ followed by projection into H^2. It is analytic if $\varphi \in H^\infty$, that is, if all the Fourier coefficients of φ with negative index vanish. Equivalently, a Toeplitz operator is one whose matrix with respect to the natural exponential basis has constant diagonals:

$$
\begin{pmatrix}
a_0 & a_{-1} & a_{-2} & a_{-3} \\
a_1 & a_0 & a_{-1} & a_{-2} \\
a_2 & a_1 & a_0 & a_{-1} \\
a_3 & a_2 & a_1 & a_0 & \ddots
\end{pmatrix}.
$$

The operator is analytic if and only if the matrix is

lower triangular.)

The answer to Problem 5 is not known, but the evidence is in the yes direction. The evidence is of two kinds: if subnormality is replaced by a suitably stronger

condition, or else if the symbol of the given Toeplitz operator is restricted to a suitably small class, then the conclusion (analytic or normal) follows.

The strongest reasonable condition that is like normality, but doesn't go all the way, is quasinormality (A commutes with A*A , or, equivalently, the polar decomposition of A has commutative factors). Amemiya, Ito, and Wong proved that every quasinormal Toeplitz operator is analytic or normal [2]. Their proof is not obvious, but it is elementary in the sense that in addition to straightforward computations, the only tools it uses are the characterization of the invariant subspaces of the unilateral shift and the characterization of commutative Toeplitz operators.

The AIW paper followed one by Ito and Wong in which the conclusion (analytic or normal) is inferred for all subnormal Toeplitz operators whose symbol φ is of a special kind: the special assumption is that there exists an inner function χ such that φ is a polynomial in χ and its complex conjugate $\bar{\chi}$ [28]. Here too the techniques are straightforward. The case $\chi(z) = z$ is typical, and, in fact, the general case can be reduced to it

(via the fact that the Toeplitz operator induced by a
non-constant inner function is a unilateral shift of some
multiplicity).

Ito and Wong mention that if "subnormal" in the
wording of the problem is replaced by "hyponormal"
($A*A - AA* \geq 0$) , then the conclusion is false. By way
of proving this negative assertion, they offer the example $A = U* + 2U$ (where U is the simple unilateral
shift). It is easy to prove that A is hyponormal but
not subnormal. A slight additional effort proves much
more: A^2 is not hyponormal. That is indeed a powerful
way to prove that A is not subnormal: if it were, then
A^2 would be also, and therefore A^2 would be hyponormal -- these implications are well-known and trivial.
The first example of a hyponormal operator with a non-
hyponormal square is very much more complicated, but it
took over twenty years before the beautiful and simple
Ito-Wong example was noticed.

The Ito-Wong work raises the question of characterizing the symbols for which the induced Toeplitz operator A is hyponormal. That question was systematically
attacked by Bruce Abrahamse [1]. His work is the deepest along these lines so far and goes furthest toward a
solution of Problem 5.

Abrahamse regards H^2 as a class of analytic functions on the unit disc (instead of a class of square-

integrable functions with restricted Fourier expansions
on the unit circle), and he uses the Nevanlinna concept
of a function of bounded type (a quotient of two bounded
analytic functions on the disc). His main theorem is
this: if A is a hyponormal Toeplitz operator with sym-
bol φ , such that either φ or $\bar{\varphi}$ is of bounded type,
and if ker(A*A - AA*) is invariant under A , then A
is analytic or normal.

It is a corollary of Abrahamse's theorem that if A
is subnormal and either φ or $\bar{\varphi}$ is of bounded type,
then the conclusion follows. The corollary has many
good special cases; the following one is perhaps the
most striking. If $\varphi = f + \bar{g}$, where both f and g are
in H^2 , and if either f or g is a polynomial in (a
finite number of) inner functions, then A is analytic
or normal. This subsumes the Ito-Wong theorem, of
course, with much room to spare.

The theory of Toeplitz operators is extensive, and
it has grown in the last ten years by more than just the
Abrahamse theorem here reported. I shall restrict myself
to mentioning only one other result, which is probably
the one of most interest to non-specialists. The assertion is due to Leonard Gambler, and it has to do with invariant subspaces. It says that if φ is (the restriction to C of) a non-constant function analytic in a
neighborhood of C , then the induced Toeplitz operator
has non-trivial hyperinvariant subspaces [19]. (Hyperinvariant for A means invariance under every operator

that commutes with A .) Like Stampfli's result about
non-cyclic operators, this assertion offers a (small?)
hope of proving another invariant subspace theorem; possibly an approximation technique might yield the result
that every Toeplitz operator has non-trivial (hyper)in-
variant subspaces.

§4. Essential normality

   The future will judge, of course, but from the
point of view of the present ,it seems clear that the
greatest single step in operator theory in the 1970's
had to do with essential normality.

   It all started in 1909, when Hermann Weyl proved
that Hermitian operators can be essentially diagonalized
even on spaces of dimension $\aleph_0$. The word "essential"
has come to have a fixed precise meaning in this context: it always means "modulo compact operators".
Weyl's theorem says, therefore, that if A is Hermitian, then there exists a diagonal operator D (meaning
one whose matrix with respect to some orthonormal basis
is diagonal), and there exists a unitary operator U
such that U*AU - D is compact. More than a quarter of
a century later John von Neumann went on from there and
solved the essential unitary equivalence problem for
Hermitian operators: if A and B are Hermitian, then
a necessary and sufficient condition that there exist a
unitary U for which U*AU - B is compact is that A
and B have the same essential spectrum. Note that the

condition is even simpler than that of the classical
principal axis theorem: the infinite-dimensionality of
the domain has done away with the need to distinguish
between multiplicities.

The theory was dormant for the next 35 years. The
possibility of generalizing Weyl-von Neumann to broader
classes of operators seems never to have been raised
till Problem 5 in 1970: is every normal operator the sum
of a diagonal operator and a compact one? Once the
question was raised, the answer was not long in coming;
in 1971 David Berg published his proof that the answer
is yes [9]. His method depended on a detailed analysis
of the spectral structure of normal operators. A year
later I called attention to a more perspicuous proof
that, in effect, reduces the normal case to the Hermitian one; the principal lemma says that every normal operator is a continuous function of a Hermitian one [23].

A related problem of some importance remained open
for a few more years and deserves to be mentioned here.
Weyl proved that every Hermitian operator is of the form
$D + K$, with D diagonal and K compact; a part of von
Neumann's contribution was a proof that K could in fact
be chosen to be a Hilbert-Schmidt operator. (Definition:
the matrix entries are square-summable.) Berg extended
Weyl's result to normal operators; could von Neumann's
strengthening be extended too? The question is not just
a specialist's technicality; it plays an important role

in, for instance, the theory of integral operators. Several people tried hard to prove the extension (and sometimes to disprove it). It turns out to be true; the result is due to Dan Voiculescu [38]. The proof is not trivial. As it now stands, it uses a modification of some of Voiculescu's earlier work (to be discussed below in another context) on representations of separable C* - algebras.

The WNB results are only a fragment of the best "essential" generalization of unitary equivalence theory for normal operators. All that was reported so far was what happens when "equivalence" is interpreted in the essential sense; to get at the deeper facts it is necessary to interpret "normal" and "unitary" in the essential sense too. The full question is this: under what conditions on two essentially normal operators A and B (A*A - AA* and B*B - BB* are compact) does there exist an essentially unitary operator U (U*U - 1 and UU* - 1 are compact) such that U essentially transforms A into B (U*AU - B is compact)?

If A_0 is normal, K_0 is compact, and $A = A_0 + K_0$, then A is essentially normal; that's the easy way. Similarly, if U_0 is unitary, K_0 is compact, and $U = U_0 + K_0$, then U_0 is essentially unitary, the easy way. The unilateral shift is a typical example that is essentially unitary (and hence essentially normal) the hard way -- it is not a compact perturbation of

a normal operator. The WNB theorem solves the hard problem (essential equivalence), but only for easy operators
(normal, or, with no extra effort, normal plus compact),
and only via easy transforms (unitary, or unitary plus
compact). The great forward step was taken by Brown,
Douglas, and Fillmore [11]. They found a usable necessary and sufficient condition in the most general case,
their condition was surprising (at least until familiar-
ity with it made it seem perfectly natural), and their
techniques (a mixture of operator theory and homological

algebra) are deep.

The first difficulty to be conquered is the distinction between essentially unitary operators and compact
perturbations of unitary ones. The distinction is genuine, but, as it turns out, in the present theory it need
not play a role. BDF prove early (with much less effort
than the major result needs, but still far from trivially) that if two essentially normal operators are essentially equivalent via an essentially unitary operator,
then they are essentially equivalent via a genuine unitary operator also. So much for that; now it is possible
to examine the main result with no further minor distrac-
tions.

The central concept is that of the (Fredholm) index
of an operator, defined as nullity minus corank. Nullity
equals dimension of kernel; corank equals codimension of
range. (Index is defined for an operator only when the

range is closed and either the nullity or the corank is finite.) Since nullity measures the deviation from one-to-one-ness and corank the deviation from onto-ness, both these numbers indicate the extent to which an operator fails to be invertible. A change in nullity is frequently accompanied by a corresponding change in corank (on finite-dimensional spaces they are always equal). The difference, the index, is, roughly speaking, the extent to which the operator is irrevocably singular.

If U is the unilateral shift and W is the bilateral shift, then both U and W are essentially

normal, and the essential spectrum of both is the unit circle. The operator W is normal; if U also were normal, the WNB result would apply, and the conclusion would be that U and W are essentially unitarily equivalent. They are not. Why not? Reason: essential unitary equivalence preserves index, and index U = -1 , index W = 0 .

The shift example does not yet reveal the full story. To get at that, it is necessary to consider the index of each scalar translate of each operator under study. Precisely: associated with each essentially normal A there is an integer-valued function, the index function, defined, for λ in the complement of the essential spectrum of A in the complex plane, as the index of $A - \lambda$. The main BDF result is that two essentially normal operators are essentially unitarily equi-

valent if and only if they have the same essential spectrum and the same index function. In a sense the index
function plays the role that multiplicity plays in the
finite-dimensional case.

The BDF theorem and its techniques of proof have
had a great effect. For one thing, they contributed to
the unification of mathematics via their connection with
topological ideas, and, at the same time, they continue
to have valuable special corollaries. Among the first
were a characterization of the essential normal operators that are compact perturbations of normal ones (A
is like that if and only if index $(A - \lambda) = 0$ for all

λ not in the essential spectrum of A), and a proof
that the set of such compact perturbations is closed (in
the norm topology). No direct proof of either of these
statements has been found yet.

§5. Quasitriangular and nilpotent

Triangular operators (that is, operators that have
an upper triangular matrix with respect to some orthonormal basis) are a natural generalization of diagonal
ones, and their "essential" generalizations have turned
out to be important. (There is, obviously, a dual theory of lower triangular operators. It doesn't matter
which one is chosen to play the central role, but, once
the choice is made, the study must remain consistent.)

The technical term for compact perturbations of triangular operators should really be "essentially triangular",
but, as a matter of historical fact, that terminological
possibility was overlooked and "quasitriangular" was
adopted.

The original prototypes of quasitriangular operators were the compact ones; a possible hope was that the
abstraction would imitate the prototype well enough to
yield invariant subspace theorems. The original prototype of a non-quasitriangular operator was the unilateral shift. The negative prototype has turned out to be
more typical than the positive one.

The unilateral shift has index -1 ; Douglas and
Pearcy observed that that's why it is not quasitriangular. More generally: if A is an operator and λ is a
scalar such that index $(A - \lambda) < 0$, then A is not
quasitriangular [16]. Soon after making this observation Carl Pearcy conjectured that the converse was true;
before long the conjecture was proved by Apostol, Foiaş,
and Voiculescu [6]. The work in which the proof first
appeared is one of a series; the proof is long and com-
plicatedly analytic. A detailed treatment, with a somewhat less complicated exposition of the proof, was given
a little later by Douglas and Pearcy [17].

An immediate corollary of the AFV result is that if
an operator is not quasitriangular, then it has a non-
trivial invariant subspace. Indeed: if index $(A - \lambda) < 0$

for some λ , then codim ran $(A - \lambda) \neq 0$, so that
$\text{ran}^{\perp} (A - \lambda) \neq 0$, which implies that ker $(A^* - \bar{\lambda}) \neq 0$.
In other words, A^* has an eigenvalue. The orthogonal
complement of the corresponding eigenspace is a non-
trivial hyperinvariant subspace for A . Tactical conclusion: the invariant subspace problem is reduced to
the special case of quasitriangular operators.

To prove something as powerful as the AFV theorem,
it is necessary to know a lot about possible ways of
constructing quasitriangular operators. Sample: by the
middle of the proof the theorem is reduced to the statement that if A and B are operators (on a Hilbert
space of dimension $\aleph_0$) such that A is normal and the
spectrum of B is included in the spectrum of A , then
the operator matrix $\begin{pmatrix} A & 0 \\ 0 & B \end{pmatrix}$ is quasitriangular. In
subsequent developments conditions that resemble the
ones of this sample statement, but are better in that
they are necessary as well as sufficient, are used to
characterize some special classes of quasitriangular operators. Sample: an operator T is biquasitriangular
(both T and T^* are quasitriangular) if and only if
it is unitarily equivalent to an operator matrix $\begin{pmatrix} A & B \\ C & D \end{pmatrix}$,
where the principal entries A and D are block diago-
nal (direct sums of operators of finite size) and at
least one of the off-corner entries C and D is com-

pact [4].

All this makes uncomfortably close contact with
Problem 7: is every quasinilpotent operator the norm
limit of nilpotent ones? It is "uncomfortable" because
I suspect that the answer to Problem 7 might be accessible by less complicated techniques, but, as of now,
there is no other way known.

The answer to Problem 7 is yes. Domingo Herrero
was the first to claim a proof [25], the details of which
appeared the year following the claim [26], but which
turned out to be wrong [27]. Next, Constantin Apostol
announced a partial solution [3]. That paper caused me
a lot of trouble (misprints, false assertions, and at
least one major step that I cannot justify yet).

Building on Apostol's partial solution, Apostol and
Voiculescu offered a complete solution [7], which makes
use of the Apostol-Foiaş characterization of biquasitriangular operators (mentioned above).

There is an even harder, but correspondingly deeper
and more informative way of solving Problem 7; it is due
to Voiculescu [36]. He proves that the norm closure of
the set of algebraic operators is exactly the set of biquasitriangular operators. (An operator A is algebraic if there exists a non-zero polynomial p such that
$p(A) = 0$.) The proof makes use of the BDF characterization of compact perturbations of normal operators. To
see that the statement includes the solution of Problem

7, it is necessary to know that every quasinilpotent operator is biquasitriangular, and to have a suitable unitary-geometric characterization of algebraic operators [22]. The characterization is simple enough: an operator A is algebraic if and only if it can be represented as an upper-triangular operator matrix

$$\begin{pmatrix} A_1 & & * \\ & \ddots & \\ O & & A_k \end{pmatrix}$$

whose diagonal entries are scalar operators
$(A_j = \lambda_j \cdot 1 , \quad j = 1,\ldots,k)$.

Suppose now that Q is quasinilpotent. Given $\varepsilon > 0$, find an algebraic operator A that is within ε

of Q in two senses: both the norm of Q - A and the spectral radius of A are to be less than ε . (This uses the upper semicontinuity of the spectrum.) If B is the result of replacing the diagonal entries of A (in the representation described just above) by zeroes, then B is nilpotent; since

$$\| A - B \| = \left\| \begin{pmatrix} A_1 & & O \\ & \ddots & \\ O & & A_k \end{pmatrix} \right\|$$

$$= \max(|\lambda_1| ,\ldots, |\lambda_k|) < \varepsilon ,$$

it follows that $\| Q - B \| < 2\varepsilon$.

Perhaps the shortest safe way of getting at Problem
7 is the one offered by Apostol, Foiaş, and Pearcy [5],
but they too use rather sophisticated tools, such as
Voiculescu's work on representations of separable C* -
algebras.

§6. Reducible approximation

Problem 8: is every operator the norm limit of reducible ones? The question sounds innocent enough. At
its most naive level it is an expression of natural topological curiosity, and one might think that it could be
answered by using not much more than the definitions.
That does not seem to be so. The answer to the question
is yes. The techniques used to get the answer were

discovered by Voiculescu; they are quite deep, and they
give much more information than the question asked [37].
Voiculescu's exposition is complicated and condensed; a
subsequent paper by Arveson examines, simplifies, and
extends the work, and is a great help [8].

A good candidate for Voiculescu's most spectacular
result is his non-commutative Weyl-von Neumann theorem.
What it says is that every non-degenerate representation
of a separable C* -algebra on a separable Hilbert space
is approximately equivalent to a direct sum of irreducible representations. I proceed to explain (1) the

terms and (2) the reason the theorem deserves its name.

A C* - algebra is an algebra of operators closed under the formation of adjoints and closed in the norm topology of operators. A representation is an adjoint-preserving homomorphism into the C* - algebra of all operators on some Hilbert space; it is non-degenerate if the image algebra has a trivial kernel, and it is irreducible if there is no non-trivial subspace that reduces every operator in the image. The really crucial concept is that of approximate equivalence. Two representations ρ and σ of a C* - algebra are approximately equivalent if there exists a sequence $\{U_n\}$ of unitary operators such that, for each A in the algebra, the differences

$$U_n^* \rho(A) U_n - \sigma(A)$$

are compact operators that tend to 0 (in norm) as n tends to infinity.

Suppose now that A is a normal operator, and let C be the C* - algebra generated by A and 1 . The normality of A implies that C is commutative. What happens when Voiculescu's theorem is applied to the identity representation ρ of C (the representation defined for each A in C by $\rho(A) = A$)? Answer: to every positive number ε there corresponds a direct sum σ of irreducible representations and a unitary operator U such that U*AU - $\sigma(A)$ is compact and has norm less than ε . The commutativity of C implies, of course, that of the image $\sigma(C)$. It is a relatively elementary fact (a consequence of the spectral theorem) that an irreducible representation of a commutative C^* - algebra is very special: except for the zero representation, the image operators must act on a 1-dimensional space. Consequence: $\sigma(A)$ is a direct sum of operators on 1-dimensional spaces, or, in other words, $\sigma(A)$ is a diagonal operator. (Quantitative information about the size of $\|U^*AU - \sigma(A)\|$ was known to Weyl and von Neumann also; for the moment it is irrelevant.) Conclusion: Voiculescu's theorem applied to the C^* - algebra generated by a normal operator includes the WNB theorem as a special case.

The non-commutative WNB theorem gives information about non-normal operators also, but for that purpose other parts of Voiculescu's work have to be used along with it. The application to a single operator always proceeds the same way: given A , form the (no longer necessarily commutative) C^* - algebra generated by A and 1 , and study its representation theory. For the next to the last result, as far as Problem 8 is concerned, the quantitative part of the non-commutative WNB theorem is crucial, but the information about compactness is irrelevant. The statement is this: to every operator A (on a Hilbert space of dimension $\aleph_0$) and to every positive number ε there corresponds an infinite direct sum B of irreducible operators such that $\| A - B \| < \varepsilon$. Conclusion: there are reducible operators arbitrarily near to A .

§7. Compact and subnormal

Among the numerous achievements of the 1970's in operator theory that did not come from the TCU problems there are two outstanding ones that have to do with invariant subspaces.

It took twenty years to extend the invariant subspace theorem for compact operators to polynomially compact ones; other desired generalizations, such as the one for commutative pairs of compact operators, remained recalcitrant even after that one succumbed. Victory, when it came, was due to a young student, Victor Lomonosov, and, to everybody's astonishment, it was short and elementary [30]. The hardest tool it used was the Schauder fixed point theorem, which has been known for about forty years, and for the heart of the matter it turned out that even that wasn't necessary.

Soon after Lomonosov's proof became known, Hilden [31] discovered a simplified version, which, to be sure, proved a little less, but which used only concepts and techniques that are commonly accessible to first-year graduate students. (The deepest tools needed are the Fredholm alternative and the Gelfand formula for spec-

tral radius.) The result is the statement that every
non-zero compact operator has a non-trivial hyperinvariant subspace. Special case: for any two commutative
compact operators (on a Hilbert space of dimension
greater than 1) there exists a non-trivial subspace
invariant under both.

Much attention has been paid in the last ten years
to the analytically important class of subnormal operators. One early and interesting news item, for instance, was the proof by Kriete and Trutt that the
Cesàro operator is subnormal [29]. (The Cesàro operator
is the one that sends each sequence in ℓ^2 onto its sequence of partial averages.) The last big news item was
the solution by Scott Brown (another newcomer) of the
invariant subspace problem for subnormal operators [12].
The statement is the obvious one: every subnormal operator (on a Hilbert space of dimension $\aleph_0$) has a non-
trivial invariant subspace. The proof is very far from
obvious.

The theory of subnormal operators has close contact
with the theory of analytic functions. To see how at
least a part of that goes, consider a finite Borel measure μ with compact support C in the complex plane,
and let $H^2(\mu)$ be the closure in $L^2(\mu)$ of the polynomials. For each bounded Borel measurable function φ
on C , let A_φ be the multiplication operator defined
on $L^2(\mu)$ by $A_\varphi f = \varphi f$. If, for instance, e is the

identity function on C ($e(z) = z$ for all z in C),

then A_e is the "position operator" on $L^2(\mu)$:

$A_e f(z) = z f(z)$. The operator A_e is normal and the

subspace $H^2(\mu)$ is invariant under A_e ; in other words,

the restriction $S(\mu)$ of A_e to $H^2(\mu)$ is subnormal.
One of the few easy statements about subnormal operators
is that the invariant subspace problem for them is equivalent to the invariant subspace problem for the special
ones just described (the $S(\mu)$'s for all possible
measures μ).

For an easy example, let μ be normalized Lebesgue
measure in the perimeter C of the unit circle. In that
case the functions e_n defined by $e_n(z) = z^n$,
$n = 0 , \pm 1 , \pm 2 , \ldots$, form an orthonormal basis for

$L^2(\mu)$, and $H^2(\mu)$ becomes H^2 , the set of all those

elements f of $L^2(\mu)$ whose Fourier coefficients
(f , e_n) vanish for all negative n . There is a natural

one-to-one correspondence between this H^2 and the set
of all those analytic functions on the interior of the
unit disc whose Taylor coefficients form a square-
summable sequence; for many purposes it is convenient to
identify H^2 with that set of analytic functions. The
original H^2 is a subset of L^2 and, therefore, its

elements are not functions but equivalence classes modulo sets of measure zero. The advantage of the analytic version of H^2 is that its elements are honest functions that have honest values at all points of the open unit disc. If $f \in H^2$, then, for instance, $f(0)$ makes sense, and the mapping $f \longmapsto f(0)$ is a bounded linear functional on the Hilbert space H^2. The kernel of that linear functional is a non-trivial subspace of H^2 and is invariant under S ($=S(\mu)$). (Note that in this case $e = e_1$.)

The assertion that for the special classical H^2 the restriction of the position operator has a non-trivial invariant subspace is a very small achievement, but the technique has a virtue: analytic functions, point evaluations, and kernels have yielded invariant subspaces (in fact invariant hyperplanes, subspaces of codimension 1) for many subnormal operators. This very virtue is at the same time a fault: it caused many workers to concentrate exclusively on point evaluations, which lead to invariant hyperplanes.

Scott Brown's idea can be described by saying that he by-passes the direct construction of sets of vectors (invariant subspaces) in favor of sets of operators. The first step is to impose some simplifying restrictions on the given subnormal operator (which means, in effect, on the given measure μ) that involve no loss of generality as far as the search for invariant sub-

spaces is concerned. (Sample restriction: the operator
is purely subnormal, that is, it has no normal direct
summand.) This sort of thing is always done. Next,
Scott Brown considers, for the restricted measures μ
that remain to be studied, the operator algebra generated by $S(\mu)$, and forms its closure with respect to a
certain operator topology, the so-called ultraweak or
σ-weak topology. Using Sarason's deep study of that topology [35], Conway and Olin have shown that the closed
algebra is isomorphic to the algebra of all bounded analytic functions on a certain open set G [14]. If p
is a polynomial, then the function p corresponds to
the restriction of A_p to $H^2(\mu)$.

The process of evaluating functions at a point and
looking at the functions for which the answer is zero is
now available again -- for each bounded analytic function
φ on G and for each z_0 in G it makes sense, of
course, to form $\varphi(z_0)$ -- but instead of sets of vectors
it yields sets of operators. A hard technical lemma is
needed, and Scott Brown provides it: if a point z_0 in
G belongs to the spectrum of $S(\mu)$, then there exist
vectors f_0 and g_0 in $H^2(\mu)$ such that

$$\varphi(z_0) = (A_\varphi f_0 , g_0)$$

for all bounded analytic functions φ on G . That's
what is needed to make the transition back from sets of
operators to sets of vectors: fix z_0 and form the span

of all those $A_\varphi f_0$'s for which $\varphi(z_0) = 0$. The result is a subspace M invariant under $S(\mu)$, and the
assumed restrictions on μ imply that $M \neq 0$. Since
$\varphi(z_0) = 0$ implies that $A_\varphi f_0$ is orthogonal to g_0,
the subspace M is not equal to the whole space $H^2(\mu)$,
and the proof of the existence and non-triviality of an
invariant subspace is complete.

§8. Epilogue

The publication date of every entry in the list of
references that follows is in the decade under discussion, that is in the closed interval [1970 , 1979].
Most of the references are to works that were explicitly
cited in the text; the mathematical and historical background can be recovered from them by iterating the bibliography operator. There are three exceptional entries
that might help to fill some of the gaps: the scholarly
and (up to its date of publication) complete monograph
of Radjavi and Rosenthal [34], the expository articles
in the Mathematical Surveys volume edited by Carl Pearcy
[32], and Pearcy's own CBMS Regional Conference Lectures
[33].

There are many more relevant publications in the
1970's, containing more, quite a bit more interesting
operator theory. I hope that despite its sins of omis-

sion this survey has conveyed the flavor and the extent
of progress in the subject during the last decade.

Will operator theory make as much progress in the
next ten years as in the last ten? I think so, and the
evidence at hand even indicates one of the possible directions of progress. More and more operator theory in-
teracts with (applies to?) other parts of pure mathematics, including number theory, topology, and, of course,
classical analysis. Hilbert space is usually the right
place to begin to look when it becomes desirable to make
something infinite-dimensional or non-commutative.
Let's wait and see. No, on second thought, let's work
and see.

References

1.   M.B. Abrahamse, Subnormal operators and functions
     of bounded type, Duke Math. J. 43 (1976) 597-604.

2.   I. Amemiya, T. Ito, and T.K. Wong, On quasinormal
     Toeplitz operators, Proc. Amer. Math. Soc. 50
     (1975) 254-258.

3.   C. Apostol, On the norm-closure of nilpotents, Rev.
     Roum. Math. Pures Appl. 19 (1974) 277-282.

4.   C. Apostol and C. Foiaş, On the distance to bi-
     quasitriangular operators, Rev. Roum. Math. Pures
     Appl. 20 (1975) 261-265.

5. C. Apostol, C. Foiaş, and C.M. Pearcy, That quasi-nilpotent operators are norm-limits of nilpotent operators revisited, Proc. Amer. Math. Soc. 73 (1979) 61-64.

6. C. Apostol, C. Foiaş, and D. Voiculescu, Some results on non-quasitriangular operators, IV, Rev. Roum. Math. Pures Appl. 18 (1973) 487-514.

7. C. Apostol and D. Voiculescu, On a problem of Halmos, Rev. Roum. Math. Pures Appl. 19 (1974) 283-284.

8. W.B. Arveson, Notes on extensions of C^*-algebras, Duke Math. J. 44 (1977) 329-355.

9. I.D. Berg, An extension of the Weyl-von Neumann theorem to normal operators, Trans. Amer. Math. Soc. 160 (1971) 365-371.

10. C.A. Berger and B.I. Shaw, Self-commutators of multicyclic hyponormal operators are always trace class, Bull. Amer. Math. Soc. 79 (1973) 1193-1199.

11. L.G. Brown, R.G. Douglas, and P.A. Fillmore, Unitary equivalence modulo the compact operators and extensions of C^*-algebras, Proc. Conf. Operator Theory, Lecture Notes in Math. 345 (1973) 58-128.

12. S.W. Brown, Some invariant subspaces for subnormal operators, Integral Equations and Operator Theory 1 (1978) 310-333.

13. J.B. Conway, A complete Boolean algebra of subspaces which is not reflexive, Bull. Amer. Math. Soc. 79 (1973) 720-722.

14. J.B. Conway and R.F. Olin, A functional calculus for subnormal operators II, Memoirs Amer. Math. Soc. 10 (1977) Number 184.

15. M.J. Cowen and R.G. Douglas, Complex geometry and operator theory, Acta Math. 141 (1978) 187-261.

16.   R.G. Douglas and C.M. Pearcy, A note on quasitri-
      angular operators, Duke Math. J. 37 (1970) 177-188.

17.   R.G. Douglas and C.M. Pearcy, Invariant subspaces
      of non-quasitriangular operators, Proc. Conf.
      Operator Theory, Lecture Notes in Math. 345 (1973)
      13-57.

18.   P.A. Fillmore, J.G. Stampfli, and J.P. Williams,
      On the essential numerical range, the essential
      spectrum, and a problem of Halmos, Acta Szeged 33
      (1972) 179-192.

19.   L.C. Gambler, A study of rational Toeplitz opera-
      tors, Dissertation, SUNY Stony Brook, 1977.

20.   R. Gellar, Two sublattices of weighted shift invar-
      iant subspaces, Indiana Univ. Math. J. 23 (1973)
      1-10.

21.   P.R. Halmos, Ten problems in Hilbert space, Bull.
      Amer. Math. Soc. 76 (1970) 887-933.

22.   P.R. Halmos, Capacity in Banach algebras, Indiana
      Univ. Math. J. 20 (1971) 855-863.

23.   P.R. Halmos, Continuous functions of Hermitian op-
      erators, Proc. Amer. Math. Soc. 31 (1972) 130-132.

24.   K.J. Harrison, Transitive atomic lattices of sub-
      spaces, Indiana Univ. Math. J. 21 (1972) 621-642.

25.   D.A. Herrero, Normal limits of nilpotent operators,
      Indiana Univ. Math. J. 23 (1974) 1097-1108.

26.   D.A. Herrero, Toward a spectral characterization
      of the set of norm limits of nilpotent operators,
      Indiana Univ. Math. J. 24 (1975) 847-864.

27.   D.A. Herrero, Erratum: Toward a spectral character-
      ization of the set of norm limits of nilpotent op-
      erators, Vol. 24 (1975), 847-864, Indiana Univ.
      Math. J. 25 (1976) 593.

28.   T. Ito and T.K. Wong, Subnormality and quasinormal-
      ity of Toeplitz operators, Proc. Amer. Math. Soc.
      34 (1972) 157-164.

29.   T.L. Kriete, III and D. Trutt, The Cesàro operator
      in ℓ^2 is subnormal, Amer. J. Math. 93 (1971)
      215-225.

30.   V.I. Lomonosov, Invariant subspaces for the family
      of operators which commute with a completely con-
      tinuous operator, Functional Anal. Appl. 7 (1973)
      213-214.

31.   A.J. Michaels, Hilden's simple proof of Lomonosov's
      invariant subspace theorem, Advances in Math. 25
      (1977) 56-58.

32.   C.M. Pearcy (editor), Topics in operator theory,
      Mathematical Surveys, Number 13, Amer. Math. Soc.,
      Providence, 1974.

33.   C.M. Pearcy, Some recent developments in operator
      theory, Regional Conference Series in Mathematics,
      Number 36, Amer. Math. Soc., Providence, 1978.

34.   H. Radjavi and P.M. Rosenthal, Invariant subspaces,
      Springer, Berlin, 1973.

35.   D.E. Sarason, Weak-star density of polynomials, J.
      reine angew. Math. 252 (1972) 1-15.

36.   D. Voiculescu, Norm-limits of algebraic operators,
      Rev. Roum. Math. Pures Appl. 19 (1974) 371-378.

37.   D. Voiculescu, A non-commutative Weyl-von Neumann
      theorem, Rev. Roum. Math. Pures Appl. 21 (1976)
      97-113.

38.   D. Voiculescu, Some results on norm-ideal perturba-
      tions of Hilbert space operators, Preprint,
      Institutul de matematică, Bucharest, 1978.

Department of Mathematics
Indiana University
Bloomington, IN 47405, USA

Submitted: August 20, 1979

Reprinted from the
INDIANA UNIVERSITY MATHEMATICS JOURNAL
Vol. 29, No. 2, pp. 293-311, Mar. 1980

Limsups of Lats

P. R. HALMOS

§0. Introduction. Question: how does Lat A depend on A? (Here A is a bounded linear operator on a complex Hilbert space H, and Lat A is the set of those closed subspaces of H that are invariant under A.) Answer: Lat is upper semicontinuous. This answer has several possible meanings, depending on how topological concepts are interpreted. The most promising meaning is this: if $A_n \to A$ (strong operator topology), then $\limsup_n$ Lat $A_n \subset$ Lat A. (The "limsup" used here is an old concept in set-theoretic topology; its definition will be recalled below.)

For good applications of the semicontinuity result, theorems about largeness are needed, that is, theorems whose conclusion is that $\limsup_n$ Lat A_n is not trivially small. The main purpose of this paper is to delimit the search for such largeness theorems by presenting (1) a weak positive assertion and (2) a strong negative one. Positive: if the sequence {rank A_n} is bounded, then $\limsup_n$ Lat A_n is "large". Negative: there exists a sequence {A_n} with rank $A_n = n$ such that $\limsup_n$ Lat A_n is trivially small. The paper concludes with a small list of possible applications of and a couple of unsolved problems about the new (semicontinuity) approach to invariant subspace theorems.

§1. Convergence. The most trivial kind of structure in which limits can be discussed, a *limit space,* is a set X with a designated set of *convergent* sequences of elements of X; the minimal axioms that convergence must satisfy are (1) every constant sequence converges to its unique term, and (2) every subsequence of a convergent sequence converges to the same limit. A subset E of a limit space is *closed* if E contains the limits of all convergent sequences in E.

Associated with every limit space X there is another one, namely the set 2^X of all non-empty closed sets in X; convergence in 2^X is defined as follows. If {E_n} is a sequence in 2^X, define $\liminf_n E_n$ as the set of all limits of convergent sequences {x_n} with x_n in E_n, and define $\limsup_n E_n$ as the set of all limits of convergent subsequences of sequences {x_n} with x_n in E_n. A sequence {E_n} in 2^X is called *convergent to the element E in 2^X* if $\liminf_n E_n = \limsup E_n = E$. (The space 2^X has a somewhat richer structure than that of a limit space; it allows a discussion of liminf and limsup as well as lim. Note, however, that the liminf and the limsup of a sequence

293

of closed sets are not necessarily closed.) These ideas were introduced by Fréchet and studied by Alexandrov and Urysohn; for detailed references and discussions see [10, Chapter II].

The collection of closed sets endows a limit space with a topology, but that topology can be quite pathological. Even if X is a good space, e.g., a separable metric space, and if convergence in X is defined in the usual manner, the limit space 2^X can still be bad, e.g., fail to be a Hausdorff space or refuse to admit any non-constant real-valued continuous functions. From the point of view of the intended applications these observations are beside the point; the only relations that matter are the combinatorial ones (such as the inclusions) that hold between certain liminfs and limsups.

If X and Y are limit spaces, a function φ from X to 2^Y is *lower semicontinuous* if

$$\varphi(x) \subset \liminf_n \varphi(x_n)$$

whenever $x_n \to x$ (*i.e.*, whenever $\{x_n\}$ is a sequence in X that converges to x); if the same condition implies that

$$\limsup_n \varphi(x_n) \subset \varphi(x),$$

the function φ is *upper semicontinuous*. The function φ is *continuous* if it is both lower and upper semicontinuous.

§2. Semistrong convergence.

Suppose now that H is a Hilbert space, and define $f_n \to f$ to mean $\|f_n - f\| \to 0$ (the usual strong convergence of vectors); the result is, of course, a limit space. The set $\mathscr{S}$ of all (closed) subspaces of H is a subset of 2^H, and, as such, it too is a limit space. (The limsup of a sequence of subspaces is not necessarily a subspace, but the liminf always is one; it follows that the limit of a convergent sequence of subspaces is a subspace.)

If $\mathscr{P}$ is the set of all projections (idempotent Hermitian operators) on H, there is a natural bijection from $\mathscr{P}$ to $\mathscr{S}$, namely $P \mapsto \operatorname{ran} P$. The set $\mathscr{P}$ is a limit space more familiar than $\mathscr{S}$; the most obvious concept of limit is the one according to which $P_n \to P$ if and only if $P_n f \to Pf$ for each vector f (the usual strong convergence of operators). It is convenient to transplant this concept to $\mathscr{S}$ and to say that a sequence $\{M_n\}$ of subspaces converges *strongly* to a subspace M,

$$M_n \to M \text{ (st)},$$

if and only if the sequence of projections corresponding to $\{M_n\}$ converges (strongly) to the projection corresponding to M. To distinguish between the two kinds of subspace convergence, the one defined first, in terms of 2^H, will be called *semistrong* (ss); in other words

$$M_n \to M \text{ (ss)}$$

means that

$$\liminf_n M_n = \limsup_n M_n = M.$$

The following result clarifies the relations between the two concepts, and, incidentally, justifies the terminology.

Theorem 1. *If* dim $H = \infty$, *then strong convergence (for subspaces) is strictly stronger than semistrong. A necessary and sufficient condition that $M_n \to M$ (st) is that both $M_n \to M$ (ss) and $M_n^\perp \to M^\perp$ (ss). If* dim $H < \infty$, *then strong convergence and semistrong convergence are the same.*

Proof. Suppose that P_n ($n = 1,2,3,\dots$) and P are projections, and write $M_n = \operatorname{ran} P_n$, $M = \operatorname{ran} P$.

(a) If $f \in \liminf_n M_n$, then $P_n f \to f$. To prove that, suppose that $f_n \in M_n$ and $f_n \to f$. Since the projection of a vector into a subspace minimizes the distance from the vector to the subspace, it follows that $\|P_n f - f\| \le \|f_n - f\|$ for all n.

(b) A necessary and sufficient condition that $M \subset \liminf_n M_n$ is that $P_n P \to P$. Indeed, if $P_n P \to P$, then, since $P_n Pf \subset M_n$ for all n, it follows that $Pf \in \liminf_n M_n$, and hence that $M \subset \liminf_n M_n$. If, conversely, $M \subset \liminf_n M_n$, then, since $Pf \in M$ for every f, (a) implies that $P_n Pf \to Pf$ for every f.

(c) If $P_n \to P$, then $P_n P \to P$, and therefore, by (b), $M \subset \liminf_n M_n$. If, on the other hand, $f \in \limsup_n M_n$, then $f_{n_k} \to f$, where $f_{n_k} \in M_{n_k}$ (for some increasing sequence $\{n_k\}$ of indices). Since $P_{n_k} \to P$, therefore $P_{n_k} f_{n_k} \to Pf$; since, however, $P_{n_k} f_{n_k} = f_{n_k} \to f$, it follows that $Pf = f$, i.e., that $f \in M$. This completes the proof that strong convergence is stronger than (or equal to) semistrong.

(d) For an example where the two convergences are different, take an infinite orthonormal basis $\{e_0, e_1, e_2, \dots\}$, let M_n be the 1-dimensional space spanned by the unit vector $(e_0 + e_n)/2^{1/2}$ ($n = 1,2,3,\dots$), and let M be the 1-dimensional span of e_0. It is easy to verify that $(P_n f, g) \to (1/2)(Pf, g)$ for all f and g, i.e., that $P_n \to (1/2)P$ weakly. If $f \in \limsup_n M_n$, $f_{n_k} \to f$, $f_{n_k} \in M_{n_k}$, it follows from (a) that $P_{n_k} f \to f$. Since, however, $P_{n_k} \to (1/2)P$ weakly, it follows that $(1/2)Pf = f$; for a projection that is possible only if $f = 0$. Conclusion: $\limsup_n M_n = 0$, and therefore $M_n \to 0$ (ss). In the strong sense, however, the sequence $\{M_n\}$ does not converge to 0, or to anything at all. Reason: if $P_n \to T$ (strong), then T must be a projection, and T must equal $(1/2)P$; that is a contradiction.

(e) If $P_n \to P$, then, of course, $1 - P_n \to 1 - P$; it follows from (c) that $M_n \to M$ (ss) and $M_n^\perp \to M^\perp$ (ss). Suppose now that $M_n \to M$ (ss) and $M_n^\perp \to M^\perp$ (ss). The auxiliary result (b) implies that $P_n P \to P$ and $(1 - P_n)(1 - P) \to 1 - P$. That is: $1 - P_n - P + P_n P \to 1 - P$, and therefore $P_n P - P_n \to 0$. Since, at the same time, $P_n P - P \to 0$, it follows that $P_n - P = (P_n P - P) - (P_n P - P_n) \to 0$.

(f) Suppose, finally, that H is finite-dimensional and $M_n \to M$ (ss). If $\{P_{n_k}\}$ is a convergent subsequence of $\{P_n\}$, with $P_{n_k} \to Q$, then, by (c),

$M_{n_k} \to$ ran Q (ss), and therefore $M =$ ran Q, or $P = Q$. That is: every convergent subsequence of $\{P_n\}$ has the same limit, namely P. Since in the finite-dimensional case the set of projections is compact, so that every subsequence of $\{P_n\}$ has a convergent subsequence, it follows on elementary topological grounds that $P_n \to P$.

Similar reasoning yields a result that sheds light on semistrong convergence and can sometimes be used to determine semistrong limits.

Corollary. *If $P_n \to T$ (weakly), then $M_n \to$ ker $(1 - T)$ (ss).*

Proof. If $f \in$ ker $(1 - T)$, then $P_n f \to Tf = f$ (weakly), and therefore $\|P_n f\|^2 = (P_n f, f) \to (Tf, f) = \|f\|^2$. It follows that $P_n f \to f$ (strongly), and therefore $f \in$ liminf$_n M_n$. In other words, ker $(1 - T) \subset$ liminf$_n M_n$.

If $f \in$ limsup$_n M_n$, then $f_{n_k} \to f$ with $f_{n_k} \in M_{n_k}$. By (a) above, $P_{n_k} f \to f$ (strongly). Since, however, $P_{n_k} f \to Tf$ (weakly), it follows that $Tf = f$, and hence that $f \in$ ker $(1 - T)$. In other words, limsup$_n M_n \subset$ ker $(1 - T)$. This completes the proof of the corollary.

The strong topology of $\mathscr{S}$ might seem more natural than the semistrong, but in fact it is less so. Reason: the definition of semistrong convergence makes sense without any change for any Banach space. Another aspect of the same comment is that "projection" is a concept of unitary geometry, and in any context in which orthogonal complementation is not intrinsic, the use of projections prejudices the issue. An example is the problem of invariant subspaces: since the orthogonal complement of a subspace invariant under an operator may fail to be invariant, to identify subspaces with projections in that context is just asking for trouble.

§3. Lat is u.s.c. Since the set $\mathscr{S}$ of subspaces of H has been endowed with the structure of a limit space (via semistrong convergence), the set $2^{\mathscr{S}}$ of non-empty closed subsets of $\mathscr{S}$ becomes one too. Most of the elements of $2^{\mathscr{S}}$ are irrelevant to operator theory, but an exact characterization of the relevant ones is not known. (A set of subspaces is "relevant to operator theory" if it is the lattice of all subspaces invariant under every element in some algebra of operators on H.) No harm done: for typographical convenience write $\mathscr{L} = 2^{\mathscr{S}}$, and leave open the possibility that in subsequent studies the role of $\mathscr{L}$ will be played by appropriate subsets of $\mathscr{L}$. The purpose of the rest of this work is to examine the properties of certain limsups in $\mathscr{L}$.

Consider the set $\mathscr{B}$ $(= \mathscr{B}(H))$ of all operators on H; endowed with the strong operator topology, $\mathscr{B}$ becomes a limit space. The function Lat maps that space into $\mathscr{L}$; the reason for the interest in the behavior of limsups in $\mathscr{L}$ is the following observation.

Theorem 2. Lat *is (strong, semistrong) upper semicontinuous.*

(The parenthetical reference to limit concepts is intended as a reminder that convergence in the domain of Lat is strong and convergence in the range of Lat is the concept induced by semistrong convergence in $\mathscr{S}$.)

Proof. Suppose that $A_n \to A$ (strong); it is to be proved that

$$\text{limsup}_n \text{ Lat } A_n \subset \text{Lat } A.$$

If $M \in \text{limsup}_n \text{ Lat } A_n$, then $M_{n_k} \to M$ (ss) for some sequence $\{M_{n_k}\}$ of subspaces with $M_{n_k} \in \text{Lat } A_{n_k}$. It is to be proved that if $f \in M$, then $Af \in M$. Since $f \in \text{liminf}_k M_{n_k}$, there exist vectors f_{n_k} in M_{n_k} such that $f_{n_k} \to f$. Since $M_{n_k} \in \text{Lat } A_{n_k}$ it follows that $A_{n_k} f_{n_k} \in M_{n_k}$. Since, finally, the sequence $\{\|A_{n_k}\|\}$ of norms is bounded, the assumed convergences imply that $A_{n_k} f_{n_k} \to Af$. Conclusion: $Af \in \text{liminf}_k M_{n_k} = M$, as desired.

§4. Change of convergence.

What happens to Theorem 2 when the convergence concepts that enter are changed? It is obvious on general grounds that if, for instance, strong operator convergence is changed to norm convergence ($\|A_n - A\| \to 0$), the resulting statement is a corollary of the present one; if convergence in the domain is strengthened (fewer sequences converge), the set of u.s.c. functions becomes larger. What if semistrong convergence in $\mathscr{S}$ is changed to strong convergence? If convergence in $\mathscr{S}$ is strengthened, then the limsup of every sequence in $\mathscr{L}$ becomes smaller, and therefore, again, the set of u.s.c. functions becomes larger. Conclusion: an upper semicontinuity theorem (about functions from $\mathscr{B}$ to $\mathscr{L}$) improves each time that the topology for $\mathscr{B}$ or for $\mathscr{S}$ is weakened. (Note: $\mathscr{S}$, not $\mathscr{L}$. The limit concept in $\mathscr{L}$ is the one induced, via liminfs and limsups, by the limit concept in $\mathscr{S}$.)

The purpose of this paper is to study the (strong, semistrong) theory; the related partially weakened versions are left unexplored. It is known (and the technique used to prove Theorem 2 proves it) that Theorem 2 remains true in the (norm, weak) sense. (The "weak" refers to defining convergence in H in the weak sense; convergence in 2^H, and hence in $\mathscr{S}$, is then defined via the induced liminfs and limsups.) No sensible u.s.c. theorem is known for weak operator convergence. The (strong, semistrong) theorem can be "applied" in the sense that it can be used to recapture an old invariant subspace theorem (the Aronszajn-Smith theorem about compact operators), and it is indicated below (§14) that it might have other applications. For the weaker theories the possibility of applications is also correspondingly weaker.

§5. Main theorem.

Theorem 2 suggests a possible way of proving invariant subspace theorems: if an operator A is the strong limit of a sequence $\{A_n\}$ such that $\text{limsup}_n \text{ Lat } A_n$ is not trivial (i.e., is not equal to the lattice Triv $= \{0, H\}$), then A has a non-trivial invariant subspace. Whether this program can be successful is not yet known. What follows is a small step in the right direction.

The completely trivial way to guarantee that a limsup is not trivial is to take each term to be the set $\mathscr{S}$ of all subspaces of H. (Note: $\mathscr{S} =$ Lat 0.) A seemingly small modification of this triviality is obtained as follows: for each subspace M, let $\mathscr{S}(M)$ be the set of all subspaces of M; take an arbitrary sequence $\{M_n\}$ of subspaces of codimension 1, and consider $\limsup_n \mathscr{S}(M_n)$. In this limsup the terms are not all equal to $\mathscr{S} (= \mathscr{S}(H))$, but they do not differ from it by much: in a certain loose sense each term contains "all but one" of the subspaces of H, or differs from $\mathscr{S}$ in one dimension only. Does it follow that the limsup is not trivial? Yes, it does, but the proof is not short. Here is a precise statement of a slightly generalized version (1 replaced by an arbitrary positive integer r).

Theorem 3. *If r is a positive integer, if* $\dim H > r$, *and if M_n is a subspace with* $\operatorname{codim} M_n \leqq r$ $(n = 1,2,3,...)$, *then*

$$\limsup_n \mathscr{S}(M_n) \not\subset \text{Triv}.$$

The limsup here referred to is the one induced in $\mathscr{L} (= 2^{\mathscr{S}})$ by strong, not semistrong, convergence in $\mathscr{S}$.

The trivial subspace 0 always belongs to $\limsup_n \mathscr{S}(M_n)$. The improper subspace H may or may not belong, but, in any event, the theorem says that some non-trivial proper subspace always does.

Since $H \notin \mathscr{S}(M_n)$ for any n, it might be puzzling how $\limsup_n \mathscr{S}(M_n)$ could possibly be *equal* to Triv: how could H get in? Easily. For an example, consider an orthonormal sequence $\{e_n\}$, and let M_n be the orthogonal complement of e_n, $n = 1, 2, 3, \ldots$. If P_n is the projection with range M_n, then $P_n \rightarrow 1$ (strong), so that $M_n \rightarrow H$ (st). Conclusion: $H \in \liminf_n \mathscr{S}(M_n)$.

The reference to strong convergence instead of semistrong makes Theorem 3 better, not worse. Recall that when convergence in $\mathscr{S}$ is strengthened, then all limsups in $\mathscr{L}$ become smaller, and, consequently, a strong assertion of non-triviality gives more information than a semistrong one.

The proof of Theorem 3 will begin in the next section.

Corollary 1. *If* $\dim H < \infty$, *and if $\{M_n\}$ is a sequence of non-trivial subspaces of H, then* $\limsup_n \mathscr{S}(M_n) \not\subset \text{Triv}$.

The corollary deserves mention, but its proof does not have to depend on Theorem 3; it can be proved directly by a trivial compactness argument. Indeed: find a subsequence $\{M_{n_k}\}$ so that the corresponding projections converge; since the corresponding traces are different from both 0 and $\dim H$, the same is true of their limit.

Corollary 2. *If r is a positive integer, if* $\dim H > r$, *and if A_n is an operator on H with* $\operatorname{rank} A_n \leqq r$ $(n = 1,2,3,...)$, *then*

$$\limsup_n \operatorname{Lat} A_n \neq \text{Triv}.$$

Proof. If A is an operator with rank $A \leq r$, then so is $A*$. (The restriction of A to $\ker^\perp A$ is injective and maps $\ker^\perp A$ onto ran A; hence rank $A* = \dim \ker^\perp A = $ rank A.) Since $\ker A = \operatorname{ran}^\perp A*$, it follows that if rank $A \leq r$, then codim $\ker A \leq r$, and, of course, every subspace of $\ker A$ is invariant under A. Since, therefore,

$$\mathscr{S}(\ker A_n) \subset \operatorname{Lat} A_n,$$

the corollary follows from Theorem 3.

When is an operator the strong limit of a sequence of operators whose ranks do not exceed r? Since rank is strongly lower semicontinuous [8], the answer is simple: if and only if its rank does not exceed r. Conclusion: Theorem 3 is not a very large step in the right direction. The invariant subspace theorems that can be inferred from it apply only in cases (operators of finite rank) where they are obvious to begin with.

§6. Special case: $\|u_n\| \to 1$. The first step in the proof of Theorem 3 is to note that there is no loss of generality in assuming that codim $M_n = r$ for all n. Indeed, if that is not true, just diminish each M_n till it becomes true. The new $\mathscr{S}(M_n)$'s are smaller than the old ones, and, consequently, the non-triviality of the limsup of the new ones implies that of the old ones.

The details of the proof are less complicated in case $r = 1$. In what follows that case will be treated first.

Find, for each n, a unit vector u_n in $M_n^\perp$. A subsequence of $\{u_n\}$ is weakly convergent. Since it is sufficient to prove the non-triviality of the limsup of the corresponding subsequence of $\{M_n\}$, assume, with no loss, that $u_n \to v$ (weakly).

The weak lower semicontinuity of norm implies that $\|v\| \leq 1$. The easiest case to deal with is the one in which $\|v\| = 1$. In that case, let M be the orthogonal complement of v, with projection P, and let P_n be the projection with range M_n, $n = 1, 2, 3, \ldots$. Since $1 - P$ is the tensor product $v \otimes v$ (that is, $(1 - P)f = (f,v)v$ for every f), and, similarly, $1 - P_n = u_n \otimes u_n$, the weak convergence $u_n \to v$ implies the weak convergence $P_n \to P$. If a weakly convergent sequence of projections has a projection for its limit, then the convergence is strong. Consequence: $M_n \to M$ (st), so that

$$M \in \liminf_n \mathscr{S}(M_n) \subset \limsup_n \mathscr{S}(M_n).$$

(The inclusion giveaway is not so generous as it might seem; recall that the sequence being discussed is perhaps only a subsequence of the original one.) Since the rank of $1 - P$ is exactly 1, so that the codimension of M is 1, it follows not only that $M \neq H$, but, since $\dim H > 1$, that $M \neq 0$.

§7. Special case: $u_n \to 0$. Another easy case is the one in which the u_n's form an orthonormal sequence. (In that case H is necessarily infinite-dimensional, of course, and $v = 0$.) Any one of the u_n's, say u_1, belongs then to the orthogonal complement of all subsequent ones; it follows that $M_1^\perp$ (which is the span of the vector u_1) is a subspace of M_n ($n > 1$), and hence that $M_1^\perp \in \mathscr{S}(M_n)$ ($n > 1$). Immediate consequence: $M_1^\perp \in$ limsup$_n \mathscr{S}(M_n)$. Since dim $M_1^\perp = 1$, it is clear that $M_1^\perp \neq 0, H$.

If $\{u_n\}$ is not necessarily an orthonormal sequence but is the image of one under an invertible operator, i.e., if $\{e_1, e_2, e_3, \ldots\}$ is an orthonormal sequence, S is an invertible operator, and $u_n = Se_n$, $n = 1, 2, 3, \ldots$, then too limsup$_n \mathscr{S}(M_n) \neq$ Triv. Reason: an invertible operator commutes with the formation of spans, and preserves dimensions and codimensions. That is, if E is an arbitrary set of vectors, then $S\left(\bigvee E\right) = \bigvee (SE)$. It follows that codim $\bigvee \{u_2, u_3, \ldots\} = 1$, and hence that the orthogonal complement of M_0 of $\bigvee \{u_2, u_3, \ldots\}$ has dimension 1. Since $M_0 \subset M_n$, or $M_0 \in \mathscr{S}(M_n)$ whenever $n > 1$, and therefore $M_0 \in$ limsup$_n \mathscr{S}(M_n)$, it follows that limsup$_n \mathscr{S}(M_n) \neq$ Triv.

What kinds of sequences are images of orthonormal sequences under invertible operators? A sufficient condition on $\{u_n\}$ and $\{e_n\}$ that guarantees the existence of an invertible S with $u_n = Se_n$, $n = 1, 2, 3, \ldots$, is that $\Sigma_n \|u_n - e_n\|^2 < 1$. Indeed: define S_0 so as to map every finite linear combination $\Sigma_n \alpha_n e_n$ onto $\Sigma_n \alpha_n u_n$; for f in the orthogonal complement of $\bigvee \{e_1, e_2, \ldots\}$ write $Sf = f$. A straightforward verification shows that S_0 is bounded, and hence that it has a unique extension to an operator S on H. A similar straightforward verification shows that

$$\|1 - S_0\|^2 \leqq \sum_n \|u_n - e_n\|^2 < 1,$$

and hence that S is invertible.

What kinds of sequences are as near to some orthonormal sequence as the preceding paragraph requires? An obvious necessary condition on such a sequence is that it tend weakly to 0. (Indeed: even if all that is known is that $e_n \to 0$ weakly and $u_n - e_n \to 0$ strongly, it still follows that $u_n \to 0$ weakly.) That necessary condition is not sufficient, but it is typical in this sense: every sequence of unit vectors that tends weakly to 0 has a subsequence that is arbitrarily near to an orthonormal sequence. Precisely: if $\{u_n\}$ is a sequence of unit vectors that tends weakly to 0, and if $\{\varepsilon_n\}$ is a sequence of positive numbers, then there exists a subsequence $\{u_{n_k}\}$ and there exists an orthonormal sequence $\{e_k\}$ such that $\|u_{n_k} - e_k\| \leqq \varepsilon_k$, $k = 1, 2, 3, \ldots$. Although this assertion seems to be widely known, it is not easy to find it in the literature; one good reference is [2] (where it occurs in the proof but not quite in the statement of Theorem 3).

Since (as was remarked before) dropping down to a subsequence does no harm in a proof of the non-triviality of $\limsup_n \mathscr{S}(M_n)$, the desired conclusion follows whenever $u_n \to 0$ weakly. Summary (so far): if $\|v\| = 1$ or 0, then $\limsup_n \mathscr{S}(M_n) \neq \text{Triv}$.

§8. Special case: $\|v\| < 1$. If $\|v\| < 1$, then $\|u_n - v\| \geq |\|u_n\| - \|v\|| = 1 - \|v\| > 0$, and it follows that $\|u_n - v\| > 0$ for all n. The differences $u_n - v$ can therefore be normalized, or in other words, the unit vectors

$$v_n = \frac{u_n - v}{\|u_n - v\|}$$

can be formed. Since, for all g,

$$|(v_n, g)| = \left| \left(\frac{u_n - v}{\|u_n - v\|}, g \right) \right| \leq \frac{|(u_n, g) - (v, g)|}{1 - \|v\|} \to 0,$$

it follows that $v_n \to 0$ weakly. Consequence (as in the preceding section): some subsequence of $\{v_n\}$ is the image of an orthonormal sequence under an invertible operator. Assume, with no loss, that in fact the whole sequence $\{v_n\}$ has that property: let $\{e_n\}$ be an orthonormal sequence and S an invertible operator such that $Se_n = v_n$, $n = 1, 2, 3, \ldots$.

Let E, U, and V be the spans of the sets $\{e_3, e_4, e_5, \ldots\}$, $\{u_3, u_4, u_5, \ldots\}$, and $\{v_3, v_4, v_5, \ldots\}$, respectively. Since codim $E = 2$ and $V = SE$, it follows that codim $V = 2$. Since each u_n belongs to $\bigvee \{v_n, u\}$, it follows that

$U \subset \bigvee \{V, u\}$, and hence that codim $U \geq 1$. The proof can now be finished

just as for orthonormal sequences. That is: find a 1-dimensional subspace M_0 orthogonal to U, note that $M_0 \in \mathscr{S}(M_n)$ whenever $n > 3$, and conclude that therefore $M_0 \in \limsup_n \mathscr{S}(M_n)$. This completes the proof of Theorem 3 when $r = 1$.

Note that the case $v = 0$ is contained in the last special case treated, $\|v\| < 1$; the reason for treating it separately was to explain the use of the "orthonormal image" technique.

§9. Special case: strong convergence. The proof of Theorem 3 for $r > 1$ is in outline the same as for $r = 1$. The zeroth step, in particular, is to make two simplifying assumptions with no loss of generality: assume that codim $M_n = r$ for all n, and, if P_n is the projection with range M_n, $n = 1, 2, 3, \ldots$, assume that the sequence $\{P_n\}$ is weakly convergent. The weak limit T is perhaps not a projection, but it is a positive contraction in any case. If $q = \text{rank} (1 - T)$, then the weak lower semicontinuity of rank [8] implies that $0 \leqq q \leqq r$.

The easiest case to deal with is the one in which $1 - T$ does happen to be a non-zero projection. In that case, let M be the range of T, and note that $M_n \to M$ (st). (Reason: $P_n \to T$ weakly and hence strongly.) Consequence: $M \in \text{limsup}_n \mathscr{S}(M_n)$. Since codim $M = \text{rank} (1 - T) = q$ and $1 \leqq q \leqq r$, it follows not only that $M \neq H$, but, since dim $H > r$, that $M \neq 0$.

The other easy extreme case is the one in which the orthogonal complements $M_n^\perp$ are pairwise orthogonal. (In that case H is necessarily infinite-dimensional, of course, and $1 - T = 0$.) The proof that $\text{limsup}_n \mathscr{S}(M_n) \neq \text{Triv}$ is the same as it was for $r = 1$. Just note that $M_1^\perp \in \mathscr{S}(M_n)$ when $n > 1$, and, consequently, $M_1^\perp \in \text{limsup}_n \mathscr{S}(M_n)$; since dim $M_1^\perp = r$, it is clear that $M_1^\perp \neq 0, H$.

If $\{M_n^\perp\}$ is not necessarily an orthogonal sequence but is the image of one under an invertible operator, i.e., if $\{L_1, L_2, L_3, ...\}$ is an orthogonal sequence of subspaces, S is an invertible operator, and $M_n^\perp = SL_n$, $n = 1, 2, 3, ...$, then too $\text{limsup}_n \mathscr{S}(M_n) \neq \text{Triv}$. Reason, as before: S preserves dimensions and spans.

To infer that $\text{limsup}_n \mathscr{S}(M_n) \neq \text{Triv}$ whenever $1 - P_n \to 0$ weakly (and therefore strongly), what is needed is a lemma that asserts (not necessarily for $\{M_n^\perp\}$ but for some subsequence) the existence of a sequence $\{L_n\}$ and an operator S with the properties described above. One way to do that is, again, to imitate $r = 1$ by first proving that some subsequence of $\{M_n^\perp\}$ is arbitrarily near to an orthogonal sequence. The idea is described in slightly greater detail in the next section.

§10. Projection approximations.

The following lemmas about projection approximations lead to the desired result and might be deemed to have some interest in their own right.

(1) If an operator is almost a projection, then it is near to a projection. That is: if $0 < \varepsilon < 1$, and if C is a normal operator such that $\|C - C^2\| \leqq \varepsilon^2$, then there exists a projection E such that $\|C - E\| \leqq \varepsilon$; in fact, E can be the value of the spectral measure of C on the set $\{z: |1 - z| \leqq \varepsilon\}$. Proof: if $p(z) = z - z^2$, then p (spect C) = spect $(C - C^2)$, and the latter is included in the closed disc with center 0 and radius ε^2. Hence

$$\text{spect } C \subset p^{-1}(\text{spect } (C - C^2)) = \{z: |z(1 - z)| \leqq \varepsilon^2\}$$

$$\subset \{z: |z| \leqq \varepsilon\} \cup \{z: |1 - z| \leqq \varepsilon\},$$

and the assertion follows from the spectral theorem.

(2) Almost orthogonal projections are near to orthogonal projections. That is: if Q and F are projections and ε is a positive number such that $\|QF\| \leqq \varepsilon$, then there exists a projection E such that $QE = 0$ and $\|E - F\| \leqq 2\varepsilon$. Proof: put $C = (1 - Q)F(1 - Q)$ and verify that $\|C - C^2\| \leqq \varepsilon^2$.

If E is the projection described in (1), then $QE = 0$. (Reason: $CQ = 0$, so that ran $Q \subset \ker C$, and therefore Q is orthogonal to the value of the spectral measure of C at any set that does not contain the origin.) The verification that $\|E - F\| \leq 2\varepsilon$ is straightforward.

(3) If $\{F_n\}$ is a sequence of projections of finite rank converging weakly to 0, and if $\{\varepsilon_k\}$ is a sequence of positive numbers (less than 1), then there exists a subsequence $\{F_{n_k}\}$, and there exists an orthogonal sequence $\{E_k\}$ of projections such that $\|F_{n_k} - E_k\| \leq \varepsilon_k$. The proof is inductive. Put $n_1 = 1$ and $E_1 = F_{n_1}$ $(= F_1)$. Suppose that $F_{n_1}, \ldots, F_{n_k}$ and $E_1, \ldots, E_k$ have been determined $(1 = n_1 < \ldots < n_k)$, with $\{E_1, \ldots, E_k\}$ orthogonal, so that $\|F_{n_i} - E_i\| \leq \varepsilon_i$, $i = 1, \ldots, k$. If $Q_k = E_1 + \ldots + E_k$, then the assumed weak convergence of $\{F_n\}$ to 0 implies that $\|Q_k F_n\| \to 0$ as $n \to \infty$. Consequence: $\|Q_k F_{n_{k+1}}\|$ is arbitrarily small for n_{k+1} sufficiently large. The problem is to find E_{k+1} so that $\|F_{n_{k+1}} - E_{k+1}\| \leq \varepsilon_{k+1}$ and $Q_k E_{k+1} = 0$. The solution is immediate from (2): as soon as $\|Q_k F_{n_{k+1}}\| \leq \varepsilon_{k+1}/2$, the existence of a suitable E_{k+1} follows. (A related result can be found in [11].)

(4) If a sequence of projections of finite rank converges weakly to 0, then it has a subsequence whose sequence of ranges is the image under an invertible operator of an orthogonal sequence of subspaces. The proof is the same, in spirit, as the proof of the corresponding statement for unit vectors. Let $\{\varepsilon_k\}$ be a sequence of positive numbers such that $\Sigma_k \, \varepsilon_k^2 < 1$; given $\{F_n\}$, find $\{F_{n_k}\}$ and $\{E_k\}$ as in (3). For each k, find an isometric mapping U_k from ran E_k onto ran F_k so that $U_k E_k U_k^{-1} = F_k$ and $\|1 - U_k\| \leq \varepsilon_k$ [9]. The rest is easy (and goes exactly as for $r = 1$): define S to be U_k on

ran E_k and the identity on $\bigvee\limits_{k}$ ran E_k, and, by computing with finite sums

of the form $\Sigma_k \, E_k f$, verify that S is bounded and invertible.

With this the proof mentioned in the last paragraph of §9 becomes complete: if $1 - P_n \to 0$ weakly, then $\limsup_n \mathcal{S}(M_n) \neq \text{Triv}$.

§11. Convergent orthonormal sets.

The most delicate part of the proof of Theorem 3 is for the case in which T is not a projection. For $r = 1$ even that part is relatively easy: if $\{u_n\}$ is a sequence of unit vectors that converges weakly to v, $\|v\| < 1$, then $\{(u_n - v)/\|u_n - v\|\}$ is a sequence of unit vectors that converges weakly to 0. In other words, subtraction and normalization reduce the in-between case to one of the extreme cases. The corresponding steps for projections seem to require more fuss.

What makes the technique work for $r = 1$ is the replacement of a subspace of dimension 1 by a unit vector, that is, by a basis of the subspace. A natural idea is to imitate that replacement for an arbitrary r: instead of subspaces of dimension r treat orthonormal sets of r vectors. What is wanted then is an orthonormal basis $\{u_n^{(1)}, \ldots, u_n^{(r)}\}$ of $M_n^\perp$, $n = 1, 2, 3, \ldots$, such that, for each j, the sequence $\{u_n^{(j)}\}$ is weakly convergent to the "right" limit. What may be assumed is that the sequence $\{P_n\}$ of projections (of corank

r) is weakly convergent. What follows is a proof that such bases do exist.

Suppose, as before, that T is the weak limit of $\{P_n\}$, with rank $(1 - T) = q \leqq r$, and let $\{v^{(1)},...,v^{(q)}\}$ be an orthonormal basis of ran $(1 - T)$ consisting of eigenvectors of $1 - T$ with non-zero eigenvalues $\lambda^{(1)}, ..., \lambda^{(q)}$:

$$(1 - T)v^{(j)} = \lambda^{(j)}v^{(j)}, \qquad j = 1, ..., q.$$

(Note: in case $q = 0$, the proof of Theorem 3 was finished in §10; in the sequel it is assumed that $q \neq 0$.) First approximations to what is wanted are the vectors

$$(1 - P_n)v^{(j)}, \qquad j = 1, ..., q.$$

They belong to $M_n^\perp$, but they are not good enough: there are only q of them (instead of r), and there is no reason why, for fixed n, they should form an orthonormal set.

Note, however, that

$$((1 - P_n)v^{(i)}, (1 - P_n)v^{(j)}) = ((1 - P_n)v^{(i)}, v^{(j)}) \to ((1 - T)v^{(i)}, v^{(j)}) = \lambda^{(i)}\delta_{ij},$$

which implies that, for large n, the vectors $(1 - P_n)v^{(j)}$ are linearly independent. Reason: their Gramian is nearly equal to the non-singular $q \times q$ diagonal matrix $(\lambda^{(i)}\delta_{ij})$.

To come nearer to what is desired, normalize; that is, form the vectors

$$v_n^{(j)} = \frac{(1 - P_n)v^{(j)}}{\|(1 - P_n)v^{(j)}\|}, \qquad j = 1, ..., q.$$

Since $\|(1 - P_n)v^{(j)}\|^2 = ((1 - P_n)v^{(j)}, v^{(j)}) \to \lambda^{(j)}$, it follows that for large n the denominators are not 0, and, incidentally, it follows that

$$v_n^{(j)} \to \sqrt{\lambda^{(j)}}\, v^{(j)}, \qquad j = 1, ..., q.$$

Since, finally,

$$(v_n^{(i)}, v_n^{(j)}) = \frac{((1 - P_n)v^{(i)}, v^{(j)})}{\|(1 - P_n)v^{(i)}\| \cdot \|(1 - P_n)v^{(j)}\|} \to \frac{\lambda^{(i)}\delta_{ij}}{\sqrt{\lambda^{(i)}\lambda^{(j)}}} = \delta_{ij},$$

the sets $\{v_n^{(1)}, ..., v_n^{(q)}\}$ are "almost orthonormal" (in the sense their Gramians are nearly the $q \times q$ identity matrix).

A good way to make the $v_n^{(j)}$'s truly orthonormal amounts, in effect, to an application of the Gram-Schmidt orthogonalization process. The formulation most appropriate to the present problem is this: if $\{v_n^{(1)}, ..., v_n^{(q)}\}$ is a linearly independent set of vectors, with Gramian V_n, and if A_n is the inverse of the (positive definite) square root of V_n, then the vectors

$$u_n^{(i)} = \sum_{k=1}^{q} a_n^{(i,k)} v_n^{(k)}, \, i = 1, ..., q$$

440

form an orthonormal set. The proof is straightforward:

$$(u_n^{(i)}, u_n^{(j)}) = \sum_{h=1}^{q} \sum_{k=1}^{q} a_n^{(i,h)} (v_n^{(h)}, v_n^{(k)}) \overline{a_n^{(i,k)}},$$

so that the Gramian of the u_n's is $A_n V_n A_n$. (Recall that A_n is Hermitian.) For the intended applications, the most important aspect of all this is that the mapping $\langle v_n^{(1)}, \ldots, v_n^{(q)} \rangle \to \langle u_n^{(1)}, \ldots, u_n^{(q)} \rangle$ is continuous: a convergent sequence of v_n's produces a convergent sequence of u_n's. If, in particular, the v_n's are almost orthonormal, so that the Gramians V_n converge to 1, then the corresponding A_n's converge to 1 also; in that case $\|v_n^{(j)} - u_n^{(j)}\| \to 0$, so that the u_n's tend to the same limits as the v_n's. (Note: this orthonormal approximation lemma could be used to give an alternative proof, for projections of finite rank, of the projection approximation lemma (1) in §10.)

The only remaining trouble is hardly any trouble at all, it is so easy to remedy. Replace the almost orthonormal sets $\{v_n^{(1)}, \ldots, v_n^{(q)}\}$ in M_n by the orthonormal sets $\{u_n^{(1)}, \ldots, u_n^{(q)}\}$ that the lemma of the preceding paragraph yields; since $v_n^{(j)} \to \lambda^{(j)1/2} v^{(j)}$, it follows that $u_n^{(j)} \to \lambda^{(j)1/2} v^{(j)}$, $j = 1, \ldots,$ q. Trouble: the set $\{u_n^{(1)}, \ldots, u_n^{(q)}\}$ may fail to be an orthonormal basis of $M_n^{\perp}$: it has q elements, and the dimension of $M_n^{\perp}$ is r. Remedy: enlarge the set $\{u_n^{(1)}, \ldots, u_n^{(q)}\}$ for each n, so as to make it an orthonormal basis $\{u_n^{(1)}, \ldots, u_n^{(q)}, \ldots, u_n^{(r)}\}$ of $M_n^{\perp}$. The new vectors $u_n^{(j)}$, for $q < j \leq r$, might not form weakly convergent sequences, but suitable subsequences of them do. Drop down to such a subsequence and therefore assume, with no loss, that $u_n^{(j)} \to v^{(j)}$, say, weakly, as $n \to \infty$, $q < j \leq r$. Since

$$(1 - P_n)f = \sum_{j=1}^{r} (f, u_n^{(j)}) u_n^{(j)}$$

for all f, and since $u_n^{(j)} \to \lambda^{(j)1/2} v^{(j)}$ for $1 \leq j \leq q$, it follows that

$$(1 - T)f = \sum_{j=1}^{q} (f, \sqrt{\lambda^{(j)}} v^{(j)}) \sqrt{\lambda^{(j)}} v^{(j)} + \sum_{j=q+1}^{r} (f, v^{(j)}) v^{(j)}.$$

But $\{v^{(1)}, \ldots, v^{(q)}\}$ is an orthonormal basis of ran $(1 - T)$:

$$(1 - T)f = \sum_{j=1}^{q} ((1 - T)f, v^{(j)}) v^{(j)}$$

$$= \sum_{j=1}^{q} \lambda^{(j)}(f, v^{(j)}) v^{(j)}.$$

Compare the last two expressions for $(1 - T)f$ and conclude that:

$$\sum_{j=q+1}^{r} (f, v^{(j)}) v^{(j)} = 0$$

for all f. Consequence: $v^{(j)} = 0$ when $q < j \leq r$.

Conclusion: it is sufficient to prove that $\limsup_n \mathscr{S}(M_n) \neq$ Triv under the added assumptions (1) the projections P_n corresponding to the spaces M_n converge weakly to an operator T, (2) $\{v^{(1)}, \ldots, v^{(q)}\}$ is an orthonormal basis of ran $(1 - T)$ such that $(1 - T)v^{(j)} = \lambda^{(j)} v^{(j)}$, $j = 1, \ldots, q$, and (3) for each n, the set $\{u_n^{(1)}, \ldots, u_n^{(r)}\}$ is an orthonormal basis of $M_n^\perp$ such that

$$u_n^{(j)} \to \sqrt{\lambda^{(j)}} \, v^{(j)} \text{ weakly}, \qquad j = 1, \ldots, q$$

and

$$u_n^{(j)} \to 0 \text{ weakly}, \qquad j = q + 1, \ldots, r.$$

§12. Subtract and normalize. The proof has reached the state in which the proof for $r = 1$ was at the beginning of §8: it is now possible (almost) to subtract and normalize, and thus to pass from $\|v\| < 1$ to $v = 0$. The obstacle in the way is a small one. The role of v is played by $\{v^{(1)}, \ldots, v^{(q)}\}$, and it is not necessarily true that the norm of every vector in that set is strictly less than 1; it is possible that $\|v^{(j)}\| = 1$ for some j. This causes no difficulty: just collect those j's and put them aside for a while.

To be precise, let J^+ be the set of those values of j ($1 \leq j \leq q$) for which $\lambda^{(j)} = 1$, and write $J^0 = \{1, \ldots, r\} - J^+$. Let Q_n^+ be the projection onto the span of the $u_n^{(j)}$'s with j in J^+; then $Q_n^0 = (1 - P_n) - Q_n^+$ is the projection onto the span of the $u_n^{(j)}$'s with j in J^0. Note that

$$Q_n^+ f = \sum_{j \in J^+} (f, u_n^{(j)}) u_n^{(j)},$$

and hence that the projections Q_n^+ converge weakly to the operator Q^+ defined by

$$Q^+ f = \sum_{j \in J^+} (f, v^{(j)}) v^{(j)}.$$

The operator Q^+ is the projection onto the span of the $v^{(j)}$'s with j in J^+. Since for j in J^+ the unit vectors $u_n^{(j)}$ converge to the unit vectors $v^{(j)}$, the convergence is strong, and, similarly, since Q^+ is a projection, the convergence $Q_n^+ \to Q^+$ is strong.

If

$$f_n^{(j)} = \frac{u_n^{(j)} - \sqrt{\lambda^{(j)}} \, v^{(j)}}{\|u_n^{(j)} - \sqrt{\lambda^{(j)}} \, v^{(j)}\|} \qquad \text{for } j \text{ in } J^0,$$

then the sets $\{f_n^{(j)} : j \in J^0\}$ are almost orthonormal; that is,

$$(f_n^{(i)}, f_n^{(j)}) \to \delta_{ij}$$

as $n \to \infty$. The verification is straightforward; all it depends on is that $\{u_n^{(j)} : j \in J^0\}$ is orthonormal ($n = 1, 2, 3, \ldots$), $\{v^{(j)} : j \in J^0\}$ is orthonormal, and $(u_n^{(i)}, v^{(j)}) \to 0$ if $i \neq j$.

The almost orthonormality of $\{f_n^{(j)}: j \in J^\circ\}$ together with the fact that $f_n^{(j)} \to 0$ weakly implies that if F_n is the projection onto $\bigvee \{f^{(j)}: j \in J^\circ\}$, then F_n converges to zero (weakly and therefore strongly). One way to prove this plausible assertion is as follows. Define a Hermitian operator C_n by writing $C_n f = \Sigma_{j \in J^\circ}(f, f_n^{(j)})f_n^{(j)}$. If $f \perp f_n^{(j)}$ for each j in J°, then $C_n f = 0$; the approximate orthonormality of the $f_n^{(j)}$'s implies that if $f \in \bigvee \{f_n^{(j)}: j \in J^\circ\}$, then $C_n f$ is nearly equal to f. Consequence: $\|C_n - F_n\| \to 0$. The weak convergence $f_n^{(j)} \to 0$ implies that $C_n \to 0$ weakly. Conclusion: $F_n \to 0$.

It is, of course, possible that all this discussion is in a vacuum: it could be that $J^+ = \{1,...,q\}$, and that therefore $J^\circ = \varnothing$. That would be fine: in that case the P_n's converge to a non-zero projection and the desired conclusion was already obtained in §9. In the remainder of the proof it is assumed, therefore, that $J^\circ \ne \varnothing$. In that case H is necessarily infinite-dimensional.

Apply now the result (4) in §10 and infer the existence of a subsequence of $\{F_n\}$ whose sequence of ranges is the image under an invertible operator of an orthogonal sequence of subspaces. Drop down to that subsequence and thus assume, with no loss, the existence of an orthogonal sequence $\{L_n\}$ of subspaces and of an invertible operator that maps L_n onto ran F_n. It is convenient at this point to exploit infinite-dimensionality and to infer the existence of a subsequence $\{L_{n_k}\}$ such that both the dimension and the codimension of its span are infinite. Consequence: both the dimension and the codimension of $\bigvee \{\text{ran } F_{n_k}: k = 1,2,3,...\}$ are infinite. Since

$$\text{ran } F_n = \bigvee \{f_n^{(j)}: j \in J^\circ\} \subset \text{ran } Q_n^\circ \vee \bigvee \{v^{(j)}: j \in J^\circ\},$$

so that ran F_n and ran Q_n° differ by not more than a fixed finite-dimensional amount, it follows that the subsequence $\{\text{ran } Q_n^\circ\}$ has the same property: both the dimension and the codimension of its span are infinite.

The proof is almost over. All that remains to be done is to put J^+ back in. For this purpose, form the span $\bigvee \{\text{ran } Q_{n_k}^\circ: k = 1,2,3,...\}$ and let P° be the projection with that range. Because of the way the subsequence $\{Q_{n_k}^\circ\}$ was constructed, both the rank and the corank of P° are infinite. Since $P^\circ \geqq Q_{n_k}^\circ \geqq 1 - Q_{n_k}^+$, the projections P° and $1 - Q_{n_k}^+$ commute, and therefore, of course, so do $1 - P^\circ$ and $1 - Q_{n_k}^+$. It follows that the product $(1 - P^\circ)(1 - Q_{n_k}^+)$ is a projection. Since that projection is orthogonal to both $Q_{n_k}^+$ and $Q_{n_k}^\circ$, so that it is orthogonal to $1 - P_{n_k}$, therefore its range belongs to $\mathscr{S}(M_{n_k})$. Since $Q_{n_k}^+ \to Q^+$ strongly, therefore

$$(1 - Q_{n_k}^+)(1 - P^0) \to (1 - Q^+)(1 - P^0).$$

Consequence: the range of $(1 - Q^+)(1 - P^0)$ belongs to $\limsup_n \mathcal{S}(M_n)$. Since $(1 - Q^+)(1 - P^0) \le 1 - P^0$, therefore $(1 - Q^+)(1 - P^0) \ne 1$; since $1 - Q^+$ has finite corank and $1 - P^0$ has infinite rank, therefore $(1 - Q^+)(1 - P^0) \ne 0$. The proof is complete: $\limsup_n \mathcal{S}(M_n)$ is never trivial.

§13. A trivial limsup.

How far can Theorem 3 be generalized? It is not likely that $\limsup_n \mathcal{S}(M_n) \ne \mathrm{Triv}$ for all sequences $\{M_n\}$ with codim $M_n < \infty$: if that were true, a positive solution of the invariant subspace problem would be an immediate consequence. (Note that every operator is a strong limit of a sequence of operators of finite rank, and apply Corollary 2 of Theorem 3 and the upper semicontinuity assertion of Theorem 2.) The following result is in fact a counterexample to many possible hopes about improving Theorem 3 by relaxing the condition of bounded coranks.

Theorem 4.   *There exists a sequence $\{U_n\}$ of partially unitary operators converging strongly to the unilateral shift U such that $\limsup_n \mathrm{Lat}\, U_n = \mathrm{Triv}$. The limsup here referred to is the one induced in $\mathcal{L}(= 2^{\mathcal{S}})$ by semistrong, not strong, convergence in $\mathcal{S}$.*

The reference to semistrong convergence instead of strong makes Theorem 4 better, not worse: a semistrong assertion of triviality gives more information than a strong one. I am grateful to L. G. Brown for calling my attention to the possibility of refining my first proof, which was for strong convergence, so as to yield semistrong convergence.

Proof.   Let H be H^2 of the unit circle, with the usual orthonormal basis $\{e_0, e_1, e_2, \ldots\}$, where $e_n(z) = z^n$, $n = 0, 1, 2, \ldots$. For $n = 1, 2, 3, \ldots$ define an operator U_n by

$$U_n e_k = \begin{cases} e_{k+1}, & 0 \le k < n - 1, \\ e_0, & k = n - 1, \\ 0, & k \ge n. \end{cases}$$

Clearly U_n is partially unitary (that is, its restriction to the orthogonal complement of its kernel is unitary), and rank $U_n = n$. If $0 \le k < n - 1$, then $U_n e_k = U e_k$; this implies that $U_n \to U$ strongly. The non-trivial part of the proof is the determination of $\limsup_n \mathrm{Lat}\, U_n$.

For each n, let ω_n be a primitive n-th root of 1, and for $0 \le j \le n - 1$ put

$$u_n^{(k)} = \frac{1}{\sqrt{n}} \sum_{j=0}^{n-1} \overline{\omega_n^{jk}} e_j.$$

A simple computation shows that $u_n^{(k)}$ is an eigenvector of U_n with eigenvalue

ω_n^k. Since U_n is the direct sum of its restriction to the span of the corresponding eigenspaces and a large 0, two consequences follow: (1) spect $U_n = \{1, \omega_n, \ldots, \omega_n^{n-1}\} \cup \{0\}$, and (2) Lat U_n is the set of all subspaces of the form $K + L$, where K is the span of a set of $u_n^{(k)}$'s (n fixed, $0 \leq k \leq n-1$) and L is an arbitrary subspace of ker U_n [6].

It is easy to prove that if a sequence $\{K_n + L_n\}$ of the kind just described converges semistrongly to a subspace M, then $K_n \to M$ (ss). (Principal reason: $L_n \to 0$ (st).) What has to be proved, therefore, is that the only possible semistrong limits of (subsequences of) sequences of spaces such as $\{K_n\}$ are 0 and H. The technique of proof is to examine the projections whose ranges are spaces of the kind in question, that is, the projections whose ranges are spanned by a set of $u_n^{(k)}$'s (n fixed).

The projection $Q_n^{(k)}$ of rank 1 whose range is spanned by $u_n^{(k)}$ is the tensor product $u_n^{(k)} \otimes u_n^{(k)}$; its matrix (with respect to the basis $\{e_0, e_1, e_2, \ldots\}$) is given by

$$(Q_n^{(k)})_{ij} = \frac{1}{n} ((e_j, u_n^{(k)}) u_n^{(k)}, e_i)$$

$$= \frac{1}{n} \omega^{jk} \overline{\omega^{ik}} = \frac{1}{n} \overline{\omega^{(i-j)k}}$$

when $i, j = 0, \ldots, n-1$, and, of course, $(Q_n^{(k)})_{ij} = 0$ otherwise.

What is important about this computation is that the $n \times n$ top left corner is a Toeplitz matrix (and, of course, Hermitian). It follows that if Q_n is a sum of $Q_n^{(k)}$'s (n fixed), then the same is true for Q_n: the top left corner is a Hermitian Toeplitz matrix.

Suppose now that $M \in \text{limsup}_n$ Lat U_n; then M is the semistrong limit of a subsequence of a sequence $\{K_n\}$. It is to be proved that $M = 0$ or $M = H$. Drop down to a subsequence, if necessary, and thus justify the assumption that the corresponding subsequence of the corresponding sequence $\{Q_n\}$ of projections is weakly convergent, to an operator T, say. By the preceding matrix observations, T is a Hermitian Toeplitz operator; by the corollary to Theorem 1, $M = \text{ker}(1 - T)$. Since the kernel of a Hermitian Toeplitz operator is either 0 or H, the proof is complete.

§14. Possible applications.

Theorem 4 shows that Theorem 2 (the upper semicontinuity of Lat) is not an automatic producer of invariant subspace theorems. That is: to prove, for some A, that Lat $A \neq \text{Triv}$, it is not enough to approximate A strongly by operators A_n with Lat $A_n \neq \text{Triv}$; the approximation must have something clever about it. The following assertion is a sample.

Theorem 5. *If A is a compact operator with ker $A = 0$, and if $\{A_n\}$ is a sequence of operators of finite rank converging to A in the norm, then* limsup_n *Lat $A_n \neq \text{Triv}$.*

445

(The "clever" part in this context is that the sequence converges in the norm.) The statement of Theorem 5 is new, but its proof is not: the conclusion is implied by the classical Aronszajn-Smith argument [1]. In outline form that argument goes like this. For each n, let H_n be a finite-dimensional subspace that "supports" A_n (that is, H_n reduces A_n and $H_n^\perp \subset \ker A_n$). Find a finite sequence of subspaces of H_n invariant under A_n and rising one dimension at a time from 0 to dim H_n. Use a suitable numerical gauge to choose, for each n, one of these subspaces "not too far from the middle". (Possibility: let u and v be orthogonal unit vectors; if M is a subspace with projection P, write $g(M) = (1/2)((Pu,u) + (Pv,v))$; define "not too far from the middle" to mean $1/4 \leq g(M) \leq 3/4$.) Use weak compactness to find a weakly convergent subsequence $\{P_{n_k}\}$ of the chosen projections; let T be their weak limit. Use the corollary to Theorem 1 to infer that ker $(1 - T)$ is the semistrong limit of ran P_{n_k}, and, consequently (by the upper semicontinuity of Lat), that ker $(1 - T)$ is in Lat A. Since $(1/2)((Tu,u) + (Tv,v)) \neq 1$, it follows that ker $(1 - T) \neq H$. The compactness of A and the norm convergence of the sequence $\{A_n\}$ (neither of which has been used so far) imply that $AT = TAT$ and from that (and the assumption ker $A \neq 0$) it follows that ker $(1 - T) \neq 0$.

That is one application of upper semicontinuity: it recaptures and very slightly strengthens the Aronszajn-Smith theorem. Another possible application might recapture the more recent spectacular breakthrough of Scott Brown about invariant subspaces of subnormal operators [4]. The idea is to use an old (but never applied) theorem of Bishop's [3] to the effect that every subnormal operator is a strong limit of normal operators. Since normal operators have highly non-trivial Lats, it seems reasonable to hope that the Lat of the limit will be non-trivial also. However reasonable the hope may be, if the approach does not propose to use any other technique, it is doomed to failure: that is what the existence of counterexamples (such as the one described in Theorem 4) shows.

A third possible application has to do with Toeplitz operators. Gambler [7] has proved that if T is a Toeplitz operator whose symbol is a trigonometric polynomial, then Lat $A \neq$ Triv. If φ is an arbitrary function in L^∞ of the unit circle, then there exists a sequence $\{\varphi_n\}$ of trigonometric polynomials that converges to φ almost everywhere and boundedly. It follows that the corresponding Toeplitz operators T_{φ_n} converge strongly to T_φ. By Gambler, Lat $T_{\varphi_n} \neq$ Triv for each n.

Problem 1. *If T is a Toeplitz operator, does it follow that* Lat $T \neq$ Triv? *Is there something that is or can be made to be true of an approximating sequence $\{T_n\}$ of Toeplitz operators whose symbols are trigonometric polynomials that implies that* limsup$_n$ Lat $T_n \neq$ Triv?

§15. Points of continuity. It is a consequence of Theorem 4 that the function Lat (from $\mathcal{B}(H)$ to $\mathcal{S}$) is not continuous at the unilateral shift, or, in other words, that U is not a point of continuity of Lat. What is?

Problem 2. *If H is infinite-dimensional, does the function* Lat *have any points of continuity?*

The problem seems to be interesting in itself, and gains additional interest from its connection with the invariant subspace problem. To see the connection, assume for a moment that there exists an operator A (on an infinite-dimensional Hilbert space) with Lat A = Triv. In that case, if $A_n \to A$ (strong), upper semicontinuity implies that $\limsup_n$ Lat A_n = Triv. Since $\liminf_n$ Lat $A_n \subset \limsup_n$ Lat A_n, it follows that Lat is continuous at A. Till such a time as somebody proves the converse implication (if Lat is continuous at A, then Lat A = Triv), Problem 3 is weaker than the invariant subspace problem. It may be more accessible, and attempts to solve it, successful or not, may yield new ideas and techniques.

It is pertinent to observe that if dim $H < \infty$, then the continuity question becomes answerable [5]. It turns out that in the finite-dimensional case the points of continuity of Lat are exactly the operators (matrices) that in classical terminology are called non-derogatory. (An operator on a finite-dimensional space is nonderogatory if and only if it is cyclic, or if and only if its minimal polynomial equals its characteristic polynomial, or if and only if each of its eigenvalues occurs in only one Jordan block.)

Acknowledgment. It is a pleasure to express my thanks to J. B. Conway for many stimulating conversations about the subject of this paper, and to D. W. Hadwin for reading it and remaining goodnatured when I accepted only most of his suggestions.

REFERENCES

1. N. ARONSZAJN & K. T. SMITH, *Invariant subspaces of completely continuous operators,* Trans. A.M.S. **68** (1950), 337–404.
2. C. BESSAGA & A. PEŁCYZŃSKI, *On bases and unconditional convergence of series in Banach spaces,* Studia Math. **17** (1958), 151–164.
3. E. BISHOP, *Spectral theory for operators on a Banach space,* Trans. A.M.S. **86** (1957), 414–445.
4. S. W. BROWN, *Some invariant subspaces for subnormal operators,* Integral Equations and Operator Theory **1** (1978), 310–333.
5. J. B. CONWAY & P. R. HALMOS, *Finite-dimensional points of continuity of Lat,* Linear Algebra and Appl. (to appear).
6. T. CRIMMINS & P. ROSENTHAL, *On the decomposition of invariant subspaces,* Bull. A.M.S. **73** (1967), 97–99.
7. L. C. GAMBLER, *A study of rational Toeplitz operators,* Dissertation (1977), SUNY Stony Brook.
8. P. R. HALMOS, *Irreducible operators,* Mich. Math. J. **15** (1968), 215–223.
9. P. R. HALMOS, *Quasitriangular operators,* Acta Szeged **29** (1968), 283–293.
10. K. KURATOWSKI, *Topology, Volume I,* Academic Press, New York, 1966.
11. C. PEARCY & N. SALINAS, *Finite-dimensional representations of C*-algebras and the reducing matricial spectra of an operator,* Rev. Roumaine **20** (1975), 1–32.

This research was supported in part by a grant from the National Science Foundation.

Indiana University, Bloomington, IN 47405

Received July 17, 1979

Reprinted from the
TRANSACTIONS OF THE AMERICAN MATHEMATICAL SOCIETY
Vol. 273, pp. 621-630, Oct. 1982

ASYMPTOTIC TOEPLITZ OPERATORS

BY

JOSÉ BARRÍA AND P. R. HALMOS[1]

ABSTRACT. An asymptotic Toeplitz is an operator T such the sequence $\{U^{*n}TU^n\}$ is strongly convergent, where U is the unilateral shift. Every element of the norm-closed algebra generated by all Toeplitz and Hankel opertors together is an asymptotic Toeplitz operator. The authors study the relations among this Hankel algebra, the classical Toeplitz algebra, the set of all asymptotic Toeplitz operators, and the essential commutant of the unilateral shift. They offer several examples of operators in some of these classes but not in others, and they raise several open questions.

What is the essential commutant of the unilateral shift? The experts are convinced that, whatever it is, it is huge. The purpose of this paper is to call attention to an asymptotic property of some operators, use that property to show that certain concrete operators that do not belong to the Toeplitz algebra do belong to the essential commutant of the shift, discuss some related examples, and pose a few unsolved problems.

The Toeplitz and Hankel algebras. The underlying Hilbert space is $\mathbf{H}^2$ of the unit circle. The unilateral shift U is defined on $\mathbf{H}^2$ by $Uf(z) = zf(z)$. The essential commutant of U is, by definition, the set $\mathbf{E}$ of all those operators T on $\mathbf{H}^2$ for which $UT - TU \in \mathbf{K}$ (where $\mathbf{K}$ is the ideal of all compact operators on $\mathbf{H}^2$).

Since U is essentially unitary (i.e., both U^*U and UU^* are congruent to 1 mod $\mathbf{K}$), it follows that $T \in \mathbf{E}$ if and only if $U^*TU - T \in \mathbf{K}$. This reformulation of the definition of $\mathbf{E}$ is convenient in matrix calculations. (For operators on $\mathbf{H}^2$, all matrices in the sequel will be formed with respect to the basis $\{e_0, e_1, e_2, \ldots\}$ defined by $e_n(z) = z^n$, $n = 0, 1, 2, \ldots$.) Since, in terms of the Kronecker delta, the matrix of U is $(\delta_{i,j+1})$, the matrix of a product TU is obtained from the matrix of T by erasing the first column, and the matrix of U^*T is obtained from that of T by erasing the first row. (Caution: "erase" means literally what it says; it does not mean "replace by 0's".) The matrix of U^*TU, therefore, is obtained from that of T by "moving one step to the southeast"; to say that $T \in \mathbf{E}$ is the same as to say that, mod $\mathbf{K}$, the matrix is not changed by the move.

The essential commutant of every operator is a norm-closed algebra. Since $\mathbf{E}$ contains every Toeplitz operator (recall a possible definition: $U^*TU = T$), it follows that the *Toeplitz algebra* (the norm-closed algebra $\mathbf{T}$ generated by the set of all

Received by the editors June 1, 1981.
1980 *Mathematics Subject Classification*. Primary 47B35, 47B37.
[1]Research supported in part by a grant from the National Science Foundation.

Toeplitz operators) is included in **E**. Question, with a not immediately obvious answer: is **E** equal to **T**? The experts' conviction (**E** is huge) means, among other things, that the answer is no; some concrete examples of operators in **E** but not in **T** will become visible presently. (The most important earlier work on a closely related problem is [2].)

In view of the role that **K** plays in the definition of essential commutativity, the relation **K** $\subset$ **E** is even more obvious than the relation **T** $\subset$ **E**. It is not only obvious: it contains no new information. Reason: **K** $\subset$ **T**. This inclusion can be inferred from a sophisticated fact about irreducible C^*-algebras [3, p. 141], or can be proved directly. [Note that since U is essentially unitary, it follows that **E** is closed under the formation of adjoints and is therefore a C^*-algebra. Since U is irreducible and $U \in$ **E**, it follows that **E** is irreducible.] Here is an elementary direct proof. Since $U \in$ **T**, therefore $E = 1 - UU^* \in$ **T**; the operator E is, in fact, the projection $e_0 \otimes e_0$ of rank 1. For arbitrary operators S and T, the product $S(e_0 \otimes e_0)T$ is equal to $(Se_0) \otimes (T^*e_0)$; it follows that if S and T are in **T**, then so is $(Se_0) \otimes (T^*e_0)$. If, in particular, p and q are arbitrary polynomials, and if $S = p(U)$ and $T = q(U)^*$, then $(p(U)e_0) \otimes (q(U)e_0) \in$ **T**. Since the set of all vectors obtained by applying a polynomial in U to e_0 is dense in $\mathbf{H}^2$, it follows that every operator of rank 1 is in **T**, and so therefore is every compact operator.

If $\varphi \in \mathbf{L}^\infty$ of the unit circle, write M_φ for the multiplication operator defined on $\mathbf{L}^2$ by $M_\varphi f = \varphi f$, and T_φ for the compression defined on $\mathbf{H}^2$ by $T_\varphi f = PM_\varphi f$ (where P is the projection from $\mathbf{L}^2$ onto $\mathbf{H}^2$). The compression T_φ is a Toeplitz operator, and every Toeplitz operator is obtained this way. If M_φ is expressed as an operator matrix with respect to the decomposition $\mathbf{L}^2 = \mathbf{H}^{2\perp} \oplus \mathbf{H}^2$, the result is of the form

$$ M_\varphi = \begin{pmatrix} T_{\tilde\varphi} & H_\varphi \\ H_{\tilde\varphi} & T_\varphi \end{pmatrix}, $$

where $\tilde\varphi(z) = \varphi(z^*)$, the diagonal entries are Toeplitz operators, and the others are *Hankel operators*. (The latter can be defined by this remark; alternatively a Hankel operator H is one for which $U^*H = HU$.) If φ and ψ are in $\mathbf{L}^\infty$, then $M_{\varphi\psi} = M_\varphi M_\psi$, and therefore (multiply matrices and compare lower right corners)

(1) $$ T_{\varphi\psi} = T_\varphi T_\psi + H_{\tilde\varphi} H_\psi. $$

What is most important about this equation is that the product of two Toeplitz operators differs from a Toeplitz operator by the product of the two Hankel operators, and every product of two Hankel operators arises in this way. A related formula with a related proof (compare upper right corners) can also be useful:

(2) $$ H_{\varphi\psi} = T_{\tilde\varphi} H_\psi + H_\varphi T_\psi. $$

Hankel operators are an essential part of Toeplitz theory. An effective way to welcome them is to consider the Hankel algebra (the norm-closed algebra $\mathbf{T}^+$ generated by all Toeplitz operators and all Hankel operators together).

Convergence. It is natural to define an *asymptotic Toeplitz operator* as an operator T such that the sequence $\{U^{*n}TU^n\}$ is strongly convergent. The limit is clearly a

Toeplitz operator, and hence of the form T_φ for some φ in $\mathbf{L}^\infty$. The function φ will be called the *symbol* of T and will be denoted by $\sigma(T)$. The simplest examples are the Toeplitz operators; the next simplest the Hankel operators.

(3) LEMMA. *If H is a Hankel operator, then $HU^n \to 0$ (strong).*

PROOF. From the matrix point of view the statement is almost obvious: the matrix of HU^n is obtained from that of H by erasing the first n columns. [Note that each entry occurs in a Hankel matrix only a finite number of times.] Alternatively, $HU^n = U^{*n}H$, and $U^{*n} \to 0$ (strong).

(4) THEOREM. *Every element of the Hankel algebra is an asymptotic Toeplitz operator.*

PROOF. The main step is to show that if $\varphi_1, \ldots, \varphi_k$ are in $\mathbf{L}^\infty$, if $T = T_{\varphi_1} \cdots T_{\varphi_k}$, and if $\varphi = \varphi_1 \cdots \varphi_k$, then $U^{*n}TU^n \to T_\varphi$ (strong). The argument is based on a telescoping sum:

$$T_{\varphi_1} \cdots T_{\varphi_k} - T_{\varphi_1 \cdots \varphi_k} = T_{\varphi_1}T_{\varphi_2 \cdots \varphi_k} - T_{\varphi_1(\varphi_2 \cdots \varphi_k)}$$
$$+ T_{\varphi_1}\left(T_{\varphi_2}T_{\varphi_3 \cdots \varphi_k} - T_{\varphi_2(\varphi_3 \cdots \varphi_k)}\right)$$
$$+ T_{\varphi_1}T_{\varphi_2}\left(T_{\varphi_3}T_{\varphi_4 \cdots \varphi_k} - T_{\varphi_3(\varphi_4 \cdots \varphi_k)}\right)$$
$$+ \cdots$$
$$+ T_{\varphi_1}T_{\varphi_2} \cdots T_{\varphi_{k-2}}\left(T_{\varphi_{k-1}}T_{\varphi_k} - T_{\varphi_{k-1}\varphi_k}\right).$$

In view of this, equation (1) implies that

$$T - T_\varphi = HH + THH + TTHH + \cdots + TT \cdots THH,$$

where each T on the right side indicates a Toeplitz operator and each H a Hankel operator; since the actual subscripts are useless, they are omitted. Multiply by U^{*n} on the left and U^n on the right; since T_φ is invariant under that operation, and since (by Lemma (3)) the right side converges strongly to 0 as $n \to \infty$, the main step is complete.

Consider next a finite product all whose factors are either Toeplitz or Hankel operators, with at least one Hankel factor present. If the rightmost factor is a Hankel operator, the asserted strong convergence (to 0) follows from Lemma (3). In the remaining cases, the first Hankel factor from the right occurs in a context HT, where, as before, the symbols H and T indicate generic Hankel and Toeplitz operators respectively. In such a case, use (2) to replace HT by $H - TH$ (subscripts still omitted), and thus replace the given operator by two others, in each of which the rightmost Hankel factor is one step nearer to the right end; the desired convergence now follows by induction.

The rest is easy. Let $\mathbf{T}_0^+$ be the (unclosed) algebra consisting of all finite sums of finite products of Toeplitz and Hankel operators. If $T \in \mathbf{T}_0^+$, convergence follows from the strong continuity of operator addition. For norm limits of operators in $\mathbf{T}_0^+$, convergence follows from the standard techniques of "$\frac{\varepsilon}{3}$" analysis.

(5) COROLLARY. *The restriction of the symbol map σ to the Hankel algebra is a contractive *-homomorphism from* $\mathbf{T}^+$ *onto* $\mathbf{L}^\infty$.

PROOF. That σ is a contraction is immediate from the strong lower semicontinuity of norm: if $U^{*n}TU^n \to T_\varphi$ (strong), then

$$\|\sigma(T)\|_\infty = \|\varphi\|_\infty = \|T_\varphi\| \leqslant \liminf_n \|U^{*n}TU^n\| \leqslant \|T\|.$$

That σ preserves sums and products in $\mathbf{T}_0^+$ follows from the main step in the preceding proof; that it preserves sums and products for all operators in the Hankel algebra follows from the (norm) continuity of operator addition and multiplication and the (just proved) continuity of σ. As for adjoints, there seems to be a difficulty; adjunction is not strongly continuous. Suppose, however, that $T \in \mathbf{T}^+$ and $U^{*n}TU^n \to T_\varphi$ (strong); the weak continuity of adjunction implies that $U^{*n}T^*U^n \to T_\varphi^* = T_{\bar\varphi}$ (weak). Since $T^* \in \mathbf{T}^+$, the sequence $\{U^{*n}T^*U^n\}$ converges strongly to something, say T_ψ. Conclusion: $T_\psi = T_{\bar\varphi}$, and therefore $\sigma(T^*) = \sigma(T)^*$.

The symbol map was originally defined for Toeplitz operators only; the existence of a homomorphic extension to the entire Hankel algebra yields a slight improvement of a curious result of Douglas [4, p. 9].

(6) COROLLARY. *If a finite sum of finite products of Toeplitz or Hankel operators is compact, then the corresponding finite sum of finite products of their symbols is zero almost everywhere.*

PROOF. If K is compact, then $KU^n e_j = Ke_{j+n} \to 0$ as $n \to \infty$ and therefore $\sigma(K) = 0$; in other words $\mathbf{K} \subset \ker \sigma$.

An important part of Toeplitz theory concerns the commutator ideal $\mathbf{Q}$ of the algebra $\mathbf{T}$ (see [3, p. 181]); the following characterization of $\mathbf{Q}$ might be useful.

(7) THEOREM. *An operator T in the Toeplitz algebra $\mathbf{T}$ belongs to the commutator ideal $\mathbf{Q}$ of $\mathbf{T}$ if and only if $U^{*n}TU^n \to 0$ (strong); equivalently $\mathbf{Q} = \ker \sigma$.*

PROOF. Suppose first that $\varphi_1, \ldots, \varphi_k$ are in $\mathbf{L}^\infty$, $T = T_{\varphi_1} \cdots T_{\varphi_k}$, and $\psi = \varphi_1 \cdots \varphi_k$. Assertion: $T - T_\psi \in \mathbf{Q}$. The proof is induction on k. For $k = 1$, the assertion is trivial. To pass from $k - 1$ to k assume, temporarily, that $\varphi_k = \alpha^*\beta$, where α and β are in $\mathbf{H}^\infty$; then

$$T - T_\psi = T_{\varphi_1} \cdots T_{\varphi_{k-1}}T_{\alpha^*\beta} - T_{\varphi_1 \cdots \varphi_{k-1}\alpha^*\beta}$$

$$= T_{\varphi_1} \cdots T_{\varphi_{k-1}}T_{\alpha^*}T_\beta - T_{\alpha^*}T_{\varphi_1 \cdots \varphi_{k-1}}T_\beta$$

$$= \left(T_{\varphi_1} \cdots T_{\varphi_{k-1}}T_{\alpha^*} - T_{\alpha^*}T_{\varphi_1 \cdots \varphi_{k-1}} \right) T_\beta$$

$$= \left(\left[T_{\varphi_1} \cdots T_{\varphi_{k-1}}T_{\alpha^*} - T_{\alpha^*}T_{\varphi_1} \cdots T_{\varphi_{k-1}} \right] + \left[T_{\alpha^*}T_{\varphi_1} \cdots T_{\varphi_{k-1}} - T_{\alpha^*}T_{\varphi_1 \cdots \varphi_{k-1}} \right] \right) T_\beta.$$

The first square bracket is a commutator, and therefore belongs to $\mathbf{Q}$. The second square bracket is T_{α^*} times an operator of the same form as $T - T_\psi$ except with $k - 1$ instead of k, and, consequently, (by the induction hypothesis) it too belongs to Q. At this point it seems necessary to use a relatively deep tool, namely the approximation theorem [3, p. 163] according to which functions of the form $\alpha^*\beta$ are dense in L^∞. With the use of that theorem the proof of the assertion is obviously complete; if $T - T_\psi \in Q$ whenever $\varphi_k = \alpha^*\beta$, then $T - T_\psi \in Q$ for all φ_k.

The preceding paragraph implies that if T belongs to the (unclosed) algebra $\mathbf{T}_0$ consisting of all finite sums of finite products of Toeplitz operators, and if $\psi = \sigma(T)$, then $T - T_\psi \in Q$. Indeed, suppose that $T = T_1 + \cdots + T_m$, where each T_j is a finite product of Toeplitz operators. It follows that $\psi = \psi_1 + \cdots + \psi_m$, where $\psi_j = \sigma(T_j)$, $j = 1, \ldots, m$, and hence that $T - T_\psi = (T_1 - T_{\psi_1}) + \cdots + (T_m - T_{\psi_m}) \in Q$.

Suppose now that T is an arbitrary operator in $\mathbf{T}$ with $\sigma(T) = 0$. Let $\{T_n\}$ be a sequence, each term of which is an operator in $\mathbf{T}_0$, such that $T_n \to T$ (norm). If $\psi_n = \sigma(T_n)$, then $\psi_n \to 0$ in L^∞ (because $\sigma(T) = 0$), and therefore $T_n - T_{\psi_n} \to T$ (norm). Since $T_n - T_{\psi_n} \in Q$ for each n (by the preceding paragraph), it follows that $T \in Q$.

What was proved so far was that $\ker \sigma \subset Q$. Since $\mathbf{T}/\ker \sigma$ is commutative, the reverse inclusion is trivial.

Examples. The condition $U^*TU - T \in K$ is (necessary and) sufficient for $T \in E$; the condtion that the sequence $\{U^{*n}TU^n\}$ be strongly convergent is necessary for $T \in \mathbf{T}$. Are these conditions sharp enough to distinguish between E and $\mathbf{T}$?

(8) EXAMPLE. The Hankel operator H whose matrix is $(1/(i + j + 1))$, $i, j = 0, 1, 2, \ldots$, (usually known as the Hilbert matrix) is a famous one; it is quite easy to see that it belongs to E. Indeed, the matrix of U^*HU is $(1/(i + j + 3))$; the difference $U^*HU - H$ has matrix

$$\left(\frac{-2}{(i + j + 1)(i + j + 3)} \right).$$

Elementary analysis shows that the sum of the squares of all the entries in this difference is finite; in other words, $U^*HU - H$ is a Hilbert-Schmidt operator. Conclusion: $U^*HU - H \in K$, so that $H \in E$.

Is H an asymptotic Toeplitz operator? The answer is yes, and the proof is easy. The necessary convergence condition is satisfied, and, for all that is visible at this stage, it could be that $H \in \mathbf{T}$.

The fact is that H does belong to the Toeplitz algebra; the proof goes as follows. Since $1/(i + j + 1) = \int_0^1 x^i x^j \, dx$, the matrix of H is a Gramian and therefore positive. The operator H^2, being the product of two Hankel operators, belongs to $\mathbf{T}$ (by (1)). Since $\mathbf{T}$ is a C^*-algebra, it contains the unique positive square root of each of its positive elements, and therefore, in particular, $\mathbf{T}$ contains the positive square root H of H^2.

The Hilbert matrix is an illuminating example, but in an attempt to get new information about E and $\mathbf{T}$, it turned out to be a failure. It is, however, not a trivial failure. It belongs to $\mathbf{T}$, to be sure (and hence to E), but not for the trivial reason; it doesn't belong to K. Proof: if f_k is the vector in H^2 whose first k coordinates are

$1/\sqrt{k}$ and all other coordinates are 0 ($k = 1,2,3,\ldots$), then f_k is a unit vector and $f_k \to 0$ (weak). Since elementary estimates show that $(Hf_k, f_k) \geqslant \frac{1}{2}$, the operator H cannot be compact.

(9) EXAMPLE. There are some near relatives of the Hilbert matrix that deserve examination. For each complex number α of absolute value 1, let H_α be the operator with

$$\left(\frac{\alpha^{i+j}}{i+j+1} \right).$$

If f_k is the vector whose initial coordinates are $1/(\alpha^j\sqrt{k})$ ($j = 0,\ldots,k-1$) and all other coordinates are 0 ($k = 1,2,3,\ldots$) then, as before, f_k tends to 0 weakly but $H_\alpha f_k$ does not tend to 0 strongly; the operator H_α is not compact. Does it belong to E? The answer depends on α. If $\alpha = \pm 1$, then $H_\alpha \in$ T; otherwise H_α doesn't even belong to E. Reason: straightforward computation shows that $U^*H_\alpha U - H_\alpha$ is a scalar multiple of H_α plus a compact operator. Consequence: $U^*H_\alpha U - H_\alpha$ is just as non-compact as H_α.

(10) EXAMPLE. The classically important Cesàro operator C is defined by the matrix

$$\begin{pmatrix} 1 & 0 & 0 & \\ \frac{1}{2} & \frac{1}{2} & 0 & \\ \frac{1}{3} & \frac{1}{3} & \frac{1}{3} & \\ & & & \ddots \end{pmatrix}.$$

Is C in E? Yes, it is. Proof (straightforward computation): $U^*CU - C$ is a Hilbert-Schmidt operator.

Since C is known to be hyponormal [1] and, in fact, subnormal [6], it follows that C is not compact. Question: is C in T_0? Answer: no. Reason: if $T \in T_0$, then $U^*TU - T$ has finite rank, but $U^*CU - C$ has a triangular matrix with all diagonal entries different from 0, and therefore has infinite rank.

The preceding two comments are evidence, however weak, that C does not belong to T. There is a bit of evidence that C does not belong to T, namely that C is an asymptotic Toeplitz operator. (In fact $\sigma(C) = 0$, which shows incidentally that $\ker \sigma \neq$ K. Cf. the proof of Corollary (6).) Is C in T? Nobody knows.

(11) EXAMPLE. Which diagonal operators are in E? Which ones are in T? (In this context a diagonal operator is not just one that can be diagonalized, but one whose matrix with respect to the standard basis is diagonal.)

The answers are easy. If $T = \text{diag}(\alpha_0, \alpha_1, \alpha_2,\ldots)$, then

$$U^*TU - T = \text{diag}(\alpha_1 - \alpha_0, \alpha_2 - \alpha_1, \alpha_3 - \alpha_2,\ldots),$$

and therefore a necessary and sufficient condition that $T \in$ E is that $\alpha_{n+1} - \alpha_n \to 0$. Since $U^{*n}TU^n = \text{diag}(\alpha_n, \alpha_{n+1}, \alpha_{n+2},\ldots)$, it follows that T is an asymptotic Toeplitz operator if and only if the sequence $\{\alpha_n\}$ is convergent. Note: if $\{\alpha_n\}$ is convergent, then $T \in$ T. Proof: if $\alpha_n \to \alpha$, then

$$T = \alpha + \text{diag}(\alpha_0 - \alpha, \alpha_1 - \alpha, \alpha_2 - \alpha,\ldots),$$

and the diagonal summand is compact. Consequence: a diagonal operator is an asymptotic Toeplitz operator if and only if it belongs to the Toeplitz algebra. Conclusion: $T \in \mathbf{T}$ if and only if $\{\alpha_n\}$ is convergent.

Here at last is a source of decisive examples: to get an operator that is in $\mathbf{E}$ but not in $\mathbf{T}$, just construct a sequence that does not converge but whose first differences tend to 0. That is easy, of course; form a sequence that oscillates between 0 and 1 more and more slowly. Concrete example:

$$0, \tfrac{1}{2}, 0, \tfrac{1}{3}, \tfrac{2}{3}, 1, \tfrac{2}{3}, \tfrac{1}{3}, 0, \tfrac{1}{4}, \tfrac{2}{4}, \tfrac{3}{4}, 1, \tfrac{3}{4}, \tfrac{2}{4}, \tfrac{1}{4}, 0, \ldots.$$

(12) EXAMPLE. Is the adjoint of an asymptotic Toeplitz operator another one? No, not necessarily.

Consider an isometry S defined on $\mathbf{H}^2$ by $Se_n = e_{2n}$, $n = 0,1,2,\ldots,$ and write $T = S^*$. It follows that $Te_{2n} = e_n$ and $Te_{2n+1} = 0$, $n = 0,1,2\ldots$. Consequence: for each k, the result of applying the "far southeast corner" $U^{*n}TU^n$ to e_k results in the zero vector. Precisely, $U^{*n}TU^n e_k = 0$ as soon as $n > k$. Conclusion: $U^{*n}TU^n \to 0$ (strong), so that T is an asymptotic Toeplitz operator. The adjoint T^* ($= S$) is not. Reason: $U^{*n}SU^n e_0 = U^{*n}Se_n = U^{*n}e_{2n} = e_n$, and the sequence $\{e_n\}$ is not strongly convergent.

(13) EXAMPLE. Is the product of two asymptotic Toeplitz operators another one? No, not necessarily. An example can be obtained by modifying (11); the first such modification was suggested by C. Foiaş.

Let S_k be the square matrix of size $2k$ defined as follows: all entries are 0 except the first k in the last row, and they are equal to $1/\sqrt{k}$ ($k = 1,2,3,\ldots$). Let S be the operator whose matrix is the direct sum of all the S_k's, and let T be the adjoint of S.

Since $Se_n \to 0$ as $n \to 0$, it follows that $\|SU^n e_k\| = \|Se_{n+k}\| \to 0$ as $n \to \infty$, and hence that $\|U^{*n}SU^n e_k\| \to 0$ as $n \to \infty$ (for each k). This in turn implies that S is an asymptotic Toeplitz operator (with $\sigma(S) = 0$). So far the exact sizes of the boxes S_k are irrelevant.

Consider next the matrix of the operator T. Since the only non-zero entry of S_1^* is in the first row of S_1^*, it follows that both T and U^*TU begin with a column of 0's, and, in fact, so does $U^{*n}TU^n$ whenever $n \geq 0$. Since the only non-zero entries of S_2^* are in the first two rows of S_2^*, it follows that $U^{*2}TU^2$ begins with two columns of 0's, and so does $U^{*n}TU^n$ whenever $n \geq 2$. Inductively: $U^{*n}TU^n$ begins with k columns of 0's whenever $n \geq k(k+1)$. Consequence: $U^{*n}TU^n e_k = 0$ as soon as $n \geq k(k+1)$ (usually sooner—the estimates are generous), so that T is an asymptotic Toeplitz operator.

The product ST is not an asymptotic Toeplitz operator. Reason: the diagonal entries of ST are 0 most of the time, but 1 infinitely often. This implies that $U^{*n}(ST)U^n e_0 = 0$ most of the time but e_0 infinitely often, and, consequently, that the sequence $\{U^{*n}(ST)U^n\}$ is not strongly convergent.

(14) EXAMPLE. Typically a projection has a diagonal matrix with diagonal entries equal to 0 or 1. Such a matrix can correspond to an asymptotic Toeplitz operator only if its rank is finite or cofinite. Are there any other asymptotic Toeplitz projections?

Yes, there are. If $\mathbf{M}$ is a subspace of $\mathbf{H}^2$ invariant under U, then the projection from $\mathbf{H}^2$ onto $\mathbf{M}$ is in the Toeplitz algebra. Reason: by Beurling's theorem [5, Problem 125] there exists an inner function φ such that $\mathbf{M} = \operatorname{ran} T_\varphi$; it follows that the projection in question is the product $T_\varphi T_\varphi^*$. (This observation is due to Sheldon Axler.)

There are asymptotic Toeplitz projections that do not seem to arise in the natural ways described in the preceding two paragraphs. Here is one. Let T_k be the matrix of size k all whose entries are equal to $\frac{1}{k}$, and form the matrix

$$
\begin{bmatrix}
T_1 & & & & & \\
 & 0 & & & & \\
 & & T_2 & & & \\
 & & & 0 & & \\
 & & & & T_3 & \\
 & & & & & \ddots
\end{bmatrix}
$$

that is the direct sum of the sequence obtained by interlacing a sequence of 0's (of size 1) with the T_k's. Clearly the operator T with that matrix is a projection. Assertion: it is an asymptotic Toeplitz projection, with $\sigma(T) = 0$. Reason: if the integer n is such that the nth column of T contains the first column of T_k, then $\|Te_n\| = \sqrt{k/k^2} = 1/\sqrt{k}$; for all larger n, the norm $\|Te_n\|$ is even smaller.

The reason the 0's were inserted into T was to make it easier to compute $U^*TU - T$. The computation has no virtues other than being easy to carry out. The result is that $U^*TU - T$ is block diagonal, and that the Hilbert-Schmidt norm of the nth block is of the order $1/\sqrt{n}$. Conclusion: $U^*TU - T$ is not a Hilbert-Schmidt operator, but it is at least compact, and therefore $T \in \mathbf{E}$. Does T belong to $\mathbf{T}$ or to $\mathbf{T}^+$? Nobody knows.

Questions. Two unsolved test problems have been posed already (see Examples (10) and (14)); each of them asks whether a certain operator belongs to $\mathbf{T}$. That seems to be the crux of the matter in much of this subject. The important question is not "what is $\mathbf{E}$?" but "what is $\mathbf{T}$?". There is, after all, a way to decide whether or not an operator T belongs to $\mathbf{E}$; just form $U^*TU - T$ and see whether it is compact. It's debatable whether this should be called an algorithm, but not even anything as good as that is known for $\mathbf{T}$. The Hilbert matrix yields essentially the only non-trivial known example of an operator in $\mathbf{T}$; all others are either in $\mathbf{T}_0$, or compact, or both. Other non-trivial examples are easy enough to construct (e.g. non-trivial continuous functions of operators in $\mathbf{T}_0$), but the experts seem to agree nevertheless that the algebra $\mathbf{T}$ is far from well understood. The four questions below are special cases or reformulations of the general problem of characterizing the Toeplitz algebra.

The important classes discussed above are: the essential commutant $\mathbf{E}$, the Toeplitz algebra $\mathbf{T}$, the Hankel algebra $\mathbf{T}^+$, and the set $\mathbf{T}^\infty$ of all asymptotic Toeplitz operators. The inclusion relations among them can be summarized by the Venn diagram, (15).

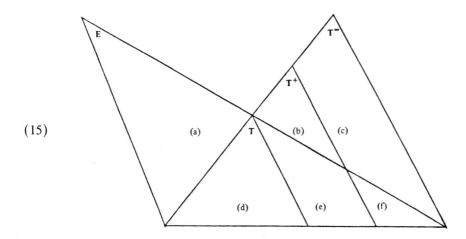

(15)

Operators corresponding to four of the indicated regions are known to exist; namely, (a) Example (11), (b) Example (9), (c) Example (12), and, for (d), any Toeplitz operator. Till now, however, no operators have been proved to belong to the classes (e) and (f).

(16) *Question. Is there an operator in* **E** $\cap$ **T**$^+$ *that is not in* **T**?

(17) *Question. Is there an operator in* **E** $\cap$ **T**$^\infty$ *that is not in* **T**$^+$?

For each operator T in the Toeplitz algebra, consider the difference $U^*TU - T$, and let **D** be the set of all such differences. Since **T** $\subset$ **E**, it follows that **D** $\subset$ **K**.

(18) *Question. Which compact operators belong to* **D**?

The reason the question is interesting is that it is a reformulation of the question "which operators belong to **T**?". That is, the set **D** characterizes **T**. More clearly said, an operator S belongs to **T** if and only if $U^*SU - S$ belongs to **D**. Indeed, if $S \in$ **T**, then $U^*SU - S \in$ **D** by definition. If, conversely, $U^*SU - S \in$ **D**, then, by definition, there exists an operator T in **T** such that $U^*SU - S = U^*TU - T$. It follows that $U^*(S - T)U = S - T$, hence that $S - T$ is a Toeplitz operator, and hence that $S - T \in$ **T**. Conclusion: $S \in$ **T**.

Example (14) describes a projection in **T**$^\infty$, and asks if it is in **T**. It would be good to know the facts in the general case.

(19) *Question. Which projections belong to* **T**?

Problems frequently become more manageable, not less, if they are embedded in a suitable enlarged context. The last question to be raised here is vague; it isn't easy to formulate a crisp, yes-or-no subquestion, but it might give a hint to a suitably general context in which Toeplitz theory can be embedded.

Begin with the observation that Toeplitz operators are the solutions of the equation $U^*XU = X$. This suggests consideration of the mapping Γ from operators to operators defined by

$$\Gamma(X) = U^*XU.$$

Toeplitz operators are the "eigenoperators" of Γ corresponding to the eigenvalue 1. Vague question: what is the spectral theory of Γ? What, in particular, can be said about eigenoperators T (generalized Toeplitz operators), $U^*TU = \lambda T$, corresponding to eigenvalues λ other than 1? What algebraic properties do they have, and what can be said about algebras generated by such operators?

REFERENCES

1. A. Brown, P. R. Halmos, and A. L. Shields, *Cesàro operators*, Acta Sci. Math. (Szeged) **26** (1965), 125–137.

2. K. R. Davidson, *On operators commuting with Toeplitz operators modulo the compact operators*, J. Funct. Anal. **24** (1977), 291–302.

3. R. G. Douglas, *Banach algebra techniques in operator theory*, Academic Press, New York, 1972.

4. _____, *Banach algebra techniques in the theory of Toeplitz operators*, CBMS Regional Conf. Ser. in Math., no. 15, Amer. Math. Soc., Providence, R.I., 1973.

5. P. R. Halmos, *A Hilbert space problem book*, Springer-Verlag, New York, 1974.

6. T. L. Kriete, III and D. Trutt, *The Cesàro operator in l^2 is subnormal*, Amer. J. Math. **93** (1971), 215–225.

INSTITUTO VENEZOLANO DE INVESTIGACIONES CIENTIFICAS, APARTADO 1827, CARACAS (101), VENEZUELA

DEPARTMENT OF MATHEMATICS, INDIANA UNIVERSITY, BLOOMINGTON, INDIANA 47405

DATE DUE

DEMCO 38-297